W9-CUD-558

The Cover

The Cover illustration is a Navajo sand painting designed for use in a ceremonial known as the Men's Night Shooting Chant. The two figures represent the earth and sky, incarnated in Navajo mythology as Nhosdzan Esdza, the Earth Mother (right), and Yadilzil Hastig, the Sky Father (left). According to Navajo legend, they were the parents of mankind. They built the seven mountains of the Navajo land, with one mountain at each cardinal point of the compass and three in the center. Through the eastern mountain they ran a bolt of lightning to fasten the land to the earth, then pushed the sky up to make room for man to live on the earth. The red and blue stripe that encloses three sides of the picture is a rainbow, and the small objects near the top are two guardians of the east, a bat to the right and a medicine bag to the left.

The black body of the Sky Father bears familiar stars and constellations: the Morning Star, the Hunter (Orion), Old Man Straddling, the Seven Sisters (Pleiades), Great Bear, Little Bear, the Pole Star, and the Milky Way, which zigzags across the chest and arms from hand to hand. The blue body of the Earth Mother bears the four sacred plants: corn at the top, beans at the right, squash at the bottom, and tobacco at the left.

The original painting is on display at the National Center for Atmospheric Research, Boulder, Colorado. (Courtesy of the Denver Art Museum.)

Frontispiece

Phobos, the larger of the two tiny moons of Mars, as viewed on November 30, 1971 by a television camera on *Mariner 9*. Note the craters, which were probably formed by the impacts of meteoroids. The picture, obtained when the range from the spacecraft to Phobos was 5,543 kilometers (3,444 miles), was subsequently processed in a computer to bring out the finer details seen here. (National Aeronautics and Space Administration)

New Horizons in Astronomy

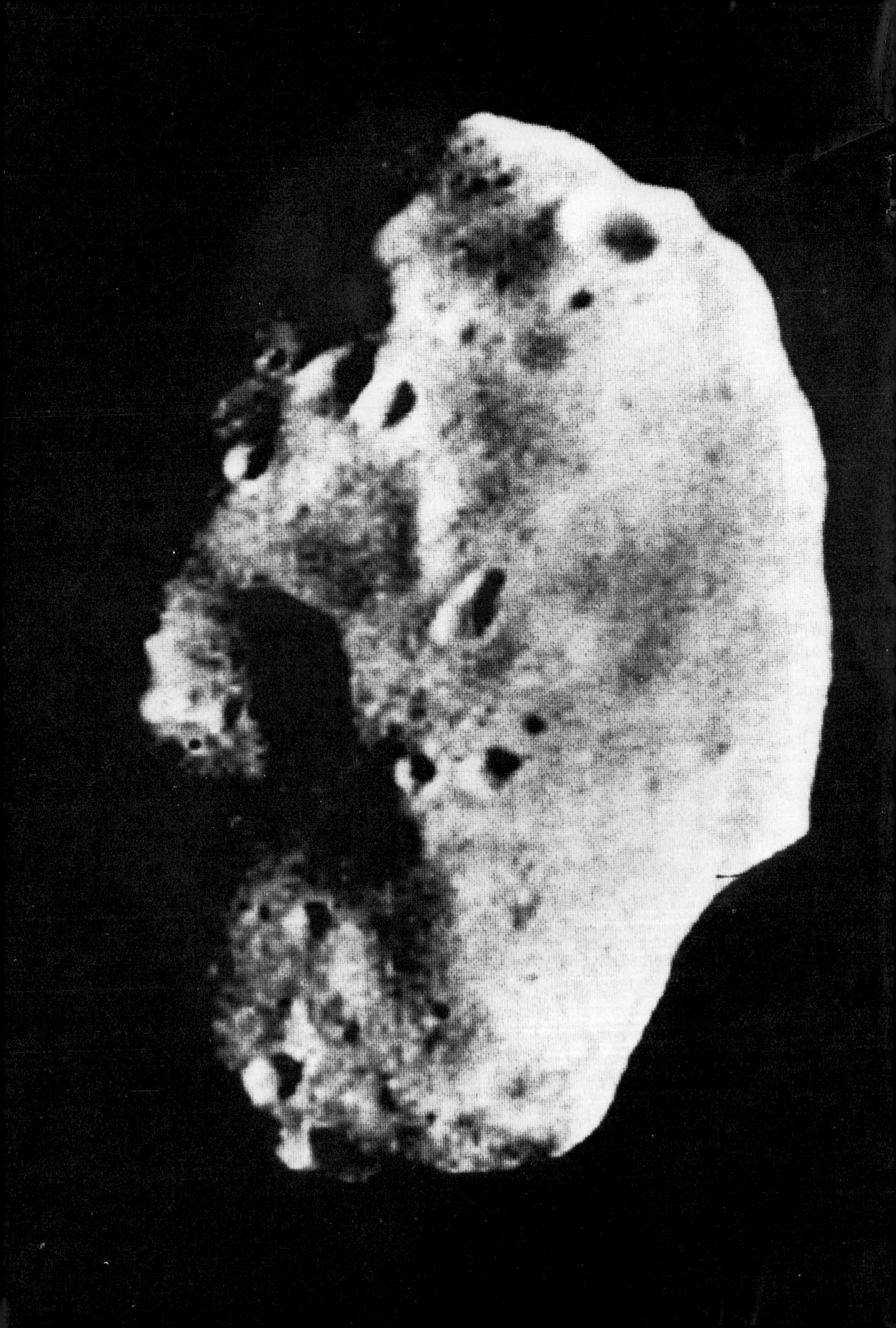

New Horizons in Astronomy

John C. Brandt
NASA-Goddard Space Flight Center,
Greenbelt, Maryland
and University of Maryland,
College Park, Maryland

Stephen P. Maran
NASA-Goddard Space Flight Center
Greenbelt, Maryland

W. H. Freeman and Company
San Francisco

A Series of Books in Astronomy and Astrophysics
EDITORS: Geoffrey Burbidge
Margaret Burbidge

Indication of the authors' affiliation with the National Aeronautics and Space Administration on the title page of this book does not constitute endorsement of any of the material contained herein by NASA. The authors are solely responsible for any views expressed.

Printed in the United States of America

Library of Congress Catalog Card Number: 74-178298

International Standard Book Number: 0-7167-0338-6

9 8 7 6 5 4

Contents

Preface

Countless examples throughout history and in our own time show that science is not an isolated activity of scholars, but rather a vital force in human affairs. In fact, the impact of science on society has never been regarded so critically, nor discussed by so many people, as it is today. In particular, astronomy, which sometimes seemed to have a largely academic character with limited applications, now, through the growth of "Big Science" and especially the space exploration programs, figures prominently in discussions of the allocation of public funds and the setting of national priorities.

Teaching an introductory one-semester course in astronomy for non-science majors, we found that no single text existed that satisfied the criteria we had set for the course. These included: avoidance of mathematics, emphasis on the relationship of life (and indeed the individual himself) to the astronomical universe, the involvement of astronomy in fundamental questions that have occupied scientific philosophers, and the development of astronomical thought in the context of the struggle for knowledge, with the adversary being sometimes Nature, and sometimes Man and his institutions.

So we have written our own text. We have devoted a good deal of it to the moon trips and Mariner-Mars encounters, to the quasars, pulsars, and other strange beasts of the twentieth-century sky—material that must inevitably obsolesce fairly rapidly. In a short text, this also means that some standard topics must be neglected. Our aim here is not to emphasize novelty for its own sake, but to portray astronomy as the living subject that it really is. Inspired by—not resting upon—the achievements of the past, astronomers are making discoveries and reaching understandings, perhaps as fundamental as those of any previous time.

We have been especially influenced by the approach to teaching astronomy developed in recent years at the University of Maryland by Professors Gart Westerhout, Donat Wentzel, and their colleagues, and we have benefited from the opportunity of participating in this teaching program and of meeting the students at College Park. We acknowledge with thanks the careful reading of the manuscript by Professor William A. Dent of the University of Massachusetts. Dr. Hong-Yee Chiu and Dr. S. Ichtiaque Rasool of the Goddard Institute for Space Studies also gave useful advice.

We thank William C. Miller and the Hale Observatories, as well as many other individuals and organizations, for their help in supplying illustrations for this book. The preparation of the typescript throughout its various stages was carried out by Dorothy Brandt, helped by Annette Diamante in an early draft; Sally Maran assisted in the proofreading. We thank them now.

John C. Brandt
Stephen P. Maran

June 1971
Greenbelt and College Park, Maryland

1

Introduction

SCIENCE AND THE ULTIMATE PURPOSE OF ASTRONOMY

How did the universe originate and how did it reach its present state? What formed the earth and how did life arise? These are among the central questions that have concerned every human culture and they obviously involve events buried in the distant past. How can we progress in understanding so as to choose among the various theories for the origin of the universe, or possibly formulate a better theory? Surely the way must lie in observing and determining as many of the properties of the universe as we can, and in looking for a common pattern in the evidence. This process of exploration is what we call astronomy, and the final purpose of astronomy is thus nothing less than to account for the origin and physical nature of the universe around us.

At every stage in history we have had a particular understanding or view of the cosmos, but as new information developed through further observations, experiments, and theoretical analyses, our picture has continually improved. This text was written shortly after the chemists and physicists began to analyze the first rocks brought back from the moon, as astronomers and planetologists studied the *Mariner 6* and 7 Mars photos, and still other scientists made the first detailed study of the birth, life, and death of an "x-ray star." Thus, many of the following chapters differ in content from the corresponding sections of books written just a few years ago, and we must accept as a certainty that future discoveries, the nature of which we cannot foresee, will in turn make this volume obsolete. Therefore, in introducing the principal concepts that underlie the present state of astronomy and some related fields, we ask you to keep in mind that "scientific truths" are often susceptible to improvement or revision, and currently accepted theories are as subject to the dictum that "this, too, shall pass away" as are the affairs of men.

ASTRONOMY AND HISTORY

Beyond the basic purpose of astronomy discussed above, astronomical discoveries have played a vital role in the history of ideas and indeed the development of intellectual freedom. Historians generally recognize three major conflicts between established authority and scientists, and astronomy relates to two of them.

The first confrontation involved the rejection of the old idea that the earth is stationary at the center of the universe (Chapter 4). The second concerned the age of the earth (Chapter 2), to which scientists assigned an increasingly greater age (now considered to be about 5 billion years) as geological evidence accumulated. This was in direct conflict with the age of several thousand years that followed from a literal interpretation of the Bible. The third of these ideological battles concerned the evolution of species theory proposed by Charles Darwin (Chapter 3). Almost all contemporary scientists believe that the concept of evolution has been fully verified, but this particular debate has not vanished from the social arena. The teaching of evolution was prohibited by law in several states until quite recently.

At the present time, science is involved in a different, perhaps more fundamental, kind of crisis. We refer to the apparent disenchantment of large segments of the public with the achievements and results of science and technology. "Progress" of the sort that was welcomed in the past is being questioned now by many people who are concerned about the deterioration of our environment, the development of weapons of mass destruction, increased automation, and so on. Astronomy is not deeply involved in these matters, although a type of pollution (Chapter 6) is affecting our capabilities of observing the stars. On the other hand, there is a clear trend toward reducing the support of basic research in favor of other important national priorities, and this has serious consequences for some aspects of astronomy that require expensive equipment, such as radio telescopes and space vehicles.

HOW THE WORLD BEGAN

It may be useful at this point to summarize briefly the history of the universe and our planet. Some aspects of this summary are the subjects of considerable controversy among scientists, although others, like the theory of evolution, are supported by an extensive array of evidence. Some of the terms used here may be new to the reader, but they are explained in the chapters that follow. Proceeding through these chapters we will explore, among other things, the full range of sizes in nature, from the atomic nucleus at 10^{-14} meters or 10^{-12} centimeters, to the distance to the farthest observed galaxies at 10^{25} meters or 10^{27} centimeters. The hierarchy of physical sizes is illustrated in Figure 1-1. If you do not know what a meter or centimeter is, or if you do not understand the meaning of 10^{-12} and 10^{27}, turn to Appendixes 1 and 2, which summarize the units and numerical notation used throughout the book.

According to our view, 10 or maybe 20 billion years ago, there was no earth, no moon, no sun, no stars, no galaxies. Just one thing existed—the enormously hot and dense primeval fireball that contained all the matter and energy in the universe. The fireball exploded into a rapidly expanding and cooling gas of protons, neutrons, and electrons, immersed in an intense sea of radiation. At first, pressure of the radiation maintained the smooth expansion, but eventually the matter, now consisting mostly of hydrogen atoms with some helium, began to form clumps. The clumps or blobs of gas continued to fly apart from each other, although the matter in an individual blob tended to contract because of its own gravitation.

In time, turbulent motions within a contracting gas blob disrupted it. One of the smaller objects produced by this disruption spun faster and faster as it contracted gravitationally, becoming flatter but continuing to condense even as it broke up into yet smaller blobs. How did this process of condensation and fragmentation come to an end? Some of the objects exploded, returning their material to the surrounding space. But,

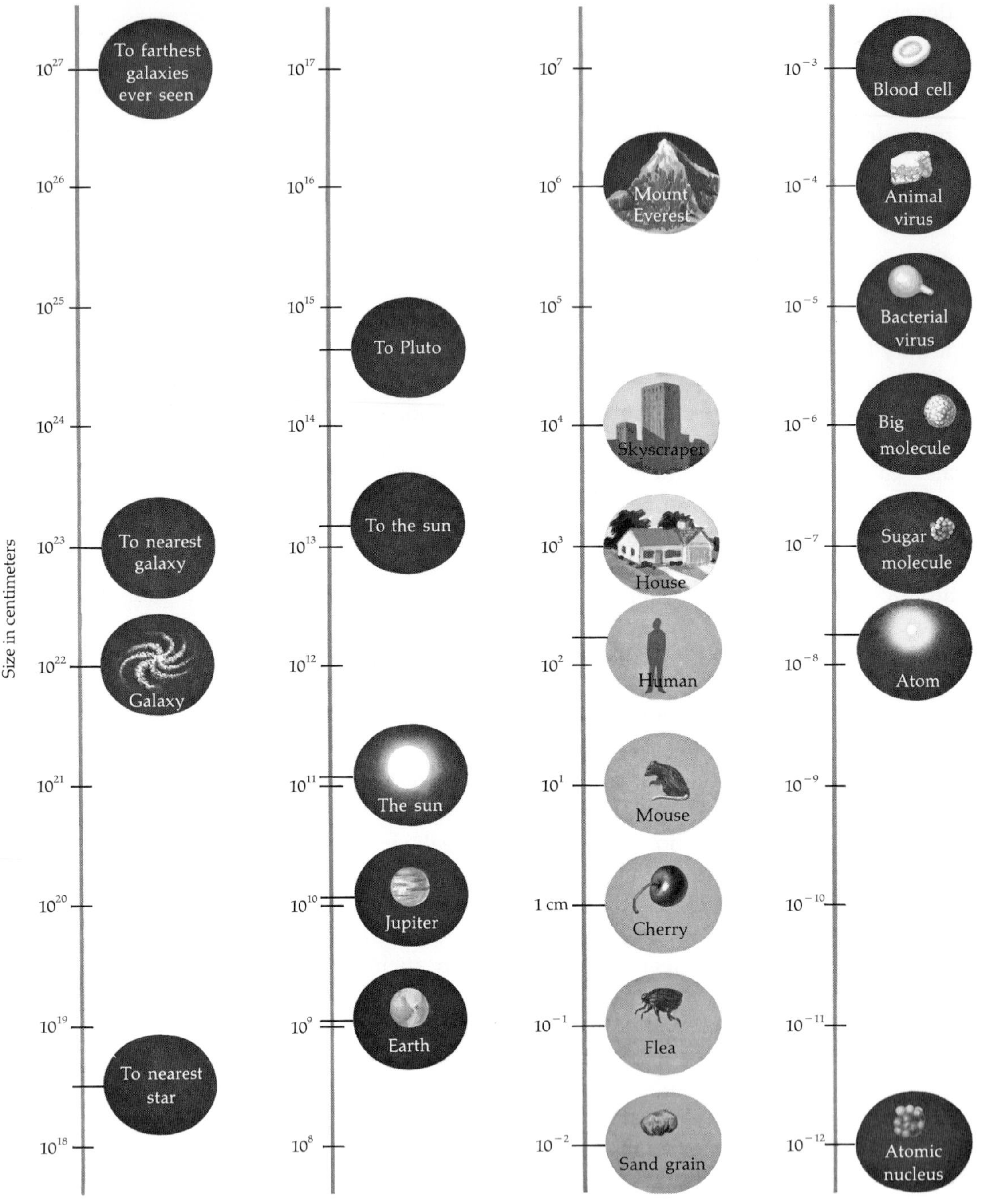

1-1 *The scale of distances and dimensions in the universe is illustrated over the range from 10^{27} cm, the distance to the farthest known galaxies, to 10^{-12} cm, the diameter of an atomic nucleus. (After* College Chemistry, *3d ed., by Linus Pauling. W. H. Freeman and Company. Copyright © 1964.)*

in one particular (and rather typical) case, the condensation produced a gaseous sphere that was sufficiently compact to resist the tendency to break into smaller pieces. Heat was released as this sphere contracted, and the temperature near the center of the object rose to the point where nuclear reactions began to occur, and these reactions constitute a source of radiant energy that has continued up to the present day. In this way the sun was born. It lay at the center of a disk-shaped array of matter, left over from the contracting cloud. The earth and other planets formed in the disk through the accumulation and condensation of this celestial debris (Figure 1-2). Thus, the origin of our own world seems almost of no consequence on the grand scale of events in the universe.

The surface of the newly made earth was quite hot, but gradually it cooled and the oceans formed as volcanic activity released water from the interior. The other basic geological

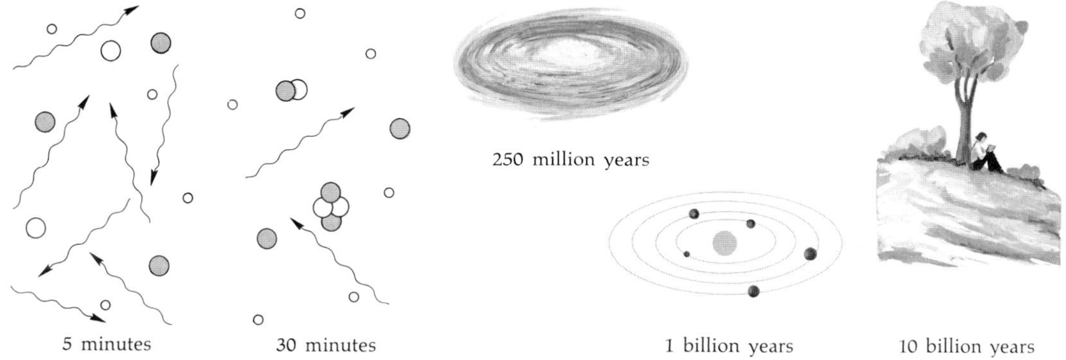

1-2 *Five stages in the evolution of the expanding universe are sketched according to a popular theory. The expansion was dominated by radiation (photons or particles of light are symbolized here by wavy arrows) during the first few minutes. The presence of matter is represented by the black dots (protons), circles (neutrons), and small circles (electrons). Within 30 minutes, however (second stage), the effects of radiation had diminished to the point that particles of matter within the expanding gas could form more complex structures, such as the deuterium nucleus (neutron and proton) and helium^4 nucleus (two neutrons and two protons). Later, most of the matter took the form of hydrogen atoms (each consists of a proton and an electron), and in the third stage, hundreds of millions of years after the expansion began, immense condensations formed in the gas—these eventually became the galaxies. Within a galaxy (fourth stage), stars formed from smaller condensations and were sometimes accompanied by planets, as is the star shown here. After several billion years had elapsed, some of the matter on the surface of a planet was organized into living forms (fifth stage). The estimates of the times are very uncertain. (After* Modern Cosmology *by George Gamow. Copyright © 1954 by Scientific American, Inc. All rights reserved.)*

processes of sedimentation, mountain building, and erosion began to operate. As the continents slowly drifted over the face of the earth, the world assumed its present general appearance.

At some time in the past, chemical compounds in the ocean water and the primitive atmosphere combined, and larger, more complex molecules were formed. Eventually, these molecules developed into the simplest forms of living organisms, which in turn evolved into more advanced marine life-forms. As evolution continued, some organisms adapted to life on the land and in the air. The numbers and complexity of animal and plant species increased and competition for food and living space intensified. As new creatures arose by mutation and climatic changes occurred, this competition led to the natural selection of those species best equipped to adapt and survive. The less efficient, less adaptable organisms were doomed to extinction.

New species continued to evolve, and an enormous, ever-changing variety of animal and plant life covered the earth. At one point, for example, dinosaurs dominated the planet for some 100 million years; finally, perhaps 3 or 4 million years ago, man-apes appeared in Africa. As brain capacity and use of tools generally increased, these hunter-gatherers evolved toward *Homo sapiens,* passing through such intermediate stages as *Homo erectus* and Cro-Magnon man. Modern *Homo sapiens* has been around for at least 35,000 years—the Bushmen of the Kalahari Desert in Africa are probably little changed from their ancestors of thousands of years ago.

Homo sapiens became the dominant species on earth, and curiosity, often reinforced by cultural and economic factors, led to a continual expansion of his horizons. Cautious inspection of the regions immediately surrounding his dwelling sites was the first step in this process, which led to the systematic exploration of the planet, and has culminated in space travel.

The voyages of Marco Polo and Magellan are especially celebrated in the annals of Western civilization. In our own country we remember the nineteenth century journeys of Lewis and Clark into the Pacific Northwest, and the expeditions of John Wesley Powell down the Grand Canyon of the Colorado River. These men traveled through parts of America that were *terra incognita* to the coastal residents. Their explorations were of tremendous value to the cartographer, for geographic information was hard to come by and was often inaccurate, as shown in Figure 1-3. Imagine how an early map-maker might have welcomed the information contained in a picture such as Figure 1-4, which shows India and Ceylon, photographed from space by an astronaut.

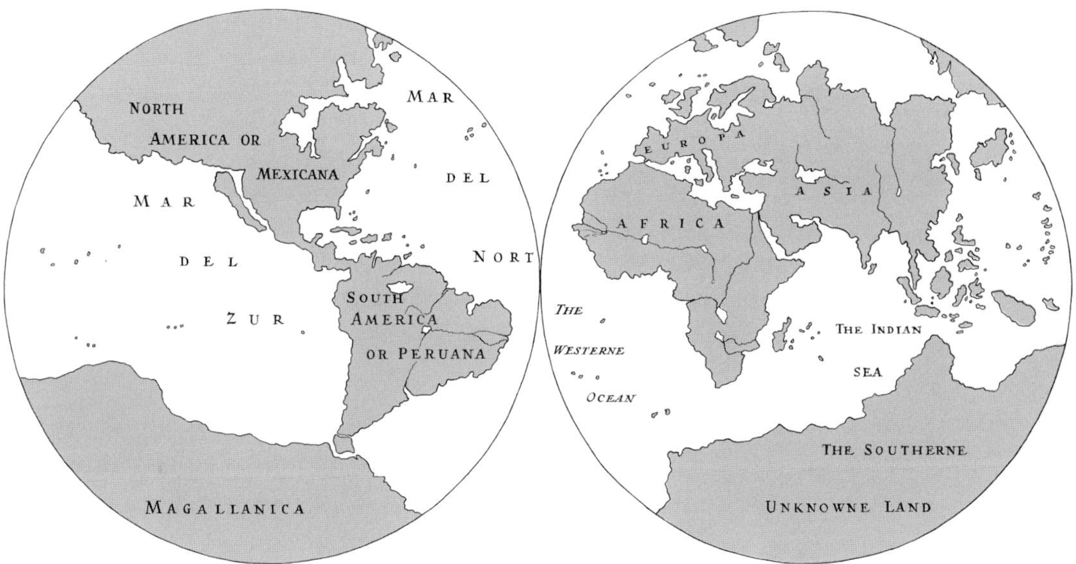

1-3 *Artist's impression of the map of the world drawn by John Speed in 1627. Note that the Mediterranean region is accurate, but errors occur elsewhere; for example, California is shown as an island.*

1-4 *India and Ceylon, photographed from the manned spacecraft* Gemini 11 *at an altitude of 655 km (410 miles). (Courtesy of National Aeronautics and Space Administration.)*

About 5,000 years ago men began to record information in an orderly and often permanent fashion. Thus, the knowledge available to succeeding generations no longer depended on direct experience or word of mouth. This stage began the cultural revolution that has accelerated so dramatically in the twentieth century. New information is being developed and published so rapidly that scholars are hard pressed to keep abreast of even the developments in their own fields, and some people believe that we are threatened by incomplete understanding and misuse of new facts and technologies.

Given man's innate curiosity and constantly expanding horizons, it was inevitable that he should turn his attention to the heavens above as well as to the earth below.

Astronomy, the study of matter and radiation in space, enables us to understand the position and nature of the earth as an element of the cosmos, and to search for an explanation of its origin. Astronomy played a basic role in the development of science and philosophy. In recent years certain astronomical subdisciplines and related fields, such as *general relativity,* have concentrated on the attempt to understand the nature of space itself (Chapter 16).

2

Our Planet Earth

Nine worlds, or *major planets*, exist in the vicinity of our sun. Among them, the earth, being nicely close at hand, offers unique opportunities for investigating the properties of a planet. The description and definition of our world has been a prime concern of science and philosophy, and the results of geological studies and exploration have given us a fair idea of its nature and the processes that determine its major surface features, including the oceans, continents, mountains, and volcanoes. As we begin the era of direct exploration of the moon by man, and of Mars and other planets by unmanned spacecraft, the techniques and knowledge developed in studies of the earth are finding immediate applications. Differences among the earth and other planets and the moon should eventually reveal their respective histories and the different forces at work among them.

In this chapter we concentrate on the surface and interior of the earth. Its atmosphere is discussed in Chapter 8, together with those of the other planets.

The shape and size of the earth are basic considerations. Aristotle (384 B.C.–322 B.C.) maintained that the earth was spherical. His opinion, although generally shared by scientists, was not universally accepted. In fact, popular sentiment for the flat earth idea existed until the actual circumnavigation of the globe in the sixteenth century dispelled any reasonable doubt that the earth was round.

MEASURING THE EARTH

Aristotle had philosophical reasons for arguing that the earth is round, but he also had evidence, in the form of the curved edge of the shadow cast by the earth on the moon (Figure 2-1) at the time of a lunar eclipse. The realization that the earth is a sphere led to the first accurate measurement of its size.

In Egypt, Eratosthenes (276 B.C.–194 B.C.) practiced geometry in its original sense—the measurement of the earth. Studying the shadow cast on a sundial at Alexandria on a certain day, he determined that the sun was not overhead at noontime but was one-fiftieth of a circle (slightly more than 7 degrees) south of the zenith.* He found that, on the same day, the sun was at the zenith at noon in the town of Syene (near modern Aswan). This was proven by the observation that the sun lit the water at the bottom of a deep well without shadowing any part of the well walls. Syene was known to be a distance of 5,000 *stadia* south of Alexandria (surveyors had paced off the distance), so Eratosthenes reckoned that if the earth were a sphere, the arc length between the two towns must be one-fiftieth of the circumference of the sphere. Then the circumference of the earth must be 50 × 5,000, or 250,000 stadia.

How good was Eratosthenes' measurement? Certainly, the main achievement lay in his reasoning; that is, the method which he originated to measure the earth, as illustrated by Figure 2-2. In fact, we don't really know how accurate the result was, because we are not sure of the length of the *stadium* unit. According to one historian's estimate, it was equal to 157.5 meters (517 feet). If this is correct, then Eratosthenes' measurement corresponded to 157.5 × 250,000 meters or about 39,400 kilometers (24,500 miles) in modern terms. This agrees very favorably with the modern value of about 39,900 kilometers (24,800 miles) for the circumference of the earth measured along a circle passing through the north and south poles. However, the good agreement must be regarded as fortuitous, not only because of our uncertainty in the value of the stadium but also because Syene was not exactly due south of Alexandria and the

*Zenith: the sky position directly above the observer.

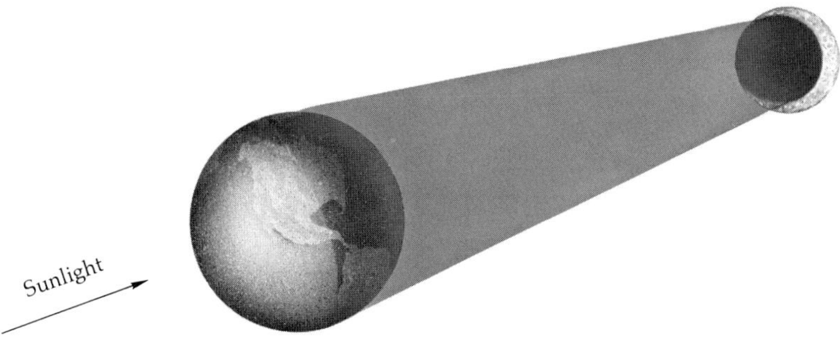

2-1 *Sketch of a lunar eclipse showing that the shadow cast by the earth is round. The simplified case shown here does not occur in nature because the earth's shadow (umbra) is about three times the size of the moon at the moon's distance. Thus, if the shadow were centered on the moon, the entire earth-facing side would be dark. The circular shape of the earth's shadow is clearly visible, however, as it moves on and off the lunar disk.*

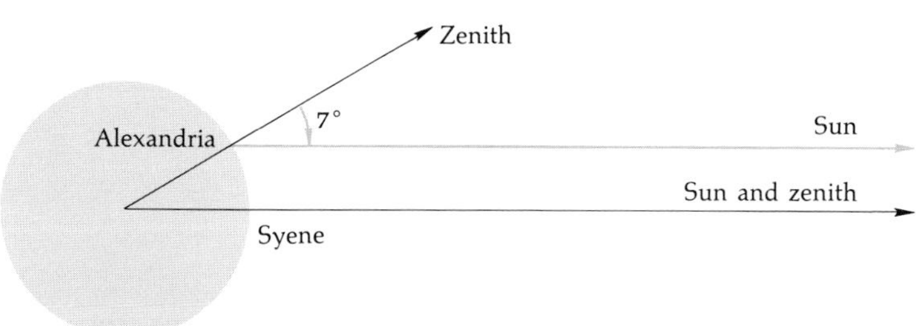

2-2 *The geometry of the sun, Syene (now Aswan), and Alexandria on June 22 as used by Eratosthenes to determine the earth's circumference.*

sundial and surveyor pacing techniques were only rough methods.

It is important to recognize that Eratosthenes' method depended on the assumption that the sun is at a very great distance from the earth, so the sun's rays are virtually parallel in the vicinity of the earth. Such an assumption is valid when the distance to the light source is much greater than the separation of the two spots on earth (Figure 2-3a). This is certainly the case for Eratosthenes' work, because the distance from the earth to the sun is now known to be about 150 million kilometers (93 million miles). We note for future reference that as the stars

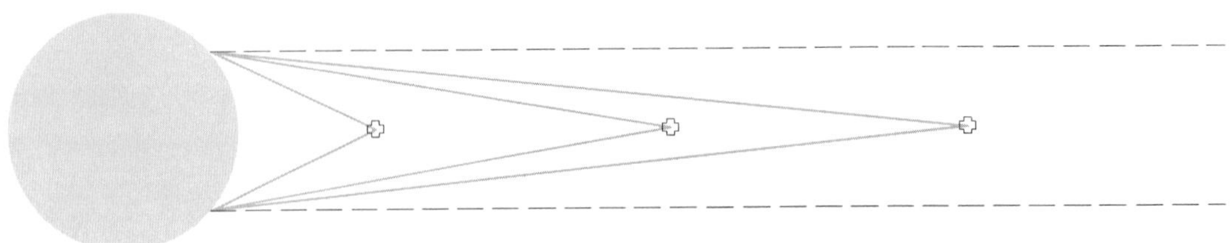

a

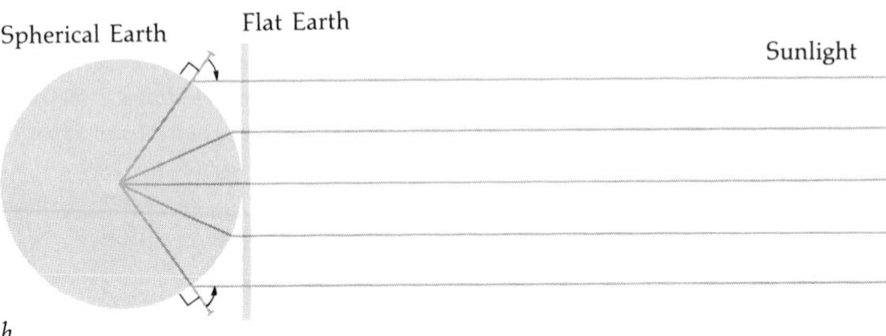

b

2-3 *(a) Light rays become more nearly parallel for objects farther and farther away. The dashed lines are parallel, corresponding to a very distant object. (b) Side view of an imaginary flat earth and a spherical earth. The sun is shown when it is overhead on the flat earth; note that it will then be overhead* everywhere. *This is* not *the case on the spherical earth and the angle between the overhead direction and sun's rays is indicated for two cases.*

are all much more distant from us than is the sun, the light rays that strike the earth from a given star are virtually parallel.

Note that if the earth were flat, then when the sun is overhead at one place, it would be overhead everywhere (Figure 2-3b). This would contradict Eratosthenes' determination that the sun was at different positions in the sky as seen from the two towns at the same time. However, despite this result, despite the teachings of Aristotle, and despite such direct evidence of the senses as the sinking of a ship on the horizon as it sails out to sea, the idea of a flat earth persisted. Drawings warned the unwary navigator against sailing too close to the edge lest he fall off to his destruction! The many voyages of exploration in the sixteenth century finally put an end to this myth.

The curvature of the earth's surface is most easily seen on photographs taken from space, such as Figures 1-4 and 17-1. In fact, the earth is not exactly spherical but has a slightly larger radius at the equator (6,378 kilometers) than at the poles (6,357 kilometers). Thus, compared to the radius at the pole, the equatorial radius is 21 kilometers, or 13 miles larger. This aspect of the shape of the earth is called the *equatorial bulge*. In addition, the earth and, particularly, the oceans are distorted by the gravitational pulls of the moon and the sun. These pulls cause *tidal bulges* and the associated phenomena of tides in the ocean. The equatorial and tidal bulges are important in discussing the precession of the earth's axis (pages 74 and 426), the history of the earth-moon system (page 424), and the circumstance that the moon keeps essentially the same face toward the earth (pages 63 and 426).

DATING THE EARTH

Modern geological science began about a century ago and the determination of the age of the earth has been one of its fundamental problems from the outset. Besides the obvious geological interest, this datum is of great importance astronomically as a milestone in the history of the *solar system* (the sun and the planets, comets, and other objects that accompany it in space), and is of concern biologically for determining the maximum time that was available for evolution of life on earth.

The age of the earth has been estimated in a variety of ways at different times. Among the early calculations was that made by James Ussher of Armagh, Ireland, in the mid-seventeenth century. He simply added up the number of years given by a literal interpretation of the succession of human generations recorded in the Bible. In this way, he found that the Creation occurred on October 23, in the year 4004 B.C. Thus, the age of the earth could be reckoned to within a day, and according to this theory it is now approaching 6,000 years.

A greater age was derived on the basis of the earth's surface temperature by Lord Kelvin in the late nineteenth century. The temperature increases as one goes down into the earth. Hence, he considered that the deep interior could be molten and that at one time in the past the entire earth was molten. From the known rate of heat loss as measured in deep mine shafts, he calculated the time required for the earth's crust to cool to its present temperature. (He assumed that there was no process that might have provided additional heat to the earth during the cooling period, and we will return to this assumption later in this chapter.) In this way, Kelvin estimated the age of the earth at 40 million years. This was a staggeringly large value as thinking went in his time. Kelvin himself considered the age a little too high because of simplifications adopted in making the computation. But, as we shall see, it was actually too low.

During the last half of the nineteenth century, geologists began to piece together an overall picture of the forces at work on the surface of the earth. There are, of course, many processes that are continually changing the appearance of the surface. The earth is ground down by erosion; wind, water, and glaciers chip away at the mountains and highlands, and much of the resulting debris is ultimately carried into the oceans in the form of silt. On the other hand, there are mountain-building processes that cause the earth's crust to be uplifted or buckled into mountain ranges. On the North American continent the Rocky Mountains have been formed relatively recently (see Table 2-1) while the much older Appalachian Mountains have been extensively eroded. Mountain building also occurs through volcanism, as in the case of the Hawaiian Islands.

As early as 1785, the Scottish geologist James Hutton had proposed the idea of *uniformitarianism* in natural processes on the earth. This theory was advocated and extended nearly half a century later by the English geologist Charles Lyell in a textbook, *Principles of Geology*, that had great influence on the subsequent development of the field. Simply stated, these men argued that the same processes of erosion and mountain building that we find now have been shaping and reshaping the earth for a very long time. This suggested that by carefully studying these processes we could infer much about the past history of the planet and the time required to bring its surface to its present state. The geological evidence indicated that most areas on the surface had been through the eroding and uplifting cycle several times. As far as the appearance of the geological features was concerned, there was no trace of a beginning and the earth's surface might well be ageless. (Modern estimates based on silt flow in American rivers indicate that, on the average, erosion lowers a region by roughly a third of a meter in 5,000 years. Thus, a plateau 3,000 meters, or 10,000 feet high, will be worn away by the processes of erosion in 50 million years.)

Hutton and Lyell's uniformitarian view of the earth's evolution, that is, that the processes at work today have been acting in the same manner over millions of years and have thereby gradually shaped the earth's surface, was not generally accepted in their time. The opponents of their view, called *neptunists*, held that the surface of the earth had remained essentially unchanged since Noah's flood. When investigation established the existence of distinct layers of rock (*strata*), the neptunists argued that such layers were sedimentary deposits from earlier catastrophic floods, of which the Noachian deluge was the last.

It was not geology, however, but the young science of nuclear physics that was to settle the question of the age of the earth. This science was born in 1896 when Pierre Becquerel discovered the radioactivity of uranium. Within a few years Lord Rutherford realized that uranium, through the process of radioactive decay (emission of particles from atomic nuclei), gradually changes into lead. The most common form of uranium decays at a rate such that one-half of the uranium in a given sample will have become lead after 4.5 billion years have elapsed. (This length of time is called the *half-life.* After 9 billion years, or two half-lives, have elapsed, only one-fourth of the uranium remains; three-quarters of it have become lead.) Thus, it is possible to determine the age of a mineral, defined as the time since it solidified, by measuring the relative amounts of lead and uranium that it contains. The oldest rocks thus far found on earth have ages of about 3.5 billion years, as determined by the radioactive dating method. There are other elements besides uranium that can be used for radioactive dating. The half-lives of these substances range from the 4.5 billion years of U^{238}, the common form of uranium, to 5,600 years for C^{14}, a rare form of carbon.* The same technique has been applied to meteorites (rocks that were in orbit around the sun before striking the earth—see Chapter 7), and has revealed ages of about 4.5 billion years. Some of the soil and rocks brought back from the moon by Apollo astronauts are as old or perhaps slightly older than this. In round figures, we can adopt 5×10^9 years as a good modern estimate for the age of the solar system.

The discovery of natural radioactivity not only led to the development of the best technique that we have for measuring the age of the earth, but it also explained why Lord Kelvin's estimate of 40 million years for this age fell woefully short of the mark. Kelvin did not include in his theory any *continuing* source of heat inside the earth. But the earth contains radioactive materials and, as these materials decay, the particles and radiation that they release heat their surroundings. Because of the heat released by this process, the earth took a much longer time to cool down than Kelvin suspected. The principal contributors to this heating are the elements uranium, thorium, and potassium.

GEOLOGICAL TIME SCALE

It is possible to deduce or reconstruct much of the earth's history by studying the traces (fossils) of ancient life forms that are found in rocks. These fossils are formed in several ways.

*The superscripts in U^{238} and C^{14} refer to the isotope atomic weights; this subject is discussed in Chapter 5.

Table 2-1 *Geologic Time Scale*

ERA	PERIOD	GEOLOGICAL EVENT	PRINCIPAL FOSSILS	LIFE DEVELOPMENT	YEARS AGO
Cenozoic (recent life)	Quaternary (an addition to a three-part classification proposed in the 18th century)	Glacial activity in the Northern Hemisphere	Animals and plants similar to those living now	Man	3 million
	Tertiary (third, from the 18th century classification)	Alps and Himalayas rise Extensive erosion in Appalachians and Rockies Climates warm, jungles widespread Appalachians reelevated	Mammals and flowering plants	Grasses become abundant Horses first appear	65 million
Mesozoic (middle life)	Cretaceous (from the Latin term for chalk)	Mountain building in Rockies Seas invade parts of North America Appalachians greatly reduced by erosion		Extinction of dinosaurs	
	Jurassic (Jura Mountains, Europe)		Conifers, dinosaurs, other reptiles	Birds and mammals first appear	150 million
	Triassic (from a time when this period was divided into three parts)	Extensive deserts in North America		Dinosaurs first appear	200 million

Table 2-1 *(Continued)*

ERA	PERIOD	GEOLOGICAL EVENT	PRINCIPAL FOSSILS	LIFE DEVELOPMENT	YEARS AGO
Paleozoic (ancient life)	Permian (after the Russian city Perm)	Appalachians complete their initial development	Amphibians		250 million
	Carboniferous (coal-bearing); often split into Pennsylvanian and Mississippian Periods	Ice age in Southern Hemisphere	Spore-bearing land plants	Coal-forming swamps Reptiles appear	300 million
	Devonian (Devonshire, England)			Amphibians appear	350 million
	Silurian (ancient British tribe, the Silures)		Fishes		450 million
	Ordovician (ancient British tribe, the Ordovices)	More than 60% of North America covered by seas	Marine plants and marine invertebrates (trilobites)	First vertebrates (fishes) Marine invertebrates form first abundant fossil record	500 million
	Cambrian (Cambria, the medieval name for Wales)	Seas invade North America			600 million
Precambrian		Mountain building in central North America		Primitive marine plants and invertebrates; one-celled organisms	3400 million
		Formation of the Earth			4500 million

Through the process of *permineralization,* the tiny voids in porous bone or other skeletal material are penetrated by mineral-bearing water; the minerals are deposited and gradually fill the voids. Other types of organic material, such as wood, may be slowly replaced by mineral deposits as the wood tissue dissolves in the water. A third type of fossil consists of an impression, such as an animal footprint or the outline of a leaf. Sample fossils are shown in Figures 2-4 and 2-5.

Rock layers that may be located far apart on earth often bear similar fossils, and thus their correlation in age can be estab-

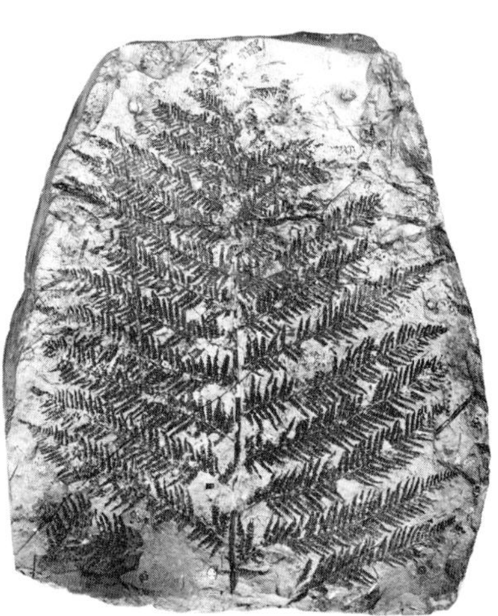

2-4 *A fossil fern. (Courtesy of the American Museum of Natural History.)*

2-5 *Trilobites, fossil marine invertebrates. (Courtesy of the Smithsonian Institution.)*

lished. For example, fossil trees are found in the coal-bearing strata that date from the so-called Carboniferous Period, whereas certain types of dinosaur fossils are found in layers of the Triassic Period (see Table 2-1).

The relative ages of rock layers are established through the principle of *superposition,* namely, that the lower layers must be older than the upper layers in any set of strata that has not been folded over by mountain-building processes. Not all strata in a time sequence are necessarily present at a given location (for example, a particular layer may have been eroded away in some regions). On the other hand, the layer that was destroyed at one place may have been preserved elsewhere. Through detailed comparison of rock strata and fossils found in various parts of the globe, it has been possible to establish the sequence in which major geologic events occurred during a substantial part of the earth's history, as summarized in Table 2-1.

The periods listed in Table 2-1 are further divided by geologists into epochs to more precisely indicate the times at which various prehistoric organisms existed and when the various rock strata were formed. Note that the Precambrian occupies seven-eighths of the total geological time scale. Unfortunately, fossils are hard to find in the Precambrian rocks; the primitive organisms of that era appear to have been mostly bacteria and algae.

The order of the periods in the geological time scale of Table 2-1 was determined from fossils and geological studies; the actual ages come from radioactive dating. Clues to the origin of the terms are given in parentheses; some of these are derived from the names of the places where particular rock layers were first studied.

The most familiar example of the physical evidence for the ancient history of the earth as portrayed by strata of different ages and origins is the Grand Canyon in Arizona. The Grand Canyon is the deepest portion of the longer and generally less spectacular canyon of the Colorado River, and it was cut by the river as geologic forces raised the terrain, a point that was first realized by the explorer John Wesley Powell a century ago.* Its numerous rock layers correspond to geological periods dating back at least 2 billion years. See Figure 2-6, where the numbers correspond to the layers labeled in Figure 2-7.

---

*Powell's deductions can be summarized like this: since the walls of the Grand Canyon are higher than those of the adjoining portions of the canyon of the Colorado, the terrain of the Grand Canyon must have been uplifted gradually after the downcutting had begun, or else the river would have had to run uphill. One can think of other explanations for the Grand Canyon, but geological studies since Powell's time show that his idea was correct.

2-6 *View of strata exposed east of Bright Angel Creek on the north side of the Grand Canyon. Numbers correspond to layers labeled in Figure 2-7. (From* Geology Illustrated *by John S. Shelton. W. H. Freeman and Company. Copyright © 1966.)*

INTERIOR OF THE EARTH

There is actually very little of the earth that we can see or touch. Although its mean radius extends nearly 6,370 kilometers (3,960 miles) from the center to the surface, our direct experience goes down only some 3 kilometers (2 miles) or so through caves, deep mines, and well drillings. Our knowledge of the interior is mostly based on inferences from earthquake measurements. The strength of a quake, the nature of the vibrations, and the times of arrival at each of many recording stations are analyzed to deduce the properties of the regions beneath the surface through which the vibrations or *seismic waves* must pass.

The earth's interior is studied by the techniques of the geophysicist, but the results are important to astronomers, because there may be some properties in common with those of other planets. In addition, the nature of the interior gives us clues about the manner in which the earth, and presumably other planets, was formed. In fact, the techniques by which earth-

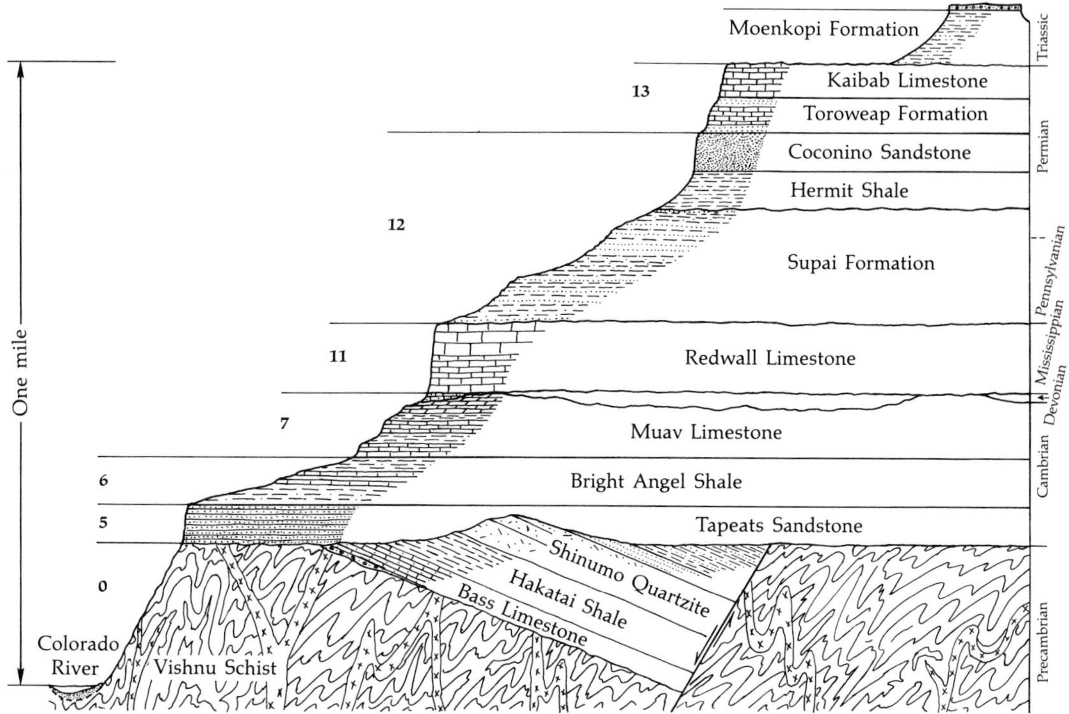

2-7 *Diagram showing labeled layers in the Grand Canyon (see Figure 2-6). Names at the far right relate the ages of the rock layers to the geological time scale discussed in this chapter. (After* Geology Illustrated *by John S. Shelton. W. H. Freeman and Company. Copyright © 1966.)*

quake phenomena are analyzed by the geophysicist are now being applied on the moon. Seismic detectors installed by the astronauts send data back to earth via radio transmission and allow us to deduce the properties of the lunar interior. In the future, similar instruments will probably be sent to Mars and perhaps other planets.

The disturbances produced by earthquakes travel across the globe in the forms of three basic types of wave: surface, shake, and push-pull. *Surface waves* are analogous to the ripples made by dropping a rock or marble into a pond or bathtub. In this case, the wave effect travels for long distances across the water's surface, but individual drops of water do not. Rather, the individual drops are vertically displaced from their rest positions as the wave passes by. (This agrees with our experience; a bather floating in the ocean bobs up and down, not moving toward shore, as a wave passes him en route to the shore. When the bather does drift, it is the result of a current, not the waves.) When the water is displaced above the average level, the *crest* of the wave is passing; when displaced below the average level, the water is in the wave *trough*. This type of wave phenomenon is called *transverse*, because the motion of the material is at right angles to the direction in which the wave is traveling. Surface waves produced by earthquakes are not important for probing the earth's interior, but their effects must be recognized and discarded from the earthquake data. *Shake waves* are produced by a side-to-side (or up-and-down) shaking motion and thus are also transverse. This kind of wave can be simulated by taking the end of a clothesline and shaking it up and down. The individual parts of the rope move up and down, but the wave itself, produced by the motion of one's hand, moves horizontally along the rope. A transverse wave can also be generated by taking one end of a rope (other end tied down) lying on a smooth surface and shaking it from side to side, rather than up and down as in the clothesline example. Unlike surface waves, shake waves can travel inside the earth, but they do not pass through liquids.

Push-pull waves are common; sound waves are a familiar example. The production of push-pull waves can be visualized through the use of a tuning fork. When it is sounded, the prongs move back and forth and the adjacent air is compressed one during each vibration of the tuning fork. This produces a series of regions where the density is alternately more or less than the original value. The motion of individual particles is back and forth along the direction of travel; hence, they are called *longitudinal* waves, and because the air is compressed, they are also often called compressional waves. As we know from experience with sound, this kind of wave can travel through solids, liquids, and through the air. The nature of shake and push-pull waves is illustrated in Figure 2-8.

The speed of a wave depends on the properties of the medium through which it travels. If the nature of the medium changes, both the wave speed and direction may be changed. Where there is an abrupt change (discontinuity) in the properties, the effect on the wave is more drastic and it may, for example, be reflected.

Waves caused by earthquakes are routinely recorded by the seismic observatories located at many spots on the surface of the earth. Consider then the time of arrival of the P (push-pull) and S (shake) waves plotted against distance from the location (epicenter*) of an earthquake. This distance is measured along

---

*The *epicenter* is defined as the point on the earth's surface directly above the actual rock movement that generates the earthquake (Figure 2-9). The measurements of seismic waves, as recorded at different locations, are used to locate the epicenter, as shown in Figure 2-10.

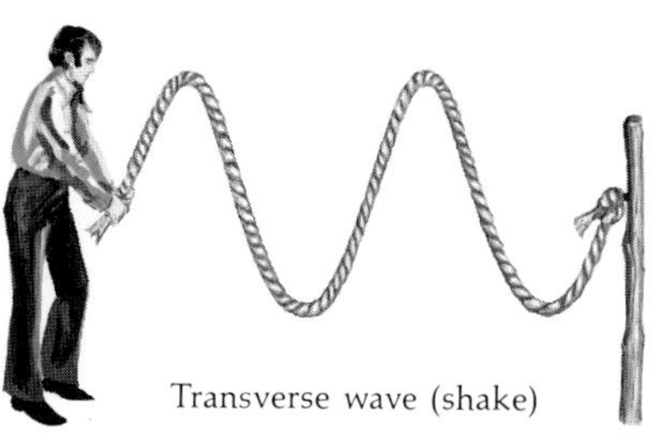

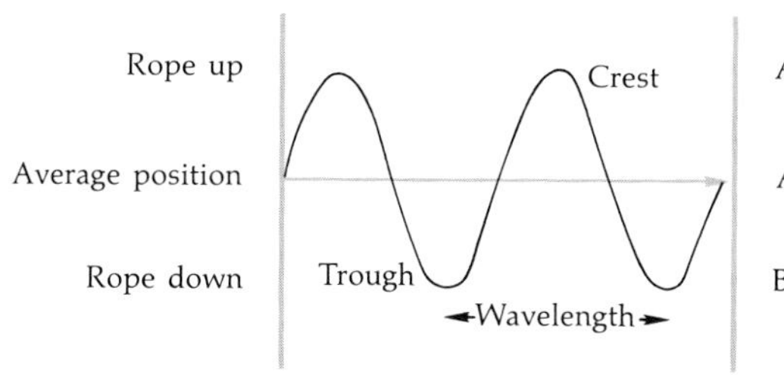

2-8 *Schematic illustrations of transverse waves (shake) and longitudinal waves (push-pull) and their graphical representation.*

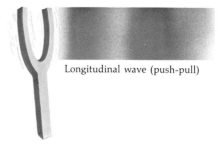

2-9 *Sketch showing the location of the epicenter (on the earth's surface) above the focus of earthquake waves caused by slippage along a fault. (From* Principles of Geology, *3d ed., by J. Gilluly, A. C. Waters, and A. O. Woodford. W. H. Freeman and Company. Copyright © 1968.)*

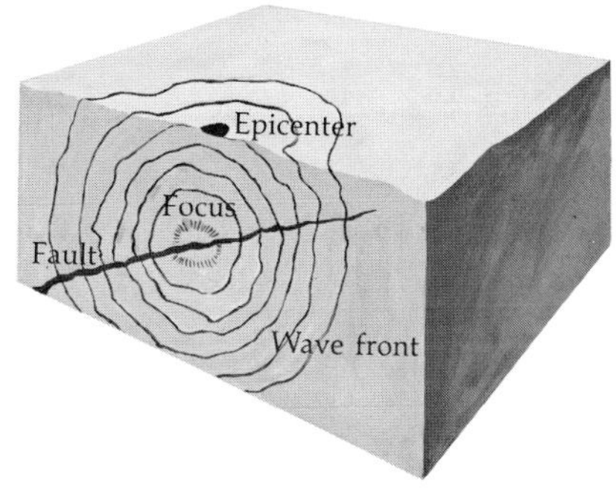

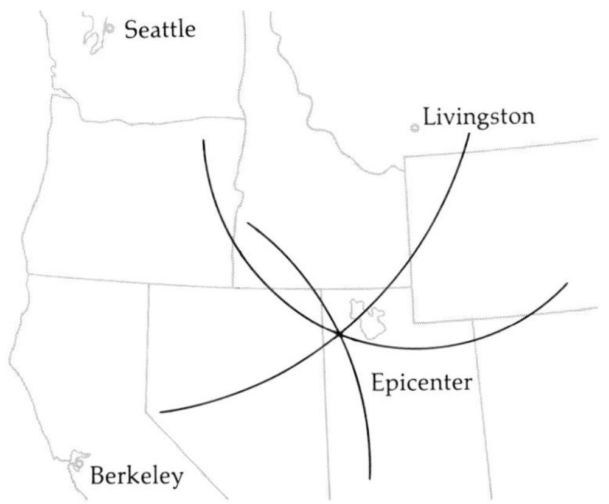

2-10 *Sketch showing how seismic observations made at three distant locations (Seattle, Livingston, and Berkeley) are used to find an epicenter in Utah. Observations at a single station determine the distance to the epicenter and, hence, the knowledge that the earthquake occurred on the edge of a circle of known radius. Data from three or more stations determine an accurate location. (From* Principles of Geology, *3d ed., by J. Gilluly, A. C. Waters, and A. O. Woodford. W. H. Freeman and Company. Copyright © 1968.)*

the surface of the earth, but the P and S waves travel through the earth. Such a plot is shown in Figure 2-11.

Both the S and P waves can normally be detected as far away as 11,200 kilometers (7,000 miles). The S wave disappears at this point, and as these waves cannot pass through liquid, we infer that a liquid region must exist inside the earth. Beyond 11,200 kilometers, the P wave is still observed, but its arrival times at various locations indicate that it has passed through regions where the wave velocities are different. These are the kinds of seismic observations that reveal the presence and rough properties of distinct zones within the earth. The geometry used in this technique is shown in Figure 2-12.

The outermost layer of the earth is called the *crust*; it extends to depths of about 35 kilometers (22 miles) beneath the continental surfaces. Under the oceans, however, the crust is only about 5 kilometers (3 miles) thick. We have never sampled the rock from below the crust, although ambitious experiments have been proposed to drill beneath the sea and obtain rock from the next lower layer.

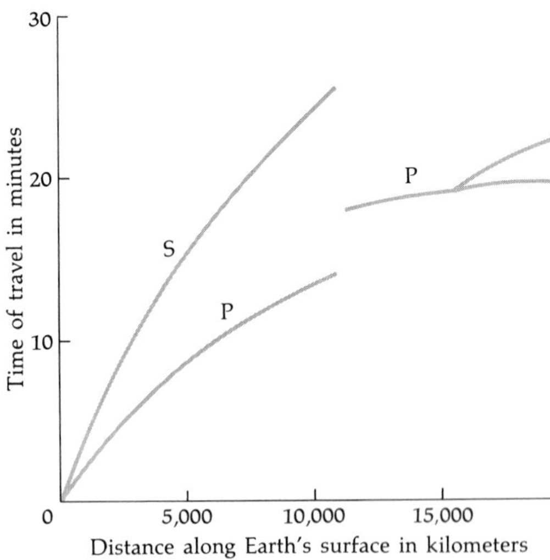

2-11 *A plot of travel times for P and S seismic waves versus distance across the earth's surface. The S waves are not recorded farther than about 11,000 km from the epicenter. The P waves observed past this distance have penetrated the core. The split in the curve for the P wave beyond about 16,000 km is caused by the earth's solid inner core.*

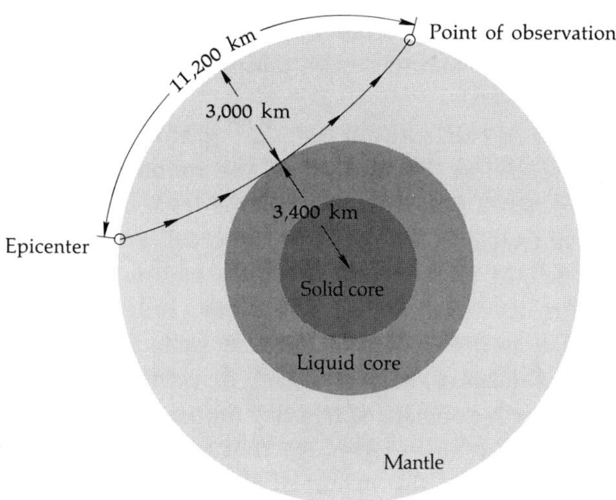

2-12 *Schematic diagram of an S-type seismic wave, as produced and observed respectively at two widely separated locations on the surface of the earth. The example, which is drawn roughly to scale, shows the wave that just grazes the liquid outer core. If an S wave were observed farther than 11,200 km from the epicenter of an earthquake, it would pass closer than 3,400 km to the earth's center. No such waves are observed (Figure 2-11). Since S waves cannot propagate through liquid, we conclude that the boundary of the liquid core is located as shown.*

The existence of a distinct layer (the *mantle*) beneath the crust was proven by the work of a Yugoslavian geophysicist, A. Mohorovičić, in 1909. He studied the P and S waves from relatively nearby earthquakes and found that their speeds changed at a certain place beneath the surface. We now call this the Mohorovičić discontinuity, or *Moho* for short. It marks the boundary between the crust and the mantle, and lies at depths of from 5 kilometers (3 miles) to more than 50 kilometers (31 miles) at different points on earth.

The mantle (see Figure 2-13) extends down to a depth of almost 3,000 kilometers (1,900 miles). The S waves do not penetrate below that depth, which indicates, as mentioned above, that a liquid zone must exist. The region below this discontinuity is called the *core*. Its outer part is certainly liquid, but a change in the traveling characteristics of P waves 2,100 kilometers (1,300 miles) below the boundary of the core suggests the presence of another zone, the *inner core*, which extends to the center of the earth and is probably solid. A cut-away model with some representative seismic wave paths is given in Figure 2-13.

The total mass of the earth, about 6×10^{27} grams (10^{21} tons), is determined by studying the motion of the moon under the influence of the earth's gravity, as explained in Chapter 4. However, when we examine the rocks of the crust, we find that they are much too light to make up the bulk of the earth and cannot account for its total mass. Therefore, much heavier material must exist at great depths, and in fact, the core is probably composed largely of iron. The discovery that the deepest part of the earth consists of heavier matter than we find in the outer regions is a result of geology that is very important in our efforts to understand the origin of the earth. It suggests that all or nearly all of the earth was once hot and fluid, so that the heavier material could sink toward the center as the lighter substances floated upward.

THE CHANGING SURFACE OF THE EARTH

As long ago as 600 B.C., the Greek philosopher Xenophanes noted the presence of a kind of clam shell on mountain tops. The shells resembled those of living clams found along the shore and so he concluded that the sea had once covered parts of the land. This view is now known to be entirely correct. It is clear that the sea position has changed in the past and that many areas that now lie far inland were once beneath the sea. It is only recently that we have been able to gain some insight into many of these changes and their causes. The earth's magnetism is fundamental to this discussion.

The practical value of the ordinary magnetic compass depends on the fact that the earth has a magnetic field. As we study the orientation of compass needles at different locations, we find that they behave as though there were a giant bar magnet inside the earth. The imaginary magnet would be tilted slightly with respect to the axis of rotation, as shown in Figure 2-14. (The basic field pattern of a bar magnet is shown in Figure 2-15.) It is well known that bar magnets and other similar devices lose their magnetism when heated. Further, the temperatures inside the earth are too high to permit a solid magnet to exist. In fact, the leading theory of the earth's magnetism asserts that it is generated by moving currents of liquid iron in the core.

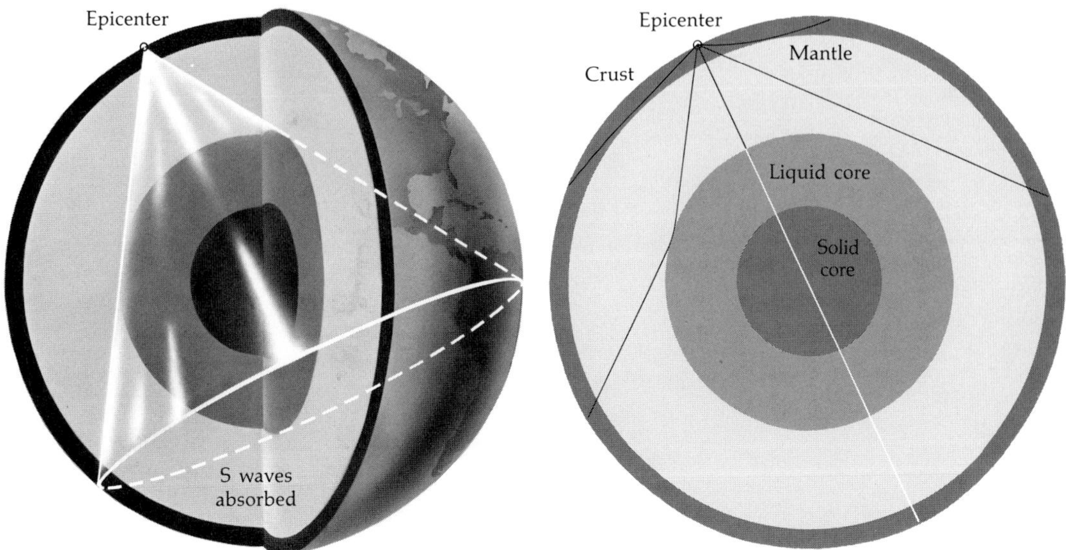

2-13 *Cut-away model of the earth and some sample seismic wave paths. (After* Structure and Change *by G. S. Christiansen and P. H. Garrett. W. H. Freeman and Company. Copyright © 1960.)*

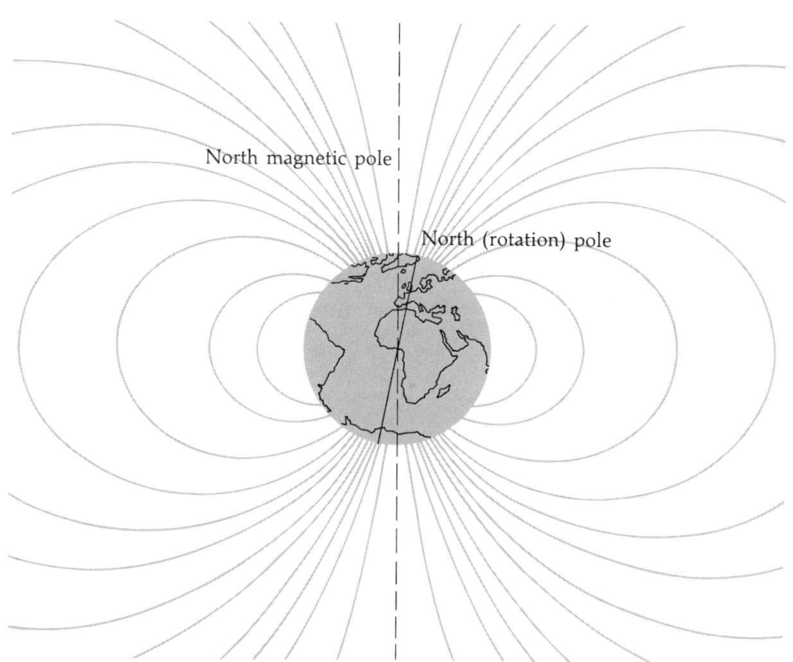

2-14 *Sketch of the earth's magnetic field lines. The magnetic and rotational poles of the earth are separated by about 12° (1,330 km or 830 miles).*

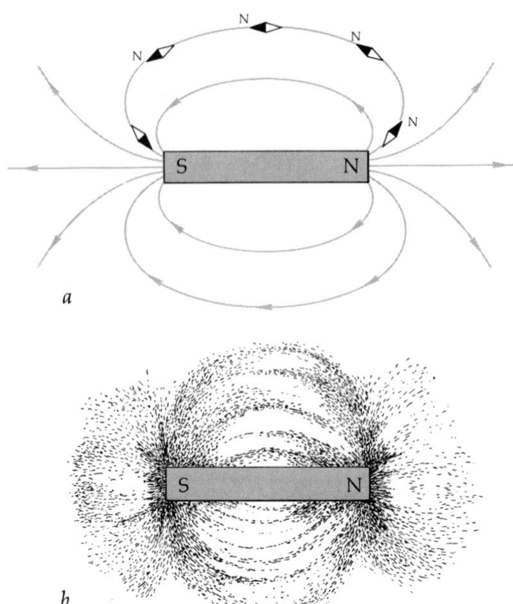

2-15 *(a) Sketch of the magnetic field around a bar magnet as revealed by a compass needle. (After* Structure and Change *by G. S. Christiansen and P. H. Garrett. W. H. Freeman and Company. Copyright © 1960.) (b) The magnetic field of a bar magnet as revealed by iron filings. (After* The New College Physics *by A. V. Baez. W. H. Freeman and Company. Copyright © 1967.)*

As rocks form,* they are magnetized in the direction of the earth's magnetic field; it is as if many tiny bar magnets within the rock line up while it is forming, and are then locked in place when it has solidified. By careful measurements of this *fossil magnetization* in rocks, one can infer the positions of the earth's magnetic poles at the different times of rock formation. The surprising result is that these poles seem to have moved around. However, an equivalent, very plausible interpretation asserts that they have remained in place, but the continents on which we find the rocks have drifted.

The idea of continental drift goes back at least to the time of the British philosopher Francis Bacon (1561–1626). It was

*Rocks form in three basic ways: (1) by the cooling of liquids that come up from far beneath the surface (as in volcanic eruptions); (2) by the action of heat, pressure and chemical effects at lesser depths beneath the surface; and (3) through the deposition of eroded material (sediments) on ocean and lake floors where the material gradually becomes cemented together.

largely stimulated by the remarkable fit between the coastlines of Africa and South America, as shown in Figure 2-16. A geologist once commented that "if the fit between South America and Africa is not genetic, surely it is a device of Satan for our frustration."

There are other kinds of supporting evidence. The presence of ancient glacial deposits suggests that South Africa once occupied a different position, and the locations of coral reefs also reveal the occurrence of major climatic changes that may have been due to continental drift. Coral grows in clear, warm, marine waters, but some fossil reefs are found in areas where the water is very cold. Likewise, fossils found in Antarctica show that it once had a much warmer climate. Finally, the close resemblance of certain fossils found on different continents indicates that the continents may once have been joined or at least were close together.

The evidence favoring continental drift is clearly widespread. Several efforts at reconstruction of the positions of the original land masses have been made. One such attempt (Figure 2-17) shows a set of hypothetical original positions.

Although the evidence for continental drift is reasonably strong, it poses a further, more difficult problem: What causes the drifts? Surely some great force must exist to move these huge land masses.

It was long thought that most of the important geological phenomena took place on the continents and the adjacent continental shelves, and that the floor of the deep oceans, beneath a thin layer of sediments, might be representative of the oldest part of the crust. In recent years, however, a variety of oceanographic observations (including the dating of sea-floor rock samples) have profoundly changed this picture, leading many geologists to conclude that quantities of fresh rock are slowly emerging from the mantle along the length of a submarine mountain range and are spreading out laterally, forming new crust that advances toward the continents at rates of up to several inches per year. It is believed that this phenomenon of *sea-floor spreading* causes the drift, and that the continents consist of blocks of crustal material (*plates*) that slowly "float" across the heavier rock of the mantle. Land masses not only move across the earth's surface, they also move up and down. In some parts of the world, such as Scandinavia, large areas formerly covered by glaciers are slowly rising. They are making an *isostatic adjustment* to the disappearance of a heavy ice layer.

The sea-floor spreading itself must originate in rock flow within the mantle. The seismic wave studies show that the

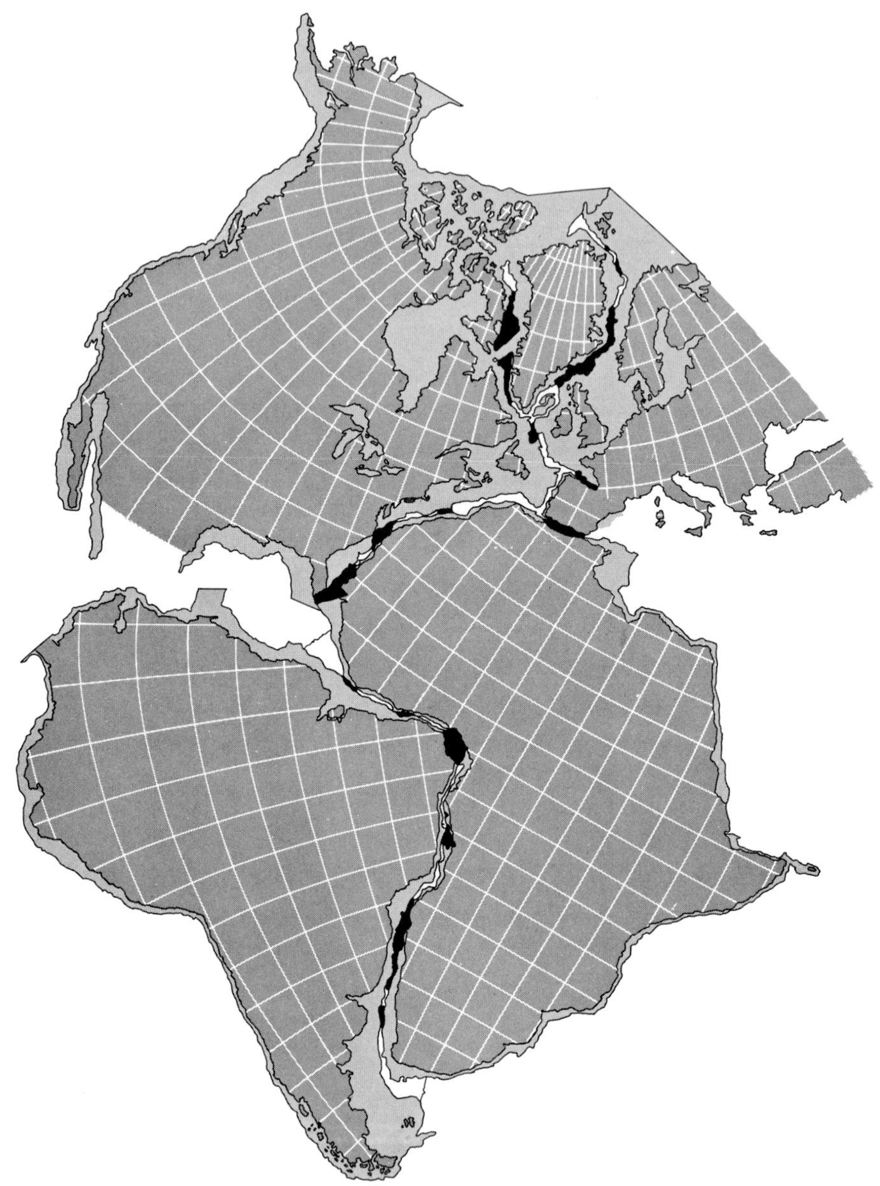

2-16 *Detailed arrangement of the continents obtained by fitting them together on a computer. The fit is excellent; the average disagreement over most of the boundary is no more than 100 km (70 miles). The regions where land masses overlap are black; gaps where there is no overlap are white. The good agreement of the fit is strong evidence for the grand scheme shown in Figure 2-17.* (*From* The Confirmation of Continental Drift *by P. M. Hurley.*

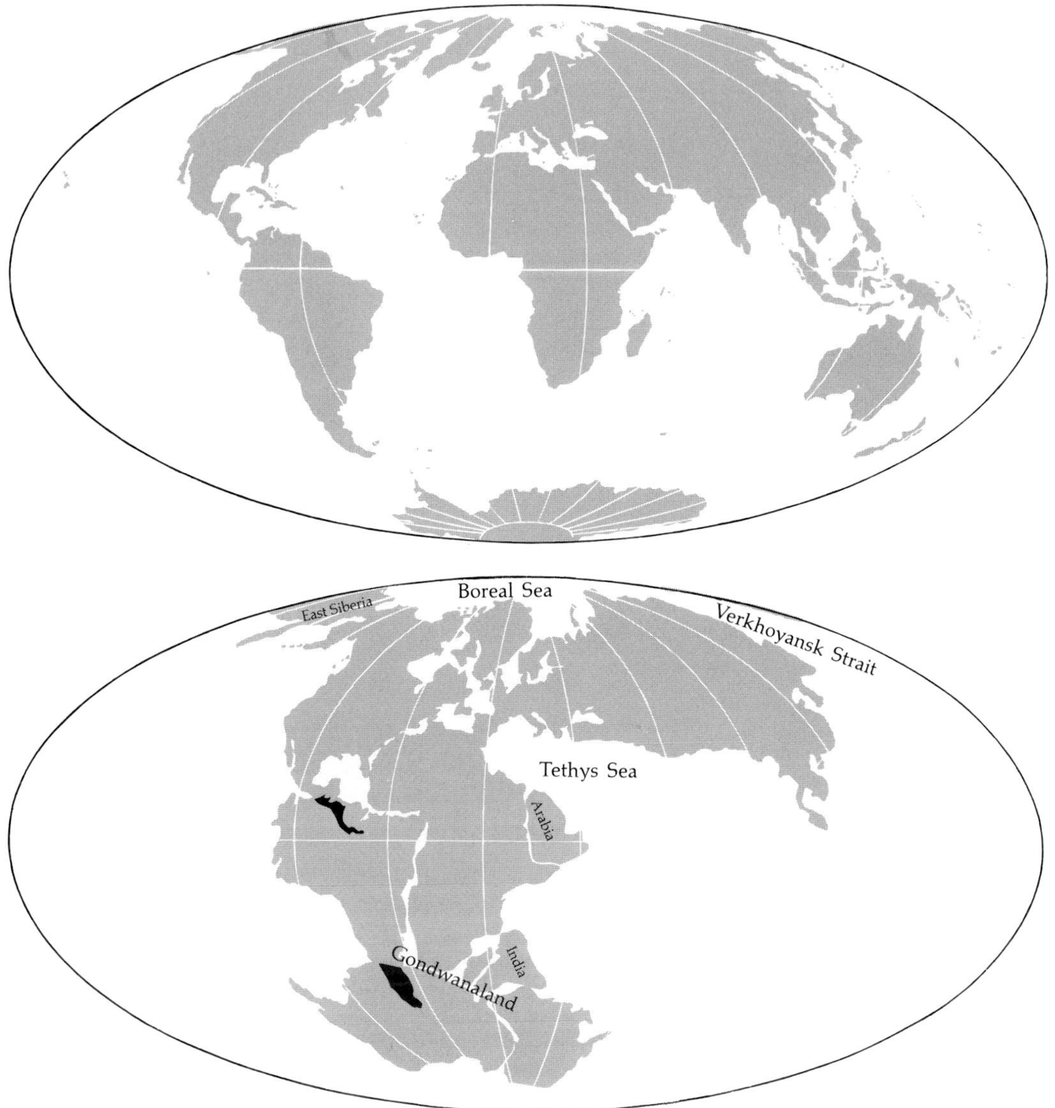

2-17 *The present-day distribution of continents (top map) as compared with the single supercontinent that may have existed some 150 million years ago. Dark areas indicate overlaps. (From* Continental Drift *by J. T. Wilson. Copyright © 1963 by Scientific American, Inc. All rights reserved.)*

material which makes up the mantle resembles a solid, and we usually think of solid rock as a rigid substance, more apt to break than to bend under force. However, rock will gradually yield under the influence of a sufficient load or force, and the rock in the mantle, subject to immense forces, has a plastic, deformable character. The influence of rock currents in the

a

2-18 *Two examples of folded strata. (a) Large-scale folds as seen looking northwest near Borah Peak, Idaho. Note the sizes of the folds as compared with the pine trees in the lower portion of the picture. (From* Geology Illustrated *by John S. Shelton. W. H. Freeman and Company. Copyright © 1966.) (b) Small scale folds near Barranca de Tolimán, Hidalgo, Mexico. (Courtesy of Kenneth Segerstrom, U.S. Geological Survey.)*

b

mantle provides a natural mechanism for mountain building. Where the mantle currents meet or sink, rock can be piled up into mountains. Alternately, the collision of continental blocks could produce mountains by squeezing and resultant buckling of the material. The Himalayas may have been formed in this way when the Indian block merged into the main body of Asia. The folding and buckling forces that build mountains act at great depths, although it is at the surface that their results are so impressive (Figure 2-18).

Earthquakes and volcanoes commonly occur in the same areas, providing dramatic evidence of the forces and events involved in shaping the crust. These areas are often marked at the surface by *fault lines*. Perhaps the best known is the San Andreas fault that passes very close to San Francisco. The plate to the west is slowly moving to the northwest with respect to the other plate. Figure 2-19 shows the effect of slippage along

2-19 *Displacements along a fault line are seen in an orange grove near Calexico, California. The view is toward the west and the trees above the fault line were displaced about 4.4 meters (14.5 feet) in a northwesterly direction. (From* Geology Illustrated *by John S. Shelton. W. H. Freeman and Company. Copyright © 1966.)*

a related fault. Earthquakes occur when the deformation of the rocks exceeds their strength. The crust then fractures and the plates shift so that the deformation decreases to an amount that the rock can tolerate. Farther to the north, along the Aleutian Island chain of Alaska, the plate boundary is marked by the presence of many volcanoes.

Volcanoes also represent a mechanism for mountain building and for altering the nature of the surface. An example is Parícutin (Figure 2-20), which was seen to form in 1943. This occurs through the build-up of cinder cones and the flow of lava or molten rock. *Magma* (the molten rock within a volcano) is believed to originate in pockets or localized regions beneath the surface. How this occurs and at what depths are still matters of conjecture.

Finally, one other major agent in changing the surface of the earth should be discussed: *glaciation*, the effect of the motion of large masses of ice across the crust. During the ice ages glaciers covered much of Europe and also the eastern North

2-20 *Parícutin erupting at dawn on February 20, 1944, one year after it first rose through the ground. (Courtesy of Tad Nichols.)*

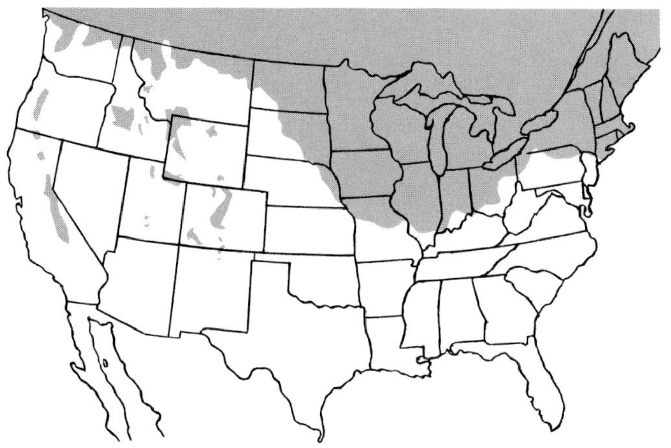

2-21 *Map showing the extent of glaciation in what is now the United States during the last 2 or 3 million years. (From* Principles of Geology, *3d ed. by J. Gilluly, A. C. Waters, and A. O. Woodford. W. H. Freeman and Company. Copyright © 1968.)*

American continent down to approximately the Missouri and Ohio rivers (Figure 2-21). The last great retreat of the ice began about 20,000 years ago and continued until about 6,000 years ago. Typically, glaciers move at about 50 meters (160 feet) per year. As they flow across the earth they alter the surface by abrasion, producing valleys, hills of debris, and other topographic features. Relatively isolated glaciers are known today in major mountain ranges, in addition to the glacial areas of Antarctica and Greenland. Glaciers also constitute a large reservoir of water, and as such, their growth or decline can cause dramatic changes in the sea level and thus the position of the seashore. We are currently in an age of relatively little glaciation, but if the remaining ice were melted, sea level would be raised by about 50 meters, enough to flood extensive coastal areas, including many major cities.

Although geologists and meteorologists have not reached a consensus on the cause of the ice ages, it seems that rather small changes in the average temperature were involved. It was colder during the last glacial epoch than at present, but probably only by about 10° F. It may be important that the average temperature over the surface of the earth increased by 1° or 2° F in the 100 years following 1850, a date that corresponds roughly to the beginning of the industrial revolution. The gaseous waste (largely carbon dioxide) produced by man's activities might be involved in raising the average temperature of the earth (see the discussion of the "greenhouse effect" in Chapter 8). If so, increased pollution or even the maintenance of the current level of pollution could hasten the melting of the glaciers and thus cause a rise in the level of the sea. However, this effect of pollutant *gas* might be offset somewhat by the *dust* released into the air by industry and agriculture. The floating dust particles reflect some sunlight back into space, thereby reducing the amount of solar energy that heats the ground. Indeed, the average temperature of the earth may have begun to *decrease* in recent years as a result of this effect. Thus, too much dust in the atmosphere could produce another ice age.

3

Evolution and Life

In ancient times, Herodotus recognized certain fossils as the remains of living organisms, and Anaximander and other Greeks discussed some concepts of evolution. Later, the interpretation of fossils as evidence of ancient life forms was strongly opposed in Europe. The existence of fossils was indisputable, and the arguments used to dismiss their significance frequently appealed to the miraculous. Fossils were described variously as accidental "sports" of nature, tricks of the Devil, or as remains of life buried in the Biblical flood of a few thousand years ago. A few men did make significant contributions to evolutionary thinking during the seventeenth and eighteenth centuries but, although they recognized the fact that evolution had occurred, they were unable to account for it or to back up their hypotheses with critical evidence. It remained for a careful observer of nature, Charles Darwin, to recognize in the properties of living organisms evidence for the nature of evolution, and thus to show (as did Hutton and Lyell in geology) that the present is the key to the past.

DARWIN AND NATURAL SELECTION

During the years 1832–1836, Darwin was an unpaid naturalist on H.M.S. *Beagle*, a ten-gun brig of the Royal Navy engaged in surveying Patagonia, the west coast of South America, and some Pacific islands. This voyage took him literally around the world and included explorations ashore in South America and in the Galapagos Islands off the coast of Ecuador. Darwin made extensive biological and geological observations which he later analyzed and used as the basis for his theory of natural selection. In fact, he thought about the evidence and accumulated additional facts for more than 20 years, and only when he learned that another scientist had formulated a similar theory was he moved to announce his own conclusions. Darwin's classic work, *The Origin of Species,* was published in 1859.

In Chapter 2, we discussed the evidence for evolution that comes from the study of fossil remains of ancient life forms. Darwin's observations during the voyage of the *Beagle* led him to recognize the process of evolution on the basis of the nature of *living* organisms. The Galapagos finches are a famous example. When a Captain Colnett surveyed the Galapagos in 1793, he found the local birds to be rather uninteresting, but years later Darwin studied the finches on these islands and found some of his best evidence for evolution.

Thirteen species of finches have been recognized on the Galapagos, and some of them behave in a very unfinchlike manner, such as the "woodpecker-finch" with a long pointed beak that preys on insects living in the bark of trees. Lacking the long tongue of a true woodpecker, it uses twigs to dislodge the insects (Figure 3-1). Other finches on the islands have short beaks and eat seeds that they find on the ground. Still another type has a beak suited to its habit of feeding on prickly pears. In fact, all of the Galapagos finch species are roughly alike, their most significant differences being in the size and shape of their beaks, and the nature of their diets. Different islands among the Galapagos are inhabited by different groups of species. Darwin concluded that "from an original paucity of birds in this archipelago, one species had been taken and modified for different ends." At one time, a single finch species came to the relatively barren Galapagos from the mainland. Living in isolation, the birds on different islands evolved by adaptation and specialization into distinct species, each capable of most effectively harvesting a particular part of the food supply. Probably, no true woodpeckers were around to feed on bark-dwelling insects, and so a finch species evolved that could

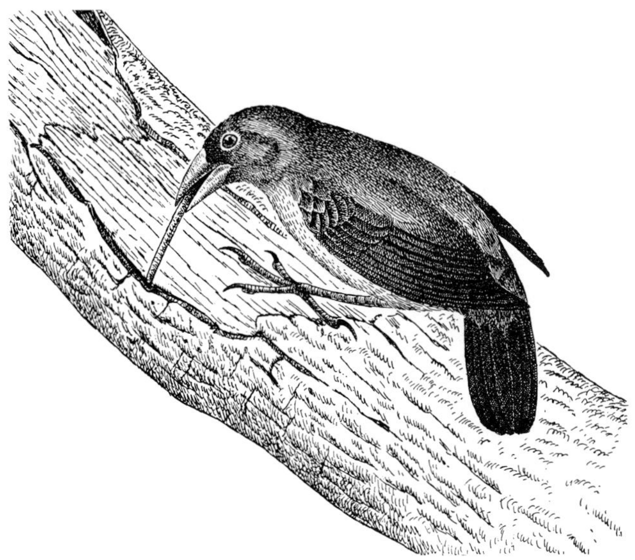

3-1 *The woodpecker-finch of the Galapagos Islands, where Darwin made a classic study. In this sketch it is using a twig to pry out insects from the bark of a tree. (From* Darwin's Finches *by David Lack. Copyright © 1953 by Scientific American, Inc. All rights reserved.)*

take advantage of this particular source of food. In the language of the biologist, they filled the *ecological niche* of a woodpecker on the Galapagos. The divergence of the original species of Darwin's finches into a variety of species, each able to survive in a different manner and thus not infringing very much on the resources available to the related species, is the classic example of the process of *adaptive radiation.*

The properties of the Galapagos that made them an excellent laboratory for evolution were isolation and a relative scarcity of animals that could effectively compete with the finches for food or prey on the birds. But these same two characteristics made the Galapagos unrepresentative of the general situation on the continents. Darwin noted that populations tend to double at characteristic intervals. The human population doubles about every 35 years; a colony of rabbits allowed to reproduce freely in the laboratory will increase at a faster rate, and many insects even faster yet. The question arises: Why isn't the earth covered with rabbits or flies or whatever? This has not happened because each species is in a delicate *ecological balance* with its surroundings. Rabbits and flies have enemies that prey on them; predators, in turn, are limited by the available food supply.

Thus, we can envision a delicate balance with competition between all species. Any slight advantage or disadvantage may be crucial for survival. The climate may change or the food supply may die out; here we encounter the general problem of adaptability to a changing environment. Success for a species is essentially the ability to stay alive *and* to reproduce. A species that can survive the rigors of the environment and the attacks of predators is still doomed if its members cannot reproduce.

Darwin reasoned that reproduction was not perfect, but that changes occur, which we now call *mutations*. The mutations are essentially random; some of them produce species of more efficient competitors (beneficial mutations) but many do not. The *principle of natural selection* states that the organisms with a higher survival probability will eventually dominate the population, as they can best feed themselves and reproduce. Thus, natural selection or the "survival of the fittest," while not the fundamental cause of evolution, determines the direction of evolution when changes occur.

Evolutionary effects generally require a great many generations to become established. Since insect lifetimes are relatively short, it is not surprising that one of the best documented cases of natural selection in action is that of an insect, the peppered moth. This moth is found in the British Isles where it flies about its business at night and rests on rocks and trees in the daytime. As its name implies, the peppered moth is, or was, characterized by a spotted pattern of light and dark markings. In 1848, when a rare, all dark mutant of this moth was first noticed, the tree trunks in England were lighter in color than most of them are today, because there was less industrial pollution. Many of the trees and rocks where the moths rested were covered with light-colored lichens (Figure 3-2). The spotted pattern of the normal peppered moths camouflaged them as they rested. The mutant dark moths, however, were easily seen and devoured by the birds. As the development of industry progressed, soot settled on and darkened the tree trunks, and frequently killed the lichens. Now the pepper pattern was easily discerned by predators, and it was the dark moths that blended in with their surroundings. The dark color had become a beneficial mutation and by the start of the twentieth century the dark moths formed the great bulk of the population.

Despite the evidence of an enormous variety of now-extinct life forms revealed by the fossil record, and despite Darwin's brilliant explanation of evolution as a result of natural selection, the theory of evolution met enormous resistance. In large part this was based on the conflict with the Biblical account of the

a

b

3-2 *Moths and survival. Each tree has both a peppered and a dark moth, but on the dark tree trunk (a) the peppered moth is clearly visible to predators. On the tree covered with light-colored lichens (b), the dark moth is easily seen. (From the experiments of Dr. H. B. D. Kettlewell.)*

Creation. There was also the rather unpleasant feeling that surrounded the idea that we were descended from the apes or related to them through common ancestors.

There were also quite reasonable scientific objections that faced Darwin and his followers. For example, evolution is clearly a slow process, and the critics doubted that enough time had been available. (Remember that it was only in the present century that we determined the true age of the earth.) Another objection was based on the fact that evolution is described as a gradual process, but the purported evolutionary lines of descent of many species lack intermediate forms. Darwin's answer was that the fossil record is incomplete. Fossils may be destroyed by heat, pressure, and chemical processes within the earth, and by erosion. In addition, the conditions necessary to form and preserve fossils may not always exist. Finally, the completeness of the known fossil record also depends on our ability to find and classify the fossils. In fact, some fossils of crucial intermediate stages in evolution have been found; a famous example is the *Archaeopteryx* (Figure 3-3). Birds are believed to have evolved from reptiles, and *Archaeopteryx* (found in Jurassic rocks) is a mixture of bird and reptilian characteristics; it has the tail, brain, and teeth of a reptile and the feathers and feet of a bird.

a

b

3-3 (*a*) *Cast of an archaeopteryx fossil. Note the feather impressions. (Courtesy of the American Museum of Natural History.)* (*b*) *Model of how archaeopteryx may have looked. (By permission of the Trustees of the British Museum—Natural History.)*

A key difficulty with the theory of evolution was its failure to explain how a mutation first occurred and how it was passed on to succeeding generations. Darwin could not answer this; the explanation was provided later by the new science of genetics. He also did not explain the origin of life, a problem that the modern chemists and biologists are now only beginning to explore.

MUTATION AND INHERITANCE

The basic rules of heredity were discovered by the Austrian monk Gregor Mendel through experiments with the reproduction of sweet peas, and they apply to all living organisms. An organism has a large set of traits; the traits are determined by genes in the cells of the organism. In a man, a pair of genes for each trait is inherited from his parents, one from the father and one from the mother. Genes may be *recessive* or *dominant*. The gene for brown eye color is dominant and blue is recessive. This means that if a man inherits a brown gene from one parent and a blue gene from the other, he will have brown eyes. Only when both genes are blue will the man exhibit the recessive characteristic of blue eyes. The situation is not quite as simple as we have indicated; for example, sometimes several genes will influence a particular trait and some of these genes may also determine other traits. Mendel's work proved that genes must exist. Much of modern biochemistry has been devoted to determining what genes actually are and how they control inheritance and pass on to succeeding generations the detailed specifications of an organism so that it will bear a clear resemblance to its parents, according to certain rules. This detailed specification is called the *genetic code*. As we have noted, Darwin was largely ignorant of the cause of mutation and the rules of inheritance. These problems were among the most serious pointed out by his critics. They could not be answered satisfactorily until some understanding of the gene and how it determines heredity was attained.

A living cell is composed of a *nucleus*, a surrounding fluid of *cytoplasm*, and the cell walls (Figure 3-4). Biologists long ago determined that structures called *chromosomes*, which are found in the nucleus, must contain the genes. More recently, it has been shown that a substance known as *DNA* (deoxyribosenucleic acid), which is contained in the chromosomes, is fundamental to the story of heredity.

The DNA molecule (Figure 3-5) is composed of subunits or building blocks called *nucleotides*. The molecule is very long (in man, DNA is composed of some 10 billion atoms) and consists

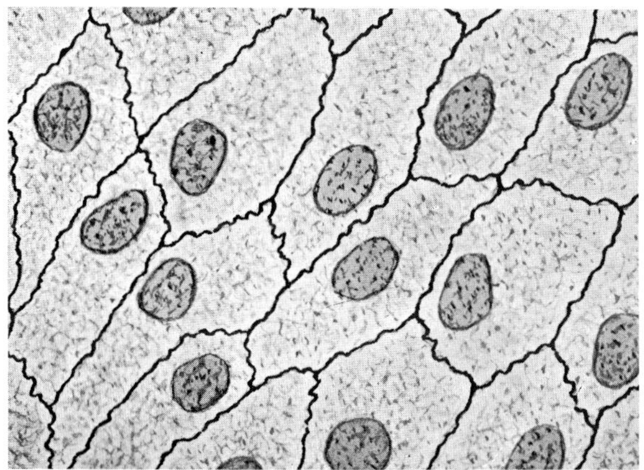

3-4 *View of human cheek cells (treated with silver nitrate) showing the nucleus, cytoplasm, and cell walls. (Courtesy of W. M. Copenhaver.)*

of two long, corkscrew-shaped strands arranged in a double helix (Figure 3-6). Relatively short strings of the nucleotides in DNA appear to determine inherited characteristics, and thus the genes, whose existence was first deduced from experiments in growing varieties of the sweet pea plant, can now be identified as chains of nucleotides in the DNA molecule.

The DNA molecule has another remarkable and vital property; that is, the ability to self-replicate. At the appropriate time the two sections of the double helix unzip or split down the middle and unwind. At this point the chemical bonds that formerly held the two parts together are unfilled. The two parts float in the cytoplasm which also contains many nucleotides. Generally, only the proper kind of nucleotide can fill a particular broken bond and thus each of the coils of the DNA molecule acts as a mold or template and can copy itself (Figure 3-7). This process of DNA replication allows a complete set of genetic instructions to be passed to each of the daughter cells.

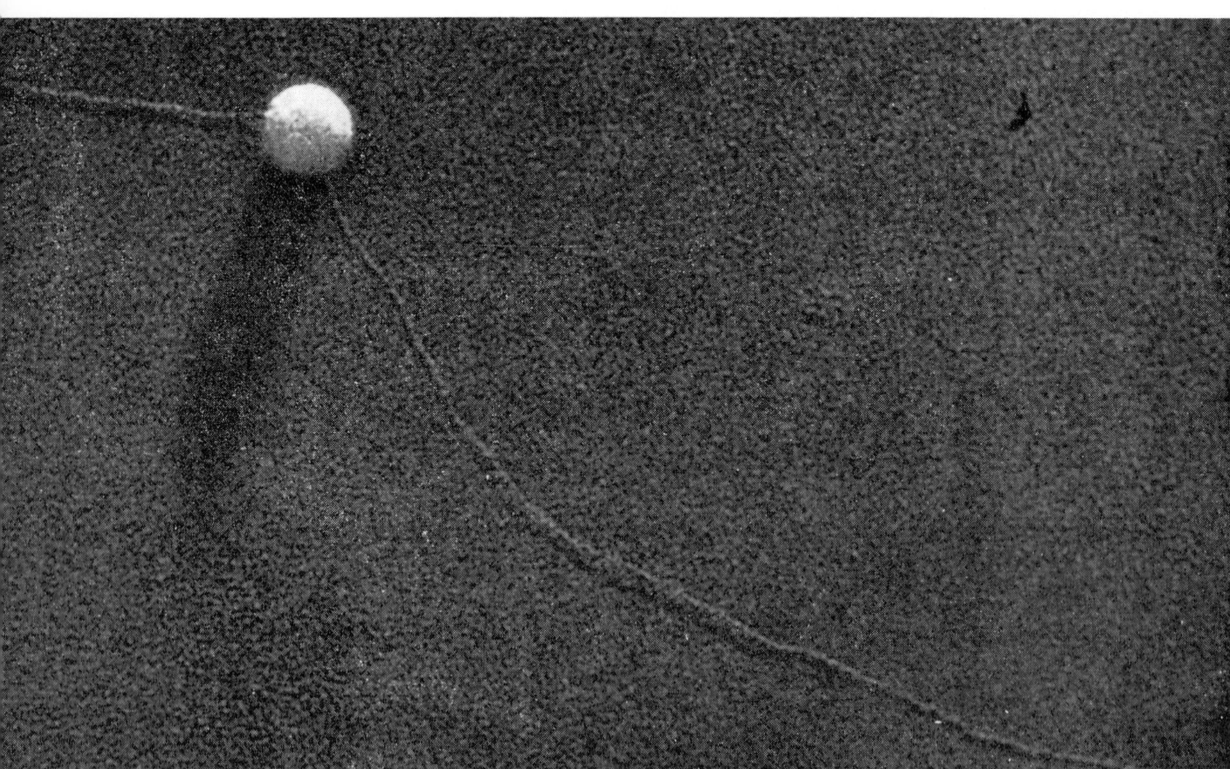

3-5 *A small part of a DNA molecule, as recorded with an electron microscope. The white sphere, a tiny piece of plastic used as a measuring stick to indicate the size of the molecule, has a diameter of* 9×10^{-6} *centimeter. (Courtesy of Cecil E. Hall, Massachusetts Institute of Technology.)*

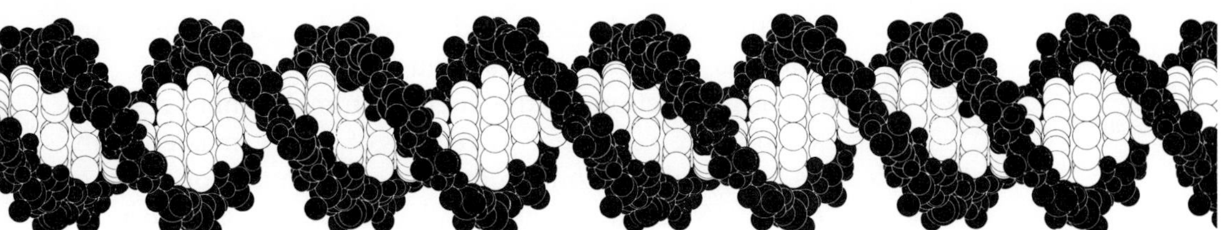

3-6 *A model of the DNA molecule. The black spheres represent the repeating units of deoxyribose sugar and phosphate that give DNA its double-helix structure. The white spheres between the two helical strands represent the nucleotides that carry the genetic code. (From* Gene Structure and Protein Structure *by Charles Yanofsky. Copyright © 1967 by Scientific American, Inc. All rights reserved.)*

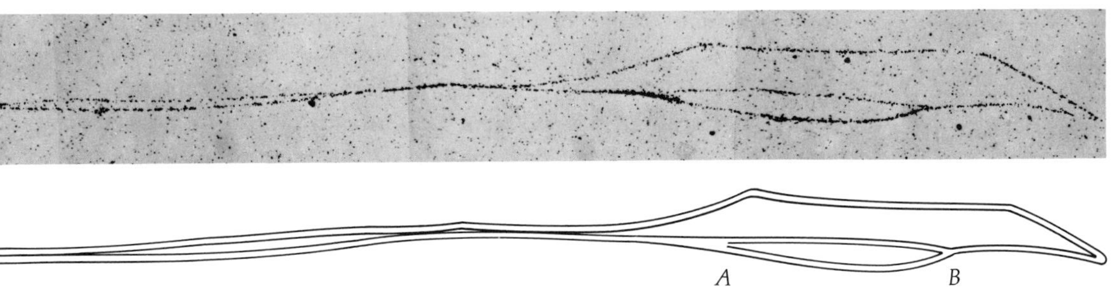

3-7 *A segment of DNA engaged in the replication process; thus far it has duplicated the portion of itself from A to B. (Courtesy of John Cairns.)*

If some event changes the arrangement of the nucleotides in a DNA molecule and that molecule is then involved in reproduction, a mutation may result. Thus, changes in the genetic code are the raw material of evolution. The mutations have a variety of causes. First, simple copying errors may occur in the DNA self-replicating process; second, the genetic material may be changed by exposure to cosmic rays,* radioactivity (both natural and artificial), and certain chemicals. The rate at which mutations occur due to the natural causes (copying errors, cosmic rays, mineral radioactivity) is estimated at one per 100 million reproductions. It appears that a small mutation rate is more advantageous from the standpoint of natural selection because most mutations are probably undesirable. Recall that most species are in a state of rather delicate balance with their surroundings. The probability of a favorable mutation can be compared with the odds that your car will be improved if a mechanic removes one part at random from your car and re-

*Fast-moving atomic particles from space; see Chapter 15.

places it with a part chosen blindly from the parts shelf. An improvement is not likely unless you have a real lemon!

Bisexual reproduction in the higher animals enhances the chance of mutation because two sets of DNA master plan molecules are involved—one in the male sperm and one in the female ovum. When the sperm and ovum unite and form a fertilized egg, the new organism develops from a combined master plan based on the DNA molecules of both parents, according to the dominant and recessive traits and the rules of inheritance. It must be realized that even recessive traits that are not evident in one parent's make-up can show up in the child if the other parent also has that recessive trait. Thus, even though not apparent in the parents, traits coded in the DNA molecules can occur in the children or in subsequent generations.

In summary, Darwin showed that the natural selection of superior characteristics governed the evolution of life. Mendel's work was the basis of the genetic theory that established the rules for the inheritance of traits. Modern biochemistry has revealed the detailed way in which inheritance of characteristics is actually controlled.

THE ORIGIN OF LIFE

Is there a scientific basis for the origin of life? Few questions have so great a philosophical importance or have so divided thinking men. Several hypotheses concerning the origin of life on earth have been made. First, a supernatural event can be invoked. This possibility is unattractive to scientists because it seems more reasonable to accept a natural process if one can be found. In any case, the supernatural is by definition outside the realm of science.

Second, it has been postulated that life came to the earth in the form of spores or microorganisms that drifted through space from some other point in the universe. This *panspermia* hypothesis is not widely believed today. Although there is some dispute over whether or not such spores could survive the harsh environment of space, the theory in any case begs the question—it does not tell us how life began but only claims that it began somewhere else. The same remark applies to what has been called the "garbage theory"—the idea that life began on earth at some time in the distant past when space travelers from another world visited our planet and contaminated it, whether accidentally or otherwise.

Third, life may have originated in common substances under physically reasonable circumstances. Modern research lends plausibility to this concept.

The cells that make up all living organisms are distinguished by the presence of four major types of substance: *carbohydrates, fats, nucleic acids** and *proteins.* The modern approach to the problem of the origin of life has been to determine how these organic substances could have come into existence. Laboratory experiments have simulated conditions that probably existed during the early history of the earth and have produced chemicals that are among the subunits or building blocks of these four kinds of organic material.

The experiments on the origin of life are based on facts and deductions from several fields of science. For example, chemistry tells us that the proteins could not have formed in the presence of much oxygen, although there is a great deal of oxygen on the earth today. On the other hand, we do not find much oxygen when we study the atmospheres of the other planets (Chapter 8), and in fact the various planets of the solar system have different atmospheric compositions. Astronomers believe that the sun and all its planets were formed from the same cloud of gas and dust, so the planets should have started out with fairly similar atmospheres. As the atmospheres are very different now, we conclude that at least some of them have changed since the planets were formed. In the case of the earth, geology and biology provide the evidence for the occurrence of this *atmospheric evolution.* It appears that most of our oxygen has been produced since life began, and is a result of photosynthesis in plants. Most of the other gases in the present atmosphere and most of the water have probably been released from the interior of the earth by volcanic activity.

The gas mixture that surrounded the earth during its early history is called the *primitive atmosphere* to distinguish it from the present atmosphere. According to one of the leading theories, the primitive atmosphere was largely composed of methane gas, with some ammonia, hydrogen, and water vapor. The fact that fairly similar mixtures exist in the atmospheres of some of the larger planets today lends credibility to the idea that the earth once had such an atmosphere.

Protein does not occur as a raw material in the crust of the earth. It is produced by living organisms and is the most complex kind of naturally occurring substance. Proteins play vital roles in the life processes of all organisms. For example, the

---

* DNA and a similar molecule, RNA.

enzymes that regulate so many of the internal chemical activities of our bodies are forms of protein. It is not surprising, therefore, that the investigations of the origin of life have emphasized the search for a way in which proteins could have been formed abiologically (without the presence of life).

The many proteins that are found in living organisms consist of distinct arrangements of only twenty subunits, or building blocks, called the *primary amino acids.* Laboratory experiments on the origin of life have not produced proteins, but they have yielded most of the primary amino acids. There have been a variety of experiments, but the basic methods are similar: a mixture of gases that simulates the composition of the primitive atmosphere is exposed to an energy source, such as an electric spark, heat, ultraviolet light, or a beam of atomic particles. This causes changes in the gas mixture, and the newly formed chemicals are analyzed; some of them have been identified as amino acids. These experiments have also produced subunits of the nucleic acids, and other constituents of living matter.

A great deal of further experimental work will be necessary to outline the methods by which life originated, but it does not seem beyond our reach to solve this problem. The present idea of the probable sequence for the development of life begins with the formation of the amino acids and other subunits of organic matter in the primitive atmosphere and oceans through the action of naturally occurring electrical discharges (lightning), heat, ultraviolet radiation (from the sun), and/or cosmic rays. The next important stage took place in water—perhaps the oceans, tidal pools, or even warm springs on land. Molecules of the amino acids came together by chance in various combinations. (Laboratory experiments show that amino acids in water will organize themselves into larger molecules.) Given millions of years, eventually the right combinations of amino acids were formed and the first proteins were created. The same kind of process over a long span of time resulted in the formation of the nucleic acids from their subunits. Eventually, the chemical activity in this *primary broth* led to the first cells. Obviously there are many gaps in this argument, but it seems to be a reasonable working hypothesis. It appears that, given the conditions astronomers and geologists believe to have prevailed on the primitive earth, life could have arisen through a long sequence of chemical processes. Once life existed, evolution by natural selection must have begun.*

---

* Also see the discussions of amino acids in meteorites (Chapter 7) and of interstellar molecules in the galaxy (Chapter 12).

Table 3-1 *Milestones in the Origin of Life on Earth*

YEARS AGO	EVENT
4.5×10^9	Formation of the earth
4.2×10^9	Occurrence of the first self-replicating entity; beginning of evolution by natural selection
3.4×10^9	Appearance of bacteria and blue-green algae
2.1×10^9	Multicellular life

The oldest known fossils, identified as bacteria and blue-green algae, have an age of about 3.4 billion years. A rough timetable of key events in the origin of life, based on the thinking of current researchers in this field, such as Carl Sagan of Cornell University, is given in Table 3-1.

THE FLOW OF EVOLUTION

The first evolutionary steps from the one-celled organisms to many-celled or relatively complex animals undoubtedly took place in the sea and would involve the specialization of activities among the different cells. Before such evolution could proceed very far, there would be a differentiation of life forms based on how food was obtained. Growth of an organism depends on the availability of nutrients; the supply produced by random chemical combinations is limited and unreliable.

A regular or dependable food supply is highly desirable. This is provided by plants and here the central figure is chlorophyll, a large molecule (but not as large as DNA) with an intricate structure. Chlorophyll molecules act as catalysts in the chemical reactions which produce carbohydrate nutrients such as sugar directly from the raw materials of light, carbon dioxide, and water; this is the process called *photosynthesis.* Oxygen is released as a by-product of this process. Nearly all the oxygen in the present atmosphere of the earth was produced by photosynthesis.

Plants provide the basic food for all animal life. In a sense, they are the only productive organisms because only they assimilate raw inorganic materials into living matter. Animals are destructive in that they require nutrients in the form of plant tissues or the flesh of other animals to grow and survive. Thus, animal life could not have arisen until plant life developed.

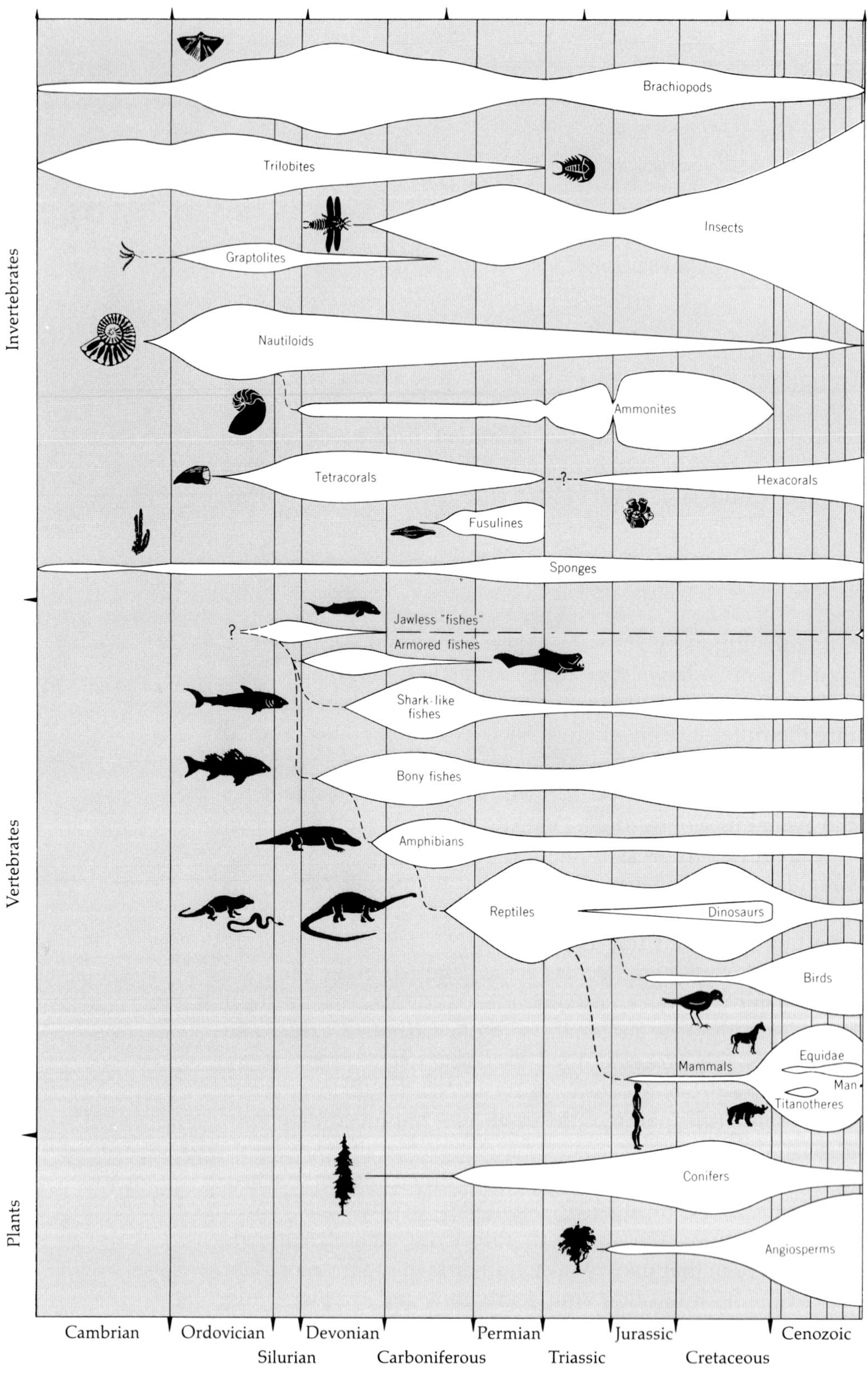

Invertebrates
Vertebrates
Plants
Brachiopods
Trilobites
Insects
Graptolites
Nautiloids
Ammonites
Tetracorals
?
Hexacorals
Fusulines
Sponges
Jawless "fishes"
?
Armored fishes
Shark-like fishes
Bony fishes
Amphibians
Reptiles
Dinosaurs
Birds
Mammals
Equidae
Man
Titanotheres
Conifers
Angiosperms
Cambrian
Ordovician
Silurian
Devonian
Carboniferous
Permian
Triassic
Jurassic
Cretaceous
Cenozoic

3-8 *Historical evolution chart showing the geologic periods of principal life forms. The widths of the areas that represent various types of plants and animals represent the extent of diversity of species. (From* Geology Illustrated *by John S. Shelton. W. H. Freeman and Company. Copyright © 1966.)*

Our knowledge of evolution is based on the fossil record and the reader should consult Table 2-1 for a brief outline of the events recorded in the rocks. Multicellular life is believed to have started in the sea and may have been similar to the present-day *phytoplankton,* the one- or several-celled aquatic green plants. The phytoplankton are the basic food for the small water animals, such as the protozoa, and may be the earth's single most important source of raw food and oxygen.

The principal mileposts of evolution are summarized in Figure 3-8, but some specific developments deserve brief mention. The fossil record for Precambrian times is scarce and consists primarily of aquatic plants. By about one billion years ago, animals had developed to fairly complex forms, such as worms, and occasionally fossil worm holes or worm tracks are found (Figure 3-9). The earliest really abundant fossils are the marine invertebrates, such as the trilobites shown in Figure 2-5. The Pennsylvanian part of the Carboniferous Period (280 to 310 million years ago) was characterized by a tremendous amount of plant growth that produced forests in swampy land. Geological processes turned the organic remains of the forests into coal over long periods of time (petroleum was probably also produced in this way); our coal supply dates largely from this period and is literally "fossil fuel."

3-9 *Cambrian sandstone, showing the tracks of large worms that crawled across the wet sand, as well as an overall ripple pattern caused by the waves. (Courtesy of the Smithsonian Institution.)*

The dinosaurs, perhaps best known of all the ancient and extinct life forms, appeared in the Triassic Period about 200 million years ago. They evolved into various species that were at home on the land, at sea, and in the air, and dominated the earth for some 100 million years. A famous example of evolution is the development of the horse over the 50 million years since its appearance as a dog-sized animal in the Eocene Epoch (56 to 37 million years ago). The modern animal is larger, faster, equipped with teeth suited to a wider variety of diet, and probably smarter than its ancient predecessor (Figure 3-10).

The Quaternary Period began 3 million years ago and includes the Pleistocene and Recent Epochs. During this period extensive glaciation and the development of man have occurred. Many of the plant and animal fossils of the Quaternary resemble the living organisms that we find today.

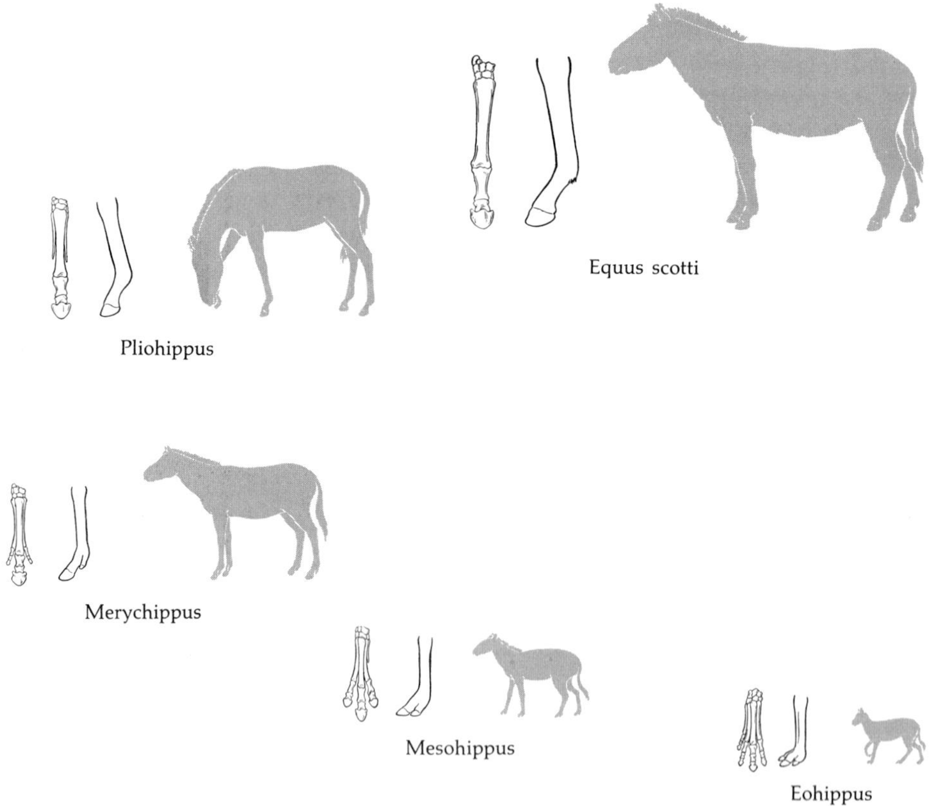

3-10 *Evolution of the horse from the Eocene (bottom pictures) to the present (top) as reflected in the relative size and development of the bones in the foreleg. (From* Introduction to Biological Science *by C. W. Young, G. L. Stebbins, and F. G. Brooks. Harper and Row, 1938, 1951, 1956. Reprinted by permission of the publishers.)*

Evolution and Life

In the preceding sections we have discussed the evidence for the evolution of species occurring on geological time scales. A similar case can be made for the evolution of man.

Fossils and contemporary anthropological evidence place the emergence of man in Africa some 3 to 4 million years ago. There appears to have been a sequence of ancient primate types that showed steady variation from ape-like properties to those found in modern man. The characteristics considered here are the shape of the hips and pelvis (to allow efficient erect walking or bipedalism), the shape of the head and tooth structure, the size of the brain, and the use of tools and fire. The use of tools was made possible by the bipedalism, good binocular vision, and the grasping hand, a remnant of tree-dwelling ancestors.

The progression can be illustrated most simply in terms of brain size. The *Australopithecus* (3 to 4 million years ago) had a brain size of about 510 cubic centimeters, *Homo erectus* (0.5 million years ago) had a brain size of about 975 cubic centimeters, and *Neanderthal Man* (35 to 110 thousand years ago) had a brain size of 1,420 cubic centimeters. This latter figure is comparable to the brain size of modern man; by comparison, the brain size of a chimpanzee is about 395 cubic centimeters.

Among the earliest primate specimens in this progression was the *Proconsul* (15 million years ago), which is classified as an early ape and is perhaps the ancestor of the chimpanzee and the gorilla. The earliest direct ancestor of man appears to have been *Ramapithecus* (13 million years ago). We note that the ancestors of both man and apes were intermingled at the time in question, about 15 million years ago. Thus, man is not strictly descended from the apes. We are "cousins," descended from common ancestors of a long time ago.

Australopithecus appeared about 4 million years ago and evolved into an advanced form that flourished between 1.0 and 1.5 million years ago; they gave rise to *Homo erectus* about 900 thousand years ago. *Homo erectus* lasted about 500 thousand years and developed into early *Homo sapiens* about 250 thousand years ago. These facts were pieced together primarily (but not exclusively) on the basis of fossil finds in Africa by Raymond Dart, Robert Broom, and Louis and Mary Leakey.

This view of "African Genesis" is widely accepted today. Modern man, for better or worse, is the result of millions of years of human evolution. Many false starts are found in the fossil record; these were dead ends which became extinct. The thrust of the development was from fruit-eating tree dweller

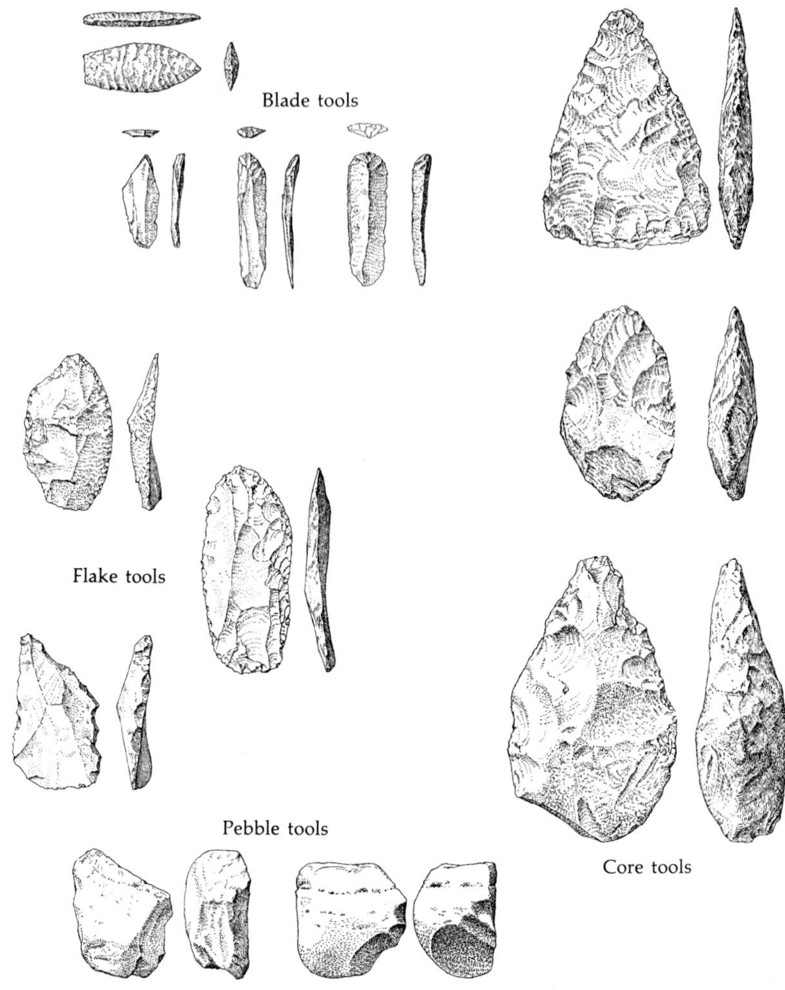

3-11 *Ancient stone tools; the earliest ones are shown at the bottom, later ones at the top.* Core tools (*right*) *were shaped by chipping off flakes;* pebble tools *were a crude form of these. The other tools shown were made from the flakes themselves;* blade tools *are flakes with nearly parallel sides.* (*From* Tools and Human Evolution *by S. L. Washburn. Copyright © 1960 by Scientific American, Inc. All rights reserved.*)

to ground-dwelling hunter-gatherer. Our fruit-eating days are mirrored in our teeth; they are not the teeth of a true carnivore. Stone tools were developed early (Figure 3-11), and enabled man to extend his own physical gifts in hunting and other activities. These tools are durable and are the chief evidence bearing on the life styles of early men. Finally, social organization permitted men to hunt physically superior animals such as mammoths.

The climatic environment of our most immediate ancestors was very difficult indeed. The Pleistocene was an era in which the earth was subjected to repeated temperature changes and ice ages. This was a time of testing. Modern man, with a large complex brain, the ability to formulate and communicate abstract thoughts, the use of tools, the ability to adapt to new situations (from a zoological point of view, to be a "specialist in non-specialization") is, to a certain extent, the product of those times; even today men are often at their best under adverse conditions. Neanderthal Man flourished between 110,000 and 35,000 years ago. Toward the end, he gave way to *Cro-Magnon Man.*

Cro-Magnon Man and related species were widely distributed over the earth and developed rapidly. They made advances, including the conversion from hunting and gathering to a crop-growing, animal-raising society, and the discovery and application of metals. The Cro-Magnons had the time or necessity for art (as preserved in the Lascaux cave paintings, Figure 3-12) and practiced ceremonial burial of the deceased. The first stirrings of religion and a curiosity about one's place in the scheme of things must have occurred by this time, if not earlier.

3-12 *Prehistoric paintings in the Hall of Bulls, Lascaux Caves. (Courtesy of French Government Tourist Office.)*

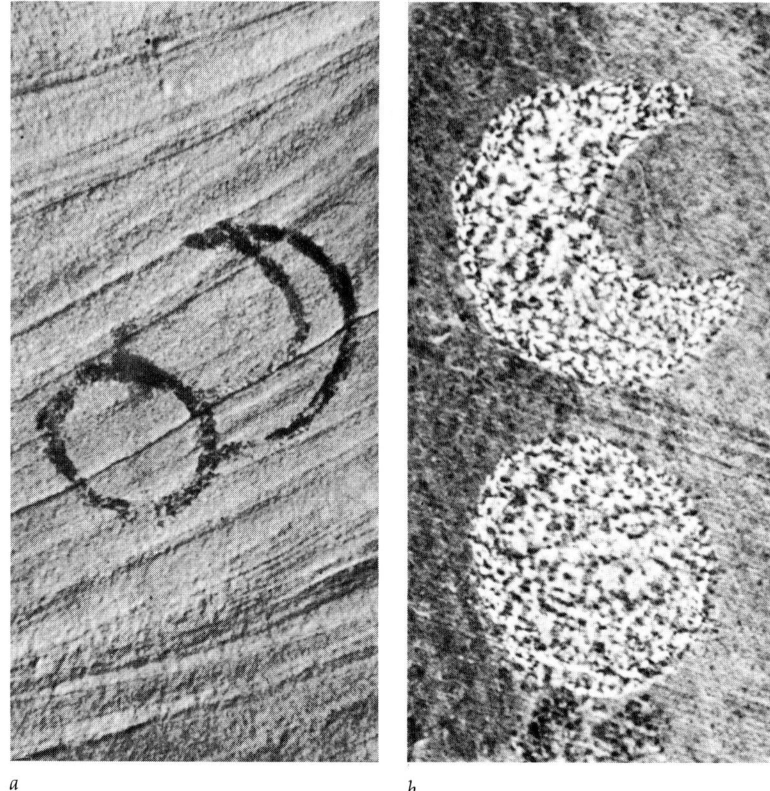

a *b*

3-13 *Ancient pictures that are believed to represent the supernova of 1054 A.D. Photographs taken (a) at White Mesa, Arizona, by William C. Miller, and (b) in Navajo Canyon, Arizona, by R. C. Euler. The crescent that depicts the moon is reversed, but this common mistake is not considered to be significant. (Courtesy of William C. Miller, Hale Observatories.)*

The development of language, and thus the ability to pass knowledge on from one generation to the next, as inferred from their development, was clearly a phenomenon of immense importance.

Where did man's interest in astronomy begin? Surely at a very early time man was aware of the sun and the moon, of the basic division into day and night. At some point in prehistory he began to speculate about the nature of the world and the origin of things. The obvious importance of the sun to all organisms and activities on earth was reflected in its incorporation as a motive power or deity in the earliest known myths and theologies of primitive civilizations. The deciphering of cuneiform inscriptions dating back to 3000 B.C.–2000 B.C. shows that the ancient Babylonians had an extensive knowledge of astronomical phenomena, and that even in those times astronomers were employed in observing the phases of the moon to establish, regulate, and periodically revise the calendar. It seems reasonable to conclude, therefore, that astronomy is among the oldest (if not in fact the very oldest) of the sciences. Observations of unusual astronomical events, such as eclipses and the appearances of comets in the sky, are found in the records of oriental civilizations. Even primitive people who lacked a written language have left us astronomical records in the form of rock paintings (Figure 3-13), found on cave and canyon walls in northern Arizona.

Before leaving the story of the development of man on earth, we consider the fascinating possibility of finding present day fossil societies; that is, people who still are living in the manner of the hunter-gatherers of the past. One such group may be the "lost stone age tribe" reported found in 1971 on Mindanao, in the Philippines.

ECOLOGY

Human development did not stop where we left it at the end of the previous discussion nor does it end today. Some other noteworthy landmarks involving cultural evolution include the Industrial Age (or Revolution) from about 1770 to 1870, and the start of the Nuclear Age in 1945. The thrust has been toward an ever-increasing development of technology and attempts to shape nature as man desires.

This has led to a situation of extreme gravity for the future of *Homo sapiens.* We have tended to make an unwarranted distinction between man and animal and have come to regard ourselves as apart from and possibly immune to natural laws. In this sense, the problems of pollution and over-population that face us today have largely arisen from ignorance and arrogance.

As the evidence of the harmful nature of many modern devices and chemicals piles up, it gives added impetus to the suddenly popular science of ecology, the discipline that views our environment as a whole.

Professor Barry Commoner, one of ecology's most persuasive spokesmen, has outlined three simple principles:

1. Everything interacts with everything.
2. Everything goes somewhere.
3. There is no such thing as a free lunch.

All of these principles are routinely violated.

The deleterious effects of Commoner's second principle are attracting the most attention today. Industrial waste has ruined Lake Erie, our tap water foams with someone else's detergent that has not been removed by the purification process, and American mothers' milk contains DDT in amounts that are illegal when found in cow's milk at the supermarket. Ecology demands that we view problems as wholes, processes as parts of closed cycles, and that we think before taking a short-cut. There is really no such thing as the dump "outside of town." Vital materials must be continuously recycled.

The interrelations of biological processes in the soil, the oceans, and the atmosphere are complex and delicate. Most species are in equilibrium with their environments, and this situation is required for long-term survival. An example of a complex problem involving oceanography, geology, biology, atmospheric physics, and astronomy concerns the oxygen and carbon dioxide content of the atmosphere. Briefly, atmospheric oxygen is produced by green plants (largely phytoplankton), and carbon dioxide is absorbed by the oceans and utilized by plants. Industrial atmospheric pollution and the exhausts of automobiles have several effects. They increase the amount of haze and dust in the atmosphere and lead to smog, which contributes to lung diseases in humans. The dust in the atmosphere reduces the solar heating and may affect the climate. Industry and automobiles also add to the carbon dioxide in the atmosphere; burning of fossil fuels between 1860 and 1960 increased the atmospheric carbon dioxide content by 14 per cent. Carbon dioxide contributes to the greenhouse effect (Chapter 8) whereby the sun's energy is trapped in our atmosphere and the earth's surface temperature is raised. Estimates vary, but (as we mentioned at the end of Chapter 2) the oppositely directed effects of the carbon dioxide and dust in the air could eventually result either in melting of the polar ice caps and a disastrous rise in sea level or in a new ice age, depending on which pollution product dominates.

The ultimate ecological blunder would be nuclear war. Many distinguished scientists have argued that there is no *technical* solution to the nuclear (or for that matter, the chemical-biological) arms race. New weapons systems beget counter-weapons systems which beget counter-counter-weapons and so on *ad infinitum.* Many of our severest environmental problems are traceable directly to the ever-increasing population and to industrial activity. Each additional person produced by the population explosion requires more materials and energy to build and run his automobiles, to clothe and feed him, and to dispose of his wastes.

In nature, population control is a normal occurrence; the system is in balance. A much studied example is that of the moose herds and wolf packs on Isle Royal in Lake Superior. Not only are the numbers of moose and wolves remarkably constant, but the predatory activities of the wolves seem to be in the long-term interest of the moose herd. This is because kills are almost always of sick or weak animals; thus, only the strong and healthy remain to propagate the herd. Many times, however, man has hunted predators such as wolves and cougars to the point of extinction. In several eastern states the deer population has no surviving natural enemies, is out of control, and must be harvested annually to prevent mass starvation. Particular species of animals have their own forms of population control, such as territorial spacing or even mass suicide (lemmings). However, we have disrupted the ecology in many places to the extent that we have doomed many species to extinction.

The details of re-establishing a stable environment on earth involve many social and political problems outside the scope of this text. In any event, a non-catastrophic solution to the problem necessitates the clear recognition and acceptance of ecological principles. Man must either acknowledge the existence of natural law and live within it or he will perish by it.

4

Historical Astronomy—Sky and No Telescopes

On a clear dark night, an observer located far from the lights and smog of the cities can perceive several thousand stars without optical aid. The men of ancient civilizations were familiar with the stars, many of which they named, and they recognized familiar patterns—the outlines of animals, people, ships—among them. These patterns are called *constellations.* Other points of light in the sky, called *planets,* were distinguished from the stars by the fact that they slowly moved across the backdrop of the constellations. (The word planet stems from the Greek *planetes,* or wanderer.) Five planets* (Mercury, Venus, Mars, Jupiter, Saturn) were recognized before the telescope was introduced, but no one thought that they too were worlds like the earth. The nightly rising and setting of the stars as they perform circular motions in the sky about the North and South Poles were ascribed to a rotation of the sky as a whole—man generally did not believe that the earth itself turned. The Milky Way was perceived as simply a hazy band

*Actually, the Greeks counted the sun and the moon as planets too, since both of them move with respect to the stars.

4-1 *The zodiacal light as seen from Mount Chacaltaya, Bolivia. (Courtesy of D. E. Blackwell and M. F. Ingham.)*

of light. Its appearance actually results from a great many faint stars spaced relatively close together in the sky, but this fact was not recognized until telescopes became available in the seventeenth century. The *Magellanic Clouds* (actually two relatively nearby galaxies, see Chapter 13) were familiar to inhabitants of the Southern Hemisphere and were recorded in the log books of sixteenth century Portuguese navigators. Resembling luminous patches of cloud, they were long regarded as isolated fragments of the Milky Way. Another hazy, dim light phenomenon is the *zodiacal light* (Figure 4-1). Now identified as the product of the scattering of sunlight by dust particles in the solar system, the zodiacal light was recorded as a cone- or pyramid-shaped glow, seen in the west just after dusk or in the east shortly before morning twilight.* Occasionally, comets were seen but their celestial nature was not admitted, and they were generally considered to be cloud phenomena in the atmosphere. The celestial origin of meteors (often called "shooting stars") was also unrecognized; they were sometimes compared to lightning flashes.

---

*The zodiacal light is dim, and one needs to be in a dark location away from city lights in order to see it.

The most obvious thing about the moon is that the shape of its illuminated surface is different from one night to the next. Throughout the month we observe a systematic progression. The thin crescent seen low in the western sky just after sunset grows in illuminated area (*waxes*) and is found farther to the east each night after sunset, becoming a half-moon (referred to by astronomers as the *first quarter* stage) and continuing to grow until it reaches the *full moon* condition when the moon rises as the sun sets. The full moon, as the month continues, decreases in lit-up area (*wanes*), passes through a half-moon stage (called *last* or *third quarter* moon), becomes a thin crescent that rises in the east just before sunrise, vanishes briefly (*new moon*), and finally becomes the thin crescent in the west at sunset that signals the start of another monthly lunar cycle. This sequence is shown in Figure 4-2, where one can also see that the moon keeps the same face toward the earth throughout the month. The term *gibbous moon* applies whenever the illuminated area is larger than at quarter-phase, but less than at full moon.

The phases of the moon are a simple consequence of two facts: (1) the moon is not self-luminous, but shines by reflecting the light of the sun; (2) the moon orbits around the earth. Since the sun always illuminates the half of the moon that faces it at any given time (except during an eclipse of the moon), the phases depend solely on what fraction of the illuminated lunar hemisphere can be seen from earth (Figure 4-3). As early as the fifth century B.C., Anaxagoras correctly explained the lunar phases by asserting that the moon shines by reflected sunlight, but for radical ideas such as this, he was convicted of impiety and banished from Athens.

Anaxagoras also understood that eclipses are simply shadow effects. The moon is eclipsed (which can occur only at full moon phase) when it moves within the shadow of the earth (see Figure 2-1). The sun is eclipsed (only possible at new moon phase) when the earth falls within the moon's shadow. Then why don't we see such eclipses every month at new moon and full moon? The explanation for this is also geometrical. There is a small tilt of about 5 degrees between the planes of the moon's orbit about the earth and the earth's orbit about the sun. At most new and full moon times the moon lies slightly above or below the sun-earth line, and thus does not eclipse the sun. But, if the moon happens to be in the plane of the earth's orbit (which is called the *plane of the ecliptic,* for obvious reasons) at new moon, then the sun, moon, and earth are in a straight line and a solar

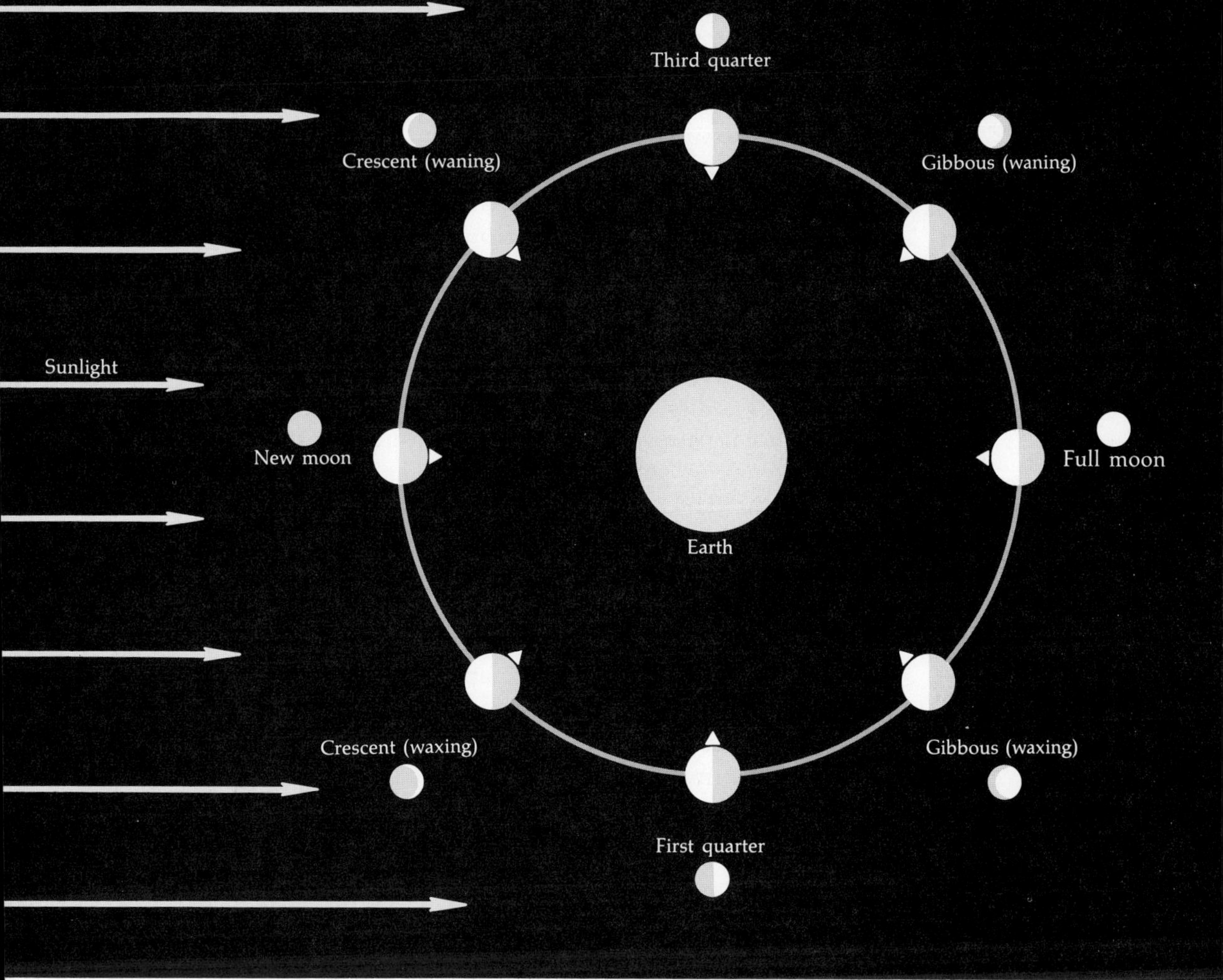

4-3 *The geometry of the moon's appearance in the sky as seen looking down on the earth's North Pole. As the moon moves in its orbit around the earth, different parts of the sunlit and shadowed lunar surface are visible from earth and this situation gives rise to the appearance of new, crescent, quarter, gibbous, and full moon as shown. The same position on the moon is marked by an arrow; note that the moon keeps the same face pointed toward the earth. Compare with the photographs in Figure 4-2.*

4-2 *The moon as seen through its monthly cycle of phases. The first example (top left) is the thin crescent at age 4 days after new moon. The additional examples, going from left to right, from top row to bottom, are age 7 days (first quarter); age 10 days (gibbous); age 14 days (full moon); age 18 days (gibbous); age 20 days (gibbous); age 22 days (third quarter); age 24 days; age 26 days. Compare with the diagram in Figure 4-3. (Courtesy of the Lick Observatory.)*

eclipse occurs. If the corresponding circumstances happen at the time of full moon, a lunar eclipse is produced.

The maximum number of eclipses possible in a year is seven (either four solar and three lunar, or five solar and two lunar); the minimum is two (both solar), and the average is four. A lunar eclipse can be seen from any place on the night side of the earth. The shadow of the earth is seen to be circular and to move across the full moon. At the time of totality, the moon seems to have a dull red or copper color, due to rays of sunlight that, passing through the earth's atmosphere, are bent so they enter the shadow and strike the moon.

A total eclipse of the sun is a much more impressive phenomenon. A solar eclipse is seen as total when the observer is in the darker part of the moon's shadow, called the *umbra;* a partial eclipse is seen when the observer is in the lighter part of the shadow, called the *penumbra* (see Figure 4-4). When a total solar eclipse occurs, the *path of totality,* that is, the region on the earth that lies within the umbra, is typically only 120 kilometers (75 miles) wide. The apparent diameters of the sun and moon are nearly equal (they are each about one-half degree) and the narrow width of the path of totality is due to the fact that the umbra barely reaches to the earth (Figure 4-4). The umbra tracks across the earth at a typical speed of 1,600 to 3,200 kilometers per hour (1,000 to 2,000 miles per hour) and the

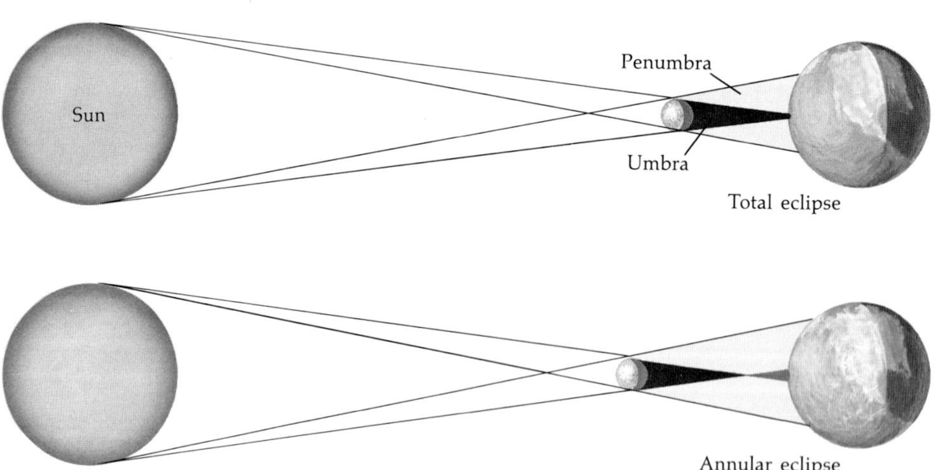

4-4 *Geometry of solar eclipses. In a total solar eclipse the umbra (from which no portion of the sun is visible) reaches to the earth's surface. Outside of the region of totality a partial eclipse occurs where the penumbra (from which part, but not all, of the sun is visible) strikes the earth's surface. In an annular eclipse the moon's shadow does not reach to the earth's surface; the result is that even in the most total phase of the eclipse, part of the sun is still visible as a bright ring around the moon. Not drawn to scale.*

duration of the total eclipse as seen from any given spot in the path of totality never exceeds eight minutes. A rather small fraction of the earth's population has witnessed a total solar eclipse; if you were to sit down at one point on the earth and wait for one, you would have a wait, on the average, of 400 years.

The duration of totality at a given spot depends primarily on the exact distance of the moon from the earth at the time of eclipse. Because its orbit is slightly elliptical, the moon is sometimes far enough away so that the earth does not lie in the umbra. In other words, the moon is too far from the observer to block out the whole sun. In this case, a thin ring of sunlight is seen around the moon's edge (*annular eclipse,* Figure 4-4).

While a lunar eclipse or a partial or annular solar eclipse is an interesting event to watch, a total eclipse of the sun is a really spectacular phenomenon. As the moon moves across the sun the observer's surroundings grow darker and colder until, at the moment of totality, the dark disk of the moon is suddenly surrounded by the glowing white *corona* or outer atmosphere of the sun (see Chapter 9). Animals of many types react to eclipses as though night had fallen, and humans have also had unusual responses. An ancient Chinese myth tells that the eclipsed sun is being consumed by a monster (Figure 4-5) so that one has to beat drums and gongs to drive the monster away and rescue the sun. So far, this procedure has always been successful!

4-5 *The monster responsible for solar eclipses according to an old Chinese myth.*

If an opportunity to view a total solar eclipse arises, it is well worth taking and, in view of the scarcity of such events at a given location, it may be a once-in-a-lifetime occasion. *Take care* not to view the partially eclipsed sun directly, as considerable infrared radiation remains and eye damage could result; it is safe to view the sun directly only during totality. When the total phase is about to end, small portions of the sun's disk emerge through the irregularities on the edge of the moon, thereby producing the diamond ring effect (Figure 4-6). This effect can also occur at the beginning of totality.

Many scientific investigations are made during eclipses of the sun. At these times the outer layers of the solar atmosphere can be studied without interference from the blinding glare of the sun's disk. Observers can search for comets and small planets near the sun, and an experimental test of the general theory of relativity can also be carried out (Chapter 16). Eclipses have been used to date the events described in ancient records. If the location at which an eclipse was seen is recorded, it is frequently possible to calculate the date, in terms of our modern calendar.

4-6 *The sun emerging from the total eclipse of March 7, 1970, showing the diamond ring effect. (Photographed by J. C. Brandt and R. G. Roosen, Goddard Space Flight Center.)*

Three natural clocks existed in the sky for ancient man: the rising and setting, or *diurnal motion,* of the sun and stars, which measures the day and night; the phase cycle of the moon, which was used to define the month; and the apparent motion of the sun, which gave rise to the concept of the year (and which seemed closely related to the seasons).

The first of these clocks, which measures the day, is due to the rotation of the earth from west to east about the axis that passes through the North and South Poles. This produces the effect that the stars appear to move in the opposite direction, namely, from east to west. The *meridian* is an imaginary line in the sky, which passes through the north and south points on the horizon and also the zenith or overhead point. One way of defining the day is in terms of the interval between two successive crossings of the meridian by the sun (*solar day*). Another way is in terms of the interval between two successive crossings of the meridian by a star (*sidereal day*). The days defined by these two methods differ in length by about four minutes. We will return to this difference near the end of the chapter.

We should note in passing that, unlike the day, month, and year, there is no basis in celestial motions for the week. Nevertheless, in many languages, the days of the week are named after the seven bright, moving astronomical bodies that were known to the ancients (this count excludes temporary events like the appearance of a comet), as shown in Table 4-1.

The second of the three clocks, on which the month was based, is the motion of the moon in its orbit about the earth,

Table 4-1 *The Days of the Week*

SKY OBJECT	DAY (ENGLISH)	DAY (OTHER LANGUAGES)
Sun	Sunday	—
Moon	Monday	—
Mars	Tuesday	Martes (Spanish)
Mercury	Wednesday	Mercoledi (Italian)
Jupiter	Thursday	Jeudi (French)
Venus	Friday	Venerdi (Italian)
Saturn	Saturday	—

which gives rise to the phase cycle shown in Figure 4-2. Similarly, the third clock, on which the year is based, is the motion of the earth about the sun. The sun appears projected against the background of different constellations as we view it from different points in the earth's orbit. Of course we don't see stars when the sun is in the sky, but we can, for example, note the point that is a given distance east of the sunset location at some fixed time after sunset, and find in this way that this point moves systematically through the constellations, returning to the same place among the stars after an interval of slightly more than 365 solar days.

A problem of considerable importance in the construction of a calendar is that the length of the year is not a whole number of days. As a practical matter, one does not want to have a fraction of a day left at the end of a calendar year, so a set of rules was formulated to give some calendar years an extra day and make the average of a large number of calendar years approximately equal to the true length of the earth's journey around the sun, 365.2422 solar days. Many of the ancient calendars were hopelessly inadequate, and when Julius Caesar came to power he found that the seasons were occurring in the "wrong" months.

Caesar extended the length of the year 46 B.C. to 445 days to make up for past errors. He also established the *Julian calendar* on the basis of advice from the Alexandrian astronomer Sosigenes, approximating the average length of the year as $365\frac{1}{4}$ or 365.2500 days. This average was obtained by adding an extra day every fourth year to a basic calendar containing 365 days. This is how the leap year originated.

While the Julian calendar was a great improvement, the calendar year and the astronomical year still differed by 365.2500 minus 365.2422 or 0.0078 days. This is a small difference, but given enough time a substantial discrepancy will accumulate. If the calendar were allowed to run for 1,600 years without adjustment, a difference of about twelve days would occur. In fact, this happened, and in the year 1582 Gregory XIII (on the basis of advice given by the astronomer Clavius) issued a papal bull designating the day after Thursday, October 4, as Friday, October 15. This proclamation established what we now call the *Gregorian calendar.* It omits as leap years the centurial years that are not evenly divisible by 400; thus, 1900 was not a leap year, but 2000 will be. On the Julian calendar there would be 100 leap years in 400 years; on the Gregorian calendar three of these are dropped. Hence, the average length of a Gregorian year is 365 plus $^{97}/_{400}$, or 365.2425 days long. This value is so close to the astronomical one that the error we make in using the Gregorian calendar is only three days in 10,000 years.

The Gregorian calendar was adopted immediately by the Roman Catholic countries, but it was opposed initially in Protestant lands. Gradually it gained acceptance (Great Britain adopted it in 1752, despite riots over "lost days") and today it is the official or *civil calendar* in use over most of the globe. The calendar change leads to an interesting historical oddity. Modern history books list the birthday of George Washington as February 22. However, calendars hanging in the American colonies at the time of his birth showed the date as February 11.

The ancients noticed that the sun was lower in the sky during winter than in summer and thus deduced that the seasons have an astronomical cause. In fact, the seasons arise from the circumstance that the earth's axis is not perpendicular to the plane of the earth's orbit around the sun, but rather is tilted some 23.5 degrees with respect to the perpendicular. The axis points north to a fixed position in the sky (North Celestial Pole) and thus, as is clear from Figure 4-7, the Northern Hemi-

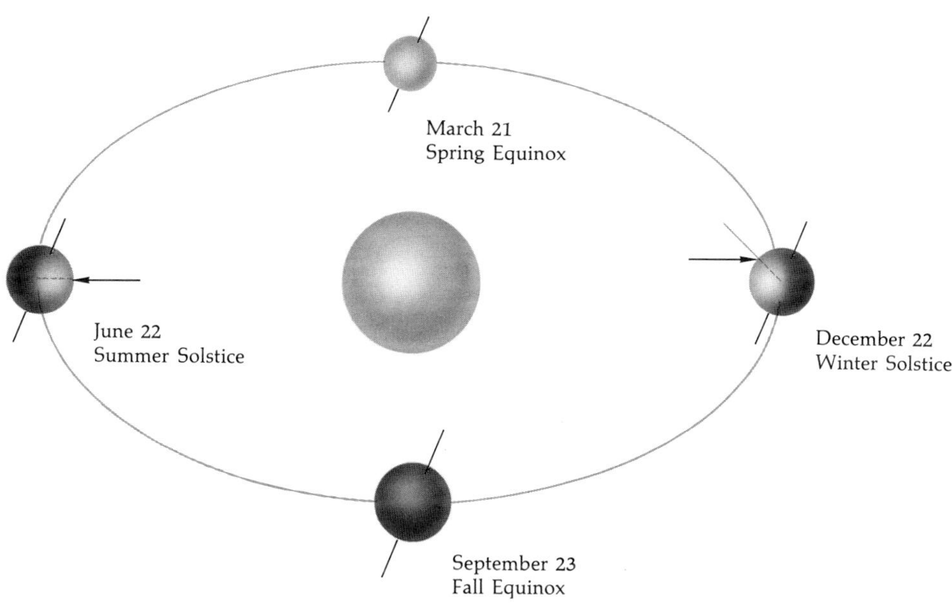

4-7 *Tilt of the earth's axis and the seasons. The earth's axis of rotation points throughout the year in the same direction in space, a direction which is tilted 23-$\frac{1}{2}$° with respect to a line at right angles to the plane of the earth's orbit. This effect causes the sun's rays to strike a given latitude on earth at different angles throughout the year. An example is indicated in the figure; at noon on June 22, the sun is overhead for latitude 23-$\frac{1}{2}$° north and heating is efficient. At noon on December 22, the sunlight strikes the same ground obliquely and heating is inefficient. See Figure 4-8.*

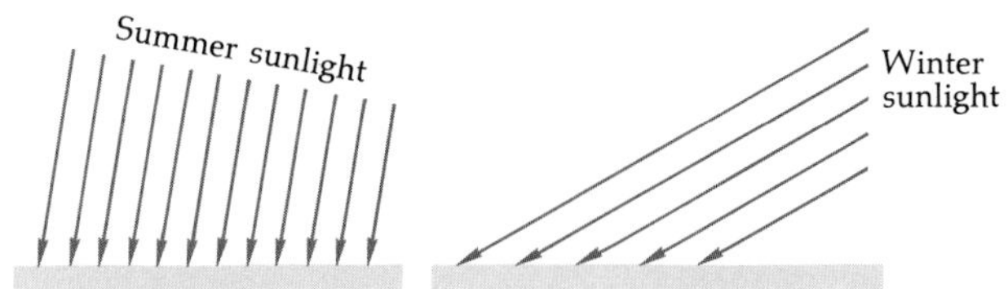

4-8 *Differences in ground heating between summer and winter. The solar intensity is equal in both cases (as indicated by the equal spacing), but because of the different angles of incidence, a given area of the ground receives much more energy in summer than in winter.*

sphere leans away from or toward the sun, depending on the season of the year. As perceived by an observer at, say, New York, the sun is higher in the sky at a given time of day (such as noon) in the summer than it is in winter. This is an effect caused by the Northern Hemisphere leaning toward the sun in summer and away from it in winter. When the sun is high in the sky it heats the observer's surroundings more effectively than when it is low in the sky (see Figure 4-8). Also important is the length of time each day that the earth is heated by the sun as compared to the length of the night when the earth loses heat. In the Northern Hemisphere the sun is highest in the sky and the day longest near June 22, but the warmest days do not occur at that time, because the earth still has not recovered from the cold of winter. It takes about a month more for the surface of the earth to reach its maximum temperature.

Note that the distance from the earth to the sun is not an important factor in determining the seasons. In fact, we are closest to the sun when winter occurs in the Northern Hemisphere; at that time it is summer in the Southern Hemisphere, which is then leaning toward the sun.

SYSTEMATIC DESCRIPTION OF THE SKY

When we study natural objects, whether they are geological formations, plant and animal species, or stars, one of the first things we want to know is: Where are they? The position of a given star can be described in terms of two quantities—its distance and its direction. We defer the discussion of star distances to Chapter 10, as the ability to measure them is a relatively recent development in the history of astronomy. On the other hand, the directions of the stars, as recorded on maps of the constellations, have been studied for thousands of years. They are conveniently specified by the *equatorial coordinate system.* Just as we can describe a location on the earth according to its *latitude* (angular distance north or south of the equator) and *longitude* (angular distance east or west of Greenwich, England), we use two coordinates, called *Right Ascension* and *Declination,* to identify points in the sky (Figure 4-9). The *Celestial Equator* is defined as the intersection of the earth's equatorial plane with

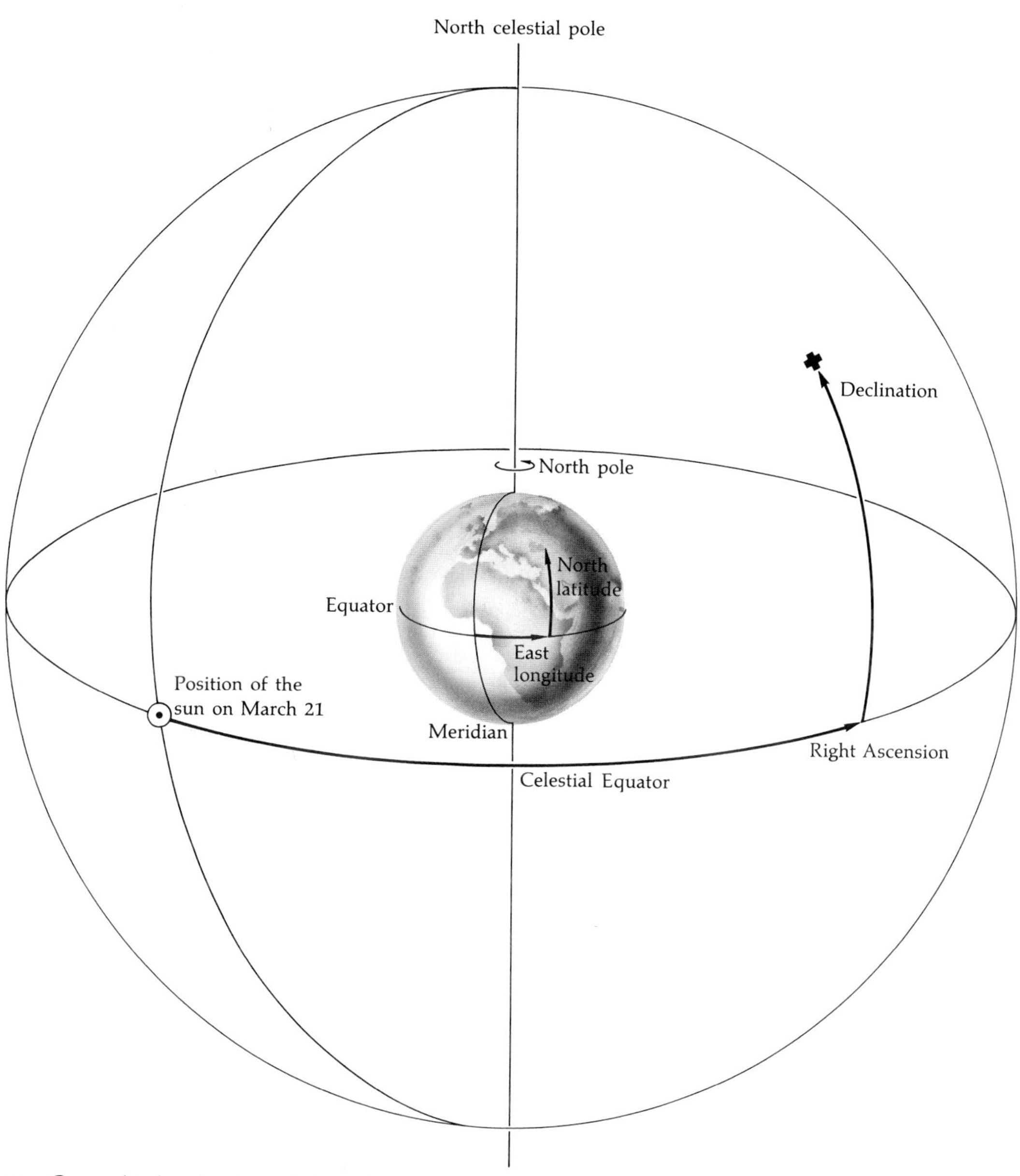

4-9 *Geographical and astronomical coordinate systems.*

the *celestial sphere,* an imaginary sphere in the sky, centered on the earth. The equatorial coordinate system can be visualized as a grid on this sphere, resembling the grid of latitude and longitude on a globe. Declination in the sky is analogous to latitude on the earth; it is measured in degrees north or south of the Celestial Equator. Right Ascension is measured eastward from one of the two points at which the earth's orbital plane (determined by plotting the sun's position on a sky map) intersects the Celestial Equator. This point, called the *Vernal Equinox,* is the location of the sun on the first day of spring* when the sun crosses the Celestial Equator, moving from the south celestial hemisphere to the northern (that is, moving from southern to northern Declinations). The path of the sun across the background of stars is called the *ecliptic;* it represents the intersection of the plane of the earth's orbit with the celestial sphere.

The Age of Aquarius

We have mentioned the three celestial clocks used by ancient men. There is still another and slower clock that requires 26,000 years for a single revolution. Although the earth's axis appears to point toward the same direction in space, actually its orientation is slowly changing, making the axis trace a circle of radius 23.5 degrees on the sky. At the present time the location to which the axis points (North Celestial Pole) is close by the star alpha Ursae Minoris, which we therefore call the North Star or Polaris, but the gradual motion of the axis, called *precession,* is moving the North Celestial Pole away from this star. As time goes on we will have other North Stars, until 26,000 years have elapsed and the axis points to alpha Ursae Minoris again. This precession motion, in which the rotation axis of the earth traces out a cone (Figure 4-10), is due to the gravitational pull of the moon and sun on the earth's equatorial bulge (page 12) and is analogous to the wobbling motion that one notes in a spinning top or toy gyroscope. The existence of precession as a slow continuous process affecting the equatorial coordinates of the stars was recognized by Hipparchus in the second century B.C. and was independently recorded in the Chinese annals a few hundred years later.

If the direction in which the axis of the earth points is changing slowly, then as the equator (by definition) always lies in

---

*Spring in the Northern Hemisphere of the earth; fall in the Southern Hemisphere.

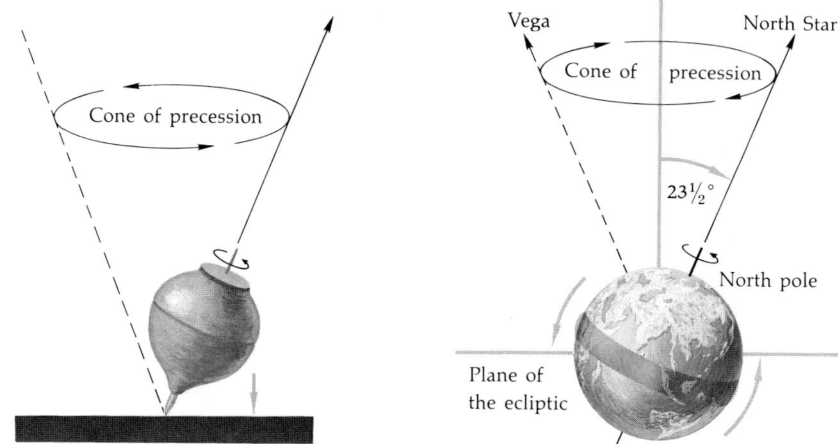

4-10 *Precession. The axis of a spinning top sweeps out a cone of precession because the top is pulled down by the earth's gravity. Similarly, the moon and sun pull on the earth's equatorial bulge (marked by the dark band) and tend to pull the earth's equator into the plane of the ecliptic. Thus the earth's rotation axis also sweeps out a cone of precession. The north half of the axis presently points approximately to the star alpha Ursae Minoris, but in A.D. 14,000 it will point near the bright star Vega, and Vega will become the North Star. 26,000 years from now the axis will point to alpha Ursae Minoris again.*

a plane at right angles to the axis, and as the intersection of this plane with the celestial sphere defines the celestial equator, the celestial equator must also be in gradual motion against the star background. The Vernal Equinox, defined above, must therefore move through the constellations. This point, where the sun lies on the first day of spring, had great significance to the ancients; it was called the "First Point of Aries" because it lay in the constellation of Aries, the Ram. More recently, it has been in Pisces, the Fish, and now it is moving into Aquarius, the Water Carrier. This is the origin of the term "Age of Aquarius" adopted by some young people to identify the present epoch.

STAR BRIGHTNESSES—REAL AND APPARENT

Continuing our study of the sky as seen with the unaided eye, we note that in some places stars seem to be located quite close together. (For example, the stars Alcor and Mizar in the Big Dipper, Figure 4-11, which can only be separated or *resolved* by

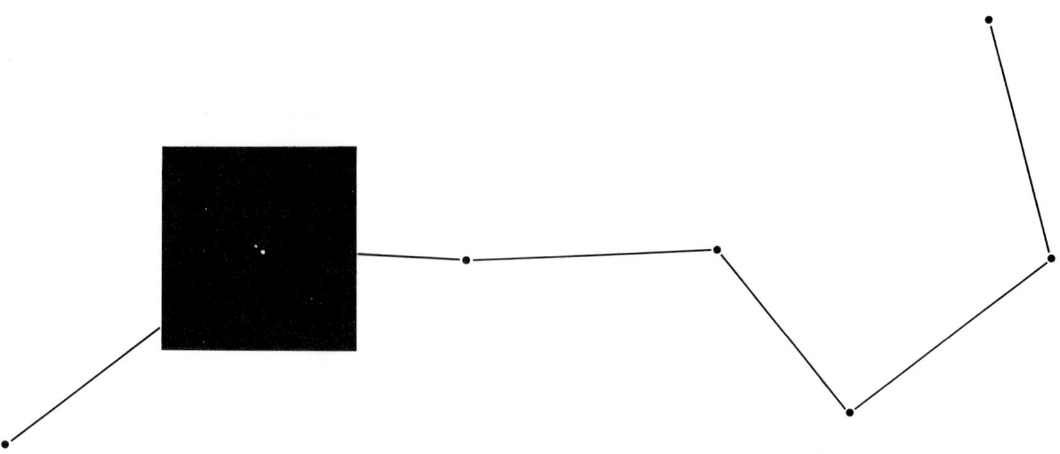

4-11 *Outline of the Big Dipper. Persons with good sight can resolve the two stars (inserted photograph) Alcor and Mizar; others see them as one.*

an observer with good vision.) This suggests that even noting the positions of the stars might not make it simple to identify them individually. Clearly, if we record a star's brightness as well as its position, we make it easier for other astronomers to determine to which star we are referring. The brightnesses can also tell us much about the stars themselves, as we shall see in some later chapters.

Hipparchus (circa 150 B.C.) is credited with preparing the first star catalogue; he tabulated almost 850 stars and introduced a brightness classification or *magnitude* system. However, this work did not survive in its original form. A later catalogue, by Ptolemy (circa 150 A.D.), contained positions and magnitudes for more than 1,000 stars and is believed to be an expanded version of Hipparchus' list. According to the modern definition of the term magnitude (which is in practice not very different from that used by Hipparchus and Ptolemy), the brighter stars are of the first magnitude, fainter stars are of the second, even fainter ones are of the third, and so on; the faintest stars visible to the naked eye are sixth magnitude. A few extremely bright stars have zero or even negative magnitude, as do bright planets like Venus and Jupiter. The faintest objects recorded with the largest telescopes are about $+24$, whereas the sun is -26. A magnitude is defined such that a 1 magnitude difference between two stars means that one star is the fifth root of 100 (about 2.512) times brighter than the other star. Thus, if two stars differ by 5 magnitudes, one of them is 100 times brighter than the other. Since the brightest star seen at night (Sirius, also called alpha Canis Majoris) has a magnitude of -1, and the faintest stars easily seen by the naked eye are of magnitude $+6$, they differ by 7 magnitudes, or a factor of about 630.

We know from experience that a lamp looks brighter when it is near than when seen from a distance. Thus, if the stars are at various distances from the earth, the magnitudes that we note for them are not necessarily intrinsic properties. For this reason, magnitudes such as those listed by Ptolemy are called *apparent magnitudes.* When we know a star's distance, we can use its apparent magnitude to determine a true measure of the intrinsic brightness of the star (see Chapter 10), which is termed the *absolute magnitude.* It is obvious to even the most casual observer that there are more faint stars than bright stars; that is, the number of stars having a high apparent magnitude (faint apparent brightness) is much greater than the number of stars with low apparent magnitude.

ORIGINAL TASKS OF THE ASTRONOMER

Time and the calendar are of great significance to us today. If one is consistently late to a 9 o'clock class one may flunk out and possibly be drafted.* If we lose track of April 15 and thereby fail to file an income tax return on time, we are likewise in trouble. Ancient man, who lived closer to the soil than we do, was equally dependent on the calendar. He had to know when to plant his crops and when to pay his tithe to the temple or the king. In Egypt he had to know when to anticipate the annual flooding of the Nile. The timing of religious occasions, as among the Hebrews, was especially important. The astronomer, who had at his disposal the natural clocks provided by the rising and setting of the stars, the phases of the moon, and the annual apparent motion of the sun, was thus provided with a livelihood. By observing the moon, sun, and stars, he determined, corrected, and kept track of the calendar. In some cultures the importance of timing religious functions was paramount, and the capacities of priest and astronomer were combined, just as in later times (Dark Ages in Europe) the all-important skill of writing was substantially confined to the clergy. Since the apparent positions of the stars depend on not only time of day and time of year, but also on the *location* of the observer, the astronomer's ability to deduce his position on earth by observing celestial positions led to the art of navigation as an outgrowth of astronomy.

---

*Selective Service laws have since been revised.

At least one of the original tasks of the astronomer, the determination of time, has remained to this day. Time signals are provided for the nation by the United States Naval Observatory in Washington, D.C.

HISTORY OF THE WORLD-VIEW

Curiosity as a characteristic of man is perhaps best exemplified by the progressive development of our world-view. The experience of early man was limited to a few square miles of foraging area in the vicinity of his lair and to a certain consciousness of the horizon and the sky above him. Thousands of years later, scientific geography began to flower among medieval men through the travels of the great explorers, typified by Marco Polo (1254–1324), who brought back to the West an account of the vast extent of China. Aristotle (384 B.C.–322 B.C.) taught that the earth is round (Chapter 2), and in the early sixteenth century, Ferdinand Magellan (1480–1521) organized, captained, and lost his life in the first successful circumnavigation of the globe. In the next few centuries (as we shall see in the following section) men discovered that the earth, long regarded as the center of the universe, in fact *moves.* It turns on an axis and also revolves around the sun. We gradually learned that there are other worlds in space, that the sun is just another star, and that the solar system is a tiny spot in an immense galaxy of some hundreds of billions of stars. There are countless other galaxies, mostly organized into clusters, that exist in space out to the limiting distances to which our largest telescopes can see. The history of astronomy documents the evolution of human thought concerning our place in the world, and the relation of the earth to the universe around us.

THE EARTH MOVES

In the previous section we referred briefly to the development of the world view. We now discuss this concept to bring out in greater detail man's long struggle toward the realization that the heavens do not revolve about him. Nearly every ancient culture possessed a world-view in which the earth was at rest in the center of the universe (*geocentric hypothesis*). The sky was the arched body of a goddess (Egyptians, Figure 4-12) or rested on the tusks of an immense elephant (Hindus), was shaped like a bell jar (Babylonians), or, somewhat more recently, was re-

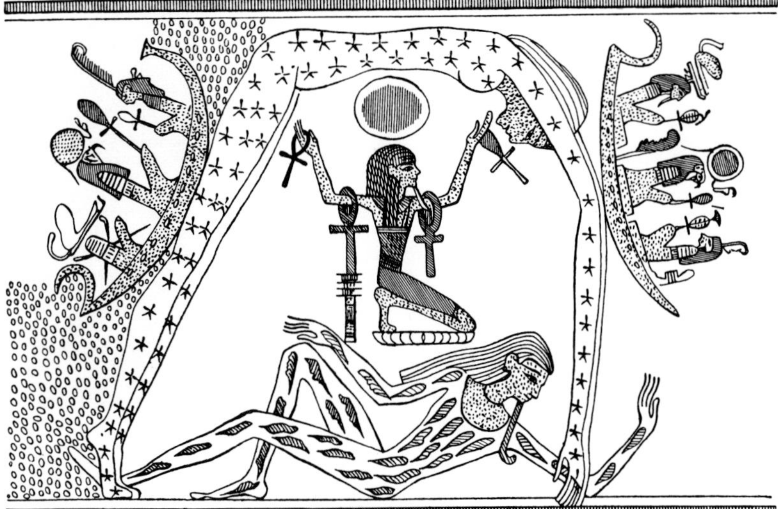

4-12 *An ancient Egyptian conception of the universe, showing the heavens represented by the goddess Nut, her starry body arched over the earth (figure adorned with leaves), and the ship of the sun in both its rising and setting positions. (From* Pioneers of Science *by O. Lodge. Dover Publications, Inc., New York, 1960. Reprinted through permission of the publisher.)*

garded as an immense tent overhead (Arabs). But in any case, the remainder of the universe rested on, or revolved about, the earth. One gets the impression that the number of cosmologies that were proposed in Greece and Alexandria was about equal to the number of philosophers who lived there. Several of these sages recognized that the earth turns; one (Aristarchus, third century B.C.) also suggested that the earth and the other planets revolve about the sun. However, the views of the Egyptian astronomer, Ptolemy of Alexandria, who compiled the star catalogue mentioned on page 76, came to dominate Western thinking for centuries.

The fundamental problem that each cosmologist faced was to explain the motions of the planets. It was observed that three of the planets known before the discovery of the telescope (Mars, Jupiter, and Saturn) occasionally traced out loops (*retrograde* or backward motion) as they wandered through the sky (see Figure 4-13). On the other hand, the other two visible planets (Mercury and Venus) never showed retrograde motion and had the curious property that they never moved very far from the sun in the sky (28 degrees in the case of Mercury and 48 degrees in the case of Venus). There was a great preference for assigning purely circular motions to the celestial bodies,

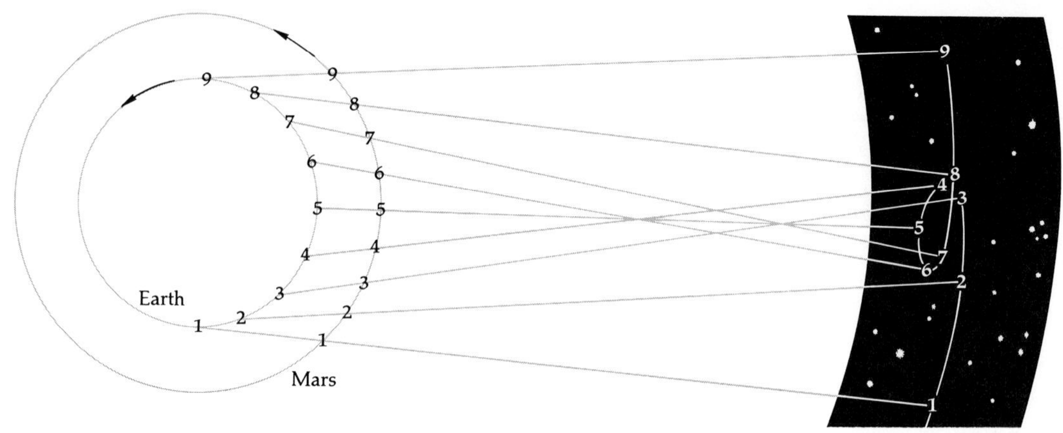

4-13 *Retrograde or backward motion of Mars. The planet's apparent track through the sky is shown at right, and the modern explanation is given at the left. For each of nine times the line of sight from earth to Mars is drawn to project the apparent position of Mars against the stars. (From* Structure and Change *by G. S. Christiansen and P. H. Garrett. W. H. Freeman and Company. Copyright © 1960.)*

since the circle was regarded as a pure or perfect geometric form. Further, as we have mentioned, there was strong sentiment for the central location of the earth. Ptolemy interpreted the planetary travels in terms of an ingenious system of circular motions centered on the earth, as shown in Figure 4-14. In the *Ptolemaic system,* the planets, sun, and moon followed circular paths about the earth. But superimposed on these paths were smaller circles (*epicycles*), each centered on a moving, imaginary point located on the larger circle (Figure 4-15). Considering the projected position of Mars against the starry background as seen from the earth (Figure 4-13), we can see how Ptolemy explained retrograde motion. Many epicycles were needed in order to reproduce the observed positions of the planets.

By postulating that the centers of the Mercury and Venus epicycles always lay on the straight line that connects the center of the earth and the moving sun, Ptolemy accounted for the fact that these planets are always fairly close to the sun in the sky.

As we have seen above, Ptolemy employed a clever geometric arrangement to explain the observed motions of the planets within the context of a philosophy that required a central location for a stationary earth and circular motions for other heavenly objects. There is no particular physical reason why the planets should move according to these geometric patterns, but

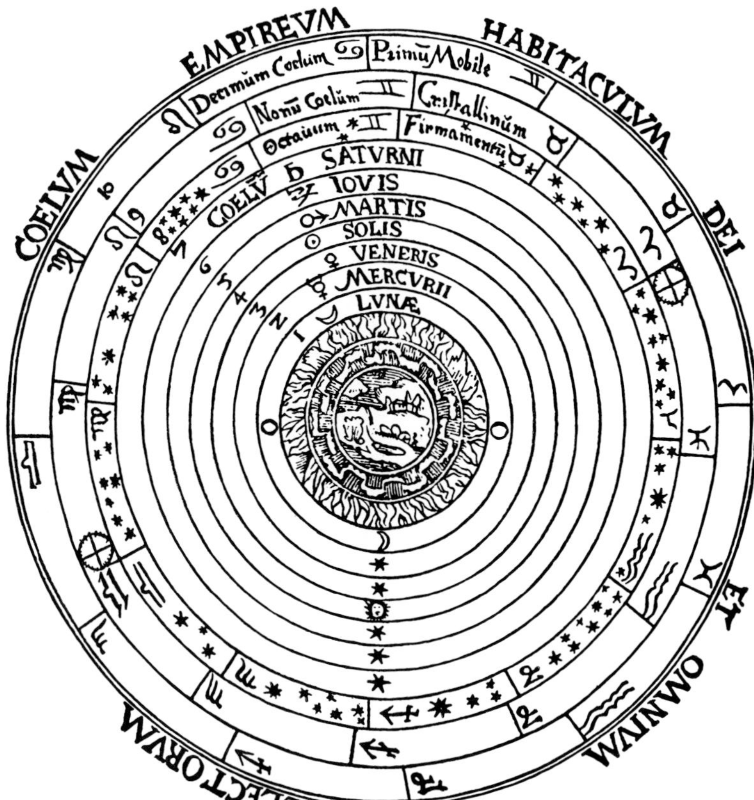

4-14 *The Ptolemaic universe as presented in Peter Apian's* Cosmographia *(1539). The earth is at the center. The moon, planets, sun, and stars occupy a series of concentric rings.*

in the age of Ptolemy few physical laws were known or recognized as such. However, the need for some degree of sophistication was apparent and a source of power to move the objects was required. Figure 4-14 shows one reproduction of the Ptolemaic system; four "elements" (earth, water, air, and fire) are represented, as are the moving astronomical bodies. Beyond them is the crystalline firmament in which the stars were fixed (Aristotle and other Greeks taught that the stars were permanent, fixed, and immutable). Still beyond was the Primum Mobile (or Prime Mover) that powered all. It seems that the existence of the Primum Mobile would excite the curious, and Figure 4-16, a medieval woodcut, shows a traveler putting his head through the vault of the sky and viewing the machinery responsible for the movement of the stars!

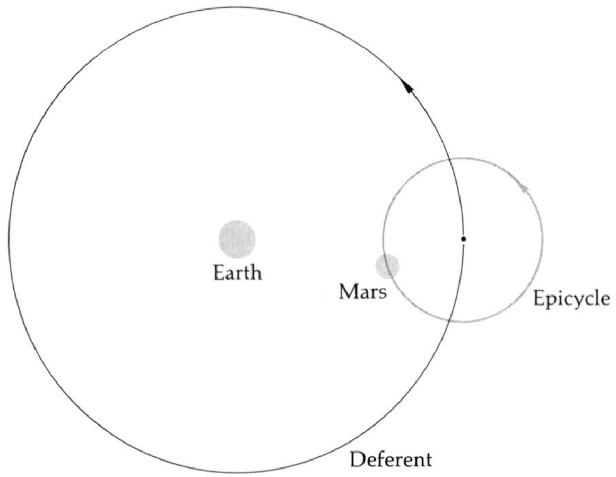

4-15 *Epicycles and retrograde planetary motion. As explained by the Ptolemaic theory, Mars was thought to move around the epicycle, as the center of the epicycle moved along the deferent, a larger circle centered on the earth. These two motions combined to produce backward loops through the sky as seen from the earth.*

4-16 *A medieval woodcut depicts a man who traveled to the end of the earth, putting his head through the sky in order to observe the machinery that moves the stars. (From* Knowledge and Wonder *by Victor F. Weisskopf. Doubleday, 1966.)*

Some psychologists believe that a person must be at least mildly egocentric to function effectively in society. Perhaps a similar need explains the fervor with which the *geocentric* (earth centered) universe of Ptolemy was adopted. In fact, it became the established and official world-view in Europe.

The credit for the revolution in science that occurred in the Middle Ages, giving us the *heliocentric world-view* (that continues to the present day in our thinking about the solar system) belongs to five men: Copernicus, Tycho, Kepler, Galileo, and Newton. Copernicus revived, extended, and discussed the theory of a planetary system centered on the sun. For 20 years Tycho (without the aid of a telescope) made precise measurements of the position of Mars. Kepler showed that the motions of Mars, deduced from Tycho's observations, were not in good agreement with the Ptolemaic theory but rather that they were in accord with certain general rules (*Kepler's Laws*) that put the sun about where Copernicus believed it must be. Among the many discoveries made by Galileo with his telescope, two in particular (the phases of Venus and the moons of Jupiter) were strongly discordant with Ptolemaic philosophy. Galileo also began the scientific study of motion. Finally, Newton's development of calculus, his formulation of the laws of motion, and especially his discovery of the law of gravity, enabled him to give a *physical* explanation for the accurate description of planetary motions that Kepler had derived from Tycho's measurements.

Nicolaus Copernicus (1473–1543), a Polish monk, studied his own and others' observations of the planets and proposed a heliocentric theory in which the earth revolved about the sun, as did the other planets, and in which the daily rising and setting of the stars was correctly ascribed to the rotation of the earth on its axis. Although the Copernican solar system (Figure 4-17) is usually remembered as a set of sun-centered circular planetary orbits, Copernicus in fact followed the practice of his predecessors in also including epicycles. Otherwise, the available observations of the planets would not have been completely accounted for by his theory. Thus, he was strongly influenced in formulating this model of the solar system by his continued belief in the circle as the ideal form for a planetary orbit.

The models designed by Aristotle, Ptolemy, and Copernicus shared one feature; they were horribly complex because they were attempting to approximate nature with basically incorrect forms. The Greek theories were so complex that Alfonso the Learned, thirteenth century astronomer and King of Castile, was moved to say that had he been present at the Creation he would have recommended a simpler plan.

Copernicus's ideas were published in 1543, literally as he lay on his death bed. Many of the questions that were asked about his heliocentric theory were of the same sort that children ask when they first become conscious of the world around them. For example, If the earth moves, why don't we fall off? But leaving aside simple-minded objections and details, his basic idea was not welcomed because established religious doctrine held that the earth did not move. Accordingly, Copernicus's book was placed on the forbidden list, the *Index Librorum Prohibitorum,* by the Church in Rome, and it was likewise denounced by Martin Luther.

Tycho Brahe (1546–1601) rejected the Copernican theory in favor of a geocentric cosmology of his own. According to this *Tychonic theory,* the sun revolved about the earth, but the other planets revolved about the sun. A more lasting contribution by Tycho was a series of very precise observations of the position of Mars, made with equipment of his own design at an observatory on the Danish island of Hven. The telescope was still unknown, and a typical Brahe instrument consisted of little more than two pointers used to sight on a star and the planet, respectively. The angle between the two pointers (and hence the angular separation of star and planet in the sky) was then read from a graduated arc.

One of the great heroes of astronomy, Tycho is also one of the more fascinating characters in its history. At the time of his discovery of a bright new star* in 1572, an observation that was a direct challenge to the accepted dogma of the immutability of the stars, he was living in splendor at Hven, surrounded by servants and research assistants. Losing his nose in a duel, he replaced it with a silver one. When Tycho's patron, the King of Denmark, died there was a falling out with the new authorities, so Tycho moved to Prague and composed a poem on Denmark's ingratitude to its greatest astronomer. In Prague he was continually plagued with difficulties such as collecting his salary on time or at all. But the final indignity was his death, after a dinner attended by imperial representatives. No one could leave the table before the V.I.P.'s were finished, and Tycho suffered a burst bladder as a result. His dying wish (not granted) was that Kepler concentrate on proving the Tychonic theory.

Johannes Kepler (1571-1630), a German, worked as Tycho's assistant and inherited his observations. Kepler's forte was not the measuring process. Centuries before the introduction of the

---

*Tycho's star was a supernova or exploding star, like the object that produced the Crab nebula (Chapter 16).

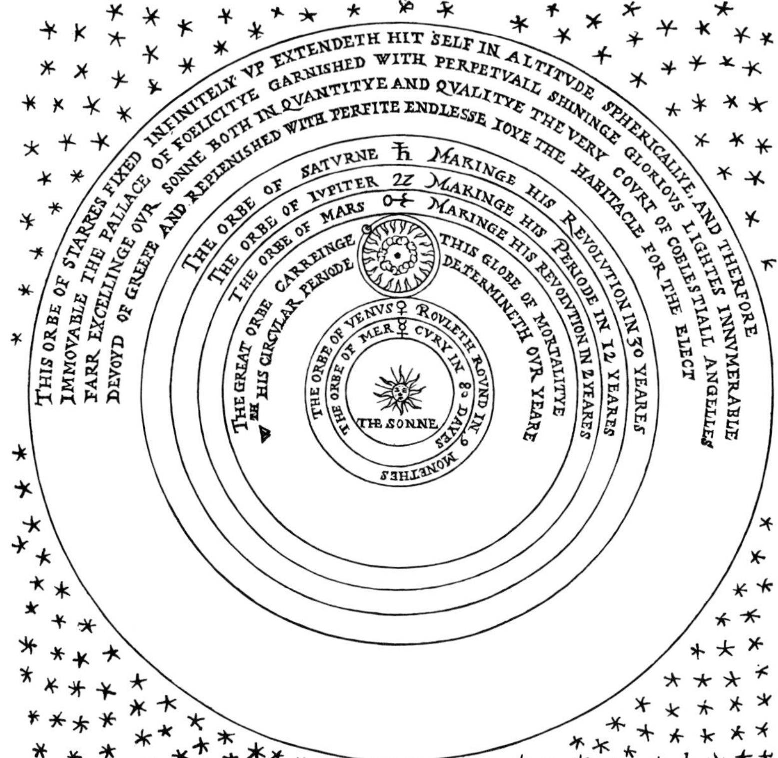

4-17 *The sun-centered Copernican universe as drawn by Thomas Digges in 1576. The earth moves about the sun, but the moon travels around the earth. This diagram contains a feature unusual for the 16th century; the stars are not in a spherical shell at a certain distance, but extend to infinity. Most heliocentric models of the time placed the stars in a shell just beyond the orbit of Saturn, just as given in the pre-Copernican, earth-centered model shown in Figure 4-14.*

electronic computer he was a master of data analysis. He analyzed Tycho's long series of observations, together with some later measurements, and found that they fit three empirical rules. Kepler showed that Copernicus was correct—the sun is the center of the planetary system. And Kepler found that Copernicus was wrong—the planetary orbits are not circles.

KEPLER'S LAWS

1. Each planet travels in an *elliptical* orbit, with the sun at one focus of the ellipse. (A description of an ellipse is given in Figure 4-18.)
2. The *radius vector* (that is, the imaginary line from sun to planet) of a planet sweeps out equal areas in equal periods of time (Figure 4-19).

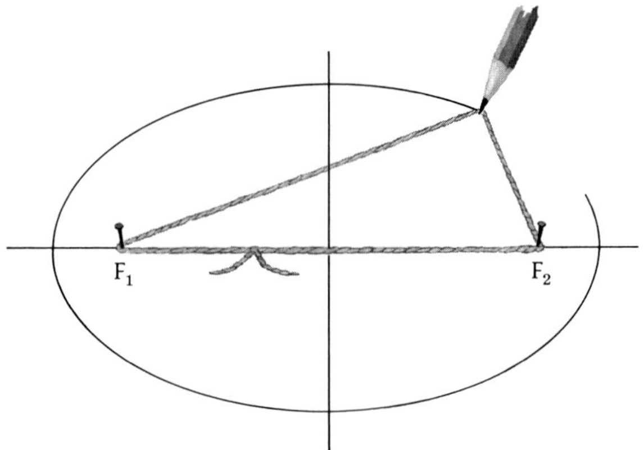

4-18 *Drawing an ellipse. If a loop of string is pinned to a drawing board as shown, the figure traced out by a pencil which keeps the string taut is an ellipse with foci at F_1 and F_2, the locations of the pins. (From* New College Physics *by A. V. Baez. W. H. Freeman and Company. Copyright © 1967.)*

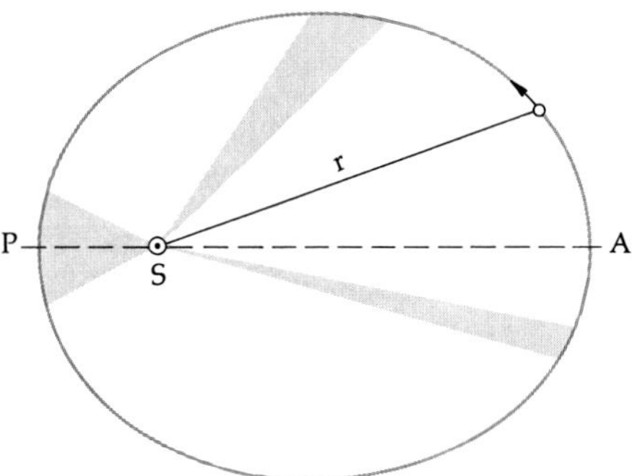

4-19 *The law of areas. Imagine a line (marked r for radius) running from the sun (S) to a planet in its elliptical orbit around the sun. As the planet travels in its orbit, the radius sweeps out an area in such a way that equal areas are swept out in equal times. The area swept out (shaded) is equal in the three cases shown, and thus the planet must travel faster in its orbit at perihelion (P) than near aphelion (A).*

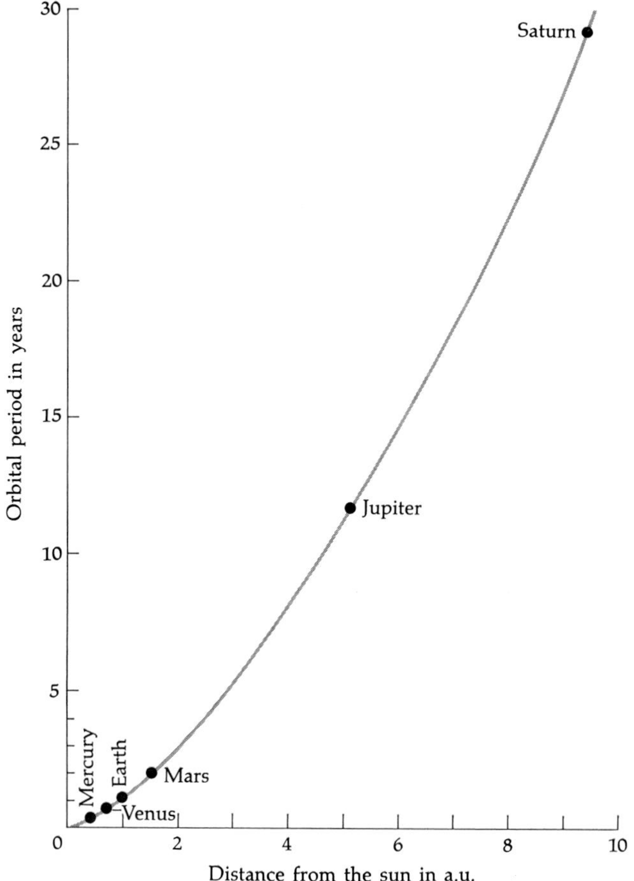

4-20 *Kepler's Third Law which states that the orbital period (in years) squared is equal to the heliocentric distance (in astronomical units) cubed. The planets out to Saturn are shown at the proper points on the graph. The curve can be used to find the period or heliocentric distance when only the other is known; for example, if an asteroid has a period of 4.6 years, its heliocentric distance is about 2.7 a.u.*

3. The distance of a planet from the sun uniquely determines the length of time required for the planet to complete one orbital revolution (the *period*). (See Figure 4-20; the distances are conveniently measured in terms of the *astronomical unit* or average sun-earth distance.) In computational terms the cube of the average distance of a planet from the sun in astronomical units is equal to the period (in years) squared. We can check this for the planet Mars. The distance from the sun is 1.52 a.u.; cubing it we get (1.52) × (1.52) × (1.52) equals 3.54. The period is 1.88 years and (1.88) × (1.88) is 3.54.

Now Kepler had no more physical reasoning to support these laws than Ptolemy had to support *his* theory. The advantage of Kepler's Laws was simply that they gave a better description of planetary motions. In particular, when calculating backward or forward in time using Kepler's Laws, the positions of the planets in the sky can be predicted more accurately than by use of Ptolemy's model.

Before proceeding to a discussion of Galileo's contributions, it is important to understand the spirit of the times. Kepler had the misfortune to observe another so-called "new star," the supernova of 1604, although established authority (despite Tycho's supernova of 1572) still maintained the unchanging nature of the heavens. As a result of this and his work on the Copernican theory, Kepler's books were regarded as dangerous and were placed on the *Index.* The philosopher Giordano Bruno fared even worse. He stated publicly not only that the earth moves, but also that the stars were other suns, surrounded by worlds similar to the earth. He was seized by the Inquisition, tried, and convicted of heresy. A lengthy attempt to make him confess error failed, and he was burned alive in 1600.

In this setting lived Galileo Galilei (1564–1642), the greatest scientist of Renaissance Italy. He originated *dynamics,* the scientific study of motion, believed in the importance of experiments, and made fundamental advances in astronomy. Galileo's discoveries contradicted existing ideas and paved the way for Newton. For example, Aristotle taught that heavier objects fall much faster than light ones. Galileo tested this theory by dropping various objects from atop the Leaning Tower of Pisa (Figure 4-21). The heavy ones barely beat the light ones to the ground. Galileo theorized that the light objects would fall exactly as fast as the others if it were not for the resistance of the air. Since he had no vacuum chamber in which to test this idea,* he performed an experiment in water. Water should have more resistance than air, so the difference in the drop times of dense and light objects would be more noticeable in water than in air. His experiment proved this point. To his students, if not to many of his fellow professors, Galileo established the superiority of experimentation to reasoning from logic alone, as often employed by Aristotle.

Galileo applied the newly discovered telescope to astronomical research in 1609. Among his many great discoveries, several concern us here. Galileo found that the planet Venus exhibits

---

*Apollo 15 astronaut David Scott, standing on the airless moon, dropped a hammer and a feather, beautifully demonstrating to the television audience that Galileo was right.

4-21 *Galileo's experiment at the Tower of Pisa. (After* Matter, Earth, and Sky, *2d ed., by George Gamow. Prentice-Hall, Inc. Copyright © 1965. Used by permission.)*

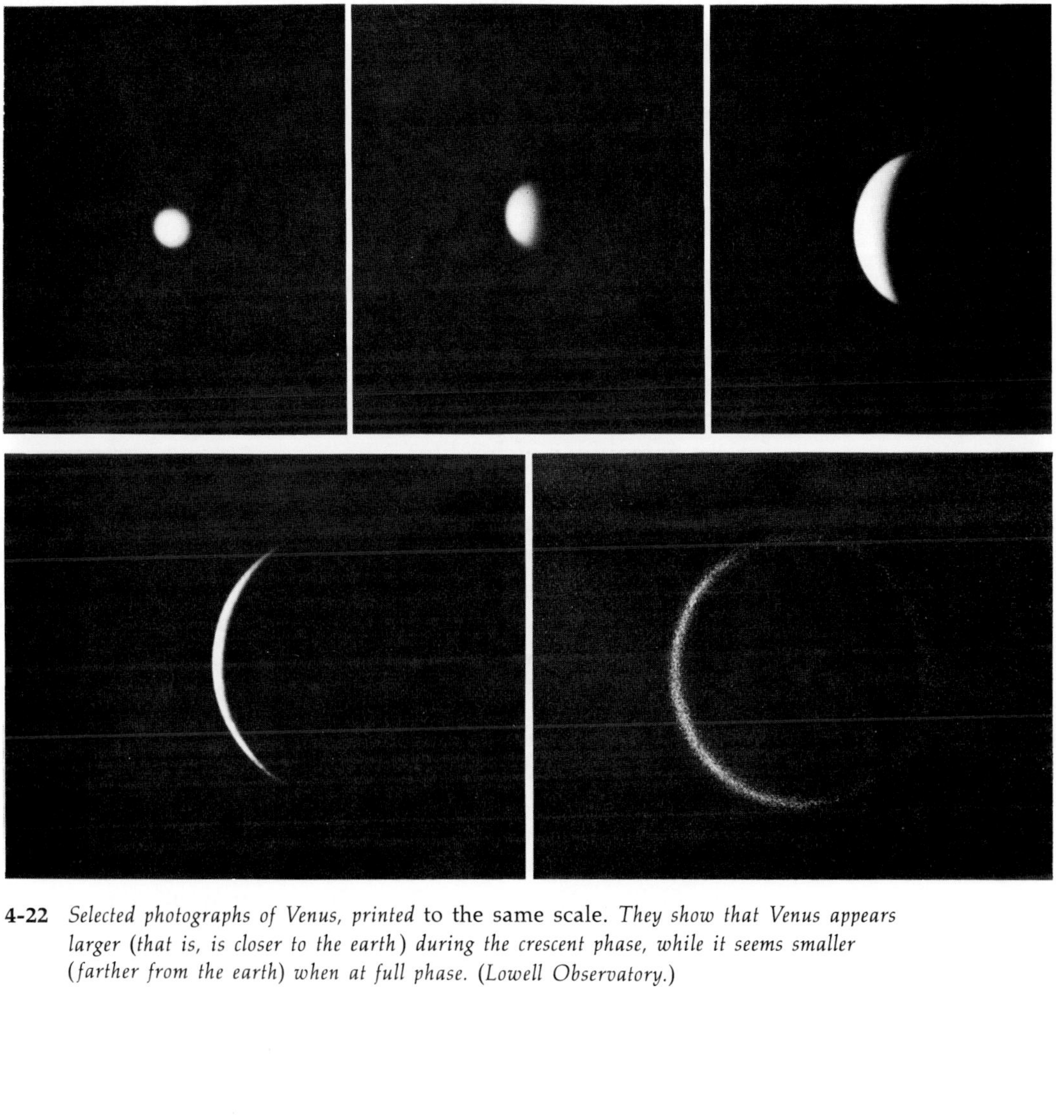

4-22 *Selected photographs of Venus, printed* to the same scale. *They show that Venus appears larger (that is, is closer to the earth) during the crescent phase, while it seems smaller (farther from the earth) when at full phase. (Lowell Observatory.)*

phases (Figure 4-22) similar to those of the moon. This proved that Venus is not self-luminous but shines by the reflected light of the sun. The observed phases agree with a heliocentric model; the crescent is seen when the planet lies close to the line that joins sun and earth; full Venus occurs when it is on the opposite side of the sun. In the Ptolemaic model, however (see Figure 4-14 again), Venus and earth are *never* on opposite sides of the sun, so Venus should never be full or nearly full.

Of lesser direct consequence to the geocentric versus heliocentric argument, but of perhaps even greater philosophical significance than the phases of Venus, was Galileo's discovery that Jupiter has four moons which orbit about that planet (Figure 6-3). Thus, the earth was clearly not the center of all motions. Later, Newton's discussion of the motions of celestial objects under the influence of gravity, a discussion which assumed no special properties for the earth or any other world, was shown to apply as well to Jupiter's satellite system as to the motions of the planets about the sun.

Galileo also found spots or "blemishes" on the sun and noted their motion (proving that the sun rotates), despite the fact that such transient markings were not permitted by the prevailing doctrine that maintained the unchanging nature of celestial bodies.

Galileo's adherence to the heliocentric world view led to trouble with the authorities, and some of his doubters (including university professors) even refused to look through his telescope. Eventually, he was summoned before the Inquisition in Rome, and was asked to recant his views on the movement of the earth. After a time he did and was sentenced to house arrest for the remainder of his life.

Galileo's house arrest caused little reduction in his scientific activity, although he later became blind, perhaps because of eye damage suffered previously when he first viewed the sun through a telescope. He invented a way of studying motion through the use of graphs, proposed the construction of pendulum clocks (but was unable to make one himself; Huygens did it later, in 1656), and composed his famous dialogues on the science of motion. Galileo died in 1642 at the age of 78. He was denied a monument in the hope that he and his offending work would be forgotten, but in that year, Isaac Newton was born, and Newton's work was to ensure the survival and triumph of the heliocentric theory.

Like Galileo, Newton (1642–1727) was a man of many talents and interests. He studied light (Chapter 5), designed the first good reflecting telescope (Chapter 6), placed the rules of motion on a precise mathematical basis, and formulated the universal law of gravitation. Although he was not the sole discoverer of the calculus, he was the one who applied it to the study of moving objects, including the planets.

NEWTON'S LAWS OF MOTION

1. A moving object on which no forces act travels at a constant speed in a straight line. (It may appear that this law is contradicted by everyday experience; for example, a ball rolled along the ground comes to a stop, but the stopping is caused by friction, a type of force exerted on the ball by the surface along which it is rolling. Experiments show that if the friction is reduced by rolling the ball on a smoother surface, the ball will go farther.)
2. A force acting on an object produces a change in its motion in the direction in which the force acts, and the amount of this change (*acceleration*) is inversely proportional* to the mass of the object. (This law also agrees with the evidence of our senses—it takes much more force to get a massive bowling ball moving rapidly than to roll a marble along the same floor at the same speed.)
3. For every action (force) there is an equal and opposite reaction (opposing force). (A familiar example: the angler rows his boat out to a good fishing spot, then throws—exerts a force on—the anchor overboard; the boat moves a slight way in the direction opposite to the throw, and this reveals the reaction—the force exerted by the anchor on the man as it is thrown.)

(In stating Newton's Laws, we have used the term mass. Mass is defined in terms of Newton's Second Law as the property of an object that determines the amount of acceleration which the object undergoes when subjected to a given amount of force. In effect, an object's mass is a measurement of the total amount of matter it contains, in terms of individual atomic particles—protons, neutrons, electrons—as described in Chapter 5.)

Given Newton's Laws, what about the motions of the planets? According to Kepler's results, they travel in elliptical orbits, but Newton's First Law says that a moving object will follow a straight line, unless a force is acting on the object. What force acts on the planets? According to a famous story, Newton, inspired by an apple falling from a tree, realized that the same force of the earth's gravity that acted on the falling apple might

*For example, if object A (mass = 10 grams) and object B (mass = 20 grams) are both acted on by the same force, the acceleration of A will be twice that of B. A gram is equal to the mass of one cubic centimeter of water at 4° C.

act on the moon, keeping it from pursuing a straight course through space. By analogy, a similar force might keep the planets in curved orbits around the sun. Other scientists were reasoning along similar lines at about the same time, but they were unable to test their theories. Newton, using the mathematical methods of calculus that he himself derived, proved that his own hypothesis of *universal gravitation* could explain the planetary motions described by Kepler. According to the universal gravitation idea, every object exerts a force of attraction on every other object, and the force between any two objects decreases as their separation increases, varying inversely as the square of the separation. Thus, the gravitational force between two objects separated by 10 kilometers is 10 squared or 100 times less than the force that would exist between the same objects if they were separated by one kilometer. Further, Newton stated that the force is proportional to the product of the masses of the two objects.

Orbital motion is achieved through a balance of the tendency for an object to keep moving in a straight line* and the force of gravity. If the gravitational attraction between earth and moon were miraculously shut off at some instant, the moon would fly off in a straight line tangent to its orbit† at that point because no force would be pulling it toward the earth. On the other hand, if the moon's orbital motion were suddenly stopped but gravity remained operative, the moon would fall directly toward the earth. The balance of these two opposite tendencies produces the actual orbit of the moon around the earth (see Figure 4-23). The moon is always falling toward the earth, but its own instantaneous motion (at nearly right angles to the direction toward the earth) produces an orbit. An analogous situation can be achieved by whirling a weight around in a circle on the end of a string. If you continue to whirl it and supply a force, the weight travels in a circle; if you stop exerting the force by releasing the string, the weight flies off; it continues moving, but not under your influence.

Newton showed that the laws of motion could explain planetary motion if the force acting on the planets were gravity. In fact, using Newton's Laws and the hypothesis of universal gravitation, it is fairly easy to derive mathematically the other laws Kepler had deduced from laborious examination of Tycho's measurements. Combining Kepler's Third Law and the law of

---

*This tendency, a consequence of Newton's First Law, is commonly called centrifugal force when motion in a curved path is under discussion.

†We have ignored the effects of the gravitational force exerted on the moon by the sun for the purpose of this discussion.

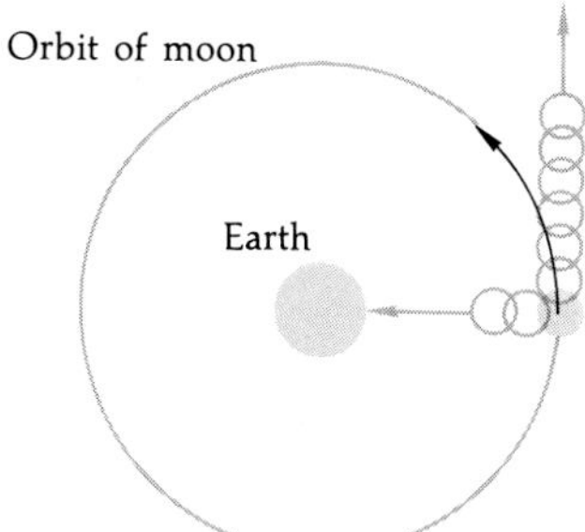

4-23 *The balance responsible for lunar orbit around the earth. If the earth exerted no gravitational pull, the moon would proceed in a straight line into outer space. If the moon were held stationary and released, the earth's gravity would pull it to the earth. These two competing processes produce an orbit. The effects of the sun are not included here.*

gravity, we find that the precise relation between the period and the distance of a planet depends on the mass of the sun (about 2×10^{33} grams). This is the basis of all fundamental mass determinations in astronomy. The orbit of the earth around the sun tells us the mass of the sun; the orbit of the moon around the earth tells us the mass of the earth. The mass of any body with a satellite can be measured by studying the orbit of the satellite.

Newton's explanation of the motions of celestial bodies was not immediately accepted, because it supported and extended the already unpopular work of Copernicus, Galileo, and Kepler, which maintained that the earth moves. Several mathematicians did make calculations that supported Newton's ideas, but the most striking confirmation came in 1758 with the return of a comet, as predicted by Edmund Halley, who based his computations on Newton's theory. In fact, a striking aspect of this successful proof of the theory of gravitation was that the comet did *not* return at the exact time predicted by Halley. Other astronomers who studied Halley's fairly simple calculations realized that, according to the law of gravity, the comet would not only be attracted to the sun, but significant forces would act on it due to Jupiter and Saturn. Taking these into account, in addition to the gravitational force of the sun, they correctly predicted the comet's return.

Halley's comet reappeared in 1758; the law of gravity dates from 1684. During the intervening period, there were nagging doubts about both gravitation and the motion of the earth. Newton could not explain *why* gravitational force exists. He hypothesized the properties of gravity and showed that they could account for the planetary motions, but he could not explain how objects exerted forces across a distance in space.

A crucial test of the earth's motion involves the *parallax* of the stars. Their directions should be slightly different as seen when the earth is on opposite sides of its orbit, because they are viewed from a different position (this does occur, see Chapter 10). Tycho had searched for it, but the effect is so small that it was not until the 1830s that it was detected. The difficulty lay in the lack of appreciation of the vast distances to even the closest stars. At that time they were thought to be just outside the orbit of Saturn (recall Figures 4-14 and 4-17).

The motion of the earth in orbit gives rise to the small difference between the solar and sidereal days, as described earlier in this chapter. This difference can be traced to the fact that during the passage of one day (as determined by the earth's rotation), the earth has moved along in its orbit and, by consequence, the sun is seen at a slightly different position in the sky. This apparent motion of the sun accounts for the 4 minute difference in the lengths of the two types of day (see Figure 4-24). Note, however, that a real motion of the sun around a fixed, rotating earth would also explain it.

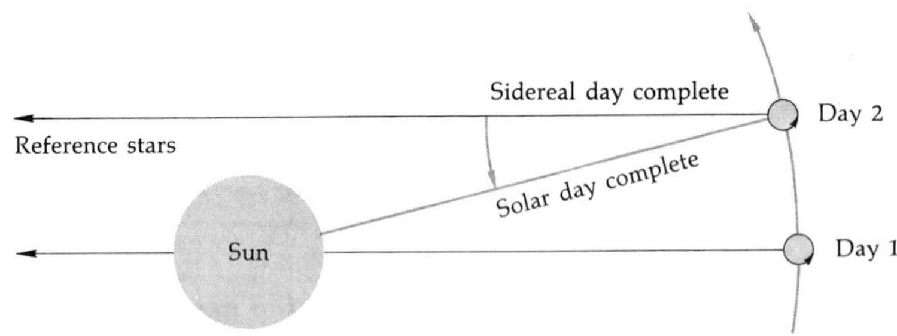

4-24 *Solar versus sidereal days. At noon on Day 1 the sun and a hypothetical reference star are directly overhead. While the earth rotates on its axis, its orbital motion also carries it to the position marked Day 2. A sidereal day has elapsed when the reference star is overhead. The sun, however, is not in the same direction at that time, and the earth must rotate through the extra angle (as marked) to complete the solar day.*

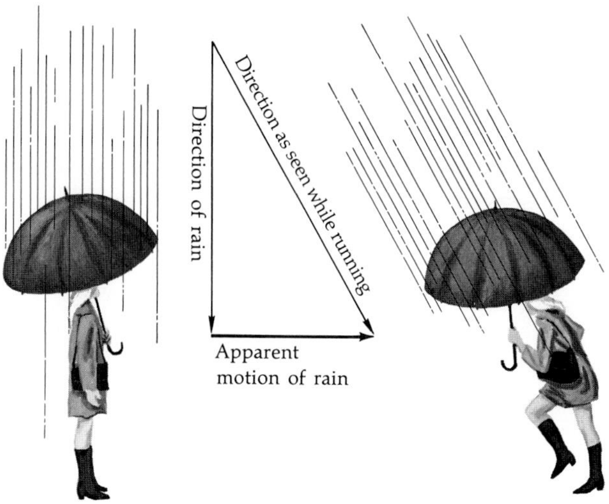

4-25 *Aberration diagram showing that the direction in which raindrops appear to fall depends on the motion of the observer.*

An additional fact easily explainable only if the earth moves is the phenomenon of *aberration,* which was discovered in 1728 by the English astronomer, James Bradley. Bradley was trying to measure the parallax of stars, but his equipment was inadequate. Instead, he found an effect that had an opposite behavior to that expected for parallax. When the earth was situated favorably for observation of the shift due to parallax (Figure 10-1), none was observed. When no shift due to parallax was expected, an easily measured shift of 20 seconds of arc on either side of the average position of a star was observed. The observations were somewhat perplexing until Bradley noticed one day that the direction in which the flag of a ship flew depended not only on the direction of the wind but also on the motion of the ship. This led him to understand that aberration was caused by the motion of the earth.

The effect of aberration can be visualized by considering the direction of falling raindrops. Suppose there is no wind; then the raindrops are falling vertically. However, if we run through the rain, we must tilt the umbrella forward to avoid getting wet because the running motion causes the rain to seem to fall at an angle (Figure 4-25). Thus, to have aberration, we must have both the motion of the objects to be observed and the motion of the observer. We see stars by means of their "light particles" or photons which travel at the speed of light, and the apparent directions from which the photons come are changed by the earth's motion, just as the raindrops' direction of origin seems to depend on the runner's motion. Thus, the aberration of starlight depends on the earth's speed and the speed of light. In fact, from the observed 20 second of arc displacement and the known speed of the earth in its orbit, it is possible to estimate the speed of light.

5

Matter, Heat, and Light

We began our survey of the universe by looking first at the earth from the viewpoint of elementary geology and relevant concepts of biology. Next we examined the earth's place in the solar system, introducing some aspects of the physics of motion (dynamics) and gravitation. Now we will present several basic ideas from physics and chemistry that are needed to understand our observations of stars, planets, nebulae, and galaxies.

HISTORY OF ATOMISM

We encounter matter in three forms in our everyday experience—as gases, liquids, and solids. Materials have different colors, different hardnesses, and so on. We sometimes see matter change its state, for example when ice melts or water becomes steam. The complexity of even the most familiar things becomes apparent when we observe them carefully. Consider a campfire—flames and smoke rise and water sizzles from the burning wood. The smoke thins out with distance from the fire, eventually merging indistinguishably into the surrounding air. A small residue of solid matter is soon all that is left. Commonplace phenomena like these were of the greatest interest to the early natural philosophers. Among the Greeks, Empedocles (490–430 B.C.) suggested that all things are composed of four "elements": air, water, fire, and earth; a Chinese doctrine recognized five: earth, fire, water, wood, and gold.

The early attempts to describe the composition of matter led to the formulation of atomism, the theory that matter is not infinitely divisible but that discrete parts or building blocks that cannot be further divided are finally encountered at some level. The idea of ultimate building blocks or *atoms* was advocated by the Greek philosophers Leucippus and Democritus in the fifth and fourth centuries B.C. The atoms were thought of as endlessly drifting through a void; an occasional combination of them formed material objects. Later, Epicurus (341–270 B.C.) taught that atoms have weight.

The concept of atomism lay dormant through the Middle Ages, then reappeared in the seventeenth century. Newton discussed and favored the theory, as did the British chemist Robert Boyle (1627–1691), who conducted experiments showing that the four elements of the Greeks were not the primary components of matter, since they could not be extracted from other substances. Later, John Dalton (1766–1844), influenced by Newton's work, formulated an Atomic Theory and conducted relevant experiments. Considering the elements as primary materials that could not be broken down further by heat or chemical processes but which did combine with each other to produce the various known substances, Dalton's Atomic Theory was based on three principles:

1. Elements consist of indivisible particles (atoms) whose properties are not affected by chemical reactions.
2. The atoms of a given element have equal weights and are also identical in their other properties, but atoms of different elements always have different weights.
3. A chemical compound is formed by the combination of specific numbers of atoms of different elements.

A familiar example of the third principle is the water compound (H_2O); each of its smallest units (molecules) consists of two hydrogen atoms and one oxygen atom. The fact that water can be broken down into two different substances (hydrogen and oxygen) establishes that it is a compound and not an element.

Dalton's theory proved capable of accounting for a wide variety of chemical experiments, and it became a keystone in the subsequent development of chemistry and physics. Today we know that although it can account for a wide variety of laboratory experiments, it is not strictly correct. The physicists have split atoms and found that they are built up from various combinations of three types of particle—the electrons, protons, and neutrons—and that a variety of other atomic particles exist as well. And it has been found that the atoms of a given element may exist in several forms (*isotopes*) that differ in weight.* In fact, there exist isotopes of different elements that have virtually the same weights but differ in their detailed makeup of electrons, protons, and neutrons. We now recognize that the atomic weights deduced for various elements on the basis of the Atomic Theory are really averages that depend on the individual weights and the relative *abundance* (frequency of occurrence) of the isotopes of each.

Certain elements have rather similar properties. For example, sodium, potassium, and lithium are all light-weight metals that readily react with other substances to form compounds. On the other hand, neon, argon, and xenon are gases that never combine with other elements under natural conditions.† Several scientists noticed that there appear to be numerical relationships between the atomic weights of elements that have similar properties. In 1863, the Englishman John Newlands suggested that (looking at the elements in order of increasing atomic weight) every eighth element was similar. The most important work on this subject was that of the Russian chemist Dmitri Mendeleev (1834–1907). He also tabulated the elements in the order of their known atomic weights. Believing in numerical relationships for similar elements, Mendeleev suggested that many elements were at the wrong places in the table and therefore their measured atomic weights must be wrong. Furthermore, he pointed out that there were gaps in the table of atomic weights where no elements were known to exist. He deduced that six such elements should exist, and he predicted their physical properties on the basis of their locations in his Periodic Table.‡ Mendeleev's criticism of the old atomic weight measurements was confirmed by subsequent experiments; five of the new elements were later found in nature, and the sixth has been created artificially in the laboratory (Table 5-1).

Mendeleev's predictions of the properties of these new elements were also verified. For example, he suggested that eka-silicon, the element of atomic weight about 72, would be

---

*An example is given on page 106.

†In 1962, chemists finally succeeded in creating compounds of xenon.

‡As often happens in science, similar ideas were developed elsewhere at about the same time; in this case, the German chemist Lothar Meyer also formulated a periodic table of the elements.

Table 5-1 *Six Elements Predicted by Mendeleev in 1871*

MENDELEEV'S NAME	YEAR OF DISCOVERY	PRESENT NAME
eka-boron	1879	scandium
eka-aluminum	1875	gallium
eka-silicon	1886	germanium
eka-manganese	1937*	technetium
dvi-manganese	1925	rhenium
eka-tantalum	1917	protoactinium

* Produced artificially.

a gray metal about 5.5 times denser than water and that it would form a white compound when heated in oxygen. These predicted properties of eka-silicon were indeed found to characterize the element of atomic weight 72.6, which was discovered in 1886 and named germanium, and which has been of vital importance to twentieth-century electronics as a key ingredient of the transistor.

The brilliant success of Mendeleev's predictions showed the importance of investigating what it is about an atom that allows its physical properties to be determined by its position on the table of atomic weights. A modern version of the Periodic Table is given in Table 5-2. The elements are not arranged by atomic weight but by *atomic number,* the number of electrons or protons in the atom.

PARTICLES IN THE ATOM

The first component particle of the atom to be recognized was the *electron,* the negatively charged particle that carries the smallest (that is, indivisible) unit of electric charge. Its name is derived from the Greek word elektron, meaning amber, or fossilized pine resin. Amber figured in an electrical phenomenon that was known to the Greeks. They found that when a chunk of it is rubbed with a woolen cloth, the amber will attract light objects, such as feathers. Later experiments showed that other objects could be charged in this way and that there are two kinds of electrical charge (called positive and negative) which can be distinguished by the circumstance that objects with like charge repel each other, whereas oppositely charged objects attract each other (Figure 5-1).

Table 5-2 *The Periodic Table of the Elements*

Hydrogen 1 **H** 1.008	Helium 2 **He** 4.003

Lithium 3 **Li** 6.939	Beryllium 4 **Be** 9.012	Boron 5 **B** 10.81	Carbon 6 **C** 12.01	Nitrogen 7 **N** 14.01	Oxygen 8 **O** 16.00	Fluorine 9 **F** 18.99	Neon 10 **Ne** 20.18
Sodium 11 **Na** 22.99	Magnesium 12 **Mg** 24.31	Aluminum 13 **Al** 26.98	Silicon 14 **Si** 28.09	Phosphorus 15 **P** 30.97	Sulfur 16 **S** 32.06	Chlorine 17 **Cl** 35.45	Argon 18 **Ar** 39.95

Potassium 19 **K** 39.10	Calcium 20 **Ca** 40.08	Scandium 21 **Sc** 44.96	Titanium 22 **Ti** 47.90	Vanadium 23 **V** 50.94	Chromium 24 **Cr** 52.00	Manganese 25 **Mn** 54.94	Iron 26 **Fe** 55.85	Cobalt 27 **Co** 58.93	Nickel 28 **Ni** 58.71	Copper 29 **Cu** 63.55	Zinc 30 **Zn** 65.37	Gallium 31 **Ga** 69.72	Germanium 32 **Ge** 72.59	Arsenic 33 **As** 74.92	Selenium 34 **Se** 78.96	Bromine 35 **Br** 79.90	Krypton 36 **Kr** 83.80
Rubidium 37 **Rb** 85.47	Strontium 38 **Sr** 87.62	Yttrium 39 **Y** 88.91	Zirconium 40 **Zr** 91.22	Niobium 41 **Nb** 92.91	Molybdenum 42 **Mo** 95.94	Technetium 43 **Tc** (99)	Ruthenium 44 **Ru** 101.1	Rhodium 45 **Rh** 102.9	Palladium 46 **Pd** 106.4	Silver 47 **Ag** 107.9	Cadmium 48 **Cd** 112.4	Indium 49 **In** 114.8	Tin 50 **Sn** 118.7	Antimony 51 **Sb** 121.8	Tellurium 52 **Te** 127.6	Iodine 53 **I** 126.9	Xenon 54 **Xe** 131.3
Cesium 55 **Cs** 132.9	Barium 56 **Ba** 137.3	Lanthanide series 57 – 71	Hafnium 72 **Hf** 178.5	Tantalum 73 **Ta** 180.9	Tungsten 74 **W** 183.9	Rhenium 75 **Re** 186.2	Osmium 76 **Os** 190.2	Iridium 77 **Ir** 192.2	Platinum 78 **Pt** 195.1	Gold 79 **Au** 197.0	Mercury 80 **Hg** 200.6	Thallium 81 **Tl** 204.4	Lead 82 **Pb** 207.2	Bismuth 83 **Bi** 209.0	Polonium 84 **Po** (210)	Astatine 85 **At** (210)	Radon 86 **Rn** (222)
Francium 87 **Fr** (223)	Radium 88 **Ra** (226)	Actinide series 89 – 103	Rutherfordium 104 **Rf** (260)														

Lanthanide series	Lanthanum 57 **La** 138.9	Cerium 58 **Ce** 140.1	Praseodymium 59 **Pr** 140.9	Neodymium 60 **Nd** 144.2	Promethium 61 **Pm** (145)	Samarium 62 **Sm** 150.4	Europium 63 **Eu** 152.0	Gadolinium 64 **Gd** 157.3	Terbium 65 **Tb** 158.9	Dysprosium 66 **Dy** 162.5	Holmium 67 **Ho** 164.9	Erbium 68 **Er** 167.3	Thulium 69 **Tm** 168.9	Ytterbium 70 **Yb** 173.0	Lutetium 71 **Lu** 175.0
Actinide series	Actinium 89 **Ac** (227)	Thorium 90 **Th** (232.05)	Protoactinium 91 **Pa** (231)	Uranium 92 **U** 238.0	Neptunium 93 **Np** (237)	Plutonium 94 **Pu** (242)	Americium 95 **Am** (243)	Curium 96 **Cm** (247)	Berkelium 97 **Bk** (247)	Californium 98 **Cf** (249)	Einsteinium 99 **Es** (254)	Fermium 100 **Fm** (253)	Mendelevium 101 **Md** (256)	Nobelium 102 **No** (256)	Lawrencium 103 **Lr** (257)

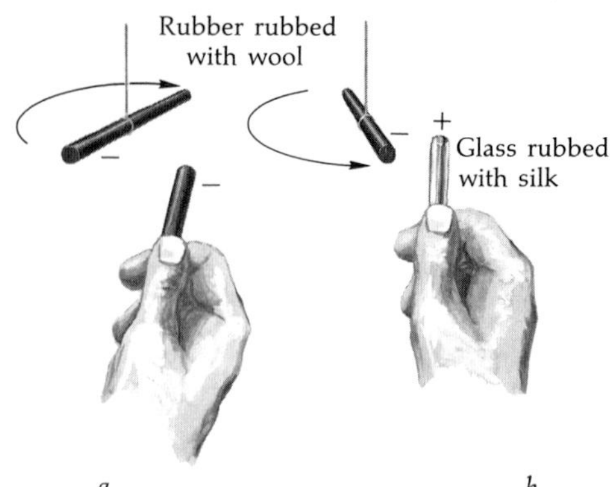

5-1 *Illustration of (a) the repulsion of like charges and (b) the attraction of unlike charges. (From* Structure and Change *by G. S. Christiansen and P. H. Garrett. W. H. Freeman and Company. Copyright © 1960.)*

The interpretation of the experiment in which the initially neutral amber becomes attractive after being rubbed is that a certain amount of negative charge was literally rubbed off the amber, leaving it with a net positive charge. For a long time a controversy existed over whether electricity was continuous or consisted of discrete indivisible particles, like the atoms of matter. During the second half of the nineteenth century, the discovery of *cathode rays* and subsequent experiments with them resolved this argument.

When each of the two terminals of a high voltage battery is connected to a separate metal plate, and the two plates are housed inside a glass tube filled with air at low pressure, electricity flows through the gas, which shines like a neon sign. Pumping out the air, a point is reached where the pressure is so low that the glowing gas is no longer seen. However, a faint glow *in the glass itself* suggests that it is being struck by particles (Figure 5-2). Experiments showed that these particles emanated from the negatively charged plate (cathode); hence they were named cathode rays. Physicists disagreed about them; some believed that the rays were really particles of light (*photons,* page 122), whereas others suggested that they were the long-sought individual particles of electric charge. About 1897, J. J. Thomson at Cambridge University measured the velocity of the particles, showed that it was less than the velocity of light, and thus demolished the first theory. Thomson also demonstrated that the cathode rays behaved in the same way regardless of what metal was used for the negatively charged plate, or what gas

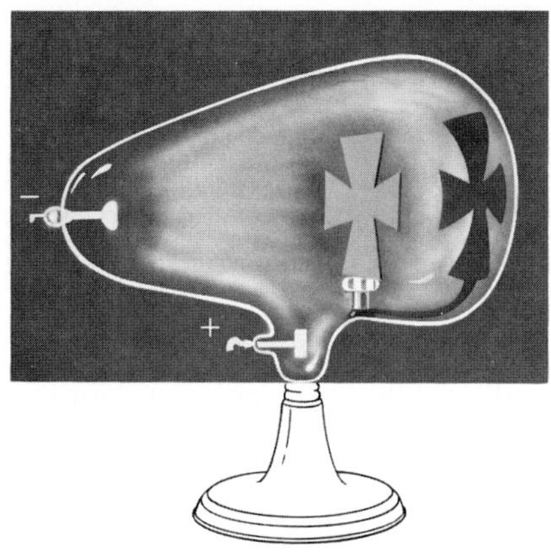

5-2 *The Crookes tube, with electrons moving from left to right. The glow is due to the electrons striking the glass. The presence of the shadow shows that the electrons move in straight lines. (From* College Chemistry *by Linus Pauling. W. H. Freeman and Company. Copyright © 1964.)*

was placed in the tube. Thus, it appeared that they were neither particles of light nor atoms of a particular element. The evidence suggested that the cathode rays (now called electrons) were indeed particles of electric charge.

It was possible, from a variety of experiments, to estimate the value of the electron's charge. But in each case, this value was found as the average of a great many particles, and it was not known if *each* particle had an identical charge. This was resolved in a famous experiment by the American physicist, R. A. Millikan (1868–1953), who developed an ingenious method to measure the electrical charges of tiny oil drops. He found that the charge values were all multiples of one negative number, and this largest common denominator was thus identified as the charge of the electron. It was also found that all electrons have the same mass, and that this is equal to $^1/_{1837}$ of the mass of a hydrogen atom.

At this point in historical development, we know that electrons are a common constituent of matter, that they are charged, and that they have a specific mass. However, ordinary matter such as wood is not electrically charged. Hence, atoms must also contain an equal and opposite (positive) charge. Thomson proposed a "plum pudding" model in which the atom consisted of a homogeneous sphere of positive charge, where the electrons were embedded like raisins or other goodies (Figure 5-3).

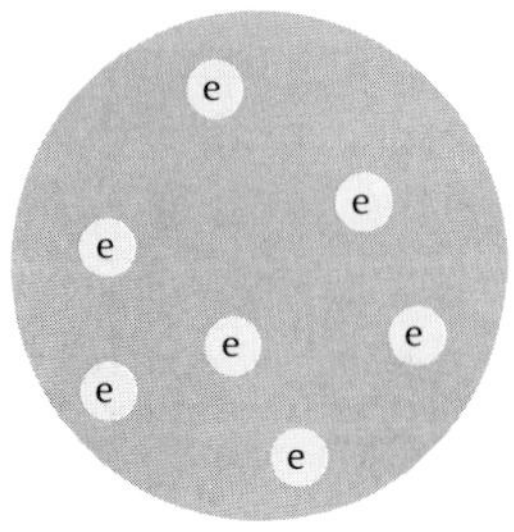

5-3 *The Thomson plum pudding model of the atom. The shaded area represents the positive charge.*

The validity of the plum pudding or any other atomic model could be checked if we had a way to probe the structure of the atom. A technique was in fact developed, making use of particles released in the radioactive decay of some atoms (recall the discussion of natural radioactivity in Chapter 2). Three kinds of ray were known to be emitted in the process of radioactive decay:

alpha rays: positively charged, also called alpha particles; now identified as helium nuclei, consisting of two protons and two neutrons.

beta rays: negatively charged; now identified as electrons.*

gamma rays: now identified as high energy photons.

The type of electrical charge of each kind of ray was determined by seeing how it behaved when passing through the field of a magnet (Figure 5-4), as positive and negative charges were known to be deflected in opposite directions by the magnetic field, whereas uncharged particles were not.

In atom-probing experiments directed by Lord Rutherford in 1911 (Figure 5-5), a thin gold foil was bombarded by a stream of alpha particles. Most of these particles passed through the gold foil, but one in 8,000 bounced off at an angle. This suggested that most of the alpha particles hit nothing or, at most, hit the much less massive electrons in the gold. But every once in a while, an alpha particle did hit a relatively heavy, positively charged object. The fact that the positive charges were encountered so rarely meant that the plum pudding model was wrong and the positive charge was probably confined to a minute fraction of the atom's volume. Thus, Rutherford was led to an atomic model in which the electrons circle about the positively charged nucleus in great orbits like planets around the sun (Figure 5-6). In the case of the simplest isotope of hydrogen (which is the simplest element), the atom consists of one electron in orbit around a particle of equal and opposite charge. This positively charged *proton* has a mass of 1,836 times the electron's mass.

Rutherford's work suggested that the positive charge in the atom was concentrated in a small fraction of the atomic volume, but this was hard to understand. Protons have like charges and therefore repel each other. What process could confine them to the nucleus? This question is answered by the existence and

---

*Electrons were called cathode rays when they were found in the electrical experiments described on page 102; they were called beta rays when they were found as a decay product of radioactive atoms; eventually it was realized that these two kinds of ray were identical.

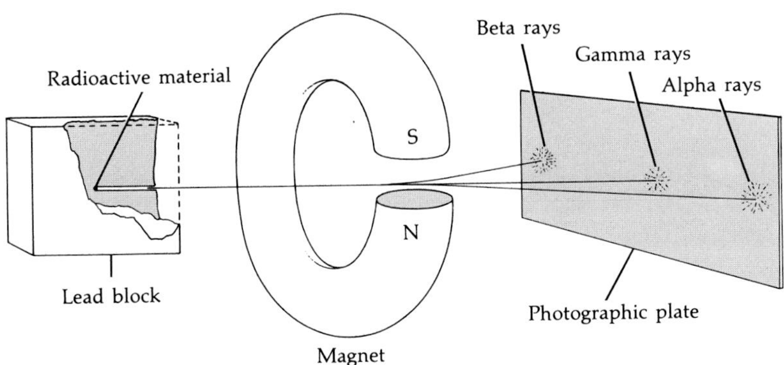

5-4 *Alpha and beta rays are deflected by a magnetic field while gamma rays are not. (From* General Chemistry, *3d ed., by Linus Pauling. W. H. Freeman and Company. Copyright © 1970.)*

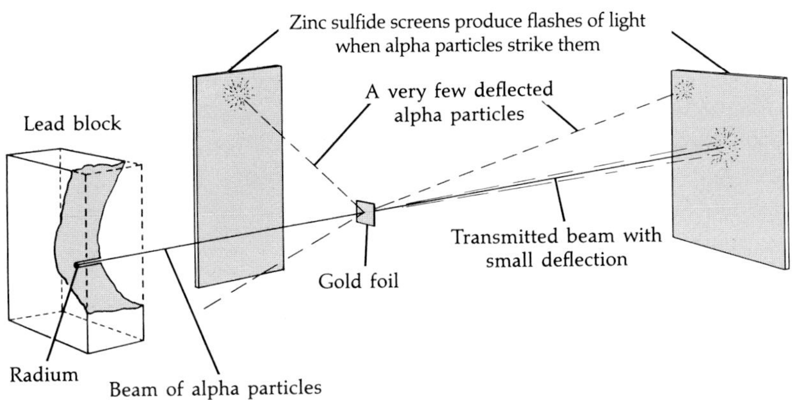

5-5 *The Rutherford experiment. The fact that very few of the alpha particles were deflected by nuclei in the gold foil showed that the nuclei in the foil atoms were small compared to the atoms themselves. (From* General Chemistry, *3d ed., by Linus Pauling. W. H. Freeman and Company. Copyright © 1970.)*

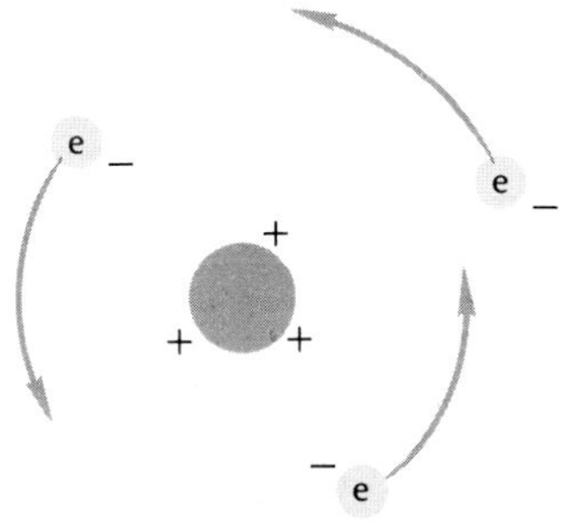

5-6 *The Rutherford planetary model of an atom, for the case where the atomic number is 3. The orbits of the electrons are achieved in a manner analogous to the planetary orbits discussed in Chapter 4. In the atomic case, the effect of gravity is very weak, and the tendency for the electron to continue in a straight line is balanced by the electrical attraction of the nucleus. At the time of Rutherford's experiments, the neutron had not been discovered. In fact, the various isotopes of the element (lithium) with atomic number 3 have from two to six neutrons in their nucleii.*

properties of the neutron (discovered in 1932 by James Chadwick), a particle with approximately the same mass as the proton. As its name implies, the neutron has no charge. Free (that is, not in a nucleus) neutrons decay spontaneously with a half-life of 12 minutes, leaving a proton and an electron. Both protons and neutrons are found closely packed in the nucleus. They are attracted to each other by a *short-range force* (also called the strong nuclear force) that is much stronger than either gravitational or electric force over tiny distances like that in the nucleus, but which falls off in strength very rapidly with distance and so is insignificant on scales much larger than the nucleus.

Several terms are used in describing the properties of atoms in relation to their different numbers of particles. The *atomic number* is equal to the number of protons in the nucleus of an atom and is the same as the number of orbiting electrons when the atom is in its normal or electrically neutral state. The *atomic mass* is approximately equal to the total number of protons and neutrons in the nucleus. The *isotopes* of a given element are atoms with the same atomic number (equal numbers of protons) but with different atomic masses (different numbers of neutrons). For example, the common form of the hydrogen atom which we described above as consisting of one proton and one electron has an atomic mass of about 1* and atomic number 1. On the other hand, *tritium,* a scarce isotope of hydrogen consisting of one proton, two neutrons, and an electron, has an atomic mass of about 3 and atomic number 1. There is an isotope of helium that also has an atomic mass of about 3: it has two protons, two electrons, and one neutron. Its atomic number is 2, indicating that it is an isotope of a different element than tritium. The *atomic weight* of an element as usually given in tables, such as Table 5-2, is an average of the atomic masses of the isotopes of that element, calculated according to their relative frequency of occurrence in ordinary matter. For example, the tabulated atomic weight of helium is 4.003, due to the fact that the vast majority of naturally occurring helium on earth is in the form of the isotope helium-four, which has two protons, two neutrons, and two electrons, and an atomic mass of 4.003 daltons. The helium isotope with an atomic mass of about 3 amounts to only ten-thousandth of one per cent of the naturally occurring helium on earth.

---

*More exactly, the atomic mass of the common form of hydrogen is 1.007825 daltons, where the *dalton* is defined as exactly $\frac{1}{12}$th of the mass of an atom of the carbon-12 isotope. Carbon-12 has six neutrons, six protons, and six electrons.

Despite the existence of the short-range force, there appears to be a limit to the number of nucleons (protons and neutrons) in a naturally occurring atom. The heaviest atom found in substantial amounts in nature is the uranium isotope of atomic mass 238 (atomic number 92). Short-lived heavy atoms, notably the plutonium isotope of atomic mass 239 (atomic number 94), are found in trace amounts in uranium mines where they are produced by reactions between the uranium atoms and neutrons. Heavier atoms have been made in the laboratory, but these *transuranium elements* all decay radioactively in times that are short compared to (for example) the age of the earth (indeed, some have half-lives of a fraction of a second). This has led some scientists to speculate that a number of the transuranium elements may have existed naturally in the past. There are also a few atoms of lesser atomic number than uranium, such as technetium (atomic number 43), that do not occur naturally on the earth. These elements also have been produced in the laboratory and they also do not have long-lived isotopes.

COMPOUNDS, GASES, LIQUIDS, AND SOLIDS

We will now discuss the question of compounds and the chemical bond on the basis of Rutherford's model of the atom in which the electrons are pictured in orbit around the nucleus. The basic properties of the model, in terms of the arrangement of the electrons in orbits and shells, were worked out by a number of scientists, notably the Danish physicist Niels Bohr. It has become known as the *Bohr atom* or *Bohr model* and is adequate for our purpose in explaining the emission and absorption of light by atoms in stars and other celestial objects, although some of the details have been superseded by more recent developments in physics.

The *noble gases* (including helium, neon, argon) are so named because they are inert and almost never form compounds with other elements. This occurs because the electrons of atoms do not occupy randomly located orbits. There appear to be specific shells, each capable of containing some specific number of orbiting electrons. The innermost shell can accommodate up to two electrons. The atom with one proton, and therefore one electron, is hydrogen. That with two protons and two electrons is helium. Hydrogen reacts fairly readily with other elements, but helium, which has a complete (filled to capacity) electron shell does not. It is found that the other elements which, like helium, rarely form compounds (that is, the other noble gases)

Table 5-3 *Some Elements of Interest*

ATOMIC NUMBER	NOBLE GASES	ATOMIC NUMBER	ALKALI METALS
2	Helium	3	Lithium
10	Neon	11	Sodium
18	Argon	19	Potassium
36	Krypton	37	Rubidium
54	Xenon	55	Cesium
86	Radon	87	Francium

are precisely those that normally have complete electron shells and no electron outside. For example, the second shell can accommodate up to eight electrons, and the element that has ten electrons (two in the inner shell, eight in the second shell) is neon, a noble gas. On the other hand, we recall the metallic elements such as lithium, sodium, and potassium (page 99) that react very rapidly with many common substances. These are exactly the elements in which the outermost electron is the sole electron in a shell, and the other electrons are in closed shells. For example, lithium has one more electron than the noble gas helium, and sodium has one more electron than the noble gas neon. Some noble gases and alkali metals are arranged in order of atomic number in Table 5-3.

The numbers of electrons that can exist in the various atomic shells are, respectively, 2, 8, 8, 18, 18, 32, and 32 (in order beginning with the innermost shell). If all seven shells of an atom were completed (that is, each contained the maximum possible number of electrons, as listed) then the atom would have atomic weight number 118, much higher than the heaviest known transuranic element (atomic weight 105).

The fact that the second and third shells each contain eight electrons, and the dependence of the chemical properties of an element on the arrangement of its electrons in shells, together explain why Newlands found some evidence for his idea that every eighth element was similar (page 99). The Periodic Table was designed so that the elements are listed along horizontal rows* in such a way that elements with similar properties occur in the same vertical columns (Table 5-2). The physical basis of the Periodic Table can now be understood from the atomic model discussed above. Each row in the table comprises those elements whose outermost electrons are normally confined to the same shell.

*The number of elements in a row was called the period.

A given chemical compound, such as ordinary table salt (sodium chloride), consists of two or more elements that have combined to form a substance whose properties are different from those of the constituent elements. We have seen that the short-range force holds a nucleus together, and an electrical force (the attraction between unlike charges, usually called the *coulomb force*) holds the atom together in such a way that the electrons exist in certain discrete shells. The formation of compounds is related to the latter process. As the formula for table salt (NaCl) shows, the smallest unit or *molecule* of this compound consists of one sodium (Na) and one chlorine (Cl) atom. The tendency to complete an atomic shell of eight electrons is fulfilled in this case; sodium has one electron in its outer shell and chlorine has seven. The NaCl molecule can be visualized as forming when the electron in the outer shell of the sodium atom is transferred to the chlorine atom. However, the chlorine atom then has one extra electron and is a *chlorine ion* with a negative charge. The sodium atom now has one less electron than normally and is a *sodium ion* with a positive charge (see Figure 5-7). The unlike charges attract and an *electrovalent* bond is established. The electrovalent bond is not directional; that is, there is no preferred orientation for the two atoms in the NaCl molecule.

A complete outer shell of eight electrons can also be achieved by sharing some or all of the electrons between two or more nuclei. Thus, the chlorine gas molecule (Cl_2) is formed by sharing two electrons between the two chlorine nuclei to complete two shells of eight electrons. Water (H_2O) is formed by sharing the two electrons from the hydrogen atoms and the six electrons from the oxygen atom to form a common cloud of

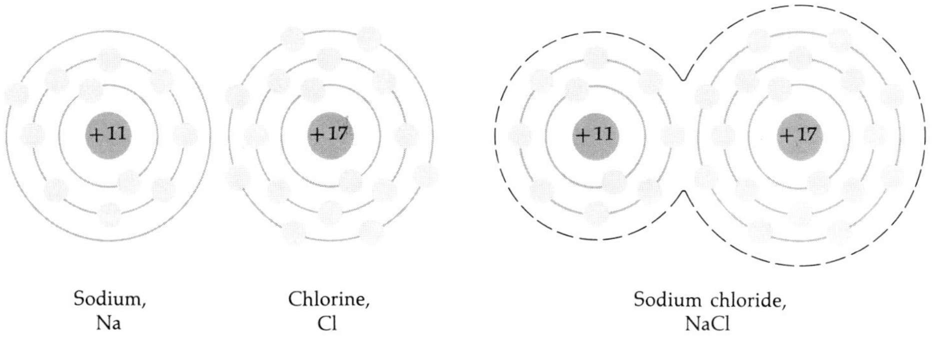

5-7 *Schematic illustration of the formation of salt, NaCl, from sodium (Na) and chlorine (Cl) through the electrovalent bond.*

electrons. In these two examples, the electrons can be thought of as forming a common cloud around both nuclei. The bonds in these molecules are called *covalent* and they have the remarkable property of being directional; that is, there are only certain possible directions of attachment. A diagram of the CH_4 molecule is shown in Figure 5-8. The large molecules encountered in biology (such as DNA) are the result of covalent bonds, as are many crystals (NaCl is an exception) with their regular arrangement and spacing. Carbon with its ability to form four equally spaced covalent bonds (Figure 5-8) is well suited to the construction of large molecules. This property of the carbon atom that lends itself to the development of large, complex molecules helps to explain why life developed from molecules based on carbon.

In a gas, the force that holds an individual molecule together is much stronger than any force that may attract one molecule to another, and so the individual molecules can move around freely. However, in a liquid the molecules are closer to one another and some of the electrons of one molecule may be attracted by the positive ion of another molecule. This bond (*van der Waals force*) is much weaker than the electrovalent and covalent bonds, but it does hold the molecules loosely together, thereby distinguishing a liquid from a gas (Figure 5-9).

The bonds in a solid are much stronger than those in a liquid, and this allows a piece of solid matter to have a definite shape and size. For example, in a metal the matter is held together by a force (*metallic bond*) that results from the general attraction between all of the electrons that are not contained in closed atomic shells and a lattice or framework composed of the positive ions. An individual free electron is thus not bound to

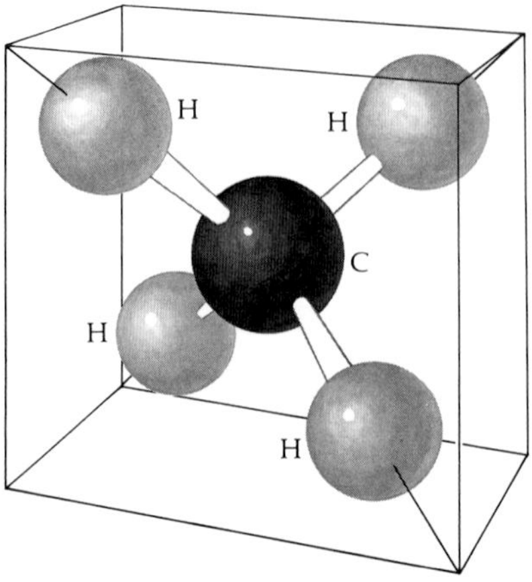

5-8 *Structure of the methane molecule, CH_4, illustrating the directional property of the covalent bond. (After* College Chemistry *by Linus Pauling. W. H. Freeman and Company. Copyright © 1964.)*

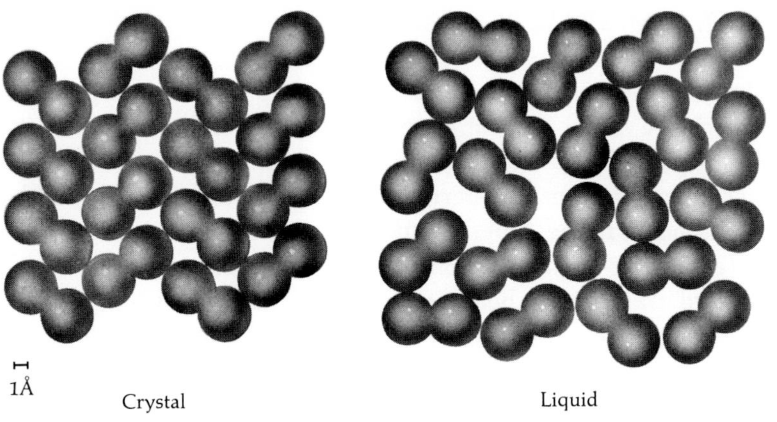

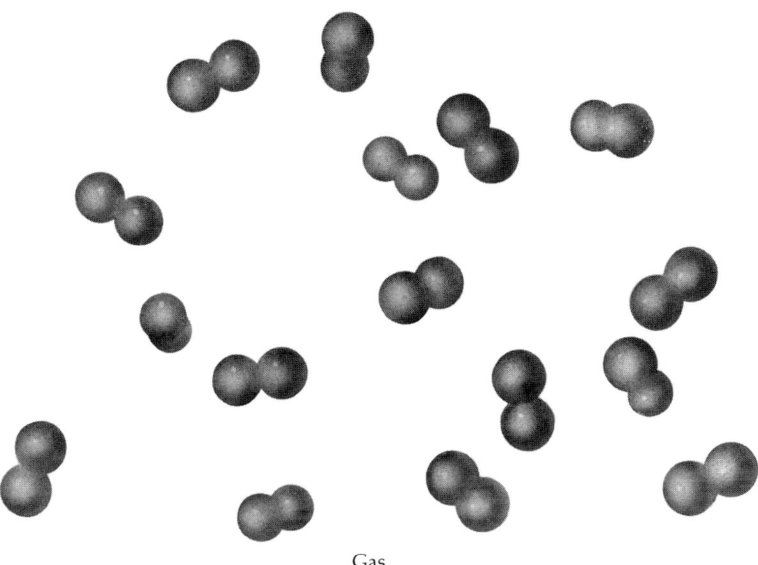

5-9 *Iodine in the form of a crystal, liquid, and a gas of diatomic molecules* (I_2). *(From* General Chemistry *by Linus Pauling. W. H. Freeman and Company. Copyright © 1970.)*

a particular ion and so it can move about, almost as a molecule in a gas. The presence of this *electron gas* causes metals to be good conductors of electricity and heat. Most solids are *crystalline;* that is, they have a regular internal structure consisting of a repeated three-dimensional pattern. Solids that lack this ordered structure are described as *amorphous*—charcoal is a good example.

Solid, liquid, and gas are the three most familiar states of matter; a fourth state, plasma, is discussed in Chapter 9.

Many of the basic facts about heat are apparent from personal experience. Heat can be transferred in several ways. A metal spoon placed in a hot cup of coffee soon becomes hot. In this case, the heat was transferred along the body of the spoon by *conduction*. On the other hand, when one boils water for the coffee, water at the bottom of the pot is heated and travels upward, transferring heat by *convection*. Finally, walking outdoors on a clear day we feel heat from the sun, transferred to us by *radiation*. Conduction is the transfer of heat by one individual particle (atom or molecule) to another; convection occurs through bulk motions of matter; radiation occurs through the transfer of photons from one object to another. In any case, the fact that heat flows from a hotter to a cooler object is clear from ordinary experience. Everyone knows that a block of ice placed in a bucket of hot water produces lukewarm water. One does not expect to place two buckets of lukewarm water together and somehow get a bucket of hot water and a bucket of ice.

The concept of heat is intimately related to the concept of temperature; things that are hotter than others are said to have a higher temperature. The temperature scales in common use have been set up in fairly arbitrary ways using identifiable points, notably the freezing and boiling temperatures of water (see Figure 5-10). The Fahrenheit scale (°F), named after the eighteenth century German physicist who invented the mercury

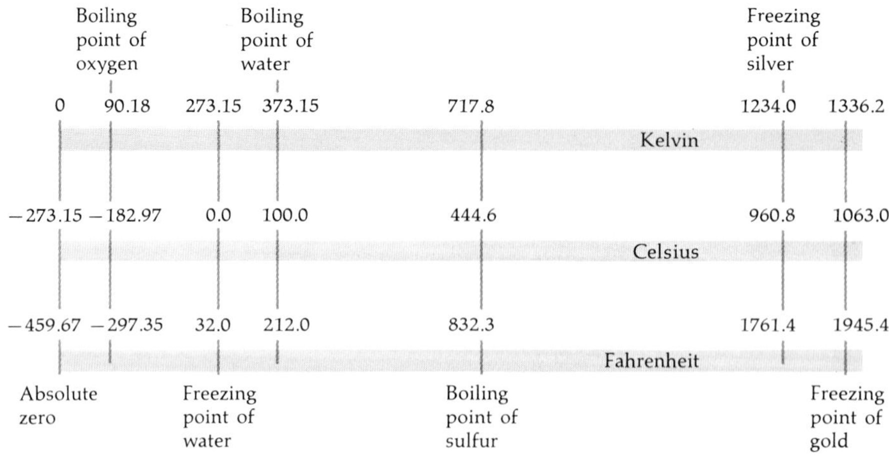

5-10 *Three commonly used temperature scales. (After* Standards of Measurement *by Allen V. Astin. Copyright © 1968 by Scientific American, Inc. All rights reserved.)*

thermometer, defines the melting temperature of ice as 32° F and the boiling point of water as 212° F. Fahrenheit attempted to set up his scale so that the normal temperature of the human body would be 100° F, but he missed by slightly more than 1° F. Since this temperature is now generally taken as 98.6° F, it has been suggested somewhat facetiously that somebody had a fever when Fahrenheit calibrated his thermometer. Although this scale is the official system in a number of English-speaking countries, most nations have adopted the nearly identical Celsius and centigrade scales (°C). The centigrade scale is set up so that the interval between the temperatures at which water freezes and boils (again, under standard laboratory conditions) is divided into 100 degrees, and the two temperatures are defined as 0° C and 100° C, respectively. The fundamental unit of heat, the *calorie,* is related to the concept of temperature through its definition as the amount of heat required to raise the temperature of one gram of water (one cubic centimeter of water under specified laboratory conditions) by 1° C.

Heat is the energy of motion of the atoms and molecules that constitute matter. As the temperature in a room is increased, for example, the average speed of the air molecules increases. As the temperature decreases, it is found that the speed of the molecules decreases. We have seen that the commonly used Fahrenheit and centigrade scales were arbitrarily defined in terms of the properties of water, and thus they are not necessarily related to the physical nature of heat. To introduce the concept of a more physically meaningful temperature scale, we imagine an experiment in which the temperature of the air in a room is constantly decreased to such an extremely low value that the motion of the air molecules would essentially stop. This hypothetical temperature is called *absolute zero,* and it is used to define the zero point of the Kelvin temperature scale (°K) that is shown in Figure 5-10. The Kelvin degrees are the same size as the centigrade degrees, and the freezing and boiling points of water are found to be 273° K and 373° K, respectively. The Kelvin temperature scale is used in physics and astronomy because of its close connection with the energy of motion of atoms and molecules. Absolute zero has not been reached in laboratory experiments, but temperatures as low as 0.0001° K have been.

All atoms and molecules in the universe are in motion. Even the atoms in the crystal lattice of a solid move back and forth in a rather restricted range. If the temperature is increased, their speeds are increased and eventually the bonds that hold the solid together are broken. This occurs at the melting point of a substance and usually produces a liquid in which the atoms or molecules can move around relatively freely, but are loosely bound together by the van der Waals force described on page 110. If the temperature is increased still more, the speeds of the molecules in the liquid increase, the van der Waals force can be overcome, evaporation occurs, and we have a gas. Molecules in a gas are relatively far apart but do collide with each other.

The fact that the molecules of a fluid are in motion can be demonstrated in simple experiments. The most famous example is the *Brownian motion,* as one terms the zig-zagging of small dust particles floating in a liquid or gas. This effect was the subject of considerable study after it was noted by the Scottish botanist Robert Brown (1773–1856). If the zig-zag motions of the dust particles were due to currents in the liquid, then the members of a small group of adjacent particles would all appear to flow in the same direction. However, each particle takes its own randomly changing path, and this is understood as the effect of its being bombarded and knocked about by the molecules of the liquid (Figure 5-11). Small particles are naturally more affected by collision with the molecules than are larger particles, and indeed they can be seen to undergo faster Brownian motion.

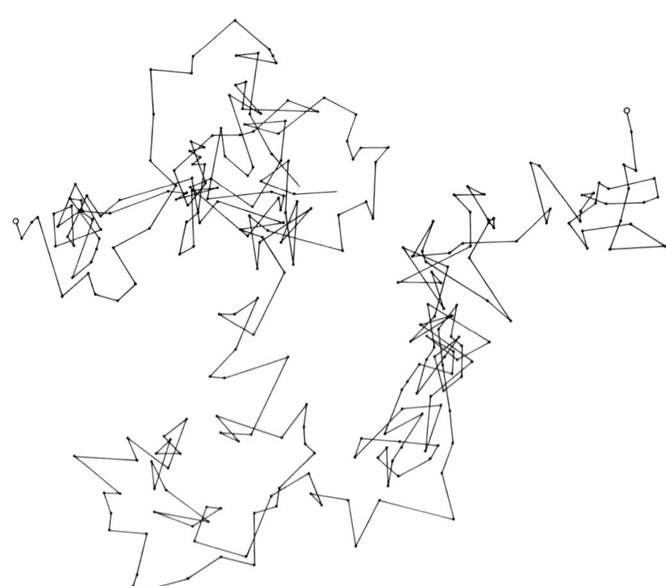

5-11 *Brownian motion. The position of a 10^{-4} cm diameter smoke particle in air as recorded at one-minute intervals by the physicist, Jean Perrin. (From* Matter, Earth, and Sky, *2d ed., by George Gamow. Prentice-Hall. Copyright © 1965. Used by permission.)*

In gases, the molecular motions have effects important in our everyday lives through the phenomenon of *pressure.* The motions of gas molecules in a container cause the molecules to strike the wall of the container and be reflected. This constant drumming of gas molecules on the walls of a container produces an average force on the container wall, and pressure is defined as the force on a unit area. Often we are unaware of atmospheric pressure because it is the same inside our bodies as outside; thus, there is a balance. But, when we specify a tire pressure of, say, 20 pounds, we are actually specifying that the pressure inside the tire be 20 pounds per square inch higher than that in the atmosphere. Such a *pressure difference* can exert a large force; in the case of an automobile tire, the pressure difference supports the automobile's weight. Atmospheric pressure at the surface of the earth is sufficient to support the weight of a column of mercury about 30 inches high. This is the figure usually quoted in weather reports.

A famous demonstration of air pressure was carried out by Otto von Guericke, a seventeenth century mayor of Magdeburg, Germany. He invented the vacuum pump and evacuated a pair of bronze hemispheres. The force holding them together was then so strong that two teams of eight horses each could not pull them apart. How did this force arise? As illustrated in Figure 5-12, the hemispheres initially were subjected to air pressure both on the inside and the outside, and hence no net

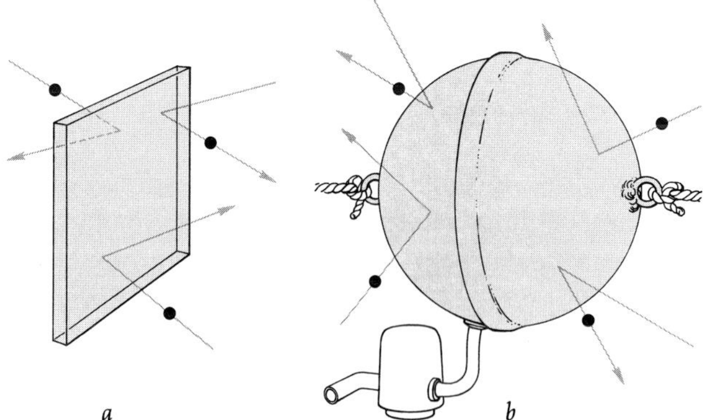

5-12 *Gas pressure and the von Guericke demonstration. If air molecules are bouncing off both sides of a slab (a), the effects of pressure may be unimportant. However, if two hemispheres (b) are sealed together and evacuated with an air pump, the air molecules bombard* only *the outside of the hemispheres. This produces a strong force which holds the hemispheres together.*

force was exerted on them. When the hemispheres were brought together and evacuated, however, the force due to the difference between the outside atmospheric pressure and the negligible internal pressure held them together.

We have mentioned that heat is simply the energy of motion of atoms and molecules; macroscopic objects also have energy by virtue of their motions. Both the microscopic and macroscopic energies of motion are known as *kinetic energy.* We can think of several other kinds of energy, such as the *nuclear energy* that powers modern submarines, and the *chemical energy* that is released in the explosion of dynamite or the combustion of gasoline. Sometimes one form of energy is converted into another, as when the chemical energy of fuel is converted into the kinetic energy of an automobile.

Related to the concept of energy is that of *work.* Work is done whenever an object is moved against a force; the work done on an object adds energy to the object. For example, if we move a book from the bottom to the top shelf of a bookcase, we increase its *potential energy* because we have moved it against the force of gravity. Giving the book a slight nudge so that it falls off the edge of the shelf and is accelerated downward by the force of gravity, we find that the potential energy is converted to kinetic energy as the book moves faster and faster toward the floor. The impact is greater for the same book when pushed from the top shelf than when pushed from the bottom shelf, because in the former case the book has more potential energy. Water stored in a reservoir behind a dam (in a hydroelectric power plant, for example) has potential energy. When the water is allowed to flow down through the dam, the potential energy is converted to kinetic energy. This energy of the moving water is in turn imparted to a water wheel or hydroturbine that drives a generator which converts the kinetic energy to electrical energy.

If we push a box along a rough floor, we are pushing against the force of friction. This work must add energy to the box; in fact, it appears as heat in the floor and the bottom of the box. Thus, kinetic energy of the moving box is changed to heat. A similar demonstration was cited by Count Rumford (circa 1798) on the basis of heat produced during the boring of a cannon. When work was done, heat was produced; these facts led to the idea that heat is a form of energy.

The principle of the *conservation of energy* states that the total energy of an isolated system is constant, although the form of the energy may change. This can be illustrated by a detailed look at a simple process such as picking up a rock and dropping it in a waste can. We can describe the energetics of this process as having started with the sun. The solar energy was transferred to the earth in the form of *radiant energy* or light. Photosynthesis in green plants converted this radiation into *chemical energy* stored in carbohydrates. A man ate some of the plants and the calories of chemical energy stored in the carbohydrates were utilized by him to lift the rock. This added *potential energy* to the rock, which was converted into *kinetic energy* when the rock was dropped. As the rock hit the can a sound was heard, and the sound waves carried away *acoustical energy*. The waste can was also *heated* slightly. When the principle of conservation of energy is combined with the equivalence of mass and energy, as stated by Einstein, there are no known exceptions to its rule.

NUCLEAR ENERGY

Nuclear reactions take place in the interior of stars and in atomic bombs, achieving the goal of medieval alchemists—the transmutation of one element into another—and at the same time accomplishing the conversion of mass into energy.

The fundamental relation governing the conversion of matter into energy was determined by Albert Einstein (1879–1955; Figure 5-13):

$$E = mc^2$$

The energy, E, liberated by the annihilation of a certain amount of mass, is equal to the mass annihilated, m, times c^2, the square of the velocity of light (all in the appropriate units).

This equation has an important application, as follows. A common sequence of nuclear reactions leads to the production of one helium atom from four hydrogen atoms; however, the total mass of four hydrogen atoms does not equal the mass of one helium atom, as can be seen from their atomic masses. Hydrogen has atomic mass 1.008 (as mentioned on page 106), and the atomic mass of helium is 4.003. The difference in atomic masses between four hydrogen atoms and one helium atom is 4.032 minus 4.003, or 0.029. Thus, in converting from hydrogen to helium, we lose slightly under one per cent of the mass. This reappears as energy in the amount given by Einstein's equation. We return to this subject in our discussion of the energy source of the sun (Chapter 9). Nuclear energy is released when the bonds due to the strong nuclear force are broken or otherwise rearranged. High temperatures (10 to 20

5-13 *Albert Einstein.* *(Yerkes Observatory photograph.)*

million degrees K) are required to do this in most cases (natural radioactivity is an exception); such temperatures occur in the central regions of stars. In nuclear reactions, energy appears to be created, but when we keep in mind that matter is at the same time destroyed, we find that a general law of *conservation of mass-energy* applies throughout the observable universe, or at least that no exceptions have been shown to exist.

LIGHT

Astronomers receive the overwhelming majority of their information about the universe in the form of light or radiant energy emitted by the celestial objects. Astronomers cannot put their hands on a star or study it in the laboratory so they must infer its properties by collecting and analyzing its light. Understanding the nature of light is therefore a matter of prime necessity to the astronomer. He is preoccupied with it and the instruments used to study it (Chapter 6).

Visible light is a specific example of what is more generally referred to as *electromagnetic radiation.* In fact, the *electromagnetic spectrum* (Figure 5-14) includes radio waves, infrared, visible light, ultraviolet, x-rays, and gamma rays. The detailed study of light has been one of the central problems of modern physics.

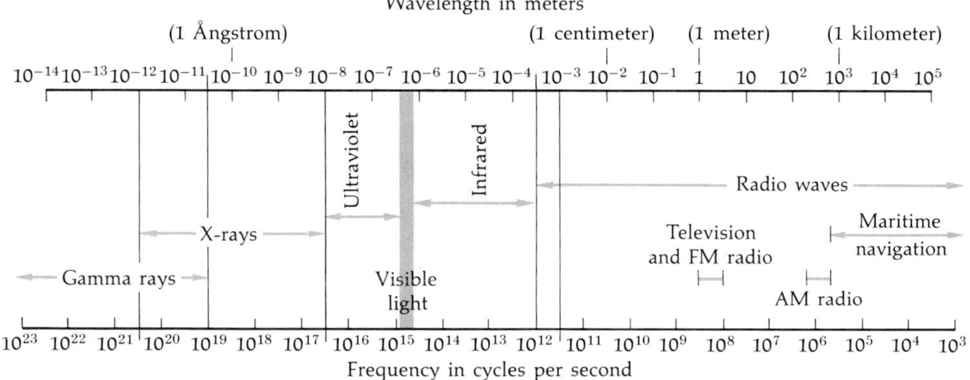

5-14 *The electromagnetic spectrum. Note the very narrow range of visible light.*

In this book we limit ourselves to summarizing the conclusions of the physicists from an operational point of view as they apply to astronomy and the analysis of light received from the cosmos.

It is often difficult to describe light in simple terms; sometimes it acts like a wave and sometimes like a particle. In the first case, it appears to behave as though it were a transverse wave, that is, a moving disturbance, with the disturbance occurring at right angles to the direction of motion. Such a wave can be characterized by its amplitude and wavelength (Figure 5-15), which are found to correspond in familiar terms to the intensity and color of light, respectively.

Light waves are able to propagate through a vacuum. For a long time this point was disputed and a hypothetical medium, the *ether,* was presumed to exist in space; the ether was disturbed by light waves just as the ocean water is disturbed by water waves.

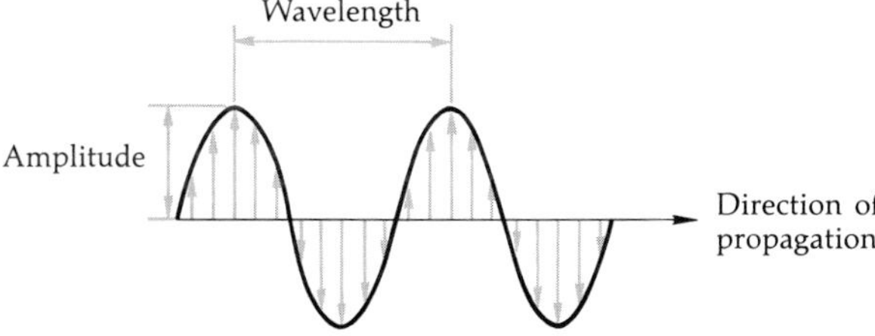

5-15 *Representation of light as a transverse wave; the electromagnetic disturbance associated with the wave is at right angles to the direction of motion.*

Two important wave aspects of light are *interference* (Figure 5-16) and *linear polarization* (Figure 5-17). The interference effect is interpreted as the reinforcement or cancellation of light waves, depending on how their amplitude peaks and valleys coincide. For example, suppose that the crest of one wave at a certain point coincides with the crest of another. The two waves can be thought of as superposed, and the amplitude of the resultant wave at that point is greater than the amplitude of either of its component waves. The result of this greater amplitude is a greater intensity of light at this position. On the other hand, suppose that the crest of a wave coincided with the trough of another wave. In this case, superposition results in a smaller amplitude. These effects are actually observed in experiments with light, and they find a natural interpretation if light is regarded as a wave phenomenon. Since the extent to which the crests and troughs of light waves coincide must depend critically on the respective wavelengths, we might expect interference phenomena to be important in the study of light of different

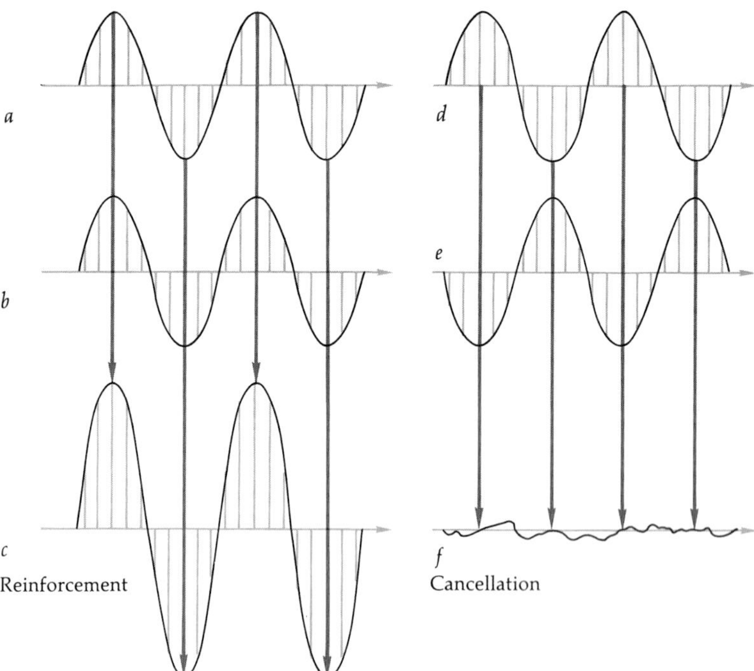

5-16 *Illustration of the phenomenon of interference through the examples of reinforcement (a and b are added crest to crest) and cancellation (d and e are added crest to trough).*

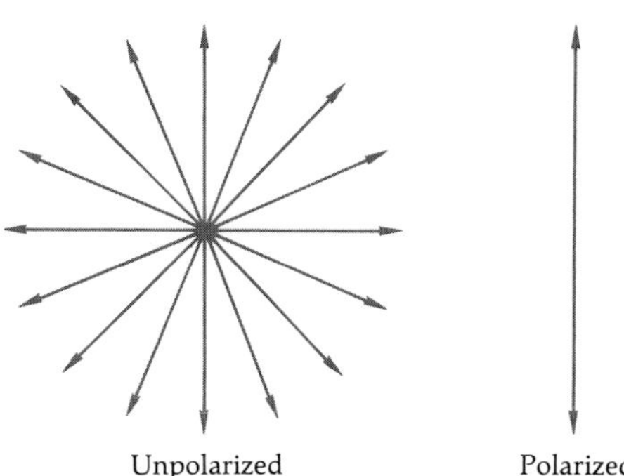

5-17 *Unpolarized and polarized beams of light, sketched "end-on" as they travel at right angles out of the page. The lines represent the direction(s) of the electromagnetic disturbances in light waves that are moving toward the reader (that is, moving at right angles to the paper). In the polarized beam, all of the waves have the same direction of disturbance or oscillation. In the unpolarized beam, different light waves oscillate in different directions.*

wavelengths or colors. The way in which light interference caused by a *diffraction grating* separates light by color is shown in Figure 5-18.

Polarization is a concept rooted in the definition of a transverse wave as a disturbance perpendicular to the direction of motion. There are an infinite number of directions (represented by lines in a plane, Figure 5-17) that are perpendicular to the straight line that represents the direction of motion. When we are dealing with a great many light waves (say the light received from a certain star) and we find that as many of them are disturbances oriented in one direction as in any other direction, then the light is *unpolarized.* However, if more of the waves are oriented in one direction than in any other, then the light is *partially polarized;* if all of the light is oriented in one direction, then it is *fully polarized.* A man floating on the ocean provides an excellent analogy—he only bobs up and down as a wave passes; he is not disturbed horizontally or at some intermediate angle because water waves are polarized in the vertical direction. There are other kinds, but this linear polarization is both the simplest and the most important type in basic astronomy. Like interference, polarization is a phenomenon that seems easily understood in terms of the wave

5-18 *A beam of white light strikes a diffraction grating and a spectrum is produced. (Courtesy of Bausch and Lomb.)* See also Plate 2 following p. 260.

nature of light. Some aspects of polarized light are illustrated by the familiar Polaroid sunglasses, which primarily pass light of a particular polarization. The sunglass makers have empirically determined the best orientation, so that little of the reflected, polarized sunlight (glare) comes through. If you look at the sky and rotate your Polaroid sunglasses slowly, it will be alternately bright and dim, depending on how the polarization of the light lines up with the preferred direction of the Polaroid material.

Sometimes, particularly when light interacts with matter, the light seems to behave as though it were composed of a stream of discrete particles (the *photons*) which have specific energies. A good example is the *photoelectric effect;* this term refers to the ejection of electrons from certain metal surfaces when light falls on them. Two basic aspects of this photoelectricity can be accounted for if we think of light in terms of particles. The first is an example familiar to astronomers, for the photoelectric effect is used to measure star brightnesses. A *photomultiplier* tube, placed at the focus of a telescope and illuminated by the light of a star, produces an electric current due to the electrons released by the photoelectric effect. When this technique is used to measure a very faint star, one finds that the current is intermittent and in fact seems to be composed of individual pulses of electricity. These are hard to understand on the basis of the theory of light as a continuous wave, but are easily interpreted on the particle theory as being due to the arrival at the photomultiplier tube of individual photons from the star. (For a bright star, there are so many photons coming in that we don't notice gaps between their arrivals.) The second aspect is the dependence of the photoelectric effect on the color of light. After the discovery of photoelectricity by the German physicist Heinrich Hertz in 1887, experiments by other scientists revealed that the number of electrons emitted depended on the intensity of the light, but the velocities of the electrons depended only on the color of the light. Blue light releases more energetic (faster moving) electrons than does red light, and infrared light may produce no electrons at all, depending on the particular metal involved. These facts are easily explained if light comes in discrete packets of energy so that the energy of a blue photon exceeds that of a red photon, which exceeds that of an infrared photon, and so on. In terms of the wave theory, the blue, red, and infrared colors correspond to increasingly longer wavelengths (see Table 5-4). It takes a certain amount of energy to knock an electron out of a metal. Thus, Albert Einstein (who received the Nobel Prize in 1921 for this explanation of the photoelectric effect) reasoned that each photon of light that strikes the metal causes the release of either one electron or no electron at all, depending on the photon energy. If the individual photons do not have enough energy to knock the electrons out of the metal, no electrons are produced. When

Table 5-4 *Correspondence of Colors and Wavelengths*

COLOR	WAVELENGTH RANGE (Å)
Violet	3,800–4,400
Blue	4,400–5,000
Green	5,000–5,600
Yellow	5,600–5,900
Orange	5,900–6,400
Red	6,400–7,500

the photons have enough energy to produce electrons, part of their energy is used in removing the electron from the metal and the remainder shows up in the speed (kinetic energy) of the electron. Blue light produces more energetic electrons than red light, and hence the shorter wavelengths correspond to more energetic photons. Observation shows that light also exerts pressure (*radiation pressure*) and this effect is also readily interpreted in terms of photons just as atmospheric pressure (page 115) is caused by gas particles or molecules.

How can light be both a particle and a wave? Perhaps we can turn the question around and ask why light *should* consist of a familiar entity like a wave or a set of particles? It is in fact a complex phenomenon that we can interpret in the familiar terms (wave, particle) that seem most helpful in particular cases.

ELEMENTARY SPECTROSCOPY

Spectroscopy, the analysis of the wavelength-dependent properties of light, was initiated by Sir Isaac Newton. While attempting to make some lenses for a telescope, he found that the images were always fuzzy. He could not correct the lenses and so questioned whether or not the problem was with light itself. In a classic experiment he passed a beam of sunlight through a prism and saw that it was broken up into colors of the rainbow, as shown in Figure 5-19. When the colors were focused on the same spot, the original character of the sunlight was restored. Newton immediately saw the reason for his fuzzy images. The bending of light by glass is used in a lens to produce a focus. His method of producing the spectrum showed that light of different colors is bent by different amounts by the glass of the prism. Thus, his lens had failed to focus sunlight sharply because the light of different colors was being focused at different distances from the lens.

The colors of light correspond to wavelengths or frequencies of oscillation (wave interpretation) or different energies (particle interpretation). Wavelengths of visible light are usually quoted in terms of the Angstrom unit (Å),* which is 10^{-8} centimeters. The approximate wavelength ranges for the major colors are given in Table 5-4.

A considerable advance in the development of spectroscopy was made in 1858 by the German physicist Gustav Kirchhoff.

*Named for A. J. Ångström, 19th century Swedish astronomer who studied the spectrum of the sun.

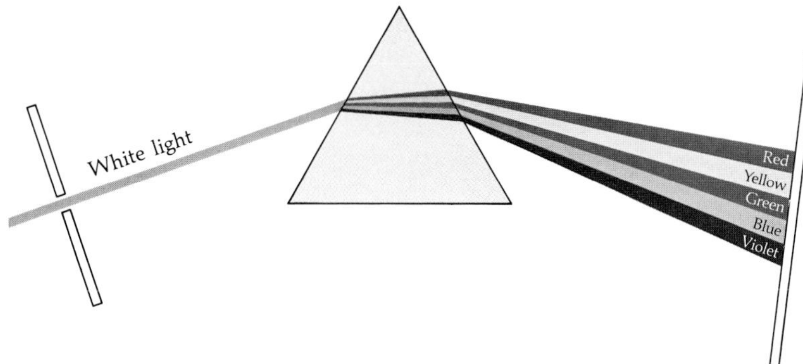

5-19 *Sketch of Newton's spectral experiment. White light or sunlight passed through a prism breaks into a spectrum.*

His experimental researches (in collaboration with the chemist Robert Bunsen) led to three important conclusions concerning the emission and absorption of light. These ideas, sometimes called *Kirchhoff's laws of spectral analysis,* are:

1. An incandescent solid or liquid, or a gas under high pressure, emits a *continuous spectrum* (that is, produces light at every wavelength over a wide range).
2. Incandescent gases under low pressure give a spectrum composed of individual bright lines, and the number and wavelengths of these *emission lines* depend on the gas involved. (By line, we mean that the emission occurs at a particular wavelength or narrow range of wavelengths, as distinct from the continuous spectrum defined above.)
3. When the source of a continuous spectrum is viewed through a cooler low pressure gas, dark lines appear in the continuous spectrum at exactly the same wavelengths as the bright lines emitted when the same low pressure gas is viewed in the incandescent state. (The dark features are called *absorption lines.*)

Kirchhoff's conclusions are illustrated by the experimental arrangement shown in Figure 5-20.

Understanding Kirchhoff's laws requires an understanding of atomic structure. Specifically, we need to know why atoms absorb and emit light in certain characteristic wavelengths to produce lines. In addition, we need to know the physical conditions that lead to their production.

In the model of the atom as developed by Bohr, the electron orbits are not located randomly—only certain specific orbits are

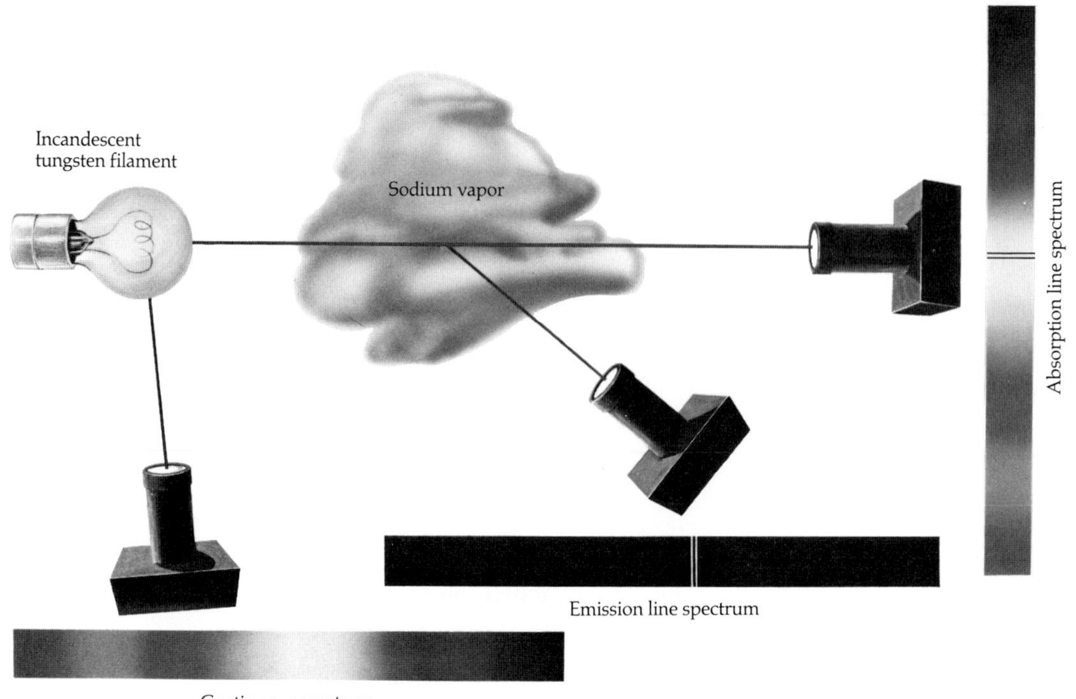

5-20 *Sketch illustrating Kirchhoff's Laws.* See also Plate 6 following p. 260.

possible (Figure 5-21). Electrical force binds the electron to the positively charged nucleus. If an electron moves from one orbit to another, it must change in energy, as in the example of the book moved from one shelf to another (page 116). When an electron moves from an outer orbit to an inner orbit, the energy decreases. (In the terminology of atomic physics, we say that the atom changed to a lower *energy state.*) Since energy cannot be destroyed, the lost energy must appear in another form. In fact, it appears as a photon of light produced by the atom (*emission process*). Since the two atomic energy states corresponding to the outer and inner orbits have specific values, their difference is also a specific amount of energy. The energy of photons is related to the wavelength of light, and so when an electron moves from a particular outer orbit to a particular inner orbit, the atom in question always radiates light of the corresponding specific wavelength. This is the origin of the emission lines described above. In like manner, an electron may move from an inner orbit to an outer one and, in this case, the atom must have gained energy. Therefore, an energy source must

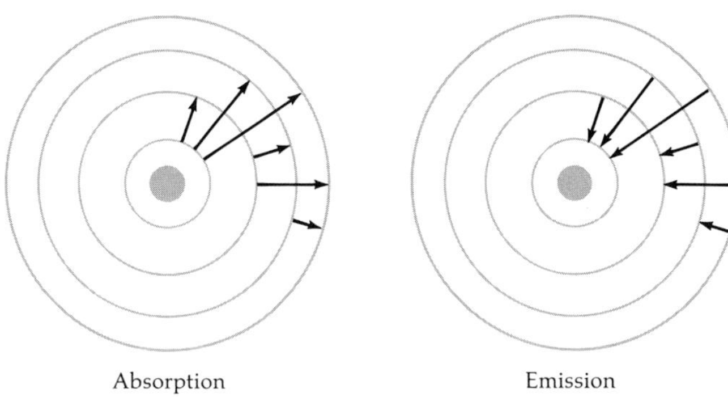

5-21 *Allowed electron orbits and transitions. (Not to scale.)*

exist, and it may be a photon of light that the atom has encountered and absorbed (*absorption process*). This photon has to have an energy just equal to the difference of the two atomic energy states. Thus, if a hydrogen atom produces an emission line of red light at wavelength 6,563 Angstroms under certain circumstances, it will also produce an absorption line at that wavelength under other circumstances, as can be proved in the laboratory. An important corollary is that when we see absorption lines at the wavelength of 6,563 Angstroms in the spectra of stars (as we frequently do), we know that these stars must contain hydrogen atoms, and that the circumstances in these stars are such that the atoms are able to exist in the two energy states involved in the production of this line. A similar remark applies to the quite common observation of an emission line at 6,563 Angstroms in the spectra of gaseous nebulae (Chapter 12); the nebulae also must contain hydrogen atoms and at least a crude idea of the nebular conditions can be gotten from the observation of this emission line. For example, by Kirchhoff's second law, nebulae that produce this emission line must contain or be composed of incandescent gas at low pressure.

(The existence of discrete energy levels in the atom is also expressed by the statement that the atomic levels are *quantized.* For this reason, photons are sometimes referred to as *quanta.* The scientific analysis of quantized atomic energy states was largely begun by the German physicist Max Planck in the early twentieth century.)

It is found that when electrons are in the higher energy levels, they tend to drop into lower levels spontaneously until they reach the lowest permissible level not already occupied by an electron. Thus, the process of absorption of a photon is eventually followed by the process of emission. A useful analogy is to think of the atom as a sort of quantum staircase in which the individual levels (steps) slope slightly forward and are not necessarily of equal heights. The steps are analogous to the energy levels in the atom and the slope is indicative of the tendency for *spontaneous emission.* Suppose we lift up a ball from the floor and place it on one of the higher steps. Due to the slight slope, the ball rolls off the step and the potential energy that we gave it by lifting it from the floor is converted into kinetic energy. In the atom, energy involved in raising an electron to a higher level reappears as one or more light quanta when the electron drops to a lower level or to a succession of lower levels.

Electrons in an atom can also be excited to a higher energy state by collisions with *free* electrons (those not bound to any atom), atoms, or molecules. The requisite amount of energy is taken from the kinetic energy of the colliding particle so that the latter particle moves at a slower speed after such a collision. This process occurs in a hot gas. The electrons in the atoms (*bound* electrons) then drop down spontaneously to produce the bright lines in the spectrum. Atoms in liquids, solids, and in gases under higher pressure, have their energy levels smeared out by electric effects of their close neighbors. Since the allowable energy values are no longer highly specified, a great many energy differences correspond to the allowable changes in energy, and hence these atoms produce a continuous spectrum.

Two other phenomena associated with light quanta and atoms should be mentioned here. Often the absorption of a photon is followed immediately by the emission of a photon of the same energy in an arbitrary direction. To an external observer the photon appears to have simply changed direction. This process is called *scattering.* If, on the other hand, the re-emission does not occur immediately and the atom suffers a collision with some other particle, the energy gained by absorption of the photon may be transferred to the motion of the colliding particle. In this case a photon goes into the atom but does not come back out because its energy reappears as kinetic energy rather than radiation. This process is called *pure absorption.* Note that in all the phenomena described in this section with electrons moving from one energy level to another, we refer only to the outermost electrons.

This brief review of atomic processes enables us to understand Kirchhoff's laws as illustrated in Figure 5-20. Photons at certain specific wavelengths are absorbed by the sodium vapor to produce the absorption line spectrum in the light from the tungsten filament as viewed through the vapor. The light absorbed in the vapor is re-emitted at the same wavelength but in all directions (scattering), causing the vapor to become incandescent and show a bright line spectrum. In this case, the energy supply for the incandescent vapor is the light from the tungsten filament. In other circumstances, the energy supply could come from the heat of the vapor itself with the bright lines excited by electron collisions.

The principles outlined above and used to explain Kirchhoff's laws enable us to identify elements in gases under observation, for example, in the sun's atmosphere. In a crude way we can think of continuous radiation from the sun's interior shining through its atmosphere, which produces absorption lines at specific wavelengths corresponding to the elements present. This is observed, and many elements have been identified (Figure 9-8) from the presence of their characteristic lines. The strength of the absorption line (how much light is removed as judged by the strength of the light at adjacent wavelengths) depends in part on the amount of the element present. Obviously, the rarer elements are less likely to produce strong lines. Spectroscopic analysis of emission and absorption lines constitutes our primary evidence concerning the abundance of the elements in the sun and the universe beyond the solar system.

RADIATION LAWS

Two important laws describe the amount and wavelength distribution of light emitted from the surfaces of objects at different temperatures. Strictly, these laws apply to an ideal radiator, originally called a blackbody because it could be approximated by a surface covered with lampblack. In practice, many things, including astronomical bodies, are fairly good approximations to ideal radiators.

The correct wavelength distribution of light from an ideal radiator was first calculated by Planck in 1901. The intensity of emission as a function of wavelength for different temperatures is shown according to *Planck's Law* in Figure 5-22. The energy source for this radiation is the motions (kinetic energy) of the atoms or molecules in the emitting material. The motions increase with temperature and such radiation is therefore called *thermal radiation.*

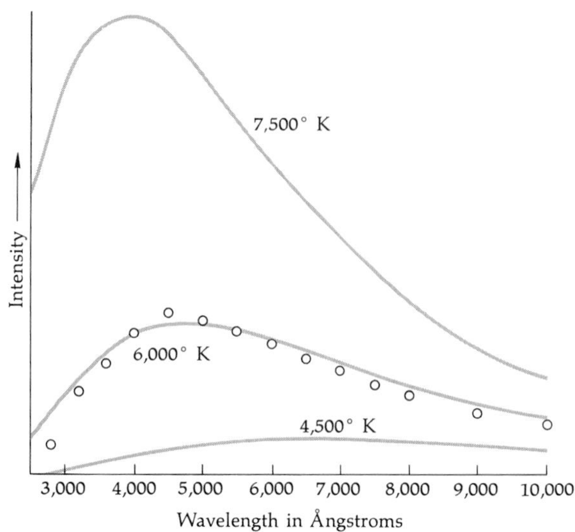

5-22 *Planck's Law, the variation of intensity as a function of wavelength for an ideal radiator at different temperatures (solid curves). The circles are the observed intensities of solar radiation and show that the solar radiation can be approximated by an ideal radiator at about 6,000° K. The graph shows that a slightly lower temperature would be a better approximation and the accurate value is 5,750° K.*

Inspection of Figure 5-22 shows two important features. As the temperature increases so does the *total amount* of radiation emitted, and the wavelength of greatest emission occurs at *shorter wavelengths.* Thus, if the temperature of an emitting surface is slowly increased, the peak of the emission will move in the direction red→orange→yellow→green→blue→violet, and so the color of an object, such as a star, depends on its temperature. (The color as perceived by our eyes also depends on the relative sensitivity of our eyes to light of different wavelengths.)

The variation of total energy output with temperature is governed by the *Stefan-Boltzmann Law.* The total energy outputs at different temperatures can be represented by the areas under the various curves in Figure 5-22. The total amount of energy radiated increases as the absolute temperature (°K) raised to the fourth power. Thus, an ideal radiator with twice the absolute temperature of another one of equal size radiates (2) × (2) × (2) × (2), or sixteen times more energy.

These laws can be applied to the determination of the temperature of the sun's visible surface or photosphere (Chapter 9).

Both the Planck Law and the Stefan-Boltzmann Law indicate a temperature of about 6,000° K for the photosphere. The agreement is not exact because the sun is not an ideal radiator.

DOPPLER EFFECT

An apparent change in the pitch (frequency) of sound waves, occurring when the sound-maker and the listener are in motion relative to one another, is a well known phenomenon. For example, the whistle of an approaching train is higher pitched when the vehicle is approaching the listener than when it is receding. This Doppler effect (explained by the Austrian physicist Christian Doppler in 1842) occurs for both sound waves and light waves.

The crux of the matter is that it takes a certain amount of time for a cycle of a wave to be emitted. Thus, if a light source is moving, the beginning and end of the wave are emitted at different locations. This causes a lengthening or shortening of the wave (as perceived by a stationary observer) depending on the direction of motion of the source (see Figure 5-23). If the wavelength is shortened (approaching source) but the velocity of light remains the same, more waves must pass the observer per second. Thus, the light appears to have a smaller wavelength (bluer color). For a receding source, the wavelength is lengthened, fewer waves pass by an observer per second, and the light seems redder. Exactly the same effects occur in the case of a stationary source and a moving observer—approach causes an apparent wavelength decrease.

In astronomical spectra, absorption and emission lines are found shifted toward the red or toward the blue in comparison to laboratory standard spectra, due to the motions of the stars and other astronomical objects with respect to the observer on earth. Since a faster motion produces a greater Doppler shift, measurement of the amount of shift in Angstroms reveals the speed at which the object and the observer are receding from or approaching each other.

VELOCITY OF LIGHT

Light travels at a very large but finite velocity. Thus, light reaching an observer was generated at some time in the past. For example, the travel time for light from the sun to reach the

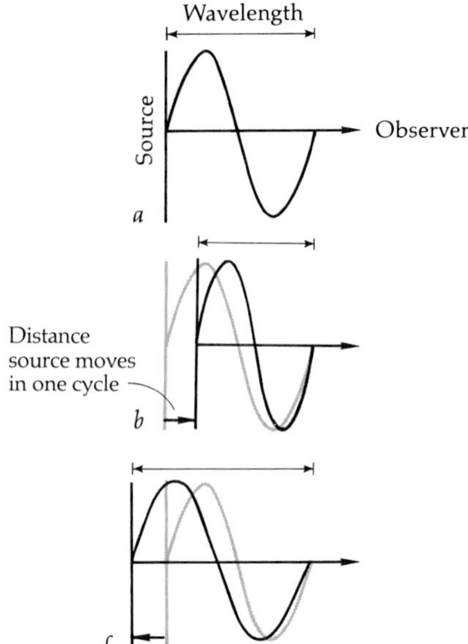

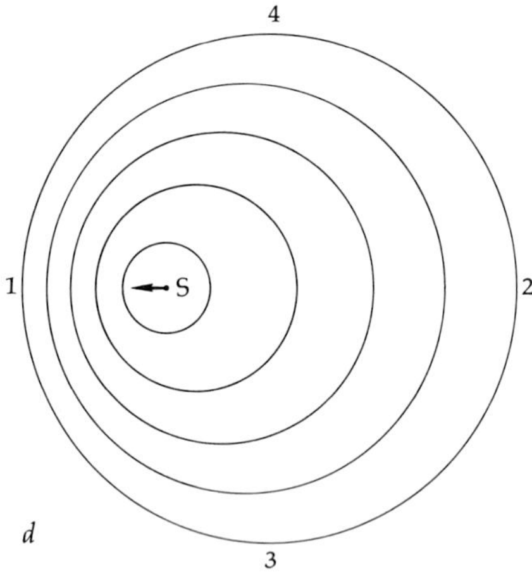

5-23 *The Doppler effect. The wavelength (and frequency) of an emitting source* as perceived by an observer *is altered by motion with respect to the observer. (a) Stationary source; (b) moving toward observer; (c) away from observer. (d) Another way to understand the Doppler effect. Sketch of the positions of crests of waves emitted by a source, S, that is moving in the direction indicated by the arrow. The most recently produced crest is represented by the smallest circle. An observer at position 1 receives Doppler-shortened waves, whereas an observer at 2 receives Doppler-lengthened waves. At 3 and 4, essentially no Doppler shift is observed.*

earth is about 8 minutes. Radio waves are a form of light, and the time required for them to travel between the earth and moon causes the slight delays in communication with the astronauts on the moon.

After the basic ideas about light (such as traveling from the object to the eye and not vice versa as some early philosophers

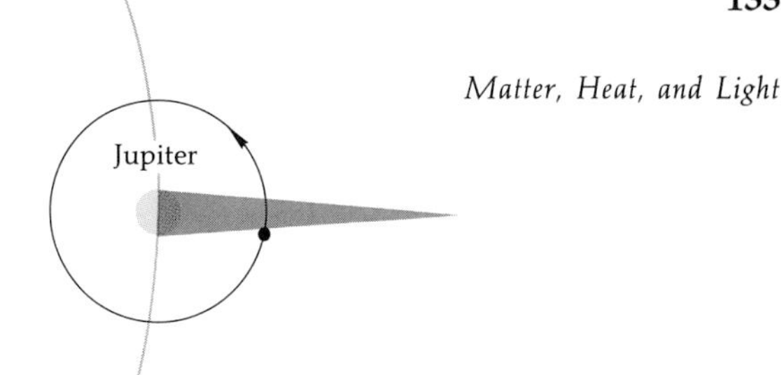

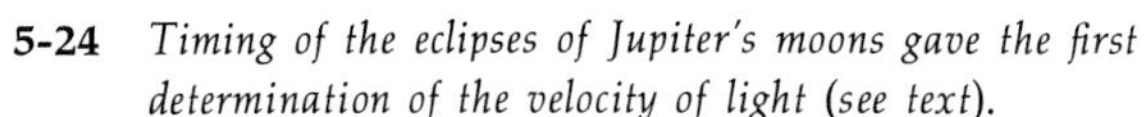

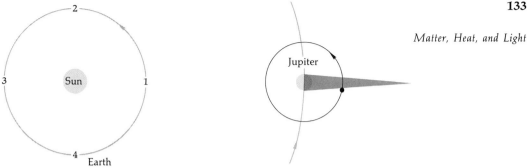

5-24 *Timing of the eclipses of Jupiter's moons gave the first determination of the velocity of light (see text).*

believed) were established, the first serious attempt to measure the speed of light was carried out by Galileo. Assistants with lanterns were stationed on successive hills. The first man uncovered his lantern. As soon as the second man saw the light from the first lantern he was to uncover his lantern; the elapsed time required for light to travel twice the distance between the observers could be measured by the first man. Galileo and his associates soon realized, however, that what they were actually measuring was the reaction time of the observers and not the velocity of light. They could only conclude that the velocity of light was very great.

The first evidence for a finite velocity of light came from observations of the moons of Jupiter. These satellites pass behind Jupiter and are eclipsed, but a curious phenomenon was found when analyzing the times at which the eclipses occurred. Referring to Figure 5-24, the eclipses were on time when the earth was at positions 2 and 4, but early when the earth was at position 1, and late when the earth was at 3.

This phenomenon was discovered and explained in 1675 by the Danish astronomer Ole Roemer. He pointed out that the irregularities in the eclipse schedule could be understood if the light took more time to reach us when the earth was farther away from Jupiter; thus, its propagation could not be instantaneous. His crude data (especially concerning the scale of the solar system) led him to a determination of the velocity of light as about 210,000 kilometers per second. This value turned out to be low by about 30 per cent, which is not bad for a determination made about 300 years ago, and most important, his method and reasoning were correct.

A little over 50 years later, Bradley's discovery of aberration (page 95), confirmed Roemer's finding that light does not propagate instantaneously. The inferred value for the speed of light was approximately the same as Roemer's. Many modern methods are available for determining the speed of light and the accuracy has been constantly improved. The current value is a speed of 299,792.5 kilometers per second (about 186,000 miles per second).

6

Telescopes and Observatories

With the exception of the occasional meteorite that reaches the earth, astronomy has been limited (until the recent advent of space travel) to *observational* studies. Unlike physicists and chemists the astronomer could not experiment with the objects under investigation. Whatever he learned about the stars came from analyzing their light from a distance. As most of the information in the remainder of this book is based on the results of studying light from the sky, it seems desirable at this point to review the instruments that are used for this purpose. In recent years increasingly complex instrumentation has become vital to advances in many areas of astronomy, but most observational results have been obtained with relatively simple optical systems, and among them the human eye is a most important example.

THE HUMAN EYE

The human eye was once the sole instrument of the astronomer. Even today, in the age of large telescopes and sophisticated optical/electronic instruments, it remains a vital tool. Looking

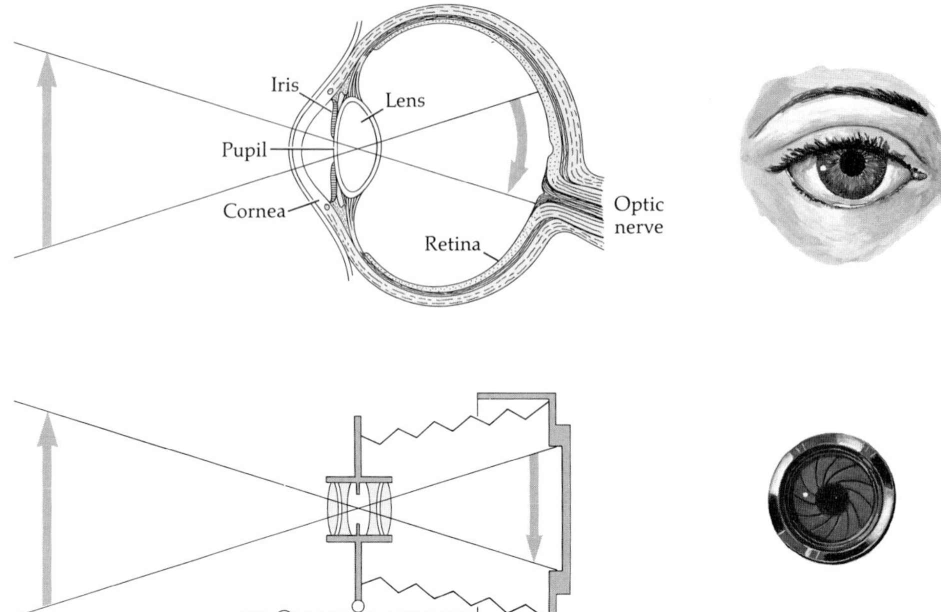

6-1 *A comparison of the optical properties of the eye and the camera.* (*After* Eye and Camera *by George Wald. Copyright © 1950 by Scientific American, Inc. All rights reserved.*)

through the telescope we use the eye to find the star, comet, or other object to be tracked by the telescope.

Light from a star or other source passes through the *cornea* and *pupil* of the eye. The size of the pupil is adjusted by muscles in the *iris* that widen it to accept more light from dim targets or contract it to exclude excessive light from bright sources. The light that enters is bent (*refracted*) by the materials that compose the various parts of the eye (Figure 6-1). Most of the refraction occurs in the cornea, but the *lens* does the fine focusing: the shape of the lens is adjusted by a muscle in order to form sharp images of objects at different distances. Thus, the iris acts like the diaphragm on a camera which the photographer adjusts to various f numbers, depending on the brightness of the scene, and the lens muscle is analogous to the focusing mechanism of the camera.

The image of the object is focused by the lens onto the *retina.* This surface contains a large number of small structures called *rods* and *cones,* which function, respectively, under conditions of dim and bright light. When light is absorbed by the molecules in a rod or cone, it produces an electrical effect that is sensed by a nerve, which transmits the information to the brain. Thus, the retina corresponds to the light-sensitive surfaces in the astronomical detectors (such as photographic plates and photomultiplier tubes) discussed later in this chapter. Under ideal circumstances, the eye can sense as few as 2 photons, absorbed in adjacent rods within an interval of 0.1 second.

CAMERAS

Most of us are familiar with simple box cameras. A lens (or lenses) at the front of the camera focuses an image on the film (Figure 6-1). A shutter opens briefly to allow enough light through to properly expose the film. The telescope used by an astronomer is not really very different; lenses or mirrors are used to form an image at a place called the *focal plane.* Depending on the purpose of the observation the astronomer may place a photographic film or plate (see page 158) at the focal plane, or he may mount some other instrument there to analyze or record the light. Cameras are generally held in the hand, but when steadiness is important the photographer uses a tripod. Steadiness is almost always important in using a telescope, since many of the stars under observation are so faint that the astronomer must take a long exposure in order to record them properly. During this time the stars move across the sky (due to the rotation of the earth), so the telescope must be appropriately mounted, with a *clock drive* to move it at the correct rate to track the stars.

THE FIRST TELESCOPES

As far as can be determined, the first telescopes were constructed in Holland, perhaps as early as 1604. They were intended, at least in part, for military applications, but within a few years rumor of these devices reached Galileo in Italy. By 1609 he had built his own telescope and made the first astronomical observations with optical aid. An explanation of the principles by which such *refractors* (telescopes that use lenses) work was published by Kepler in 1611; by the mid-seventeenth century they were in use throughout Europe, although some astronomers continued to build new lensless instruments patterned after those of Tycho (Chapter 4). In 1671, Newton built

a

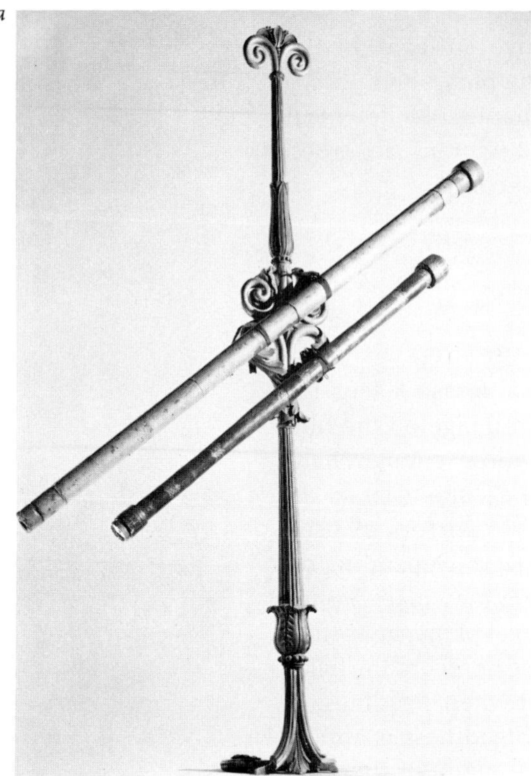

b

6-2 *Early telescopes: (a) Galileo's refracting telescopes (courtesy of Prof.ssa Dott.ssa Maria Luisa Righini-Bonelli, Istituto e Museo di Storia della Scienza, Firenze, Italy); (b) Newton's reflecting telescope (courtesy of the Royal Society).*

the first *reflector* (telescope that uses mirrors) in order to overcome some of the problems with glass lenses, including the poor focus of light of different colors, as noted in Chapter 5. (See Figure 6-2).

GALILEO, THE TELESCOPE, AND THE SCIENTIFIC METHOD

For centuries the ideas of Aristotle, Ptolemy, and other ancient Greeks had dominated Western scientific thinking. As we have seen in Chapter 4, the work of Tycho and Galileo in obtaining new facts by observation and measurement led to the adoption of a revolution in thinking—the heliocentric world view. In fact, Galileo, in his additional capacity as a laboratory physicist, conducted a variety of experiments that tested and sometimes revised previous ideas. Perhaps Galileo's greatest contribution was not a particular experiment or discovery, or even the sum of all his discoveries—it was his life-long fight against the dogmatic assertion (and meek acceptance) of scientific "truths" unsupported by experiment. His great legacy is thus the active use of the *scientific method,* an approach to research that is now the basic procedure in all fields of science. Confronted with a set of "facts", consisting of data or measurements on some aspect of nature, a scientist proposes a theory or model that can explain them; the characteristics of the model enable us to predict further properties of the subject being studied (whether it be a star or a human blood cell); new observations and measurements are then made to test these predictions. The last step in turn usually leads to a slightly (or greatly) improved theory from which new predictions are made and the cycle continues. We generally find that the hard part in this scheme is to choose and perform a *critical* experiment, namely, one that really distinguishes between rival theories. An important adjunct to the scientific method is the test of *reproducibility;* that is, we are not prone to accept the results of one scientist or group if they cannot be duplicated by others who repeat the experiment under similar conditions.

Some of Galileo's astronomical discoveries were mentioned in Chapter 4. Several others are described here to illustrate the point, which has been repeatedly demonstrated throughout the history of science, that a powerful new instrument capable of making observations or measurements not previously feasible will inevitably change our view of the nature of things. Just as this was true for Galileo, it is true today. The first radio telescope (built by Karl Jansky at the Bell Telephone Laboratories in 1931) revealed unsuspected emission from the Milky Way galaxy (Chapter 12); the first attempt to detect x-rays from the moon (by Ricardo Giacconi of American Science and Engineering, Inc., in 1962) resulted in the discovery of the first example of a new class of objects, the galactic x-ray sources (Chapter 14); the 200-inch telescope at Mount Palomar was used in the discovery of the enormous red shifts of the quasars (Chapter 16). Galileo's observations with the first crude astronomical telescope in the sixteenth century were bound to be fundamental to the progress of science.

Galileo turned his telescope to the moon and found not the unblemished surface cherished by some earlier philosophers, but a *world* of mountains, craters, and vast dark plains (the *maria,* Chapter 15). He looked at the sun, as did many of his contemporaries, and watched the sunspots move across the solar disk. Sunspots were known even in pre-telescopic times, as some were occasionally big enough to be seen by the unaided eye. However, they were generally thought to be planets located between the earth and sun that were simply being seen in projection

a

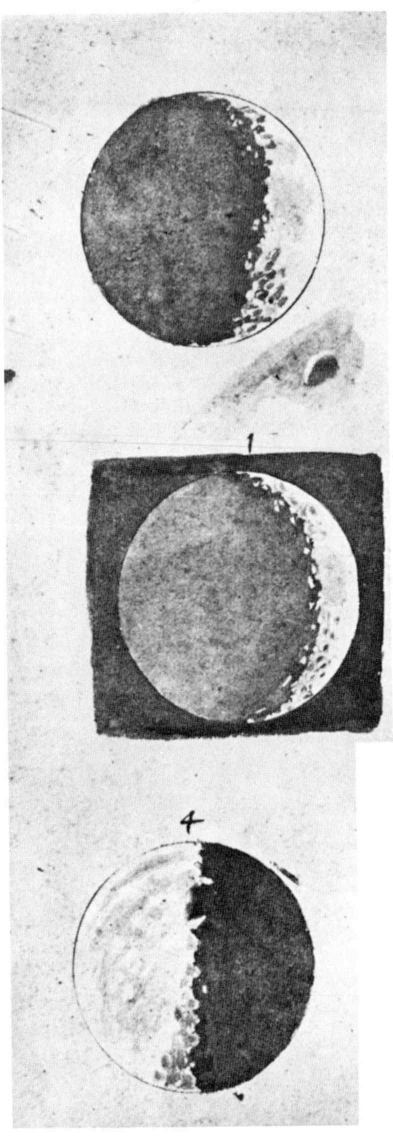

b

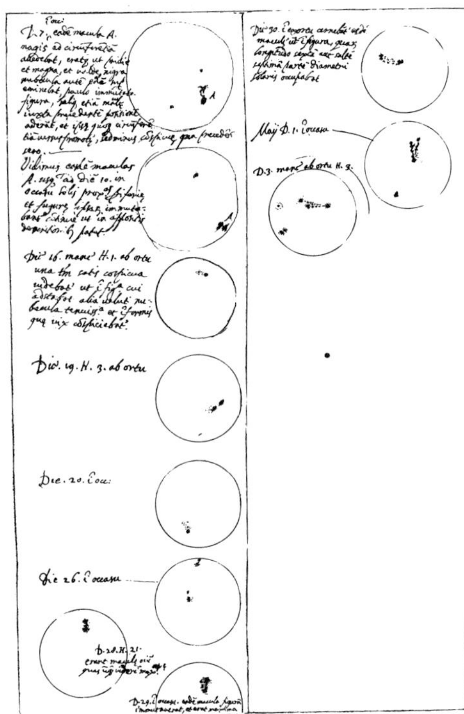

c

6-3 *Galileo's observations. Drawings in his notebooks show (a) the surface of the moon, (b) sunspots, and (c) Jupiter and its four largest satellites at different times. (Yerkes Observatory photographs.)*

against the solar disk. Galileo correctly surmised that they were fixed to the solar surface and he used them to measure the solar rotation rate, and thus to prove that the sun turns. He also found apparent appendages of Saturn that subsequent astronomers, with better telescopes, discovered to be rings (see Chapter 7). He looked at the smooth surface of the Milky Way with his telescope and realized that it was nothing less than the assemblage of a myriad of faint stars. As we mentioned in Chapter 4, he also made fundamental discoveries about Jupiter and Venus, which had extreme impact on the development of the heliocentric world view. Some of these observations, as recorded in Galileo's notebooks, are shown in Figure 6-3.

REFRACTING AND REFLECTING TELESCOPES

The purpose of a telescope is to collect light and to focus it on a small area where it can be conveniently inspected, photographed, or measured. This purpose can be achieved through the use of lenses (refracting telescope or *refractor*) or mirrors (reflecting telescope or *reflector*), as shown in Figures 6-4 and 6-5. The diameter of the light-collecting mirror or lens is called the *aperture* of a reflector or refractor, respectively. When we speak of a bigger telescope, we mean one with a larger aperture and not necessarily one that is longer.

Why is a big telescope better than a small one? Figure 2-3a shows that the light received from a distant point (such as a star) is essentially parallel, a fact that Eratosthenes realized long ago. Now consider the parallel light from a star as shown in Figure 6-6; clearly a telescope with a larger light-collecting lens (or mirror) intercepts more of the starlight, and therefore it

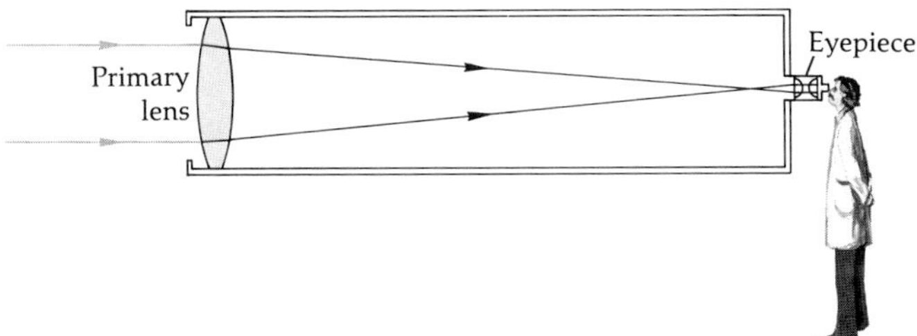

6-4 *Sketch of a refracting telescope. In practice, the "primary lens" is usually composed of more than one individual lens.*

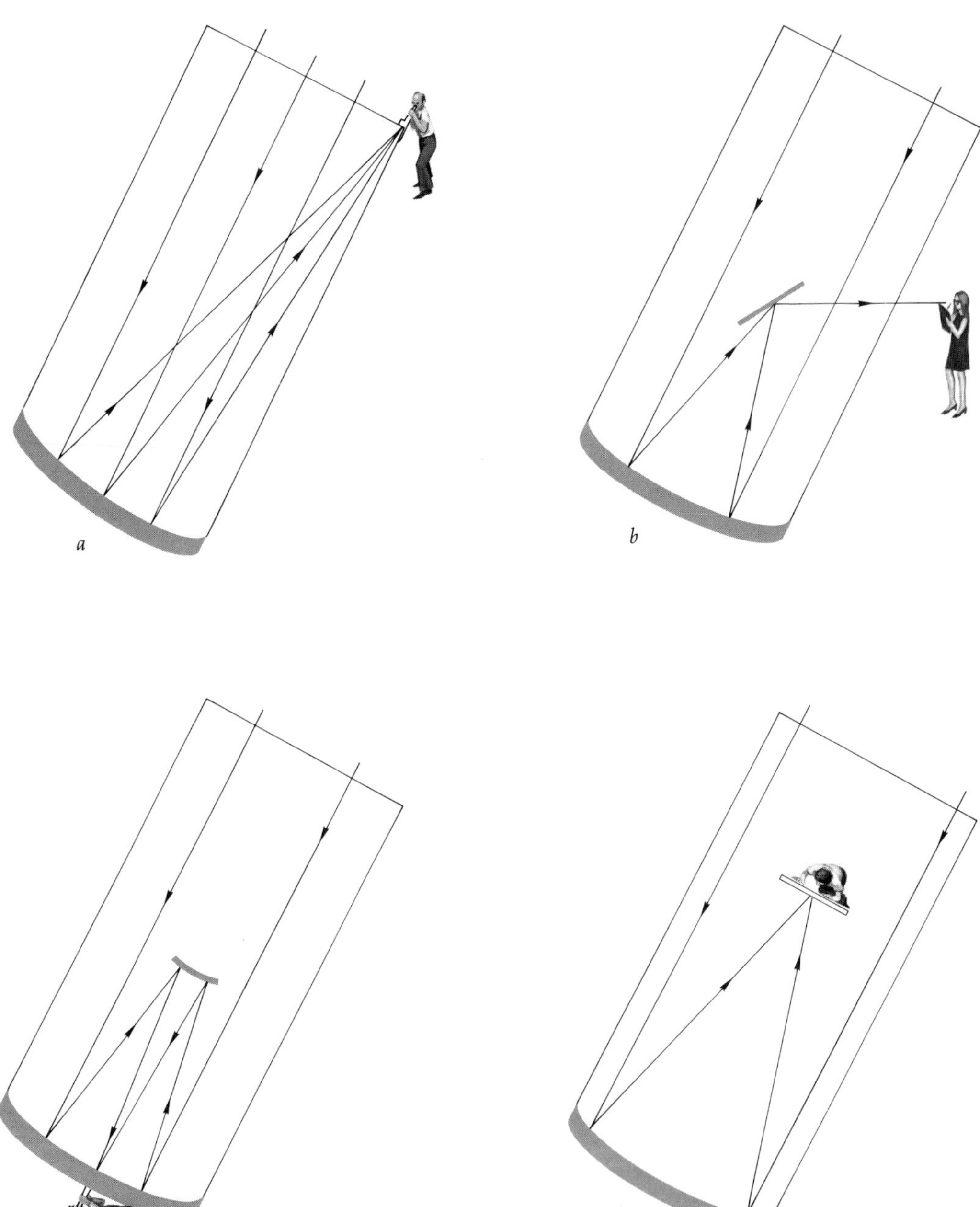

6-5 *Schematics for reflecting telescopes showing four different ways of viewing the image:* (*a*) *method employed by Herschel;* (*b*) Newtonian *method;* (*c*) Cassegrain *method, the most common arrangement in modern telescopes:* (*d*) prime focus *method used with very large telescopes. Not shown is the Coudé arrangement, in which additional small mirrors are used to divert the light to a large stationary spectrograph. (a, b, c are after* Matter, Earth, and Sky, *2d ed., by George Gamow. Prentice-Hall, Inc. 1965. Used by permission.)*

produces a brighter image of a given star. When photographing the stars the use of a larger telescope enables an observer to obtain an equivalent photograph in a shorter exposure time than if he had used a smaller telescope. Because fainter objects can be recorded with larger telescopes, and a distant star appears fainter than a similar star that happens to be close to us, larger telescopes allow us to see farther into space.

Many astronomical investigations require detailed analysis or measurement of an image. For example, when we observe the moon or Mars we want to see the fine details of their surfaces and not merely measure the intensity of their light. Larger telescopes can provide sharper images enabling us to *resolve* finer details (Figure 6-7) in the images. Unfortunately, turbulent motions of the earth's atmosphere cause a blurring of telescopic images, so that the full resolving power of the larger telescopes is rarely attained. This is one reason why astronomers are now conducting space experiments, with telescopes located above all or most of the atmosphere.

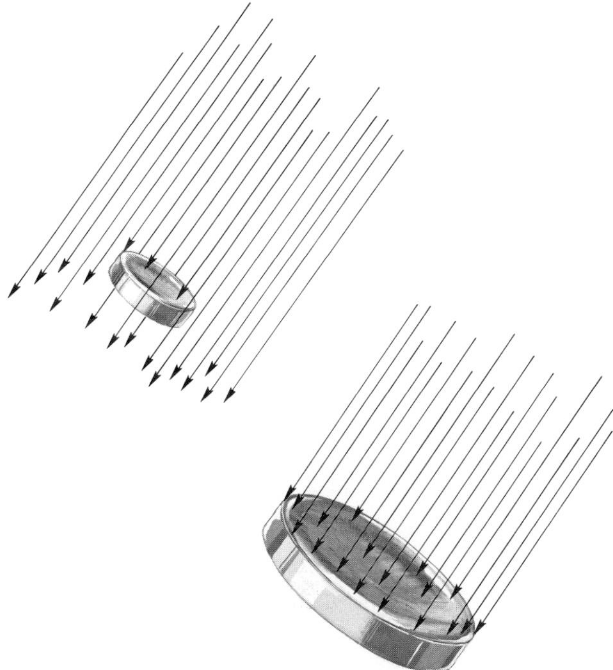

6-6 *The larger telescope intercepts (and therefore can collect) more light than the smaller telescope.*

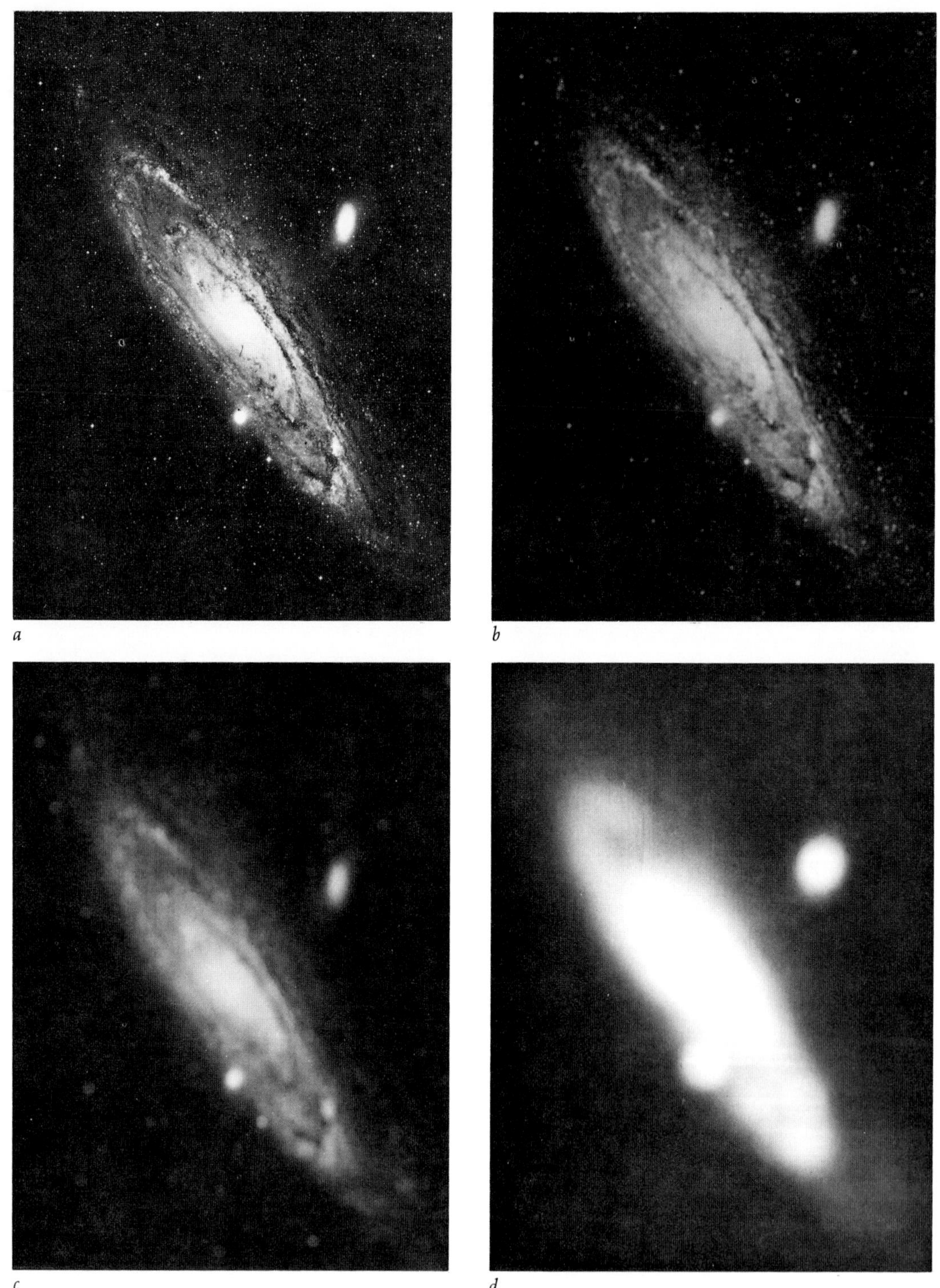

6-7 *Resolution and the quality of an image. The spiral galaxy M 31 is shown as it would appear (a) as photographed with a large optical telescope, or with resolutions of (b) 1 minute of arc; (c) 3 minutes of arc; and (d) 12 minutes of arc. (Courtesy Leiden Observatory.)*

Historically, major new telescopes have led to significant advances in the astronomical knowledge of their times. Two examples are given here; in one case the size of the telescope was of prime importance; in the other case the quality of fabrication was paramount.

72-inch Reflector in Ireland*

In 1845, the third Earl of Rosse built by far the largest telescope of his time. The main mirror was fabricated from a metallic alloy called speculum; the great light-collecting and resolving powers of this 72-inch (183 centimeters) mirror enabled Lord Rosse to discern the spiral nature of several galaxies (Chapter 13) and to discover the filaments in the Crab nebula (Chapter 16). The origin of spiral structure remains as one of the great unsolved problems in the study of galaxies; observations of the Crab filaments later provided dramatic confirmation that the nebula is expanding into space as the result of an ancient explosion.

Eighteenth Century English Reflector

Sir William Herschel (1738–1822) attained great skill in shaping and polishing telescope mirrors. Although he began as an amateur astronomer, he was soon making telescopes that surpassed those of the Greenwich Observatory in optical performance. The sharp images obtained with one of his smallest telescopes enabled him to recognize the unusual nature of what might otherwise have seemed to be an ordinary star. Viewed through Herschel's telescope this object had a slightly larger image than the other stars, and the image increased in size with greater magnification. Viewed on subsequent nights it was found to have moved against the background of stars and was soon shown to be a planet. Herschel had discovered Uranus, the seventh planet of the solar system and the first new world of the telescopic age. The importance of his telescope's quality in resolving the disk of the planet cannot be underestimated. Going back through the older records of other astronomers it has been found that Uranus was observed many times before Herschel saw it, but his predecessors noted it as just another "star."

---

*In this chapter we depart from our practice of always giving precedence to metric units; most of the telescopes discussed here have been named in English units.

The largest telescopes are reflectors; they include the 200-inch (508 centimeters) telescope on Mount Palomar and the 120-inch (305 centimeters) telescope at Lick Observatory, both located in California. Large reflectors being completed or under construction at the present time include a 234-inch (594 centimeters) telescope for a Russian observatory in the Crimea and several instruments of about 150-inch (381 centimeters) aperture that are located in Arizona, Chile, Australia, and elsewhere. Some modern telescopes are shown in Figures 6-8 and 6-9.

A special type of telescope, the Schmidt camera, uses both a primary mirror and a smaller lens, called the corrector plate (see Figure 6-10). It is used to photograph a large region of the sky at one time. The light-gathering power of this telescope is determined by the size of the corrector plate because, although the primary mirror is larger than the corrector, it is the latter object that intercepts the light from a star. Thus, the large Schmidt instrument on Mount Palomar (Figure 6-11) which was

6-8 *The 200-inch (508 cm) Hale Telescope located on Mount Palomar. (Courtesy of the Hale Observatories.)*

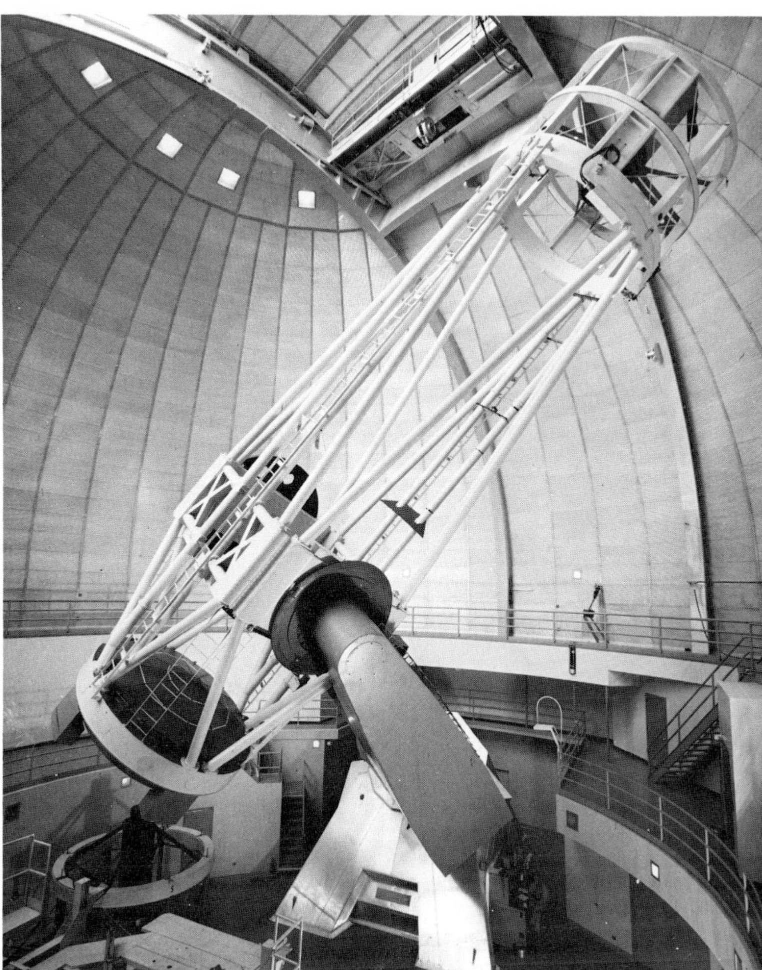

6-9 *The 120-inch (305 cm) telescope. (Lick Observatory photograph.)*

used to make a photographic sky atlas (*Palomar Sky Survey*, see page 166) is referred to as a 48-inch (122 centimeters) telescope, although the primary mirror has a diameter of 72 inches (183 centimeters).

The largest refractor, a 40-inch (102 centimeters) telescope at the Yerkes Observatory in Wisconsin, was built during the late nineteenth century. Since that time all telescopes of comparable or greater aperture have been reflectors (or Schmidt telescopes, in which the largest optical component is also a mirror). There are several reasons for this. Among them is the fact that the operation of figuring (grinding and polishing) the

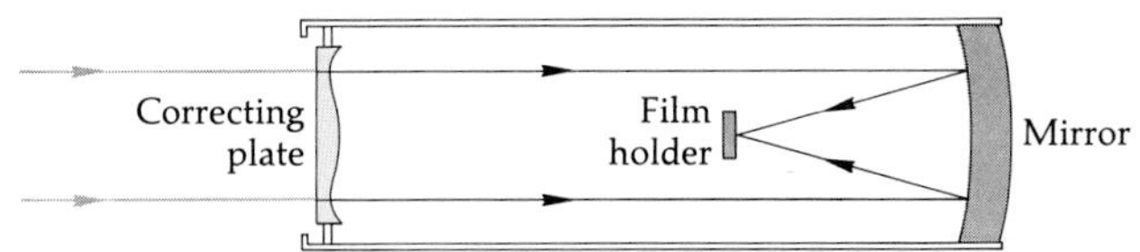

6-10 *Schematic of a Schmidt telescope.*

6-11 *The 48-inch (122 cm) Schmidt telescope at Mount Palomar. (Courtesy of the Hale Observatories.)*

surface of a large lens or mirror to the necessary optical precision can require a year or more of expert labor. This work must be done on both sides of a lens; however, it is required on only one side of a mirror because the light does not pass through the glass. By the same token, the difficult (and expensive—a modern 150-inch mirror costs about one million dollars) problem of casting a large piece of high quality, bubble-free glass is reduced for a mirror (compared to a lens) since the important region is limited to a thin layer near the reflecting surface. Finally, the lens in a large refractor sags and changes shape slightly as the telescope is tilted at different angles to view stars in different parts of the sky, and this degrades the optical performance of the telescope. A mirror, on the other hand, can be mechanically supported not only around its rim (as is done for a lens) but also across its rear surface. Thus, it is possible to mount even the largest mirrors so that they produce sharp images regardless of the telescope orientation. Still another major consideration is Newton's discovery that a lens focuses light of different wavelengths (colors) at different distances. Due to this effect of *chromatic aberration,* while one color is in focus, all the other colors will be out of focus and the recorded image will not be as sharp as one might expect from the aperture of a refractor. On the other hand, the focus of a mirror does not depend on the wavelength, and thus reflectors are not affected by chromatic aberration. Finally, new forms of glass and quartz developed in recent years make it possible to fabricate large mirrors that are lighter in weight and less sensitive to temperature changes. The primary mirror for a large modern telescope is shown in Figure 6-12.

Telescopes of the future are unlikely to continue the trend toward larger and larger primary mirrors, although engineering studies of an "X-inch" (giant reflecting telescope of as-yet-unspecified diameter) have been recommended. For one thing, the brightness of the faintest objects that can be recorded with existing telescopes is severely limited by the existence of faint light in the night sky and by air turbulence which smears out the images of stars. In fact, it has been shown that the faintest objects that could be detected with a 400-inch telescope would be only some 40 per cent fainter than those recorded by the 200-inch telescope, although the money needed to build the larger telescope would pay for at least five 200-inch ones. For this and similar reasons there is considerable work underway in studying techniques to use telescopes of conventional size more effectively. Such methods include automation, combination of observations made simultaneously by several telescopes, and creation of more sensitive detectors for use with existing telescopes. In the last category are included the development of better photographic plates and eventually the replacement of photography by electronic image recording techniques.

6-12 *The primary mirror of the 120-inch (305 cm) telescope. (Lick Observatory photograph.)*

RADIO TELESCOPES

Light, as the term is used in astronomy, is not limited to the radiation that our eyes can perceive; it includes all the forms of electromagnetic radiation that we mentioned in Chapter 5. Each of these (radio, infrared, visible light, ultraviolet, x-rays, gamma rays) describes light within a certain range of wavelengths. (Figure 5-14 depicts the electromagnetic spectrum.) In fact, among the most fundamental advances in twentieth century astronomy has been the observation of the cosmos in radio waves, x-rays, and other wavelength regions not accessible with ordinary visible-light telescopes of the sort just described. The great importance of these observations has been that as each new wavelength region has been investigated, astronomers have discovered celestial sources of such radiation that have a completely different character than the familiar stars, planets, nebulae, and galaxies of visible-light or optical astronomy.

The most familiar radio telescopes are the "big dishes" that resemble radar antennae; they are simple reflecting telescopes for radio waves (Figure 6-13). However, the elaborate electronic techniques that have been developed for processing radio signals also make it possible to operate several dish antennae simultaneously as one instrument (radio interferometer, Figure 6-14). Furthermore, the *elements,* or individual antennae, of some radio telescopes are not reflectors but remind us of those commonly used with short wave and television receivers. In any case, the function of a radio telescope, like that of an optical telescope, is to collect electromagnetic radiation and feed it to devices that detect, analyze, and record it.

The laws of optics show that a desired telescope performance in terms of *angular resolution* (the ability to discriminate fine detail) depends on the ratio of telescope aperture (diameter) to

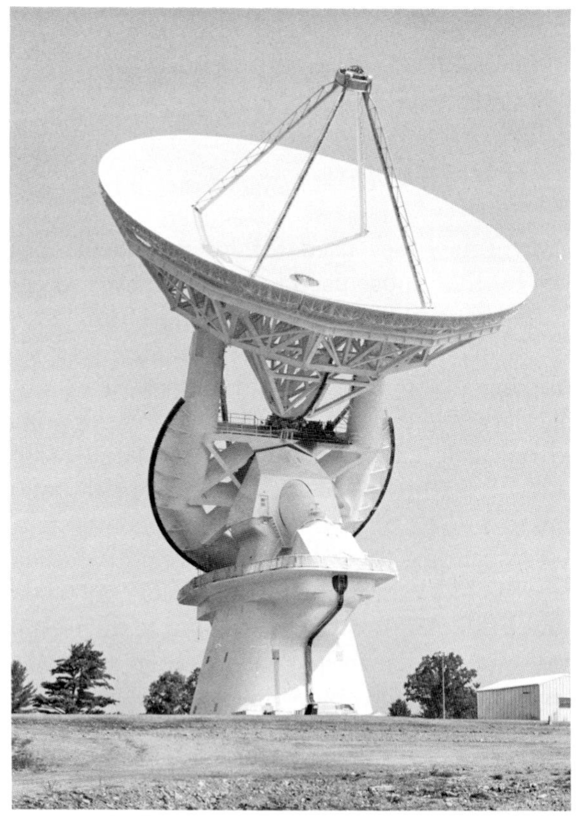

6-13 *The 140-foot (43 meter) dish at NRAO. (Courtesy of National Radio Astronomy Observatory, Green Bank, West Virginia.)*

6-14 *The radioheliograph at Culgoora, New South Wales, Australia. The instrument is composed of 96 individual antennas 12 meters (39 feet) in diameter arranged around a circle 3 kilometers (2 miles) in diameter. The individual antennas show as white dots around the circle in the panoramic view and the detail of one antenna is shown in the inset. This instrument allows snapshots of the sun in radio wavelengths to be taken with good time resolution. (Courtesy of J. P. Wild, Division of Radiophysics, C.S.I.R.O.)*

radiation wavelength. Telescopes used for longer wavelength radiation must be larger than those used to observe short wavelengths in order to resolve equally fine detail. Therefore, since radio waves are enormously longer than optical light waves, radio telescopes are usually much larger than ordinary telescopes (they are also larger because the cosmic radio emissions are very weak and because it is practical to build large radio telescopes). For example, a 10-inch (25 centimeters) optical telescope has a resolving power of about 0.5 arc seconds at a wavelength of 5,000 Angstroms. In order to resolve equally fine detail at the wavelength of 21 centimeters, which is over 400,000 times longer than the 5,000 Angstrom wave, a simple dish-type radio telescope would have to be more than 60 miles (97 kilometers) in diameter. However, the radio telescope at Arecibo, Puerto Rico, which is the largest existing single-dish antenna, has a diameter of only 1,000 feet (305 meters; Figure 6-15).

The great sensitivity of radio telescopes means that inter-

6-15 *The 1,000-foot (305 meter) diameter telescope at Arecibo, Puerto Rico. (Courtesy of the Arecibo Observatory.)*

ference from man-made signals, such as radio, radar, and television, is a serious problem. By international agreement, broadcasting on certain frequencies is prohibited to allow radio astronomers to receive cosmic signals with a minimum of interference. Our planet is so noisy due to radio transmissions that some people believe that life on other worlds may discover *us*.

OBSERVATORY SITES

Until fairly recently, observatories were located on convenient hills near their parent organizations, generally universities or government agencies. The objective was simply that the view of the sky be unobstructed by natural or artificial structures. Unfortunately, the development of modern civilization and its concomitant urban technologies has resulted in the many forms of environmental pollution that face us now. Among these, and perhaps least appreciated (except by the astronomers!), is *light pollution.** The increasing brightness of the night sky, due to atmospheric scattering and reflection of artificial light, limits our ability to observe faint stars at night, just as the scattering of sunlight in daytime prevents us from seeing even the brightest stars. This is a prime reason for the location of all major new observatories at considerable distances from urban centers. In a dispatch from Tucson, Arizona, dated November 21, 1970, the Associated Press reported that "Directors from five astronomical installations asked the City Council to pass an ordinance that would keep the night skies dark by making the city's lights reflect to the ground rather than upward."

Other factors besides sky brightness are important in the selection of an observatory site. Obviously, the frequency of clear skies is important, so one studies the weather and cloud cover records; a recent, very effective method for doing this is the analysis of weather photographs (Figure 6-16) obtained by meteorological satellites. Another criterion is the astronomical *seeing,* a term that describes the steadiness and size of star images, as affected by turbulent motions in the atmosphere. (Seeing includes the twinkling effect familiar to anyone who has ever looked at the stars.) Clearly, the perfection of the lens- and mirror-grinder's skill is of little avail if the telescope that he produces is mounted in a location where the atmospheric conditions ensure that the stars will usually be blurred by seeing or obscured by clouds. The seeing criterion is a principal reason for the location of observatories on mountain tops, where they are often above a good fraction of the turbulent air and the air-borne dust. In fact, the hypothetical 10-inch telescope mentioned on page 152 could rarely attain its theoretical resolution of 0.5 arc seconds, because the atmosphere is seldom stable enough. The water vapor in the atmosphere (see Chapter 8) is a very effective absorber of the infrared radiation that is of increasing interest to astronomers; this is still another reason for choosing a high and dry location.

In short, an observatory site should have dark and dust-free skies, low cloudiness, steady dry air, and unobstructed horizons. For radio astronomy observatories the dark sky criterion becomes one of low man-made interference. Also, low cloudiness

---

*Recently conservationists have also recognized the problem of light pollution. In May 1971, the Sierra Club's Board of Directors adopted as a policy that the ". . . Club opposes unnecessary night lighting in both urban and suburban areas. This practice is a waste of electrical energy, destroys the aesthetics of the nighttime sky, and seriously interferes with astronomical research."

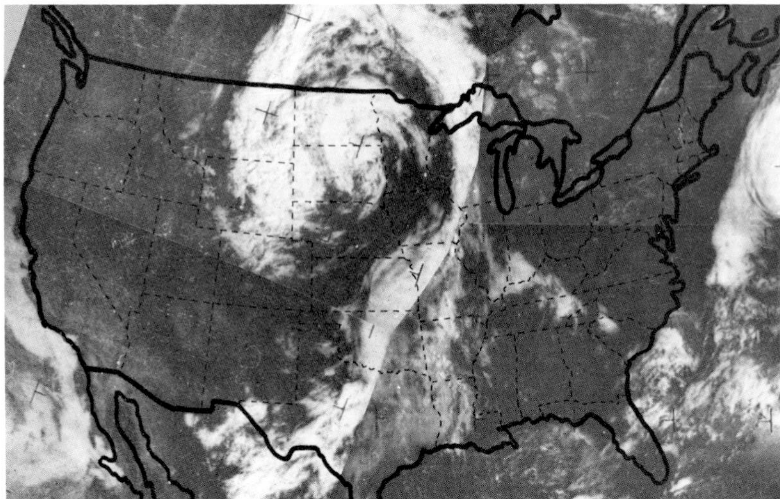

6-16 *A composite satellite photograph showing cloud cover over the United States on September 14, 1967; also visible off the New England coast is Hurricane Doria. (National Environmental Satellite Service.)*

and dry air are important at short radio wavelengths (millimeter and centimeter waves). Good optical observatory sites are generally on mountain tops; the better radio observatory sites are in valleys (for protection against interference—at the longer wavelengths of from tens to thousands of centimeters) although a few (for which low water vapor is important—millimeter wavelengths) are on mountains.

Keeping in mind all the above criteria, the best American observatory locations appear to be in the southwestern states, including southern California, and in Hawaii, but such sites are getting fewer every year due to pollution and urban expansion. The Kitt Peak National Observatory near Tucson, Arizona, provides a variety of telescopes and instruments at a good mountain-top location for use by astronomers and graduate students who lack access to comparable facilities at their own institutions. Potential users of the Kitt Peak telescopes send descriptions of their proposed research programs to the Observatory, which assigns telescope time on the basis of scientific merit. A similar facility for radio telescopes, the National Radio Astronomy Observatory, has most of its antennae at Green Bank, West Virginia, with one dish, for millimeter wavelength studies, located on Kitt Peak. In cooperation with the University of Chile, the Kitt Peak organization also maintains an Inter-

Table 6-1 *Some Major Optical Observatories*

NAME	LOCATION	CHIEF INSTRUMENTS
Hale Observatories	Mt. Palomar, California	200″ (508 cm) reflector, 48″ (122 cm) Schmidt
	Mt. Wilson, California	100″ (254 cm), 60″ (152 cm) reflectors
Lick Observatory	Mt. Hamilton, California	120″ (305 cm) reflector
McDonald Observatory	Fort Davis, Texas	107″ (272 cm), 82″ (208 cm) reflectors
Kitt Peak National Observatory	Tucson, Arizona	84″ (213 cm) reflector 60″ (152 cm) solar telescope 150″ (381 cm) reflector (under construction 1972)
Cerro Tololo Inter-American Observatory	La Serena, Chile	60″ (152 cm) reflector 150″ (381 cm) reflector (under construction 1972)
Lunar and Planetary Laboratory	Tucson, Arizona	Several 60″ (152 cm) reflectors
Steward Observatory	Tucson, Arizona	90″ (229 cm) reflector
Mt. Stromlo Observatory	Canberra, Australia	74″ (188 cm) reflector
University of Hawaii Observatory	Mauna Kea, Hawaii	84″ (213 cm) reflector
Crimean Astrophysical Observatory	Nauchny, USSR	104″ (264 cm) reflector
Sacramento Peak Observatory	Sunspot, New Mexico	30″ (76 cm) solar telescope

American Observatory at Cerro Tololo near La Serena, Chile, for studies of the southern stars. Several other observatories are being developed in the Chilean Andes, where the weather and seeing conditions are superb.

Some major observatories are listed in Tables 6-1 and 6-2.

DETECTION METHODS

The astronomer uses a telescope to select and track a target for observation and to collect light from the target. At the telescope focus he uses an instrument to record the light or analyze its properties. This may be a camera to photograph a portion of the sky, a *photometer* to measure light intensity, a *spectrograph* to study the distribution of light at the various wavelengths, or a *polarimeter* to determine the polarization properties of the light. Each instrument contains optical components (lenses, mirrors, filters, prisms, and the like) necessary for the measurements, but its key component is the light *detector* or sensor that records the light (as on a photographic plate) or produces a signal that measures the light (as, for example, a photoelectric cell or

Table 6-2 *Some Major Radio Astronomy Observatories*

NAME	LOCATION	CHIEF INSTRUMENTS
Arecibo Observatory	Puerto Rico	1,000-foot (305 m) fixed spherical antenna
National Radio Astronomy Observatory	Green Bank, West Virginia	300-foot (91 m) meridian transit reflector 140-foot (43 m) fully steerable reflector 3-element interferometer, with 1-mile (1.6 km) baseline; 1 portable element
National Radio Astronomy Observatory (field station)	Tucson, Arizona	36-foot (11 m) fully steerable reflector, usable to 1 mm wavelength
Jodrell Bank Experimental Station	Manchester, England	250-foot (76 m) fully steerable reflector
Mullard Radio Astronomy Observatory	Cambridge, England	Array of antenna elements 3-element, 1-mile (1.6 km) interferometer
Radiophysics Laboratory CSIRO	Parkes, Australia	210-foot (64 m) fully steerable reflector
Molonglo Radio Observatory	Molonglo, Australia	1-mile (1.6 km) cross-shaped interferometer
Algonquin Radio Observatory	Lake Traverse, Ontario, Canada	150-foot (46 m) fully steerable reflector
Northeast Radio Observatory Corporation	Tyngsboro, Massachusetts	120-foot (37 m) steerable reflector in radome
Radio Observatory of Nançay (Observatory of Paris)	Nançay, France	1,000-foot (305 m) partially steerable (meridian transit) array
Max Planck Institute for Radio Astronomy	Effelsberg, Germany	328-foot (100 m) fully steerable parabola

photomultiplier tube). A fundamental property of a sensor is its efficiency at producing a measurable signal or record from a given amount of light. For example, if a new kind of film requires less light to produce a recognizable image of a given star, then (other things being equal) we may photograph stars that are fainter than those that could be recorded previously with the same telescope.

PHOTOGRAPHY

The photographic detector is usually a glass plate or a celluloid film coated with a light-sensitive emulsion. Photons of light that strike the emulsion produce chemical reactions in the grains* of emulsion material. On the developed film or plate (a negative), the areas that have been struck by light are dark and the images of the brighter stars are larger and darker than those of the dimmer stars. The glass plates are usually preferable to film because they are less apt to shrink or bend, and thus measurements of the positions of stars are more reliable and a more permanent record is obtained. In addition to sensitivity differences, emulsions differ in such matters as grain size and spectral properties (response to light of different wavelengths). An emulsion with finer grains can yield sharper images, although it may not record stars that are as dim as those photographed with a more sensitive emulsion. A plate with high red sensitivity is better for photographing the red portion of stellar spectra, although it may be useless in observing blue light, and so on. In 1839, Louis Dauguerre announced the invention of photography in France, and within a year a dauguerrotype of the moon was made by John W. Draper of New York.† The first great astronomical triumph of this invention came in 1860 when photographs of an eclipse of the sun proved that the red *prominences* (Chapter 9) then observable only during eclipses were solar features, and not lunar phenomena as some astronomers had suggested. Early photography was crude, slow, and often quite awkward. Some early processes literally required that the photographic plate be wet during the exposure. But the techniques were improved and the development of emulsions

*When viewed with a magnifying glass, the emulsion of a photographic plate or film is seen to consist of many tiny clumps of material. If you look at a photograph enlarged considerably from the original negative, you note that it seems less sharp. This happens because the enlargement allows us to notice the grainy pattern of the film.

†Dauguerre actually perfected and popularized a process that was invented by J. N. Niepce in 1822.

with sufficient sensitivity to record dim stars produced a revolution in stellar map-making. In earlier times, astronomers charted the stars by making laborious measurements of their positions, one star at a time; now it became possible to record hundreds or thousands of stars on a single plate. Furthermore, information on both the positions and brightnesses of the stars is preserved. Comparison of several plates of the same sky region is a valuable tool for discovering the *variable stars* (Chapter 10) that change in brightness with time, or moving objects, such as asteroids (Chapter 7), that appear at different locations on plates obtained on different nights (or as streaks on individual plates). In addition, the photographic plate allows one to accumulate the effects of many photons during a long exposure and thus produce images of fainter and fainter objects, as illustrated in Figure 6-17.

In a great many cases, the astronomer does not photograph a star—he photographs its spectrum. These pictures, called *spectrograms,* are studied to determine the temperature, chemical composition, radial velocity and other properties of the star (Chapter 10). Recall Figure 5-19 (page 125); the instrument used to photograph spectra is similar in concept to Newton's experiment. A simplified diagram of such a *spectrograph* is given in Figure 6-18.

PHOTOELECTRIC DETECTORS

The photoelectric effect, in which light shining on a metal or other substance causes electrons to be ejected from the surface, was important in understanding the quantum nature of light (Chapter 5). Because the flow of electrons from a metal surface constitutes an electric current, it is possible to use the photoelectric effect to determine the brightness of a star by focusing its light on a suitable metal surface and using the appropriate electronic instruments to amplify and measure the resulting electrical current. Modern electronic technology has produced a variety of these sensors, known generally as photoelectric detectors. The type most widely used by astronomers is the *photomultiplier.* This is a vacuum tube in which, after light strikes the sensitive metal surface (photocathode) the ejected electrons strike another metallic surface (dynode), where each incident electron causes additional electrons to be released. These secondary electrons in turn strike another dynode where again each incident electron produces many more particles. In this way, the number of electrons produced at the photocathode by a

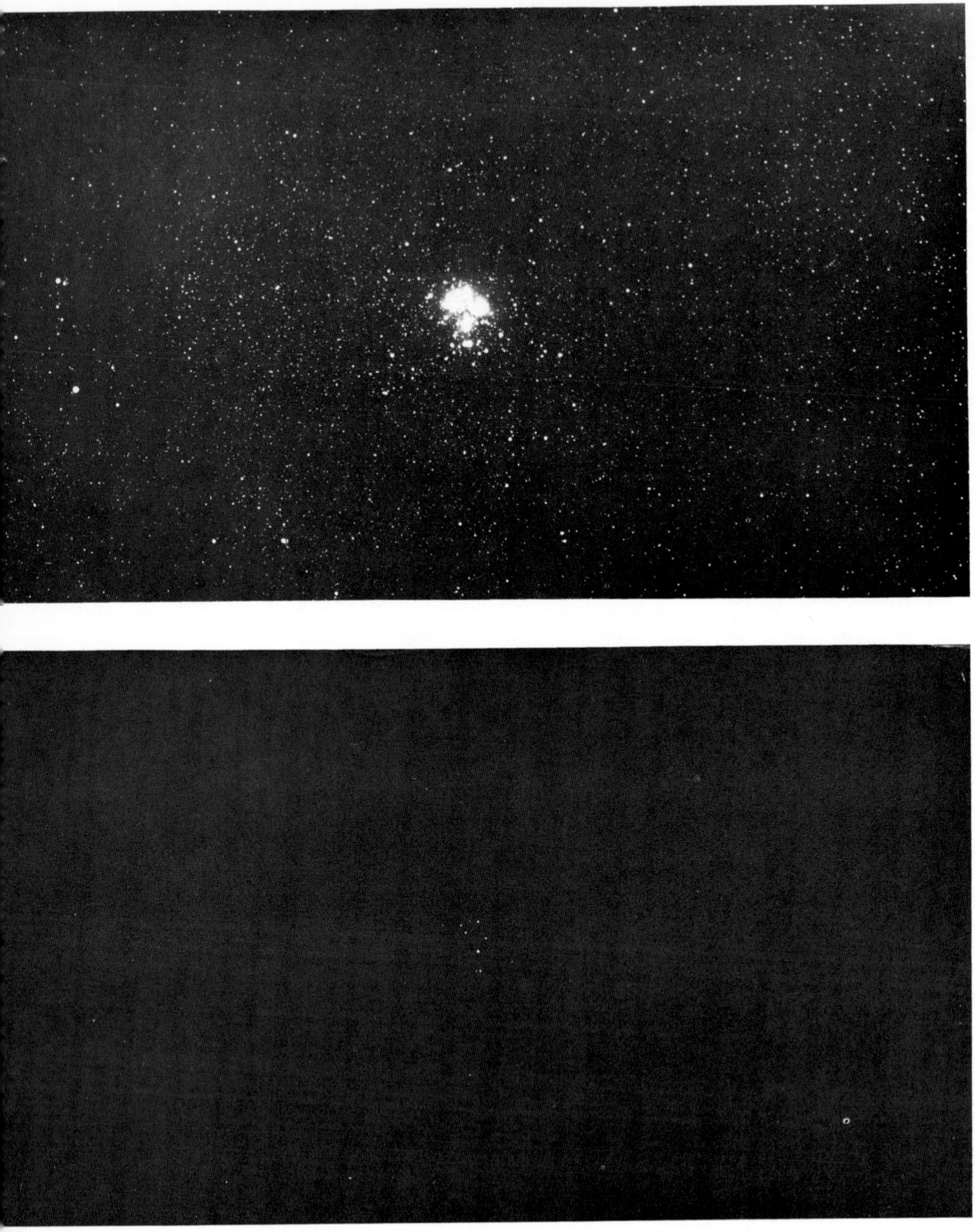

6-17 *Photographs of the Pleaides star cluster taken with different exposures: bottom, 6 minutes; top, 600 minutes. (Yerkes Observatory photograph.)*

given amount of light is effectively multiplied (in fact, by a factor in the millions) so that a larger, more easily measured electric current is produced (Figure 6-19). In some applications, particularly when the incident light is very dim so that the number of electrons produced is rather small, the output of a photomultiplier is measured by counting the individual bunches of electrons, each of which corresponds to the arrival of one photon (*pulse counting technique*).

Photomultipliers are being used increasingly in modern spectroscopic instruments. These devices, called *spectrometers* or *photoelectric spectrum scanners,* may employ a prism to produce the spectrum, as does the spectrograph of Figure 6-18, or they may use a diffraction grating (recall Figure 5-18). The example sketched in Figure 6-20 was used in an artificial satellite to record spectra in the ultraviolet wavelength range. Since light of different wavelengths is reflected from a grating at different angles (Figure 5-18), it follows that only one narrow range of wavelengths falls on the photomultiplier in Figure 6-20 at a given time. However, rotating the grating (note curved arrow) causes the different wavelengths to sweep past the photomultiplier. When the signal from the photomultiplier is amplified and displayed on chart paper, the result is a graph of the intensity of the spectrum as a function of wavelength. An example of such a graph (*spectral scan*) is given in Figure 6-21. Before the intensities on a spectral scan are used in calculations or theories, they have to be adjusted for such effects as the absorption of light by the earth's atmosphere and the sensitivity of the photomultiplier, both of which depend on the wavelength.

Photomultiplier tubes are used to measure intensity of light. However, there are several types of imaging photoelectric detectors that use the photoelectric effect to produce pictures.

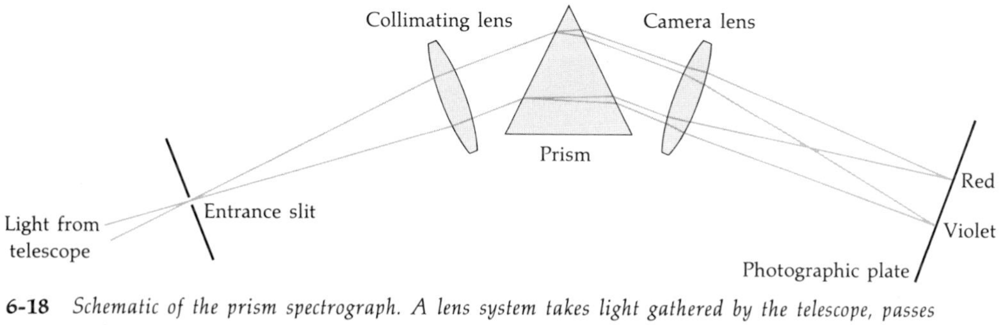

6-18 *Schematic of the prism spectrograph. A lens system takes light gathered by the telescope, passes it through a prism, and brings it to a focus. The fact (illustrated in Figure 5-19) that violet light is bent more than red light in passing through a prism produces separate images in the different colors on the photographic plate. The bending in the camera lens is exaggerated for clarity.*

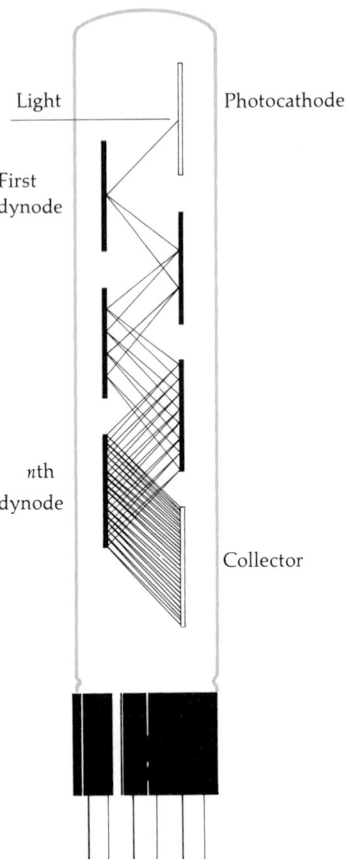

6-19 *Schematic of a photomultiplier tube. Light strikes the photocathode and an electron is emitted through the photoelectric effect. Electric fields are used to accelerate and focus these electrons onto successive dynodes where these energetic electrons knock out still more electrons at each step. In practice, the electric current at the collector is a million or more times the current at the first dynode.*

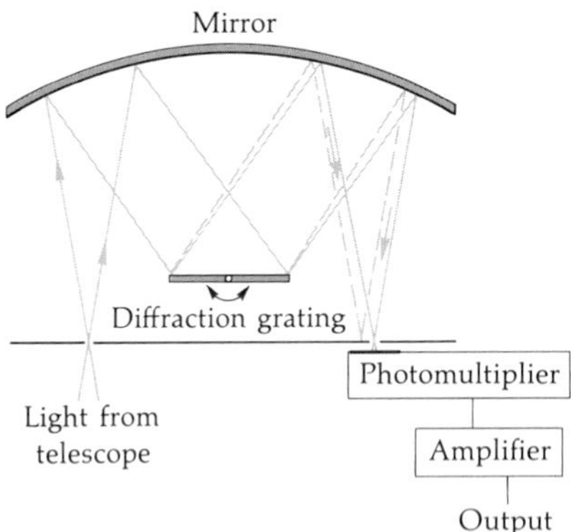

6-20 *Schematic of the grating spectrometer. The different colors are supplied to the photomultiplier by tilting the diffraction grating to produce scans of a spectrum; an example is given in Figure 6-21.*

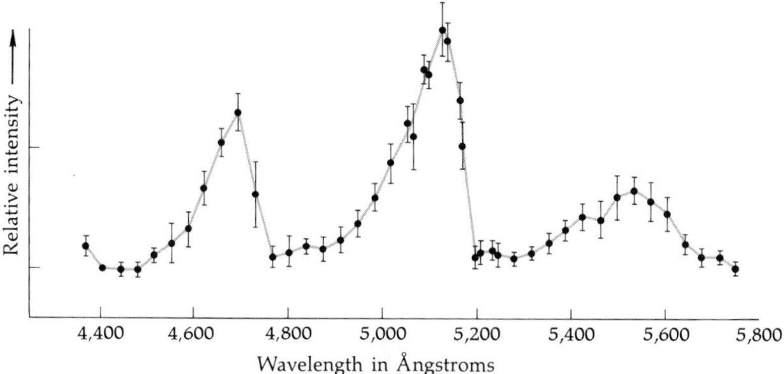

6-21 *Scan of the spectrum of Comet Honda (1968c) made with a grating spectrometer. The relative intensity shows the characteristic variation of the* C_2 *molecule. (Courtesy of W. L. Gebel, State University of New York at Stony Brook.)*

In a common application of such *electronic image tubes* the image formed by a telescope is focused on the photocathode; at the opposite end of the tube a much brighter duplicate of the incident image is produced on a glowing phosphor screen (Figure 6-22). This brighter image is recorded on a photographic plate. The advantage of this method is that the output image of the tube is much brighter than the input image formed by the telescope. Therefore, using plates or film of a given sensitivity, it is possible to photograph much fainter stars than otherwise, or to reduce the exposure time required to record a star of a given brightness. Since astronomical photography frequently involves exposure times of several hours, the possibility of reducing the exposures through the application of new electronic techniques is very attractive. In another modern development, *vidicon* tubes, which apply the television process, have been used in conjunction with telescopes (Figure 6-23). The pictures taken with the vidicons can be preserved on video tape and should be relatively easy to feed into a computer for analysis.

OTHER DETECTORS

The photographic plates and photomultiplier tubes described above are useful for detecting light in the visible wavelength region. Various types of devices are used in observing at other

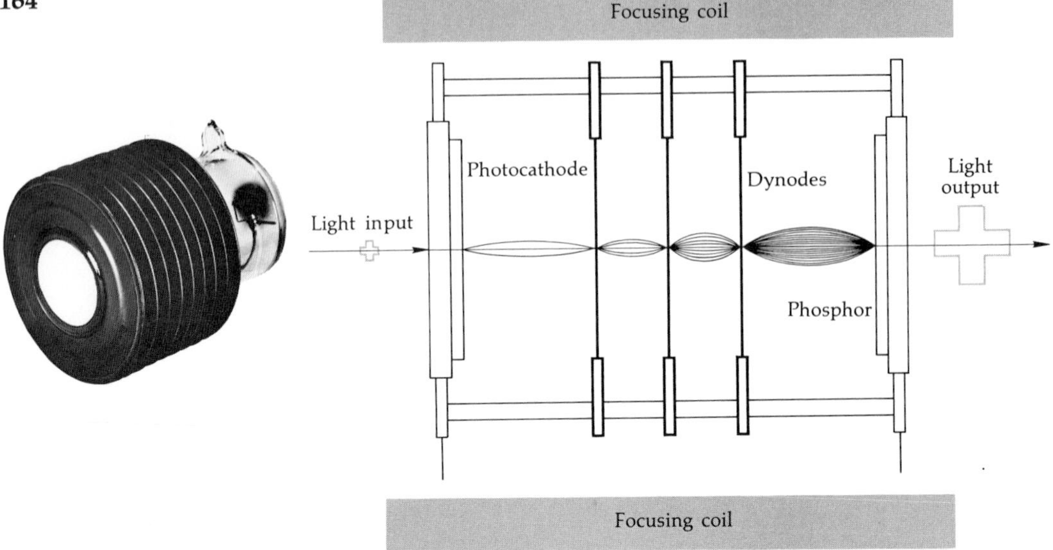

6-22 *Sketch and photograph of an image converter tube. The basic principle of electron multiplication is the same as for photomultipliers (illustrated in Figure 6-19), but focusing properties are provided to produce an image, which is converted to light by the phosphor screen. (Courtesy of Westinghouse Electric Corporation.)*

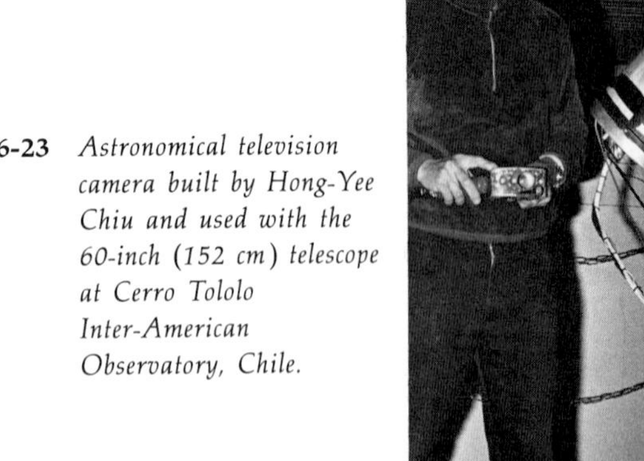

6-23 *Astronomical television camera built by Hong-Yee Chiu and used with the 60-inch (152 cm) telescope at Cerro Tololo Inter-American Observatory, Chile.*

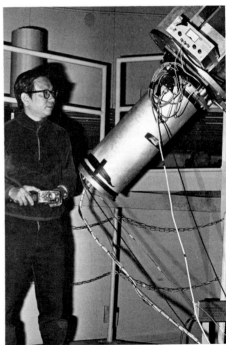

wavelengths. For example, *Geiger tubes* and *proportional counters* are flown on rockets to measure celestial x-rays, and *Germanium bolometers* are used to study infrared radiation. Although the operating principles of these other devices are different from those described above, in each case an electrical signal that corresponds to the light collected by the telescope is finally produced and measured.

SKY SURVEYS

Astronomy, as other sciences, seeks to describe and explain a certain class of natural phenomena. The first step in this process is to record and classify the subjects of investigation. From this, a collection of measurements and observations is obtained which is useful in selecting particular targets for future investigations and which provides a suitable basis for statistical analysis.

In studying the earth, the geologist uses maps of the topography, lists, descriptions, and locations of the different mineral and rock forms, and catalogues of the various types of fossils and their ages. Similarly, astronomers use maps of the sky and catalogues of the stars and their properties.

The construction of star maps and catalogues began in antiquity with the recording of the brighter stars visible to the unaided eye. In more recent times sky surveys have been compiled by observing the positions of stars with the *meridian circle,* a special kind of telescope designed for precision measurements. Such a catalogue, or sky survey, is called a *Durchmusterung,* after the German term adopted by the creator of the first major survey, Friedrich Argelander, who worked at the Bonn Observatory in the mid-nineteenth century. Over a period of seven years Argelander and his assistants, using the eye, the telescope, and the clock, mapped the northern half of the sky, plus a small zone just south of the celestial equator. Their maps and catalogue, which show the positions and approximate brightnesses of 324,000 stars, are known as the *Bonner Durchmusterung* (or B.D.), and are still in regular use today. Later, other astronomers, working in such places as Argentina and South Africa, extended the surveys to cover the southern half of the sky and to include fainter stars.

By recording the stars according to position and brightness and by giving each one a B.D. number, Argelander provided a convenient means for astronomers to locate individual stars and to identify those of special interest for further observation by their colleagues. Thus, the Durchmusterung is equivalent to a crude census that tells where people live. As the demands of the government statisticians have grown, so have the number of questions in the census, so that information is gradually accumulated on the personal characteristics, property, and so on, of the residents. Likewise, the next step after the Durchmusterung was to catalogue the individual properties of the stars listed at the various positions. Because the spectrum of a star (Chapter 10) gives valuable information on the chemical composition, surface temperature, and other properties, catalogues of the stars classified according to their spectra have been of fundamental importance to astronomers. Chief among these surveys has been the *Henry Draper Catalogue* (H.D.), which was compiled at the Harvard College Observatory around 1920 under the leadership of Annie Jump Cannon. The rudimentary state of astrophysics in those days did not yet allow reliable estimation of physical properties from the stellar spectra, so Miss Cannon classified the stars according to the appearance of their spectra (H.D. types). Modern investigations have developed the physical interpretations of the various H.D. types so that the nine volumes of the H.D., listing brightnesses and spectral types of 225,000 stars, are a major tool of the astronomical researcher.

With the advent of large telescopes, observations of stars and galaxies have been extended to such faint brightnesses that the job of cataloguing the millions of fainter stars by Argelander's visual techniques would exceed both the patience and the lifetime of any astronomer. The production of photographic sky maps was the only answer to this dilemma, and several star atlases have been compiled. These have culminated in the great survey made with the Schmidt camera on Mount Palomar which is used by all astronomers who work with the major telescopes.

In its original form the *Palomar Sky Survey* comprises 1,870 photographic plates, each 14 inches (36 centimeters) square, arranged in pairs. Each pair consists of an exposure taken with a red filter and one taken on a blue-sensitive plate. Thus, the examination of the plates reveals positions and brightnesses of stars, nebulae, galaxies, etc., and by comparison of the two plates in each pair, one can determine the approximate colors of the objects.

The Palomar Survey covered the northern half of the sky plus a good portion of the Southern Hemisphere (down to declination $-33°$). Later, partial surveys extended the sky coverage and it is hoped that eventually the entire sky will be mapped with equal quality. (This would require the installation of a major

Schmidt camera in the Southern Hemisphere.*) An enormous effort was required to obtain the 935 plate pairs: clouds, airplanes, poor telescope operation, bright moonlight, and the occasional heart-breaking dropped plate, all interfered with the survey; over 3,200 plates were taken before the desired complete set of high quality photographs was obtained. The effort and expense involved explain why the original plates are stored in a vault in Pasadena with a high quality duplicate set kept on Mount Palomar. Additional copies, on glass and on photographic paper, are in use at universities and observatories throughout the world. A sample of a Palomar Survey plate is shown in Figure 6-24.

Radio astronomers have likewise surveyed the sky. Their early catalogues contained relatively few radio sources, each of which bore a name in the style Cygnus A, Cygnus B, and so on (first and second sources found in constellation Cygnus). Soon, larger antennae at such places as Cambridge University were in use, and individual research groups were listing as many as several hundred sources found in their observations. One of the most important of these studies was the *Third Cambridge Catalog of Radio Sources.* Many of the quasars (Chapter 16) and other radio sources are still known by their "3C" numbers as listed in this catalog. By 1964, a *General Catalogue of Discrete Radio Sources,* which consisted of a summary and cross-tabulation of the individual observers' lists, was published. It included nearly 1,300 radio sources both within and beyond our galaxy, and since then, more sensitive radio telescopes have detected many more.

Figure 6-25 is a map which contains the same portion of the sky shown on the preceding sample Palomar Survey picture. This figure, however, shows the distribution of individual radio sources (represented by circles, triangles, and squares) in this portion of the sky, and the contour lines show the background radio emission of our galaxy.

CONTINUING EFFORTS

As modern technology advances and we observe with improved sensitivity, or especially as we look out in new regions of the electromagnetic spectrum (for example, ultraviolet observations of the stars from orbiting observatory satellites), the first step

*A large Schmidt camera is being put into operation at the European Southern Observatory in Chile.

6-24 *Palomar Sky Survey plate (taken in red light) of the area marked by the square in Figure 6-25; the area covered by this plate is 6.°6×6.°6. An overexposed image of the Crab nebula (see also Figure 16–4) is shown (indicated by the arrow) as well as many stars and some wisps of nebulae. (Courtesy of the Hale Observatories. Copyright © by the National Geographic Society—Palomar Observatory Sky Survey.)*

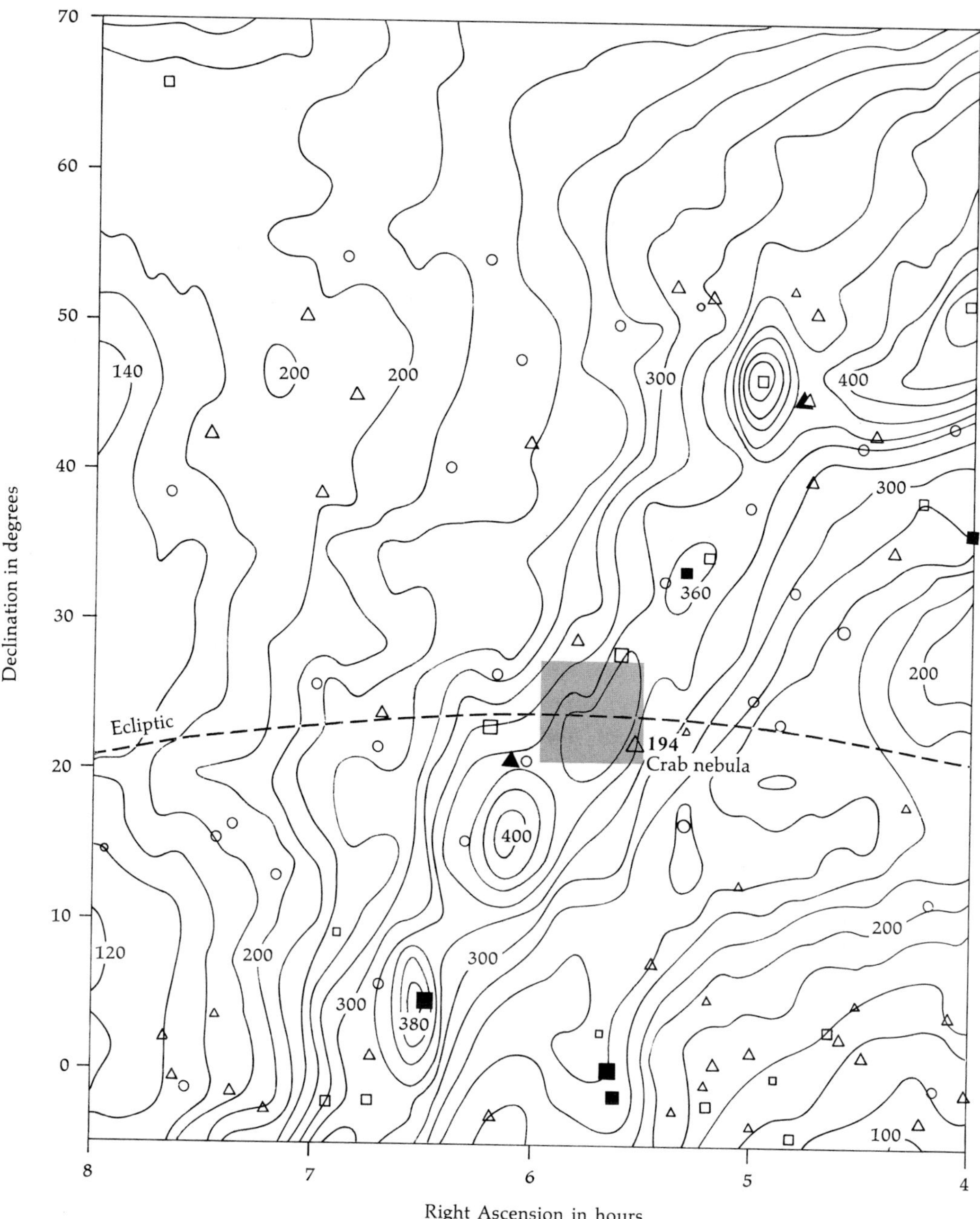

6-25 *Radio map of a large portion of the sky which includes the area photographed in Figure 6-24 (indicated by the square). The contour numbers indicate the brightness of the galactic background radio radiation. The circles, triangles, and small squares represent individual sources such as the Crab nebula or Taurus A radio source which is No. 194 in the General Catalogue of Discrete Radio Sources. Note that the Crab nebula is a strong radio source, but the bright star near the Crab nebula on the Palomar plate is not. (Drawn after a map by W. E. Howard, S. von Hoerner, H. D. Aller, and L. R. Walker, National Radio Astronomy Observatory.)*

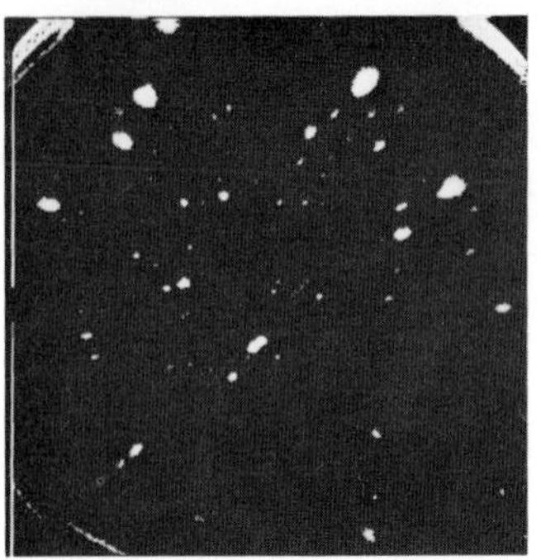
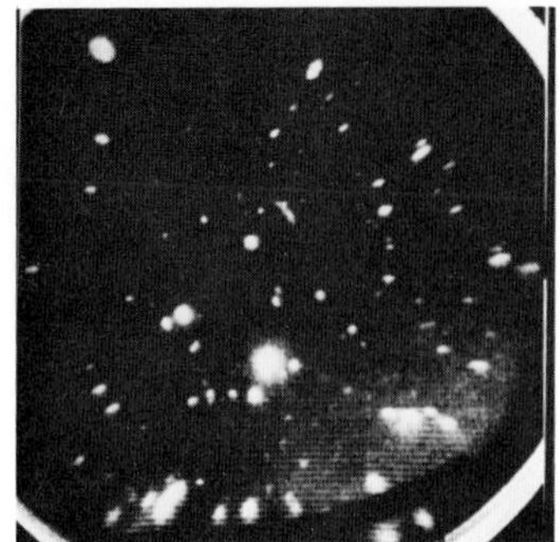
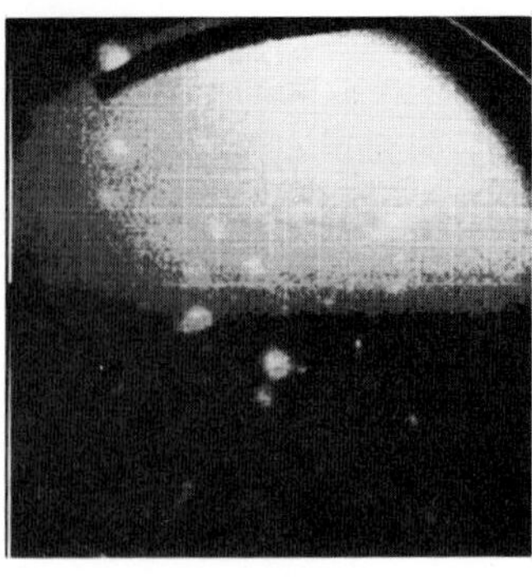

6-26 *A star field as viewed with television cameras on the OAO-2 satellite through filters covering the wavelength range from 1,050 Å to 3,000 Å. Notice the different relative brightnesses as seen through the different filters. Each view actually contains two different filters; this is clearly shown in the last view where the top part appears to be foggy. The "fog" is sunlight (wavelength 1,216 Å), scattered by hydrogen atoms in the vicinity of the earth. (Courtesy of Project Celescope, Smithsonian Astrophysical Observatory.)*

is generally the same as Argelander's: we begin by mapping the skies. The first ultraviolet star atlas, recorded with a vidicon camera on the OAO-2 (second Orbiting Astronomical Observatory), does not provide full sky coverage, but does include pictures of many of the most interesting regions. A sample ultraviolet sky picture is shown in Figure 6-26. As we shall see in Chapter 14, the observations in relatively unexplored regions of the spectrum are revealing dramatic new phenomena that enlarge and revise our concept of the astronomical universe.

7

The Solar System—An Overview

The concept of the *solar system* with the sun at the hub of an array of orbiting planets, comets, asteroids, and meteoroids is so firmly engrained in our Space Age consciousness that the term has entered our language as a metaphor for a center of power attended by lesser luminaries. Nevertheless, as we have seen, the road to acceptance of a heliocentric system operating under the force of gravity was a long and hard one, and the recognition of the other planets as worlds was also slow in arriving.

The outermost of the nine known major planets, Pluto, was not discovered until 1930. Adding it to the list, the planets are (in order from the sun): Mercury, Venus, Earth, Mars, Jupiter, Saturn, Uranus, Neptune, and Pluto. Those who like to commit things to memory can use any one of several silly mnemonic devices to recall this list, such as: "Many Volcanoes Erupt Mulberry Jam Sandwiches Under Normal Pressure", or "My Very Eager Mother Just Served Us Nine Pizzas." Uranus and Neptune were discovered in 1781 and 1846, respectively. Searches for other major planets of the sun have not been successful. For a while it was thought that a planet, Vulcan, lay between Mercury and the sun, but the observations were apparently erroneous. Also, searches for trans-Plutonian planets have been fruitless.

The orbital period of a planet, that is, the time required for one complete circuit about the sun, is easily established by observing the motion of the planet through the sky. In this way, for example, we find that the period of Mars is 687 days, or about 1.9 earth years. In the case of the earth the orbital period, one year, was determined by watching the apparent motion of the sun projected against the background of stars, as discussed earlier (page 70). Given the periods of 1.9 and 1.0 years, we can find relative sizes of the orbits of earth and Mars using Kepler's Third Law (page 87). The average distance of Mars from the sun is found in this way to be 1.5 times the average earth-sun distance of one *astronomical unit.* Table 7-1 (down to the double line) gives the approximate orbital periods and average planet-sun distances (from the Third Law) for the earth and the five other planets that are visible to the unaided eye and thus were known in antiquity. The more recently discovered planets are listed below the double line.

In 1772, J. D. Titius found an intriguing numerical relationship which he published in a little-noticed footnote. It later gained prominence through the work of J. E. Bode in Berlin, and has become known as *Bode's Law.* The recipe is as follows: consider the simple series 0, 3, 6, 12, 24, 48, 96, 192, etc. Except for 0 and 3, the numbers in this sequence are obtained by

Table 7-1 *Approximate Orbital Periods and Average Planet-Sun Distances*

PLANET	ORBITAL PERIOD (EARTH YEARS)	AVERAGE PLANET-SUN DISTANCE (ASTRONOMICAL UNITS)	BODE'S LAW NUMBERS
		VISIBLE TO UNAIDED EYE	
Mercury	0.24	0.39	0.4
Venus	0.62	0.72	0.7
Earth	1.00	1.00	1.00
Mars	1.9	1.5	1.6
			2.8
Jupiter	11.9	5.2	5.2
Saturn	29.5	9.5	10.0
		NOT KNOWN BEFORE INVENTION OF TELESCOPE	
Uranus	84	19.2	19.6
Neptune	165	30.1	38.8
Pluto	248	39.4	77.2

doubling the preceding number. If we then add 4 to each number in the series, we obtain: 4, 7, 10, 16, 28, 52, 100, 196, etc. If we continue by dividing each of these numbers by 10, we now obtain the series of numbers in the last column of Table 7-1. Bode's Law certainly gave a reasonable representation for the distances of the six planets known in 1772, and it also predicted something beyond—at 19.6 a.u.

HERSCHEL AND URANUS

In our age of training and specialization, a Ph.D. degree is usually required for full membership in the American Astronomical Society, and is, by itself, insufficient qualification for participation in the International Astronomical Union, and it is also widely held that the physicist must make his mark before the age of 35. Astronomy and other sciences are fairly well-paying occupations today, although a male student may need a deferment or a suitable draft lottery number to complete his training before he can enter one of them. The situation of the scientist was quite different in the 18th century, and it seems almost fantastic to us that one of the handful of astronomers throughout history who has been privileged to discover an unknown planet, a fundamental innovator in telescope design, a leading observer of nebulae, and the first man to apply statistics to the study of our galaxy, was a professional musician who had crossed international borders to escape military service, and who first took up astronomy as a hobby when he was approaching middle age! On March 13, 1781, William Herschel was looking through a reflecting telescope of his own manufacture (as mentioned in Chapter 6) and noted a faint "star" that differed from the others in that it seemed to have a perceptible diameter, like the head of a faint comet, instead of appearing as just a point of light. When he increased the magnification, the stars still appeared as points (since they were unresolved), but the unusual image increased in size. It was being resolved. Repeating the observations on subsequent nights, he discerned a slow movement of this object across the background of distant stars. Mathematicians applied the rules of Newtonian mechanics to Herschel's observations and discovered that the newly found object was not moving in the elongated orbit of a comet (such as Halley's; see page 190). Situated at 19.2 a.u. from the sun, twice the distance of Saturn, and with an orbital period of 84 years, *Uranus* was moving in a nearly circular elliptical orbit. It was indeed a planet, the first new world discovered since antiquity. The occasion was fortunate for Herschel and for astronomy. In recognition of this accomplishment, King George III awarded him an observatory and a stipend that enabled him to concentrate on research.

The discovery of Uranus had come nine years after Titius published his original formulation of Bode's Law. Uranus' average solar distance of 19.2 a.u. was within 2 per cent of the prediction of 19.6 a.u. It is unique among the planets in that its axis of rotation lies nearly in the plane of its orbit (see Appendix 5).

THE NEPTUNE EPISODE

After the discovery of Uranus, old records were examined for pre-discovery observations of the planet. Sure enough, it had been recorded as a star on many occasions during the 90 years prior to its discovery (including a naked eye sighting by Tycho). Immediately, however, difficulties arose with computing its orbit. An ellipse that fit one set of observations did not seem to accurately predict the future motion of the planet. The discrepancy between the observed and predicted positions grew worse as time passed.

A first attempt at a solution was to discard the older observations on the grounds that they were likely to be in error. A new orbit for Uranus was calculated in 1821, but the problem had been merely postponed and not solved. The planet Uranus systematically departed from the improved orbit and the discrepancy continued to increase.

The problem could then be resolved in at least two ways. One approach questioned the validity of Newton's law of universal gravitation under certain circumstances. Perhaps it must be modified at large distances from the sun. If so, understanding the universe beyond the solar system might prove to be extremely difficult. An alternative to this undesirable solution was to postulate that the discrepancies in the orbit of Uranus were due to the gravitational attraction of another, as yet undiscovered, planet beyond Uranus. The motion of Uranus would, of course, be controlled primarily by the sun, but the presence of another planet would *perturb* the orbit and cause the observed discrepancies.

This problem was tackled in 1841 by John Adams, a 22-year-old student at Cambridge University. His solution was, to a certain extent, one of trial and error. There were essentially two unknowns—the size of the hypothetical planet's orbit and its mass. The procedure was then to try different values of the orbit size and planetary mass until a combination was found that could account for the observed perturbations of Uranus. Other information could be used as a guide. When the discrepancies between the predicted and observed positions of Uranus were largest, Uranus and the unknown planet were probably closest together so that the gravitational perturbations were greatest. The relative directions of the two planets could be inferred by noting whether Uranus was being accelerated or retarded with respect to its nominal orbit.

After a great deal of work, Adams had a satisfactory solution. His new planet was at a distance of 36 a.u. from the sun and had a mass about 25 times that of the earth. This work was completed in October of 1845 and Adams took his computations to an astronomer at Cambridge University in the hope that the hypothetical planet could be observed. The Cambridge astronomer was "not interested" and passed the buck to the Astronomer Royal at Greenwich who was "not interested" either. We have placed quotation marks around the words "not interested" to point up a common situation. Almost any astronomer would be delighted to discover a new planet, but then, as now, scientists were besieged with new theories and revolutionary discoveries. Even when they originate with established authorities in the field, a fair amount turn out to be wrong, and those that come from unknown persons are often classed as crank letters or bothersome inquiries. The astronomers who ignored Adams' requests are considered to be the villains of this story, but the reader should bear in mind the justification for at least initial skepticism.

The same theoretical problem was tackled by Urbain Leverrier in France, and by the summer of 1846 he had presented his results to the French Academy. By August, the news had made its way back to England and a search was begun and mishandled. Despite the close agreement between the predictions of Adams and Leverrier, the survey was performed in a most leisurely fashion. The missing planet was actually seen twice by an astronomer in Cambridge during the month of August, 1846, but it was unrecognized! Meanwhile, Leverrier sent his predicted position to the Berlin Observatory because star charts for this region had just been prepared there (remember that neither the B.D., page 165, nor any other comprehensive star survey existed at that time). The location was examined on September 23, and an object apparently having a disk was noticed. When the same field was examined on the next night, the object had moved. Thus Neptune was discovered. Detailed observations of this planet and its satellites show that Neptune is 30 a.u. from the sun and has a mass of about 18 times that of the earth (somewhat less than predicted).

The research by Adams had not been published when the discovery of Neptune, made on the basis of Leverrier's work, was announced. When Adams' claim was advanced, the French regarded it as an attempt to steal Leverrier's discovery. Bitter words were exchanged and it was only after some years that calmer heads prevailed. The predictions of Adams and Leverrier were remarkably similar and were made entirely independently of each other. Both of these men are therefore usually regarded as the discoverers of Neptune.

Neptune is, in fact, closer to Uranus and to the sun than predicted. Its closeness is partially balanced by its smaller mass; this combination produces approximately the same gravitational perturbation. Although Adams and Leverrier had calculated the orbit and mass of a planet that would produce the observed perturbations on Uranus, there was no guarantee that their solution was the only one possible. The triumph was a substantial one; a new planet had been discovered using Newton's theory. Further probing of the universe using these laws was now regarded as very promising.

THE SEARCH FOR PLANET X

Following the discovery of Neptune, investigation of the outer solar system was continued because some small, but persistent, discrepancies existed in the orbits of both Uranus and Neptune. This suggested that another planet might lie beyond Neptune.

In 1906, Percival Lowell began a search for the trans-Neptunian planet at the observatory in Flagstaff, Arizona, that now bears his name. In retrospect, it appears that the sought-for planet was slightly south of the region recorded on Lowell's closest photograph, and so it remained undiscovered for some time. Meanwhile, Lowell published his calculations of the likely properties of the hypothetical planet.

Although Lowell died in 1916, the search continued. The great difficulty was that thousands of stars appeared on each plate taken in search of the new planet. Plates of a given region taken at different times would show these numerous, essentially stationary, star images as well as (hopefully) the trans-Neptunian planet, which would show movement (Figure 7-1).

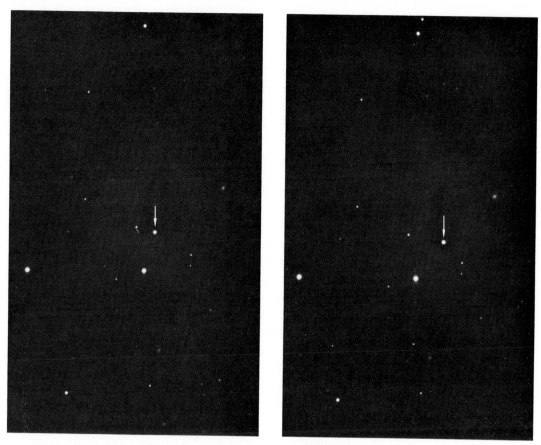

7-1 *Photographs of Pluto taken one day apart with the 200-inch telescope, showing Pluto's motion with respect to the stars. (Courtesy of the Hale Observatories.)*

No appreciable disk was expected in view of Planet X's great distance. To examine large numbers of plates without mechanical aid thus would be hopeless. A device called the *blink comparator* simplified the problem. Each of the two plates to be compared is mounted in the instrument, so that both can be examined through a common viewing device, but only one at a time. The instrument can be adjusted so that the images of the stars appear in the same locations on the two plates; then the view field is rapidly switched back and forth from one plate to the other. When this is done, any object that has shifted its position between the time the two plates were taken appears to jump back and forth while the stars stay fixed.

By 1919, W. H. Pickering made independent calculations of the likely position and brightness of Planet X, and photographs were taken at Mount Wilson Observatory to check his prediction. However, only the parts of the plates that were most likely to contain Planet X (according to the calculations) were carefully examined. And so it was missed because, although actually photographed, it was not in any of the carefully searched areas. In 1929, 23-year-old Clyde Tombaugh joined the Lowell Observatory staff and went to work with a new 13-inch telescope, especially designed to record sharp images on large (14 × 17-inch) plates. Using one-hour exposures, he began to systematically survey the predicted region in the constellation Gemini and other possible but less likely locations. Experience with the blink comparator showed that there were always some possible planets on a plate pair, usually due to defects in one plate or the other. So Tombaugh took *three* plates of each region. Then, finding a possible planet by blinking two of the plates, he could immediately check its reality on the third plate of the same area. Depending on how far the photographed region was from the Milky Way, a given plate contained from 50,000 to 300,000 star images, and it took Tombaugh several days to complete the blink comparison of a plate pair. There were several thousand false alarms that were eliminated by the third-plate technique. But finally, he was studying two plates of the region near the star delta Geminorum when,

> . . . one-fourth of the way through this pair of plates, on the afternoon of February 18, 1930, I suddenly came upon the image of Pluto! The experience was an intense thrill, because the nature of the object was apparent at first sight. The shift in position between January 23 and 29 was about right for an object a billion miles beyond Neptune's orbit.
>
> In all the two million stars examined thus far, nothing had been found that was as promising as this object.

The planet was named Pluto for the ancient Greek god of the underworld and also because the first two letters were Percival Lowell's initials. Because the gravitational perturbations caused by Pluto are very small, its mass is not known very accurately, but it appears to be smaller than assumed by the searchers and hence its discovery was probably fortuitous. Its disk has only been resolved with the 200-inch telescope at Mount Palomar. Although Pluto is 39.5 a.u. from the sun on the average, it has a noticeably elliptical orbit, part of which lies within the orbit of Neptune (Figure 7-2) and, in fact, Pluto is now in this part of its orbit.

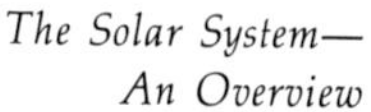

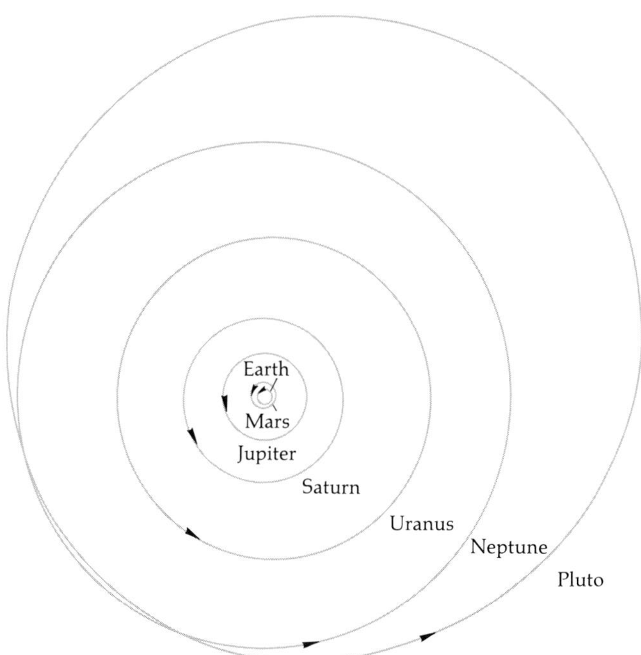

7-2 *Orbits of Neptune, Pluto, and other outer planets as plotted on the plane of the ecliptic. Actually, there are slight tilts between the planes of the different orbits. (After* The Solar System Beyond Neptune *by Owen Gingerich. Copyright © 1959 by Scientific American, Inc. All rights reserved.)*

ASTEROIDS

A review of the orbits of the nine major planets and their relation to Bode's Law (Table 7-1) can now be made. Except for the gap at 2.8 a.u., the law fits rather well in all but the outermost reaches of the solar system. But if Bode's Law were correct, then there ought to be a planet out at 2.8 a.u. from the sun. In 1801, such a planet was discovered by the Italian monk Giuseppe Piazzi. Named *Ceres* after the guardian goddess of Sicily, the planet's mean solar distance was indeed 2.8 a.u. Things seemed to be falling into place, but then a year later, *Pallas* was discovered, also near 2.8 a.u., and in 1804 *Juno* (2.7 a.u.) was found. *Vesta* (2.4 a.u.) was discovered in 1807. Now there seemed to be an embarrassment of riches—four planets where one was expected! In fact, all four are tiny; each is less than 800 kilometers (500 miles) in diameter. Now referred to as *asteroids* or minor planets, several thousand have been identified, most of them located in the *asteroid belt* between Mars and Jupiter, and most of them much smaller than the first four discovered. The appearance of asteroids on photographs is illustrated in Figure 7-3.

The success of Bode's Law in predicting the distances of Uranus and the asteroids from the sun raised the question of whether the Law was really nothing but a numerical curiosity. Perhaps it does have a physical explanation. Bode's Law predicted only one planet where the thousands of asteroids occur. On the other hand, the asteroids (their total mass is believed to be about one-tenth of one per cent of the earth's mass) may represent the debris of one or more somewhat larger ancient planets disrupted by collisions or by some cause that we can only guess at, or they may have somehow formed as the small bodies that they are today. Recent studies do suggest a sound scientific basis for Bode's Law and imply that the spacing of the planets is not accidental. We will return to this subject in Chapter 11, when we discuss the origin and evolution of the solar system.

7-3 *Asteroids photographed in motion (short lines) against a background of stars. (Yerkes Observatory photograph.)*

THE ASTRONOMICAL UNIT

An important early use of asteroid observations was in the determination of the astronomical unit (a.u.). The distance to an asteroid passing close to the earth could be determined by triangulation from different locations on the earth and the distance so determined was used to scale the solar system.

The distances of the planets from the sun listed in Table 7-1 are all relative distances; that is, they are not given in miles or kilometers, but (from Kepler's Third Law) only in terms of the average sun-earth distance, the astronomical unit. Until the a.u. is determined we are in the position of having an atlas of the solar system that lacks a scale. An analogous situation would be a state map with no mileages marked on it and the calibration statement of (say) "one inch equals ten miles" left off. In both cases, we can establish the scale by determining one distance on the map.

The basic idea is to determine the distance to an object passing close to the earth by direct triangulation from opposite sides of the earth (the principle is shown in Figure 2-3a). Photographs of objects close to the earth taken from different geographical locations, but at the same time, should show the object in projection at different positions among the stars.

Mars was observed in 1862 and 1877, but it has a substantial disk and the position shift was relatively hard to measure. A smaller star-like image would be much better. The asteroid Eros filled the bill, coming to within 0.27 a.u. of the earth in 1900-1901 and to within 0.17 a.u. in 1930–1931. Many observations were made at those times to determine the a.u.

A new, more direct method for calibrating the distance scale of the solar system involves bouncing a radar pulse off a planet, such as Venus. The pulse travels at the speed of light and the distance is determined with great accuracy simply by measuring the time required for the round trip. The modern value for the astronomical unit is 149,600,000 kilometers (92,956,000 miles).

COMETS

Comets are beautiful, frequently unpredictable, and still rather difficult to understand. For centuries they were shrouded in superstition and ignorance. Aristotle held that comets were fiery things in the upper layers of the earth's atmosphere. Their supposed proximity made them likely scapegoats for various calamities, including some of the plagues that swept medieval

Europe. Men sometimes prayed for deliverance from the evil of the comet.

If comets were, in fact, atmospheric phenomena, as was then believed, they would show a parallax, that is, they should have a different position with respect to the stars when viewed from various localities on the earth. The subject was investigated by Tycho Brahe, who observed the comet of 1577. Tycho could not find any evidence for parallax and concluded that the comet was quite distant, at least beyond the moon (60 earth radii away). These observations should have established the comets as *bona fide* solar system objects but, as usual, not everyone was convinced.

Observations of Halley's comet eventually led to the conclusion that comets were indeed solar system objects subject to the laws of motion, just like the planets. Newton's laws seem very reasonable to us today, but after all, when they were introduced they only accounted for *known* properties of planetary orbits, as summarized by Kepler. As mentioned in the discussion of the scientific method, the real test of a theory is not how well it fits the existing facts, but rather, how well it predicts as yet unobserved effects.

The Florentine astronomer, Paolo Toscanelli, a contemporary and correspondent of Columbus, was foremost in scientific observations of the comet of 1456, which had a tail that extended across one-third of the sky, causing fear and consternation in Europe.* Centuries later, Toscanelli's estimates of the changing position of the comet among the stars were used to determine its orbit. Similar bright comets were observed in 1531, in 1607, and in 1682. In England, Edmund Halley studied the observations of the comet of 1682 to test one of Newton's conclusions. According to Newton's theory of gravitation, comets could travel in orbits of closed form (ellipses) or in open orbits (parabolae and hyperbolae). Halley's study showed that it must have an elliptical orbit and that based on the theory of gravitation, it should have an orbital period of about 75 years. Looking back through the old records of when comets had appeared, he noted that

$$
\begin{aligned}
1682\text{–}1607 &= 75 \text{ years}\\
1607\text{–}1531 &= 76 \text{ years}\\
1531\text{–}1456 &= 75 \text{ years}
\end{aligned}
$$

In Halley's time it had not previously been suspected that a comet, once gone from sight, returns again. Although Newton's

*It was even blamed for the fall of Constantinople to the Turks, although that event had occurred in 1453!

theory showed that the ellipse, a closed orbit, was a possible orbit for a comet, Newton's personal preference was for an open orbit which provides only one trip past the sun. Thus, when Halley predicted that the comet of 1682 would return in 1758, he was really putting his reputation on the line. Well after his death, his hypothesis was vindicated when the comet indeed returned late in 1758. In fact, the date when the comet reached perihelion (closest approach to the sun) was in accord with the more exact calculations of the French mathematician, Alexis Clairaut, who improved Halley's work by allowing for gravitational perturbations of the comet by Jupiter and Saturn. It was a great triumph for the young science of celestial mechanics, which described motions in space as the result of gravitational forces. Any doubt that may have remained was dispelled by the discovery of Neptune. Halley's triumph also swept away much of the superstitious feeling about comets, as they were now shown to be governed by known physical laws.

Later students, checking archives, have found extensive references to past appearances of what we now call Halley's comet. In the Chinese records in particular there were reports of every one of its apparitions during almost thirteen centuries. However, lacking the necessary theory, and hampered by the fact that one man rarely lives to see two appearances of the comet, the Chinese astronomers had not realized that it was the same one that they continued to record. The records of other cultures also show traces of Halley's comet; for example, its apparition in 1066, on the eve of the Battle of Hastings, is depicted in the Bayeux tapestry (Figure 7-4).

7-4 *Halley's comet as shown on the Bayeux tapestry. (Yerkes Observatory photograph.)*

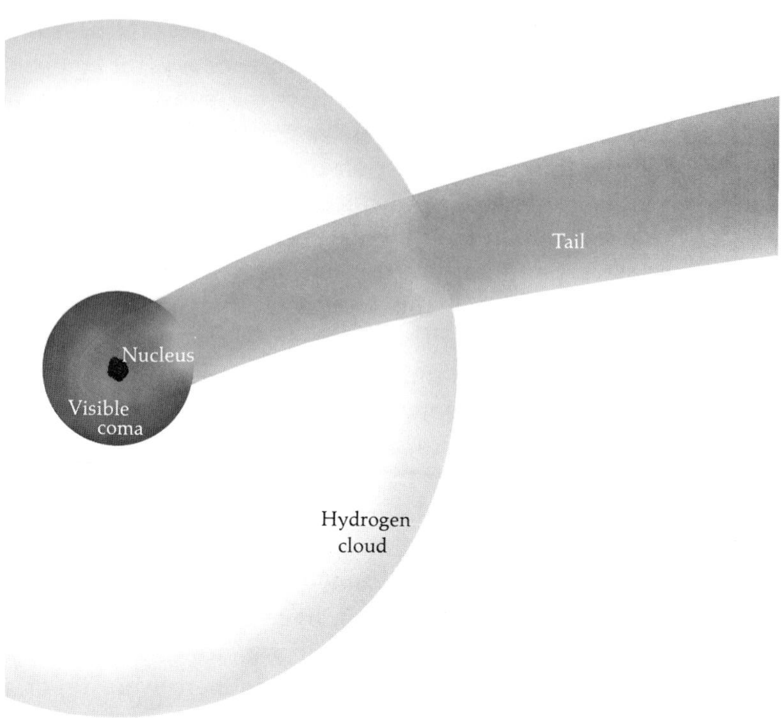

7-5 *The principal parts of a comet.*

The earliest known observation of a comet is, in fact, of Halley's comet in 467 B.C. It last appeared in 1910, and the authors hope that both they and their readers will be around for the next visit in 1986, and that we can all meet on a mountaintop, have a good look at the comet, and raise a toast to the great mathematicians of the past!

The basic components of cometary forms are shown schematically in Figure 7-5. The central *nucleus* is believed to be solid and composed of "ices" (frozen gases, such as water, methane, and ammonia) and dust particles. This structure, which has been likened to a large, dirty snowball, is not observed in every comet. The *coma* is an essentially spherical cloud of gas and dust, centered on the nucleus. The radius of the coma can reach 100,000 kilometers (62,000 miles). The most spectacular part of the comet is the *tail,* which can be as long as 10^8 kilometers (6.2×10^7 miles). In 1970, observations of two bright comets from above the atmosphere disclosed a new feature—a cloud of hydrogen atoms about 10^7 kilometers (6.2×10^6 miles) in diameter (see page 384).

The different types of tails are well illustrated in the photographs of Comet Mrkos (Figure 7-6). The long straight tail with

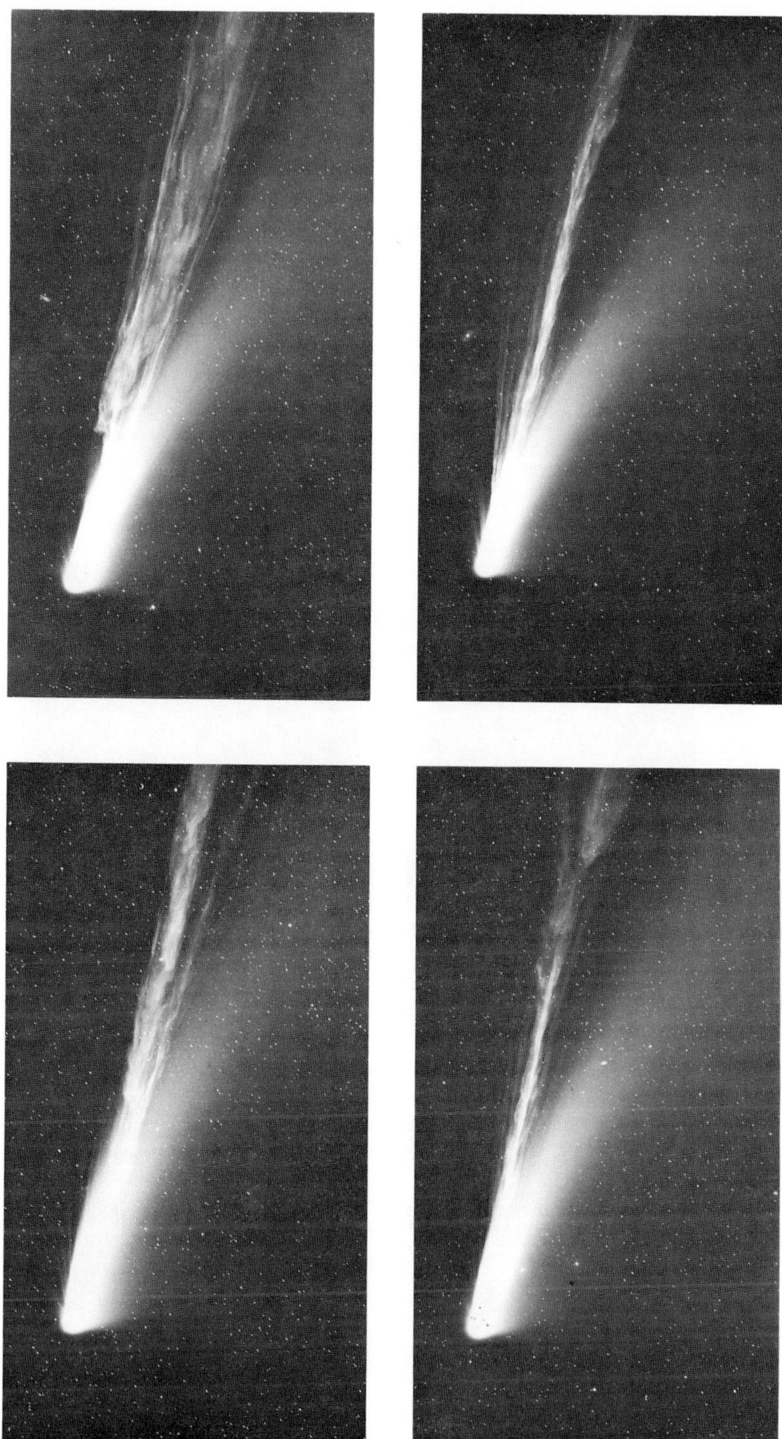

7-6 *Comet Mrkos photographed on four nights in 1957 with the 48-inch Schmidt telescope on Mount Palomar. (Courtesy of the Hale Observatories.)*

a great deal of filamentary structure is composed of ionized molecules (mostly carbon monoxide that has lost one electron, CO^+); this ion tail (by contrast, the coma is not ionized) is pushed away from the sun by a continuous outflowing of gas from the solar corona, called the *solar wind* (Chapter 9). The shorter, smoother, and curved tail is composed of dust particles with sizes of about 10^{-4} centimeter (3.9×10^{-5} inch); these particles are pushed away from the sun by the radiation pressure of sunlight (Chapter 5). The ionized tails are blue (because of the emission of light by the CO^+ molecules), while the dust tails are yellow (because of reflected sunlight). This effect is shown in a color picture of Comet Bennett (Plate 11). The two different types of tails can occur separately or together. A

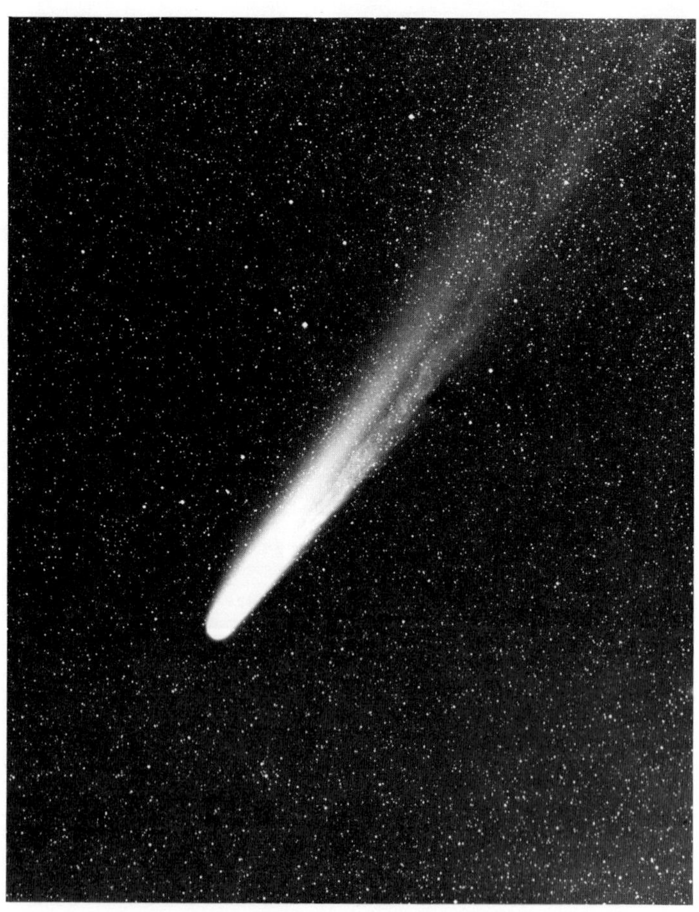

7-7 *Comet Bennett photographed on April 16, 1970. (J. C. Brandt, R. G. Roosen, and S. B. Modali, Goddard Space Flight Center.)* See also Plate 11 following p. 260.

7-8 *Comet Cunningham photographed on December 21, 1940. (Courtesy of the Hale Observatories.)*

comet with an ionized tail only is shown in Figure 7-8. When comets pass close by the sun, the emission of light by atoms (notably by sodium) appears in addition to the molecular light mentioned above. Material in the tails is blown away and lost forever; a large comet, such as Halley's, probably has enough material for about 100 close approaches to the sun.

Little is known about the origin of comets. According to a leading theory, they were produced out beyond the present orbit of Pluto as a by-product of the process that formed the solar system, and a vast swarm (numbering 100 billion or more) of comets exists out there today. Their average distance from the sun is estimated by the Dutch astronomer J. Oort at 150,000 a.u. Since the nearest star is only about twice this far from the sun, it is clear that such remote comets, although orbiting the sun, would be perturbed by other stars. Sometimes these perturbations could enable a comet to escape from the solar system (that is, from a closed orbit around the sun), but occasionally a comet would be perturbed into the central part of the solar system where it could be captured in a highly eccentric* orbit that would make it periodically visible from the earth.

---

*The point of closest approach to the sun is called *perihelion,* that of farthest removal is called *aphelion.* When the perihelion and aphelion distances are nearly equal, so that the elliptical orbit is nearly circular, it is called a low eccentricity orbit; when the distances are very different, so the orbit is cigar-shaped, the eccentricity is high.

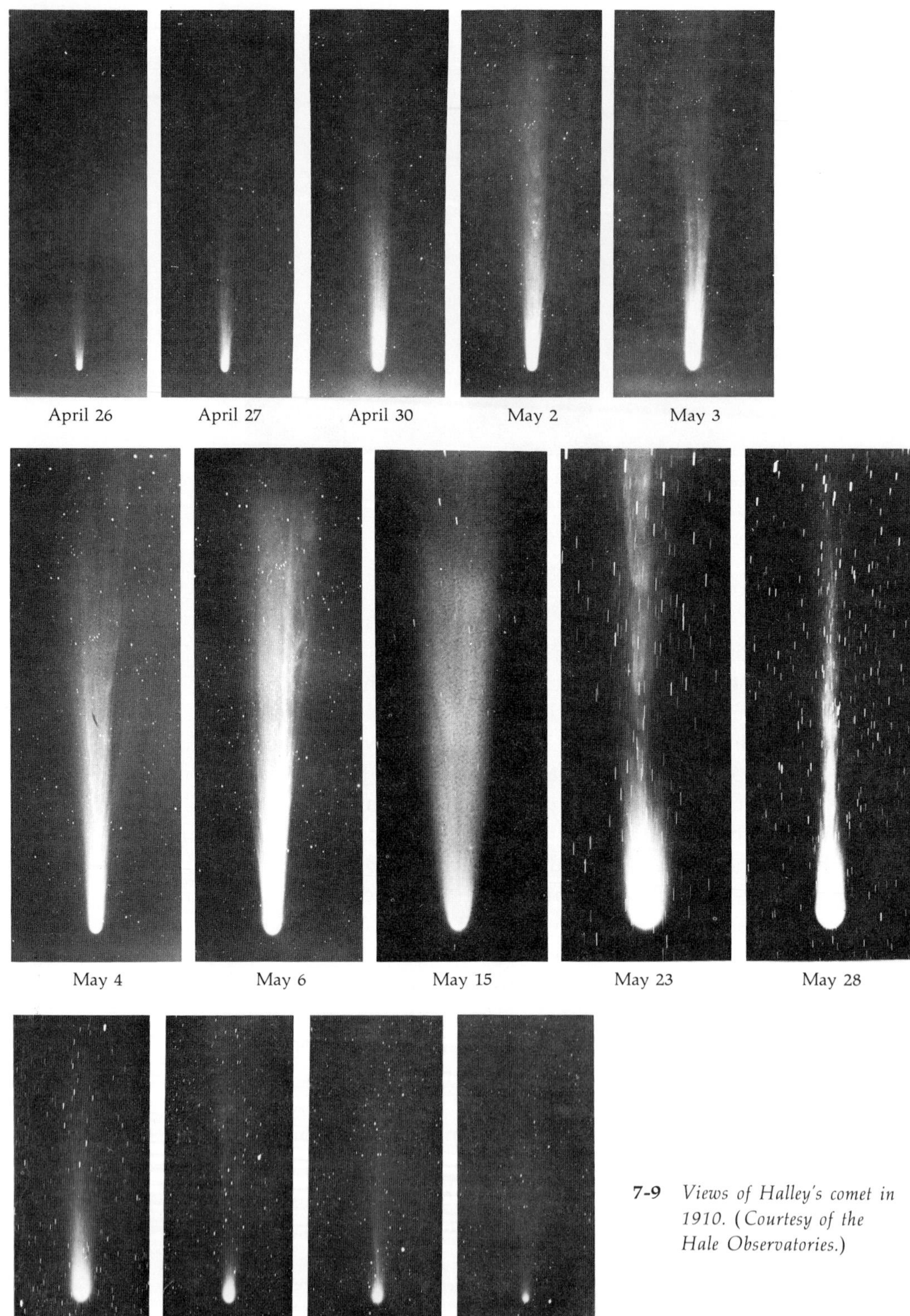

7-9 *Views of Halley's comet in 1910. (Courtesy of the Hale Observatories.)*

As the cometary nucleus approaches the sun, the frozen material is heated by solar radiation and releases gases that form the coma and hydrogen cloud. If ionization (the stripping of outer electrons from atoms) occurs, an ionized tail is formed and pushed away from the sun by the solar wind. As the ices are vaporized, the dust embedded in them is also liberated and blown into the dust tail by the solar radiation pressure. As the comet moves away from the sun, the heating decreases, the liberation of ices and dust ceases, and the nucleus goes into cold storage.

When the comet is so far from the sun that its ices are not vaporized, it no longer has a tail. A new tail forms as the comet returns to the vicinity of the sun. The changing appearance of a comet as it moves in orbit is illustrated (Figure 7-9) with photographs of Halley's comet taken in 1910.

Comet orbits are mostly quite elongated, and their orientations in space are randomly directed, so they may cross the plane of the earth's orbit at quite steep angles. This is very different from the case of the planetary orbits, which are all tilted less than 8 degrees to the earth's orbit (except Pluto at 17 degrees). Occasionally, comets pass close to Jupiter and are perturbed into paths that are nearly circles. These comets have shorter periods of revolution and are observable through a larger fraction of each orbit, as compared with those in elongated orbits like Halley's comet (Figure 7-10).

Bright comets that can easily be seen by the unaided eye are discovered from time to time, frequently by amateur astronomers and sometimes by airline pilots; three were discovered this way during the first half of 1970. Comets are usually named for their discoverers, although the most famous of all is an exception—Halley's comet was named after the man who first suggested that it had a closed orbit and predicted the date of its return.

METEORS AND METEORITES

Meteors are light emission phenomena in the earth's atmosphere, once popularly called shooting stars. For the amateur they are great fun to observe from a comfortable horizontal position under dark country skies.

A photograph of a meteor is shown in Figure 7-11; it appears as a streak of light in the sky. The production of meteor light is caused by rocks (*meteoroids*) that were in orbit around the sun before they entered the earth's atmosphere. At the typical entry

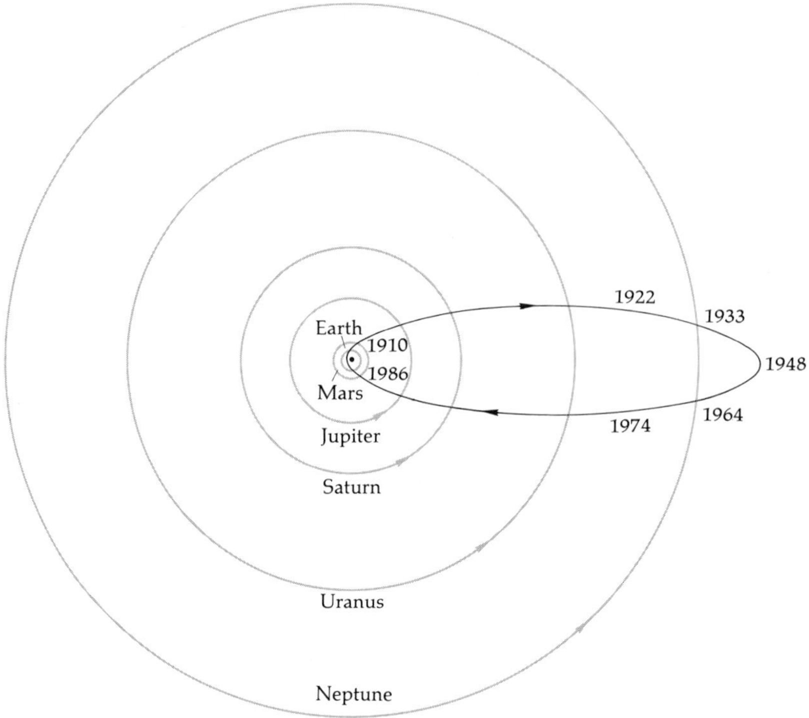

7-10 *Orbit of Halley's comet compared with the orbits of the planets. The planetary orbits are nearly circular; the orbit of Halley's comet is a highly eccentric ellipse and the comet travels around the sun in the direction opposite to that of the planets.*

speeds of 30 kilometers per second (19 miles per second), the rocks are heated by friction with the air molecules.* This heating can melt or actually vaporize the rock; the vast majority of shooting stars are caused by meteoroids no larger than gravel or grains of sand which are fully vaporized and never reach the ground. (Some meteoroids are so tiny that the amount of friction they cause is not enough to make them melt or vaporize. These *micrometeorites* drift down through the atmosphere like dust.) The collisions between a meteoroid and the air molecules also heat the air in the immediate surroundings. This hot gas emits the light we see as the meteor. If the initial meteoroid is sufficiently large, it can survive entry and reach the ground; such rocks of celestial origin found on the earth are called *meteorites* (Figure 7-12). If a meteorite is found immediately after falling to the ground it will be hot to the touch.

---

*This friction process was the basis of the re-entry problem that had to be solved for manned spaceflight.

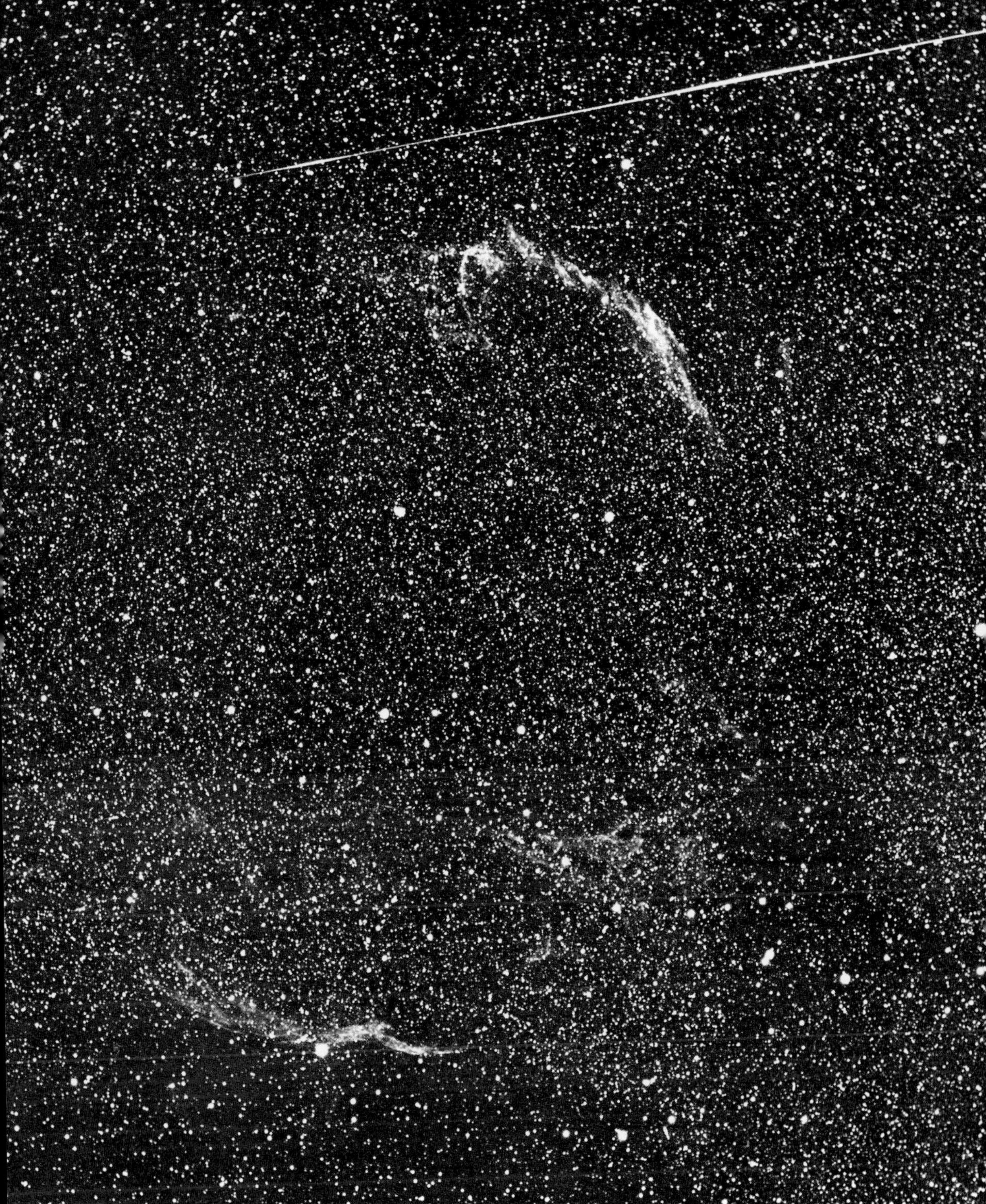

7-11 *A meteor trail seen against a background of stars and a nebula in the constellation Cygnus. (Yerkes Observatory photograph.)*

7-12 *A small meteorite found at the Arizona Meteor Crater. (Yerkes Observatory photograph.)*

Photographs of a meteor from different locations on the earth's surface enable a determination to be made of the meteoroid's orbit in the solar system prior to encountering the earth. Sometimes many are found to have essentially the same orbit—they constitute a meteoroid *stream.* This can be explained by material distributed along an orbit as shown in Figure 7-13. In some cases, the stream orbit is known to resemble the orbit of a comet that was observed in the past but has faded out. Perhaps the stream is the debris remaining from the breakup of the cometary nucleus. When the earth passes through a meteoroid stream, a large number of meteors are seen for a few days (*meteor shower*). The Perseid shower, which occurs each year at about August 12, is one of the most spectacular, with visible meteor rates often reaching 60 per hour. Other meteoroids, termed *sporadic,* appear to have unique orbits, not associated with streams.

The frequency with which we see meteors varies throughout the night; more are seen after midnight. This effect has been known for centuries and has a simple explanation. Unlike the planets, which all go around the sun in the counterclockwise direction (as seen from a point above the earth's North Pole), some meteoroids are going in the same direction as the earth and some are going in the opposite direction. On the side of the earth where time is before midnight, the meteoroids that can enter the atmosphere are going in the same direction and hence must catch up to it. On the side where the time is after midnight, the earth encounters meteoroids that are moving in the opposite direction. They do not need to catch up and are, in fact, swept up by the earth. Many more meteors occur on the morning side because of this effect. This situation is some-

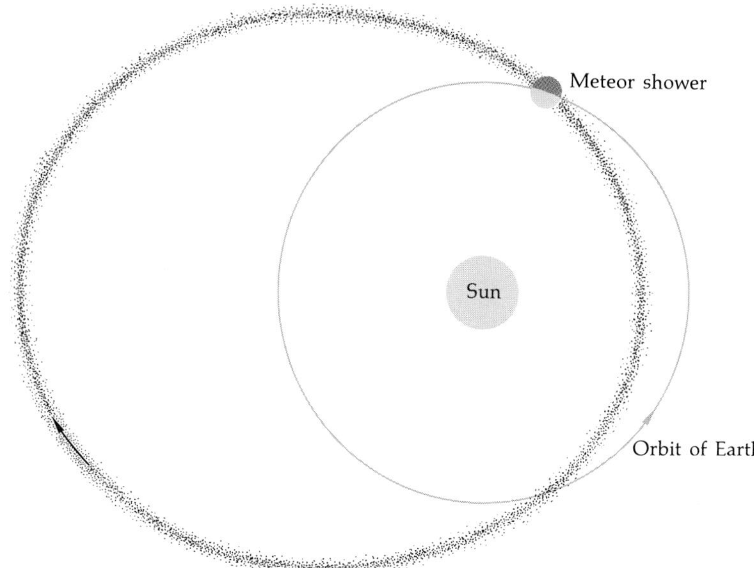

7-13 *Sketch explaining the origin of annual meteor showers. Usually the orbit of the stream particles is tilted with respect to the earth's orbit and there is only one intersection of the two orbits.*

what analogous to that of the hapless motorist who enters the San Diego Freeway via an exit ramp—he is going against the traffic and is more likely to meet a spectacular end than if he were moving with the traffic.

The rocks that actually strike the ground, the meteorites, are of great scientific interest as celestial material that can be analyzed in the laboratory. Some are stony and some are metallic (mostly iron). The reader may recall (Chapter 2) that the oldest known earth rocks are approximately 3.5 billion years old. The age of 4.5 billion years or slightly more that we quoted for the oldest rocks in the solar system was first revealed by studies of radioactive decay in meteorites. Presumably, the environment of the meteorites in interplanetary space has been less destructive than the erosive processes and mountain-building processes on earth. Recently, rocks and soil have been brought to earth from the moon by the crews of several Apollo missions, and by automated Soviet spacecraft. These samples also have been dated by radioactive means. The oldest soils and rocks are about as ancient as the meteorites (Chapter 15).

There are several important reasons for studying meteorites in the laboratory. Historically, the radioactivity dating experiments were of great interest, because they gave our first direct evidence for matter older than the oldest known rocks on earth. Secondly, chemists analyzed meteorites to determine the abundances of the elements in them. By combining information from the meteorites with data on elements detected in the spectrum of the sun and from studies of the material in the crust of the earth, a fair idea of the abundance of the elements in the solar system has been obtained. Clearly, theories of the origin of the solar system will have to explain why the elements are present in these amounts. Finally, just as we have recently been able to examine the lunar rock samples for possible biological material, meteorites have been studied for years in hopes of finding traces of organic matter from outside the earth. Occasionally, such traces seemed to have been discovered, but the evidence was not very convincing. The problem was that they might have been contaminated by organic matter *after* reaching the earth. For example, studies of some meteorites revealed tiny amounts of amino acids,* but the particular assortment of amino acids that was found turned out to be the same as is present in fingerprints. The first convincing evidence came from several fragments that fell near Murchison, Australia, in September, 1969. Eight amino acids were detected in an assortment *not* resembling those found in fingerprints or other products of living matter on earth.† The simplest conclusion was that these amino acids existed in the Murchison meteorite before it struck the earth. This does *not* prove that there is extraterrestrial life—remember that amino acids can form by nonbiological processes, as mentioned in Chapter 3—but it does mean that some of the complex molecules that are needed for life as we know it are present in space.

Occasionally, very large meteoroids that may even have been asteroids have struck the earth, with spectacular effects. The retarding effect of the atmosphere is unimportant for such massive bodies, and so they impact with great force, producing *meteor craters.* The most famous of these is the one in Arizona (Figure 7-14). This crater is 1.2 kilometers (4,000 feet) across and 170 meters (570 feet) deep, with a rim that rises about 50 meters (160 feet) above the surrounding countryside. Attempts have been made to mine the region beneath the crater floor for the primary body, which may have been a solid chunk of iron and nickel; these efforts have been unsuccessful. On the other hand, iron fragments are found throughout the countryside

---

*Living protein is composed of amino acids; see Chapter 3.

†In 1971, amino acids were found in another meteorite.

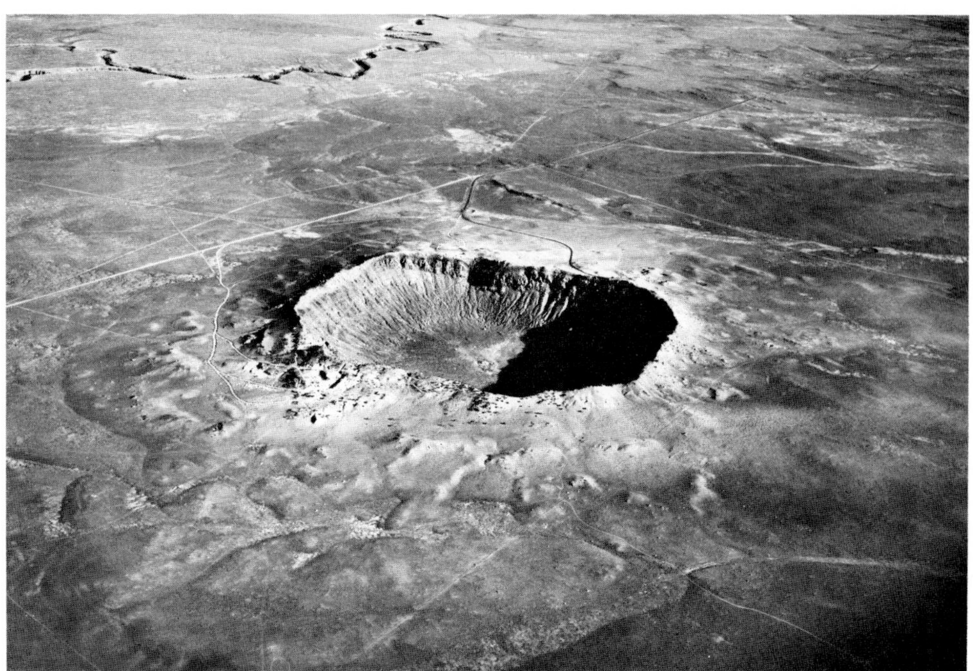

7-14 *Arizona Meteor Crater, located 18 miles west of Winslow, near Route 66 (view to the northwest). (From* Geology Illustrated *by John S. Shelton. W. H. Freeman and Company. Copyright © 1966.)*

surrounding the crater and not elsewhere in the region. Apparently, the meteoroid was disrupted by the impact and these iron fragments are parts of it. Geological evidence indicates that the impact occurred about 20,000 years ago.

About thirty proven meteorite craters are known on earth, and many other round geological features are suspected to be of meteoric origin, including craters as large as the 60 kilometer (37 miles) Manicouagan Crater in Quebec. A small comet or asteroid may have entered the atmosphere and devastated the forest around the Tunguska River, Siberia, in 1908, but no actual fragments or craters were found in the disaster area. In 1947, also in Siberia, the mid-air explosion of a meteorite produced about 100 craters, and a considerable number of iron fragments were recovered.

The identification of certain meteor streams with comet orbits establishes a cometary origin for some meteors; others may come from the asteroid belt between Mars and Jupiter. The relative numbers of meteoroids derived from these sources have not been definitely determined.

SATELLITES

The moon is the sole proven natural satellite of the earth, although tiny natural satellites have sometimes been suspected. It is also the largest satellite in the solar system, not in absolute size but judged in terms of relative sizes of a planet and its satellites. Our rudimentary knowledge of the satellites of the other planets is derived mostly from observations with the telescope. On the other hand, a fair idea of the nature of the lunar surface has developed in recent years through manned and automated space probes. This is a separate subject in itself and we defer a description of the physical properties of the moon until we discuss space exploration in Chapter 15.

Studies of the satellites of other planets began in 1610 with Galileo's discovery of the four major moons of Jupiter (Figures 6-3 and 7-15). These are now called the Galilean satellites and in order of their distance from Jupiter are Io, Europa, Ganymede (Figure 8-14), and Callisto. They are all roughly the size of our moon but, considered next to Jupiter, they are tiny, whereas our moon has about one-fourth the earth's diameter. Observations of the Galilean satellites of Jupiter led to the first successful attempt to measure the speed of light (Chapter 5). The discovery of satellites of other planets contributed greatly to the idea of a "plurality of worlds."

The planets Mercury and Venus have no known satellites, but Mars has two (Deimos and Phobos) which were discovered in 1877 by Asaph Hall of the U.S. Naval Observatory in Washington, D.C. They both are very small and very close to Mars. In addition, Phobos appears to be slowly spiraling in toward Mars—possibly the result of atmospheric resistance or *drag.* This happens to artificial satellites of the earth and in fact the Soviet astrophysicist I. S. Shklovskii once speculated that Phobos and Deimos may be artificial satellites placed in orbit

7-15 *Jupiter and the four Galilean moons showing the orbital motions of the satellites. The image of Jupiter was overexposed, in order to record the dimmer satellites. (Yerkes Observatory photograph.)*

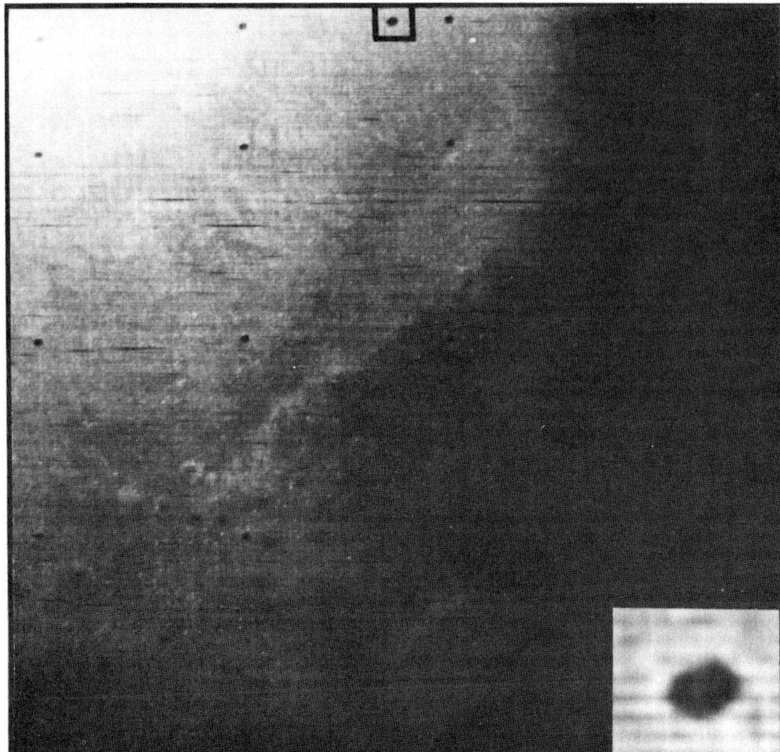

7-16 *The* Mariner 7 *spacecraft obtained this photograph of Mars' satellite Phobos. The elongated disk is more characteristic of asteroids than satellites and has prompted speculation that Phobos could be a captured asteroid. (National Aeronautics and Space Administration.)*

by an ancient Martian civilization. Recent observations by Mariner space probes show that both are oblong rather than spherical (see frontispiece and Figure 7-16).

Jupiter has eight smaller moons, in addition to four large satellites observed by Galileo. Four of these were discovered by the American astronomer, Seth Nicholson; the last one was found in 1951. At least one of Jupiter's moons (Io) seems to be related to the radio bursts observed from that planet (Chapter 8). The chief use of observations of these satellites and those of the other planets has been to determine their orbital properties, and thereby to measure the masses of their respective planets. Four of the small outer moons have retrograde orbits.

Saturn has ten moons, one of which (Janus) was discovered as recently as 1966 by the French astronomer, A. Dollfus. The biggest, Titan, is probably larger than the planet Mercury, and

7-17 *Saturn as photographed by the 100-inch telescope on Mount Wilson. (Courtesy of the Hale Observatories.)*

is noteworthy as the only satellite in the solar system known to have an atmosphere; molecular bands of the gas methane appear in its spectrum. The most distant moon, Phoebe, has a retrograde orbit.

Saturn also has a large number of satellites in the form of particles that make up the beautiful rings (Figure 7-17). Spectroscopic observations interpreted through the Doppler effect show that the particles in the rings obey Kepler's Laws for their various distances from Saturn. Thus, the rings are neither solid nor do they revolve as a rigid body; the particles are probably ices. It has been suggested that the rings are the remains of an ancient moon that disrupted. They lie in the equatorial plane of Saturn and are exceptionally thin (less than 10 kilometers, or 6 miles, thick); in fact, when viewed edge on, they are nearly invisible.

Uranus and Neptune round out our survey of planets with satellites; they have five and two moons, respectively. Uranus and some of its satellites are shown in Figure 7-18; the satellite orbits lie roughly in the plane of Uranus' equator, and hence are nearly perpendicular to its orbital plane. Their orbital motions take place in the same sense as the rotation of Uranus. Neptune's two satellites, Nereid and Triton, have normal and retrograde orbits, respectively.

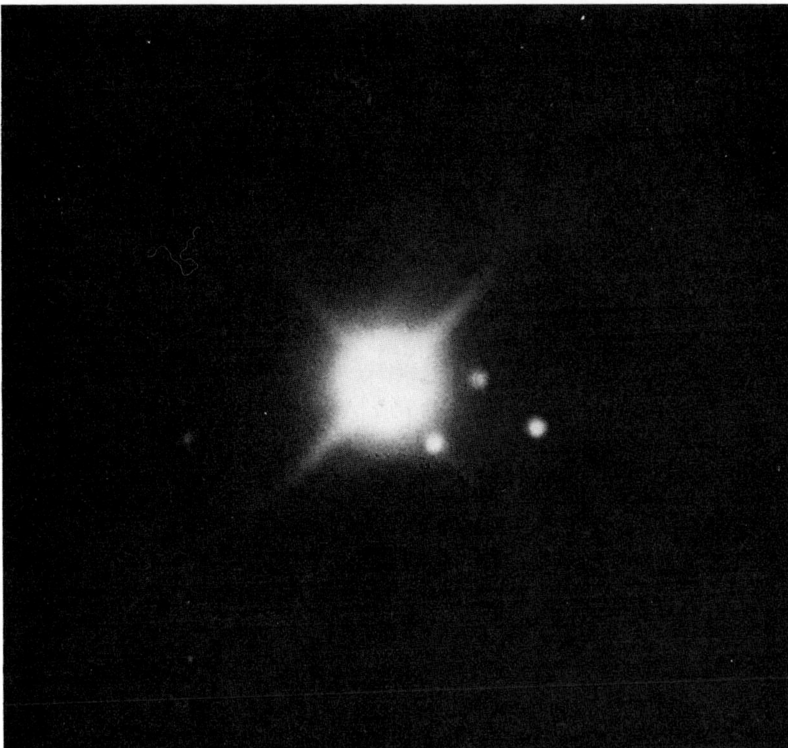

7-18 *The planet Uranus and some of its satellites. (Lick Observatory photograph.)*

Pluto has no known satellites, but astronomers are searching for some now, because observations of a satellite would allow us to use Newtonian theory and determine the mass of Pluto much more accurately than we can do at present.

With the exception of our moon, very little is known about the physical properties of the satellites. Some of the data are summarized in Appendix 5.

SCALES AND REGULARITIES

What are the facts about the large-scale structure and motions in the solar system? Some of the properties of the planets can be explained as results of atmospheric evolution (Chapter 8). The scale of the planetary orbits is shown in Figure 7-19. We have already noted the regularity of their spacing, as expressed by Bode's Law.

Of great interest in studying these broad properties of the

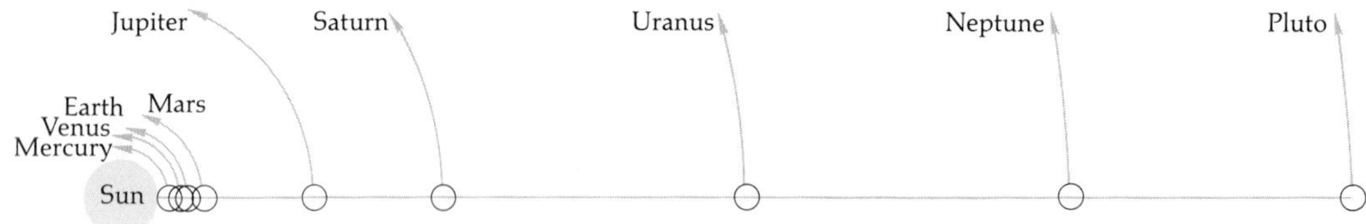

7-19 *The planetary orbits to scale.*

solar system is the regularity in the planetary motions. All the planets revolve around the sun in the same direction that the sun rotates (it is in the counterclockwise sense, as seen from a point north of the earth's orbit). At least six of the planets also rotate on their axes in this same sense; Venus rotates the opposite way (retrograde direction), Uranus rotates about an axis nearly perpendicular to its orbital motion and in the plane of its orbit so that it appears to roll along the orbit, and little is known with certainty about Pluto's rotation.* Also, as mentioned before, the orbits of the planets are nearly in the same plane. The chief exceptions are Pluto and Mercury, with orbits tilted at about 17 degrees and 7 degrees, respectively, to that of the earth. Thus, the planetary regularities are: (a) regularities of motion; (b) regularities of spacing; (c) orbits nearly in the same plane; and (d) most orbits nearly circular.

Some other fundamental properties of the solar system appear to be: (1) cometary orbits occur at all orientations to the plane of the earth's orbit; (2) possible large cloud of comets beyond the orbit of Pluto; (3) asteroids, chiefly between the orbits of Mars and Jupiter; (4) satellite orbits lie roughly in the equatorial planes of their respective planets, and the great majority of the satellites move along their orbits in the same sense as the rotation of their planets.

These general properties of the solar system provide much food for thought. A comprehensive theory of the origin of the solar system would have to account for each of them and would indeed be an impressive achievement.

---

*The light from Pluto appears to vary in brightness in a repeatable way. This may be due to different surface features that reflect sunlight in varying amounts as Pluto rotates. If so, the period is about six days.

8

The Planets and Their Atmospheres

Once it was recognized that the other planets in the solar system are worlds in the same sense as the earth, questions arose. How similar are they? Could life exist on Mars or some other planet? Investigations of the planetary atmospheres, largely by analysis of the spectra of the planets, has revealed a striking fact: no other planet has an atmosphere like the earth's. On the other hand, we have already mentioned (Chapter 3) the likelihood that at an early stage in its history, the earth was surrounded by a primitive atmosphere of very different composition. Indeed, we believe that it was rather similar to the atmospheres of the larger planets (Jupiter, Saturn, Uranus, Neptune). Thus, these planets are not only of interest in their own right, they are also of concern because they may bear clues to the history of the earth. In recent years, our interest in the planets has been renewed and increased by the possibility of making close-up investigations with space probes. Probes have already sent back valuable data from the vicinities of Mars and Venus (the Mars results are discussed in Chapter 15), and it now appears likely that automated spacecraft will be sent to study some of the more distant planets as well.

The basic planetary facts—sizes, masses, and densities—are obtainable in a straightforward manner. The masses of planets that have satellites are determined by applying the laws of motion developed by Kepler and Newton to the observed satellite orbits. This method has given us accurate values of the masses of Mars, Jupiter, Saturn, Uranus, and Neptune. Very little is known about Pluto and it is omitted from most of the discussion in this chapter. Before space probes became available, the masses of Venus and Mercury (which have no known moons) could only be ascertained by the relatively difficult procedure of studying their gravitational perturbations on the orbits of the other planets. The mass of Venus has now been more accurately determined by analysis of its effect on the motions of the Mariner space probes that have traveled to its vicinity.

The measurement of planetary sizes is straightforward in principle and they are shown to scale in Figure 8–1. The size refers to the visible or apparent surface. Mercury and Mars have a solid visible surface, but Venus, Jupiter, Saturn, Uranus, and Neptune have thick, cloudy atmospheres, and thus little direct evidence concerning the surfaces of these planets is available. The sizes are calculated by trigonometry, using the measured angular sizes of the planets and the known scale of distances in the solar system. In the case of Venus, a more accurate diameter and some idea of the surface features have been determined by radar mapping (Figure 8–4).

Inspection of Figure 8-1 immediately suggests two basic classes of planet—*terrestrial* (earth, Mercury, Mars, Venus)* and *Jovian* (Jupiter, Saturn, Uranus, and Neptune). The diameters of the Jovian ("resembling Jupiter") planets in round numbers range from about 47,000 kilometers (29,000 miles) to almost 143,000 kilometers (89,000 miles), while the four planets definitely classified as terrestrial have diameters from about 4,000 kilometers (3,000 miles) to almost 12,800 kilometers (7,900 miles).† However, the key distinction is not one of size, but of *density.* (Density is defined as the ratio of mass to volume.) The terrestrial planets must contain much rock and metals, because their densities range between 4.0 and 5.5 times the density of water, in contrast to the Jovian planets, which have densities between 0.7 and 1.7 times that of water. Clearly, the cloud-

*Pluto may also be a terrestrial planet.

†Data on the radii and other physical properties of the planets are tabulated in Appendix 5.

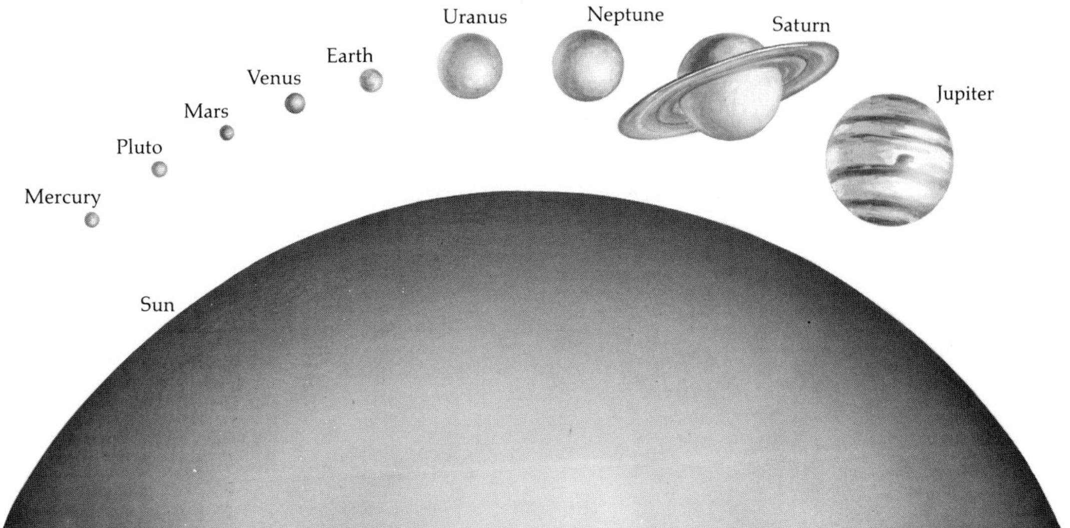

8-1 *Planetary sizes to scale. (After* Principles of Geology, *3d ed., by J. Gilluly, A. C. Waters, and A. O. Woodford. W. H. Freeman and Company. Copyright © 1968.)*

shrouded Jovian planets have a higher fraction of lighter matter than do the terrestrial planets. It would be of great interest to learn the diameters of the solid bodies of the Jovian planets, and to determine the density of this solid matter.

To speculate on the origin of the Jovian planets, we must anticipate some ideas (Chapter 11) concerning the birth of stars and the origin of the solar system. According to current thinking, the sun and the rest of the solar system formed from the *solar nebula,* a contracting gas cloud, flattened and rotating, with the same chemical composition that is found in the surface layers of the sun today: mostly hydrogen and helium with trace amounts of the heavier elements, including carbon, nitrogen, oxygen, and iron. The precise way in which the sun condensed at the center of this cloud, leaving the planets, satellites, and asteroids distributed through circumsolar space, perhaps surrounded at a very great distance by an enormous cloud of comets, is far from certain and indeed a subject of considerable debate. If the composition of the solar nebula was uniform, then the original planetary bodies or *protoplanets* would have begun with the same chemical composition as the sun.

The American geophysicist Harrison Brown has divided the substances that probably existed in the solar nebula into three reasonably distinct groups (*gases, ices, earths*) based on their melting points. The *gases* consisted of hydrogen and the noble gases (mostly helium). They are all characterized by very low melting temperatures—close to absolute zero. The *ices* were composed of carbon, nitrogen, and oxygen in various combinations with hydrogen—for example, ammonia (NH_3), methane (CH_4), and water (H_2O)—with melting points of roughly 273° K or less. The *earths* were made up of silicon, magnesium, iron, and the like, often in chemical combination with oxygen. These have very high melting points, around 2,000° K. Judging from the present composition of the sun's surface layers, the gases constituted 99 per cent of the mass of the solar nebula, the elements that combined with hydrogen to make ices made up about one per cent of the mass, and the elements that made up the earths amounted to only some 0.2 per cent.

The estimated abundances in the solar nebula (and hence, presumably, in the protoplanets, as described above) can be contrasted with the present planetary abundances, derived by studies of the earth and of the spectra of other planets and some comets. These compositions are given in Table 8-1. (Some of the rows in Table 8-1 add up to slightly over 100 per cent because the numbers are approximate.)

Thus we can see that three kinds of bodies exist now. The two largest planets, Jupiter and Saturn, are mostly gases and not unlike the sun in composition. The outer planets, Uranus and Neptune, and the comets are largely ices. The terrestrial planets are predominantly composed of earths. If all were in fact born in a nebula of uniform composition, then the histories of these three kinds of object must be quite different.

Planets tend to lose matter from their atmospheres by the process of *evaporation,* which we discuss in more detail on page 211. Evaporation proceeds more rapidly for planets with higher atmospheric temperatures and lower surface gravities, and more slowly for those with larger surface gravities and lower temperatures. We expect that planets located nearer to the sun will have higher atmospheric temperatures and *vice versa.* Mercury has a low surface gravity and is close to the sun. It undoubtedly cannot retain any substantial amount of atmosphere, and observations confirm this. Similarly, Mars, earth, and Venus have lost nearly all of their lighter (gas and ice) material and are composed almost entirely of earths. (Note that the total mass of our atmosphere is negligible compared to that of the solid earth.) The Jovian planets are much farther from the sun; thus the combination of their high gravity and low temperature allowed them to retain the lighter material in the form of gases and ices. In particular, the distances of Uranus and Neptune from the sun are sufficient to ensure that there is rather little solar heating and thus the lighter elements were more likely

Table 8-1 *Approximate Composition by Mass in the Solar System*

	HYDROGEN AND NOBLE GASES	"ICE SUBSTANCES"	"EARTHS"
Terrestrial planets	≪ 1%*	< 1%*	100%
Jupiter	90%	10%	<1%
Saturn	70%	30%	<1%
Uranus	10%	80%	10%
Neptune	10%	70%	20%
Comets	<1%	85%	15%

*The symbol < means "less than"; ≪ means "much less than".

to remain in the form of ices on these two planets. In this respect, Uranus and Neptune are similar to the comets.

The details of the solid surfaces of the planets have been observed only for Mercury, Mars, and the earth. Pluto is too far away; all of the other planets are covered by cloud. Mercury is difficult to observe because of its small size; nevertheless, its rotation was determined by the Italian astronomer Giovanni Schiaperelli in the late nineteenth century. He made drawings of streaky markings on Mercury's surface, and comparing the drawings made at different times, he concluded that Mercury rotates on its axis once in 88 earth days, the same length of time that it requires for one orbital revolution. This meant that it always kept one face toward the sun, just as the moon always keeps one face to the earth. However, some of the other observers disagreed on this point, and it seems that they were right to question it, although it was eventually accepted and has appeared in textbooks until very recently.

The modern technique for measuring Mercury's rotation is by *radar astronomy.* Unlike the radio astronomer, who receives whatever signals are coming in from the cosmos, the radar astronomer beams a pulse of radio waves of known properties at his target, then records the reflected signal. Comparison of the original and returned signals yields information about the target. For example, the time required by the pulse to travel to the target and return gives a measurement of the distance; the Doppler shift of its wavelength tells the relative speed at which the earth and the target are receding or approaching each other; *Doppler spreading* of the pulse reveals the rotation rate of the target. Doppler spreading appears in the following way. The initial signal produced by the radar transmitter is concentrated in a very narrow range of wavelengths, like an emission line in visible light. When it is reflected by the planetary target, the radar pulse is Doppler shifted to a new wavelength, depending on the relative velocity of approach or recession of the earth and the planet. But if the planet is rotating, one edge (*limb*) of it will be turning toward the earth, while the opposite limb will be turning away from us. The radar waves reflected from the approaching limb will have a slight Doppler shift toward the shorter wavelengths; those reflected from the receding limb will be shifted toward longer wavelengths. When the radar astronomer records the returned signal, he finds that it is spread over a range of wavelengths due to the planet's rotation, and the rotation velocity can be determined from the amount of this Doppler spread (Figure 8-2). The details of this procedure are more complicated than we have indicated and careful attention has to be paid to the geometry of the situation, but by observing a planet over a length of time as it moves around its orbit, the radar astronomer can determine not only its rotation rate, but also the orientation of its axis, and the direction in which it is turning. (For example, if a Martian radar astronomer observed the earth, he would find that the rotation rate is about 0.5 km/sec at the equator, or one turn per 24 hours, that the

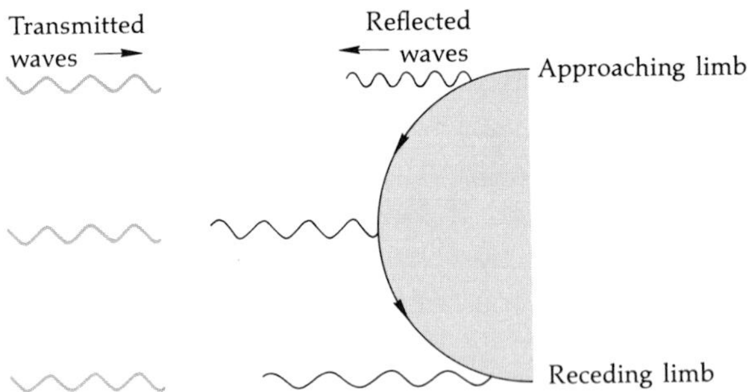

8-2 *Reflection of radar waves by a rotating planet. The arrowheads indicate direction of rotation. The radar waves approaching the planet from the direction of the earth (transmitted waves) all have the same wavelength. The wave reflected from the upper limb, which is turning toward the earth, is Doppler shifted to a shorter wavelength; the wave reflected at the bottom limb, which is turning away from the earth, is Doppler shifted to a longer wavelength. The difference in the wavelengths of the upper and lower reflected waves is determined by the rotation velocity of the planet and by the direction in which its axis points. As shown in this sketch, the axis points out of the paper, toward the reader; an additional wavelength shift due to the relative motion of the planet with respect to the earth is omitted. (Adapted from* Radar Observations of the Planets *by Irwin I. Shapiro. Copyright © 1968 by Scientific American, Inc. All rights reserved.)*

orientation of the axis is toward the star alpha Ursae Minoris (our "North Star") and that the direction of the earth's spin is counterclockwise as seen from that star. Of course, his results might be expressed in other words, since kilometers, seconds, and alpha Ursae Minoris are arbitrary terms invented by earthlings, and possibly Martians run their clocks counterclockwise by our definitions.)

In any case, radar astronomy observations do give precise measurements of planetary rotations, and in this way it was found that the old conclusions from visual drawings of Mercury were wrong. It does not rotate once per 88-day orbital period as Schiaperelli believed, but rather once in 58.7 days. In fact, Mercury makes *one and one-half rotations per orbital period.* As shown in Figure 8-3, this means that a certain diameter through the planet always points toward the sun *at the time of perihelion* (closest approach), although the opposite ends of this line (opposite faces of the planet) point toward the sun at alternate perihelia. A reasonable explanation of this phenomenon is that Mercury is not a perfect sphere but is somewhat elongated along one diameter, which is the same one that lines up with the sun at perihelion.

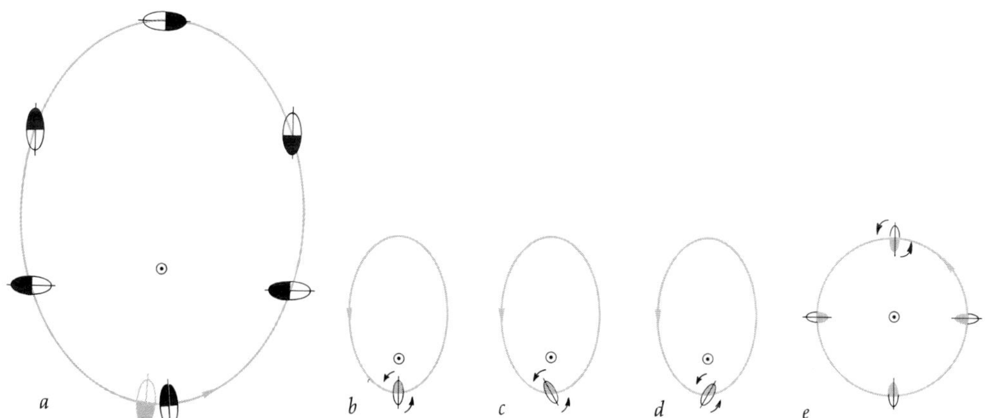

8-3 (*a*) *Sketch of the rotation and orbital motions of Mercury. Both the orbit and the shape of Mercury are exaggerated for purposes of illustration. Suppose Mercury, rotating at the observed rate, with its orientation at perihelion as shown in* (*b*), *rotated slightly faster, so that it appeared as in* (*c*) *at the next perihelion passage; then the sun's greater pull on the near half than on the far half would retard the rotation. The opposite happens in* (*d*), *where Mercury has rotated slightly slower than the normal rate. If the orbit were circular, the long diameter of Mercury would always point to the sun* (*e*).

According to the theory, the slightly greater pull of the sun on the near half of Mercury than on the far half controls this alignment, and hence the exact value of the rotation period. For if the planet turned slightly faster (or slower) than 1.5 times per orbit, the long axis would not point to the sun at perihelion time and this gravitational effect would then act to slow or hasten the rotation, respectively. The relatively large ellipticity of Mercury's orbit is also involved, because this gravitational effect is much stronger at small distances (near perihelion) than at longer ones. If the orbit were a circle, the long diameter of the planet would point at the sun continuously, in which case the rotation and orbital periods would be equal.

Radar observations show that Venus' rotation period is 243 days, or 18 days more than its orbital period; furthermore, the rotation is *retrograde.* As seen from the direction of our North Star, the earth and Venus both move counterclockwise around their orbits, but whereas the earth also turns on its axis in the counterclockwise way, Venus rotates clockwise. On the earth, the difference between our solar and sidereal days is only about 4 minutes, an effect arising from the fact that the earth moves a short way along its orbit during the time that it rotates once on its axis (see explanation on page 94). However, the orbital period of Venus is 225 days. Thus, during a *Venusian sidereal day* of 243 days, Venus moves completely around its orbit and then some. As a result, the *Venusian solar day* amounts to 117 days. (If we forget about the clouds for a moment, so that we can imagine seeing the surface of Venus, or imagine being on Venus and being able to see the stars, then the meaning of the two Venusian days is as follows: an observer on a distant star would see that Venus turns once on its axis every 243 days; an observer on Venus would see the sun rise once every 117 days.)

It is also possible to map the surfaces of planets by radar astronomy techniques. For example, as Venus rotates, it is found that some radar pulses take a little more or a little less time than expected to return from its surface. This indicates depressions and highlands, respectively. Further, the intensities of reflected signals corresponding to different areas of the planet are different, indicating differences in the smoothness (and hence the reflecting power) of the terrain. This method does not give us a very detailed picture (see Figure 8-4) of the surface of Venus, but it is of value as the clouds of Venus prevent us from seeing or photographing its surface.

A simple theory has been used to calculate the expected temperatures of the planetary surfaces. The sun shines on a

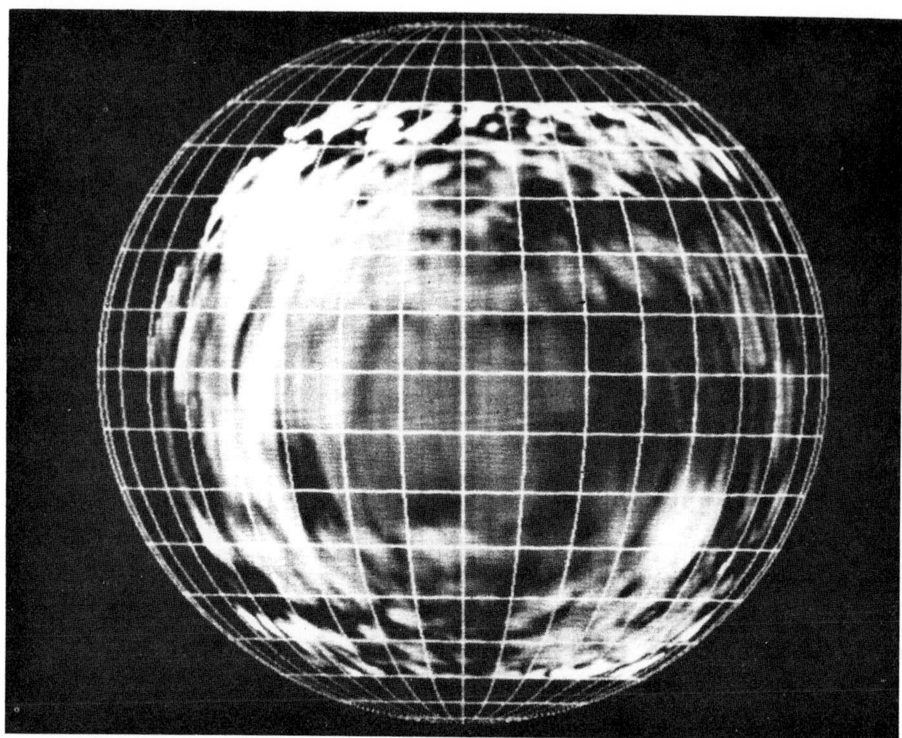

8-4 *Radar map of the surface of Venus. The bright areas are regions where radar signals are strongly reflected. (Courtesy D. B. Campbell, Cornell-Sydney University Astronomy Center.)*

surface, providing an energy input. This input is equal to the total amount of solar energy received, minus the amount that reflects back into space. We can easily calculate the solar energy received by the planet, since we know its distance from the sun. The amount that is reflected back into space is determined by observing the planet with a photometer.* The solar energy that is not reflected is absorbed by the surface and heats it. The surface must radiate energy according to the Stefan-Boltzmann Law (Chapter 5) for its temperature. After a while, the amount of energy radiated by the surface will equal the amount of energy that it absorbs. Otherwise, if (for example) the surface radiated less energy than it absorbed, it would keep heating up indefinitely. In fact, it heats up until it is just hot enough to radiate away an average amount of energy equal to the average input which it receives from the sun. It cannot heat up further, because then it would be radiating more energy than

*The fraction of incident solar radiation that is reflected into space by a planet is called the *albedo*.

it receives, and this would violate the principle of conservation of energy (Chapter 5). Thus, the energy output of the surface is equal to the energy input. We already know the input, and so we know the output, and can use the Stefan-Boltzmann Law to find out what the surface temperature must be in order to produce that output.

The average planetary surface temperatures calculated according to the simple theory outlined above are much less than the temperature of the sun. (For example, the calculated temperature of Mars is 250° K and the temperature of the visible solar surface is about 6,000° K.) We recall a property of the Planck Law discussed in Chapter 5: the wavelength at which the maximum radiation occurs is longer for cooler objects. Thus, the maximum solar radiation occurs in the wavelength range of visible light, but the planets, being cooler, radiate most of their energy in the infrared and radio wavelengths. For this reason, we use infrared and radio astronomy observations to measure the actual surface temperatures of the planets. The dependence of intensity on wavelength is determined by measuring the radiation from a planet at several wavelengths. This can be compared with a set of Planck curves corresponding to various temperatures, to find the best agreement, and hence the observed planetary temperature. In studying the observations to deduce the average temperature we have to take account of the fact that different parts of the planetary surface are at different temperatures; the part of the planet where the sun is overhead will be hotter than the other side, where it is night.

The observed average surface temperatures of Mercury, the earth, the moon, and Mars agree rather well with the values calculated by the above theory. However, the radio and infrared observations of Venus and Jupiter show that these two planets are significantly hotter than predicted by the theory. For Venus the observed temperature is about 700° K, while the theoretical value lies between 325 and 375° K. The higher surface temperature apparently results from the greenhouse effect discussed later in this chapter (page 223). For Jupiter, the observed temperature is near 150° K while the theory predicts 102° K; here the explanation is based on our ideas about Jupiter's interior. The best measurements and calculations indicate that Jupiter is radiating about four times *more* energy per second than the planet is receiving from the sun. The energy must be generated at the planet; thus, Jupiter is *self-luminous* and in that sense could be considered a very cool, dim, small star. However, the temperature is not high enough for nuclear reactions to occur, so Jupiter's internal energy source is not the same as that of the sun or other normal stars. The thermal characteristics of Saturn resemble those of Jupiter; it appears to radiate about twice as much energy as it receives from the sun. Uranus and Neptune are also hotter than expected.

EVOLUTION OF THE EARTH'S ATMOSPHERE

What is the origin of the air we breathe? The answer as we know it involves a fascinating combination of astronomy, geology, and biology. We believe that the original clump of gas (and perhaps dust) that ultimately evolved to form the earth we know was much more massive than the earth. This largely gaseous protoplanet was held together by its own gravitational attraction. How did it evolve?

The heavier elements or compounds sink to the center (this point appears in Aristotle's thinking) and the lighter atoms escape into space through a process called *evaporation.* The air molecules near the surface of the earth are in a state of constant motion, with their average speed dependent on the temperature. However, at the high densities (in the sense of the number of molecules per unit volume) near the earth's surface, particles travel only a very short distance before colliding with other ones. If we tabulate the average distance traveled between collisions (called the *mean free path,* or m.f.p.) as a function of increasing height in the atmosphere, we find that the m.f.p. gets longer as we go higher up in the atmosphere. At heights of about 500 kilometers (310 miles) and above, the density of air molecules is so low that an atom or molecule moving upward has a good chance of traveling a very great distance without a collision. If this is the case, the atom (if its speed is great enough) could escape the earth and go into an orbit around the sun. Recall our discussion of orbits in Chapter 4 and the fact that they can be thought of in terms of a balance of gravitational attraction and centrifugal force. The (always outward-directed) centrifugal force increases with the speed. Thus, three possible cases are of interest for an atom moving upward at heights of 500 kilometers or above. (1) Particles traveling relatively slowly do not have enough speed to overcome gravity and therefore they fall back (like a ball thrown upward). (2) Particles with an intermediate speed can achieve an orbit around the earth but this orbit usually carries the atom back into the atmosphere. (3) Atoms traveling sufficiently fast will be subject to a centrifugal force too large to be balanced by the earth's gravity and the atoms in question fly away, never to return. The minimum speed at which an atom moving upward can leave the earth in this way is called the *escape speed,* and it is about 11 kilometers per second (7 miles per second). The escape speed does not depend on the mass of the object; it is the same for an atom of the air as for a space vehicle.

Thus, atoms in the outer atmosphere of the protoearth could have escaped by evaporation. The hotter the temperature of the atmosphere, the faster are the speeds of the gas atoms and molecules—more of them can then exceed the escape speed. In a gas at a given temperature the less massive molecules move faster than the heavier ones. Detailed calculations show that the earth is likely to have lost essentially all of its original hydrogen (hydrogen molecules are the lightest of all) and possibly all of its original helium (the second lightest gas) by evaporation. (Compare the concept of astronomical evaporation with the evaporation of water; the basic ideas are the same but in the latter case the water molecules escape only from the body of water to the atmosphere.)

Thus, the major constituents of the solar nebula, namely hydrogen and helium, were probably lost from the earth by evaporation. It is quite likely that the original atmosphere (particularly the hydrogen and helium) was also similarly lost from Mercury, Venus, and Mars. The heavier substances, together with any dust or larger solid bodies that may have been present in the prototerrestrial planets, consolidated under their own gravitation and chemical action to form the solid bodies of these planets. On the other hand, the giant planets, Jupiter and Saturn, have much larger gravities (and hence much larger escape speeds) and much cooler atmospheres (because of their greater distances from the sun), so they have probably retained most of their original atmospheres, which may account for the resemblance of their compositions to that of the sun.

If the above discussion of the evaporation process were the whole story, then the atmospheres of the earth and other terrestrial planets ought to be quite similar. However, this is not the case. The composition of the earth's atmosphere is summarized in Table 8-2.

A vital constituent of the atmosphere, not listed in Table 8-2, is the water vapor (H_2O) that is so intimately involved with climate and weather on the earth. Depending on the conditions, a sample of air at sea level may contain as much as 2 per cent by mass of water vapor. The concentration of carbon dioxide is also variable. Various other substances, including both naturally occurring gases and the products of industrial and automotive exhausts, are also present in trace amounts.

Table 8-2 *Chief Constituents of Dry Air at Sea Level*

GAS	PER CENT BY MASS
Nitrogen (N_2)	75.5
Oxygen (O_2)	23.2
Argon (A)	1.3
Carbon dioxide (CO_2)	0.05

The composition of the earth's atmosphere is thus entirely unlike the sun and the giant planets, as expected from our discussion of evaporation. Further, it is unlike that of Mars and Venus. The unmanned spacecraft that have traveled to the vicinities of these two planets sent back data indicating that their atmospheres are mostly composed of carbon dioxide, with little if any free oxygen. In fact, it appears that the oceans and much of the earth's present atmosphere were exhaled from the crust, as we have already mentioned in Chapter 3. Specifically, the emission of gas probably took place through volcanoes, and there are indications that volcanic activity was more extensive in the past. Direct measurements on contemporary volcanoes have been made by brave geologists who find that the volcanoes exhale steam (H_2O), carbon dioxide (CO_2), molecular nitrogen (N_2), and sulfur dioxide (SO_2). The sulfur dioxide is emitted in small quantities and readily combines with rocks. The oxygen in our atmosphere has been produced almost entirely by photosynthesis in plants (Chapter 3). A continuous source of oxygen is necessary, as it has been estimated that if the production of oxygen stopped today, the existing oxygen in the atmosphere would be used up in, at most, several tens of thousands of years, chiefly as a result of oxidation in the weathering of rocks. Oxygen is also used up in the respiration of animals and in combustion.

The volcanic processes that gave rise to much of our atmosphere probably produced much more carbon dioxide than we find in the air. However, carbon dioxide is removed from the atmosphere by several processes: (1) the photosynthesis in green plants; (2) chemical reactions in the weathering of rocks; and (3) absorption by the oceans. The weathering of rocks is actually the most important long-term removal process. Both oxygen and carbon dioxide can dissolve in sea water, a circumstance that permits both animal and plant life to exist there. In fact, the sea is a great reservoir for carbon dioxide and presently contains about sixty times as much CO_2 as does the atmosphere. The oceans actually help to stabilize the atmospheric CO_2 content because they absorb it faster when its concentration in the atmosphere increases.

Carbon dioxide is produced by animals and by combustion. Under normal circumstances, the various atmospheric, biological, and chemical processes are in equilibrium. However, even the oceans may not be able to absorb sufficient CO_2 as the amount produced artificially continues to increase.

In summary, the history of the earth's atmosphere probably began with a protoearth composed of material with essentially solar composition. The heavier substances concentrated in the center and the lighter elements escaped from the protoplanet. About 4.5 billion years ago the earth achieved its present mass and composition. Formation of the crust took place about 4.0 billion years ago. The original atmosphere had largely evaporated at this point and the secondary or exhaled atmosphere would have been produced by 3.5 billion years ago. Sufficient hydrogen from the original atmosphere was stored in the crust in the form of compounds (water, hydrogen sulfide, ammonia, methane, etc.); these were emitted into the atmosphere during the exhalation process. The oceans had developed to about their present volume around the time that the exhaled atmosphere was formed, or perhaps a little later. The combination of the compounds of hydrogen mentioned above and the water provided the raw materials for the primary broth that may have led to the origin of life as outlined in Chapter 3. Starting about 2 billion years ago, substantial amounts of oxygen began to be produced by plant life. By approximately 1 billion years ago the oxygen content of the atmosphere had increased to its present value. The present atmosphere as we know it dates from this time. Note that 1 billion B.C. precedes the geological times for which there is generally agreed-upon identification of strata and absolute dating (Table 2-1). The classification and dating of earlier strata will undoubtedly contribute greatly to the fund of knowledge concerning evolution of life, the atmosphere, and the earth's crust.

CHARACTERISTICS OF THE EARTH'S ATMOSPHERE

The composition of the earth's present atmosphere was given in Table 8-2. Solar radiation affects the bulk of the atmosphere indirectly; it heats the ground and the atmosphere is then warmed by convection as air near the ground is heated and rises.

Some parts of the atmosphere, for example, the ionospheric layers near 160 kilometers (100 miles) altitude, are heated directly by the absorption of solar radiant energy. The pressure decreases with height in the atmosphere as shown in Table 8-3; the variation of temperature is also shown. These are average values around which there are considerable fluctuations, which, at the lower altitudes, we experience as "weather." The properties of the atmosphere at high altitudes were determined from many rocket and satellite studies.

A layer of ozone (O_3) molecules exists near 30 kilometers (19 miles) altitude in the atmosphere. This layer is a highly effective absorber of solar ultraviolet radiation for wavelengths shorter than 3,000 Angstroms. For wavelengths near 3,000 Angstroms, about 7 per cent of the light passes through the ozone layer. At the absorption peak in this wavelength range (near 2,600 Å), only 10^{-32} of the incident sunlight passes through the ozone layer. The small amount of ultraviolet radiation that does reach the ground is responsible for suntans and sunburns. The intensity of this radiation at the earth's surface if the ozone layer

Table 8-3 *A Model Terrestrial Atmosphere*

HEIGHT ABOVE SURFACE				
KM	MILES	REMARKS	PRESSURE, PER CENT OF SURFACE VALUE	AVERAGE TEMPERATURE (°K)
0	0	Surface	100	280
1.5	1	Altitude of Denver, Colorado	85	273
3	2		70	266
6	4	Tops of high mountains	50	255
10	6	Cruising altitude of jet-liners	30	238
16	10		10	210
30	20	Ozone layer	1	250
50	30		0.1	290
80	50	Aurora and meteors near 100 km	10^{-3}	170
160	100	Ionosphere	10^{-6}	600
480	300	Atmospheric evaporation takes place	10^{-13}	1200

were not present would be far more than uncomfortable—it would be lethal. On the other hand, in the early history of the earth, this radiation may have been an important source of energy for the chemical reactions in the primary broth before the ozone shield was formed from atmospheric oxygen.

The absorption of solar radiation by ozone is an energy source for the atmosphere near 30 kilometers and above, and this energy is responsible for the higher temperatures found at 30 and 50 kilometers as seen in Table 8-3. Similarly, the temperatures increase above 100 kilometers because of the ionization* of atmospheric constituents by solar radiation. This region of the atmosphere, called the *ionosphere,* contains substantial numbers of free electrons and ions, although most of its atoms and molecules are neutral. It consists of several layers and reflects radio waves with wavelengths of a few meters or greater. Multiple reflections between the ionosphere and the earth can send radio signals all the way around the earth. The principal layers of the ionosphere are called the D region (altitude 90 kilometers, or 56 miles), the E region (110 kilometers, or 68 miles), and the F region (200 to 300 kilometers, or 120 to 190 miles). The D layer has the smallest *electron density* (number of electrons per cubic centimeter), and the largest density of neutral gas molecules. This combination makes the D layer weakly reflecting and strongly absorbing, so far as radio waves are concerned. The electrons and ions in the D region recombine to form neutral atoms and molecules shortly after sunset, but particles in the higher ionospheric layers recombine more slowly. Therefore, the D region vanishes and radio propagation by ionospheric reflection (Figure 8-5) is most effective at night. Since the ionosphere is produced by x-ray and ultraviolet radiation from the sun, variations in the intensity of this radiation due to solar flares and other events (Chapter 9) create temporary changes in the ionosphere, and can thus affect radio communications.

Our own experience with the weather and the seasons tells us that the atmosphere is not in a static and uniform state. The combined effect of the heating of the earth by the sun and the rotation of the earth is to produce a global circulation pattern (Figure 8-6). Many facets of this pattern were known empirically to the navigators of sailing ships. The fairly steady *trade winds* are a result of large-scale atmospheric circulation patterns as are the *horse latitudes* characterized by light and undependable winds. These examples are representative of the surface circu-

---

*The heating occurs because the energies of the photons absorbed in the ionization process are partially converted to kinetic energy of the electrons released from the atoms.

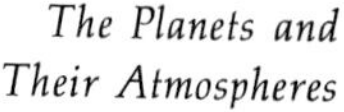

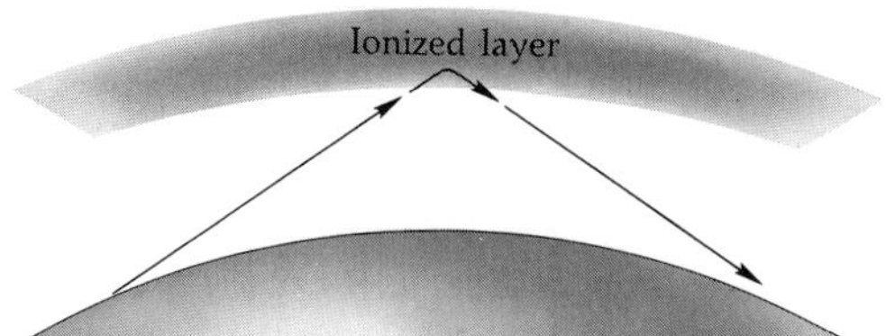

8-5 *Sketch showing that reflection by an ionized layer in the atmosphere permits radio transmission over large distances on the earth.*

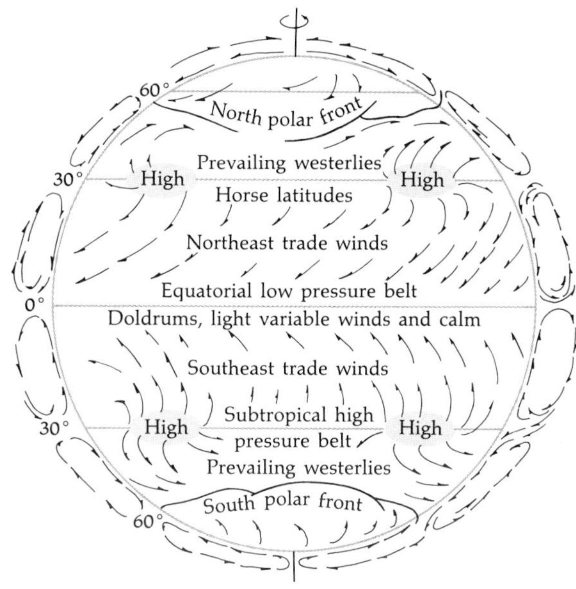

8-6 *Major patterns in the circulation of the earth's atmosphere. (From* Principles of Geology, *3d ed., by J. Gilluly, A. C. Waters, and A. O. Woodford. W. H. Freeman and Company. Copyright © 1968.)*

lation pattern. The general atmospheric circulation is also largely responsible for the gross pattern of currents in the oceans.

Other features of the circulation are found at different heights. The most familiar example is the *jet stream,* a thin current of air that flows from west to east near an altitude of 10.5 kilometers (35,000 feet). The wind speed in the jet stream can reach 320 kilometers per hour (200 miles per hour). As a result, for example, it takes longer to fly east-west across North America than it does to fly west-east; airline schedules reflect this fact and the cruising altitude is normally chosen to avoid the stronger parts of the jet stream when going west and to take advantage of them when going east.

Besides the gross changes of pressure, temperature, and circulation patterns, the atmosphere plays a vital role in the movement of water between the oceans and the land masses—the *hydrologic cycle.* Water moves through the cycle as illustrated in Figure 8-7, evaporating into the atmosphere principally from the oceans. It condenses into the clouds, which are transported by the wind, and is deposited as rain and snow on the continents. Eventually it reaches the ocean again, and the cycle is repeated. The same water flows continually through these stages and we say the cycle is a closed one. Impurities in water blocked at some stage of the cycle will accumulate there; an example is the continual buildup of salt in the oceans as minerals are washed away from the land.

Clouds often contribute to a beautiful sunset, but certain features of the atmosphere, without clouds, can be seen in the setting sun and blue color of the sky. As the sun sets, its color turns first to orange and then red. This effect is caused by the molecules in the atmosphere. When the molecules are much smaller than the wavelength of light, the sunlight is affected

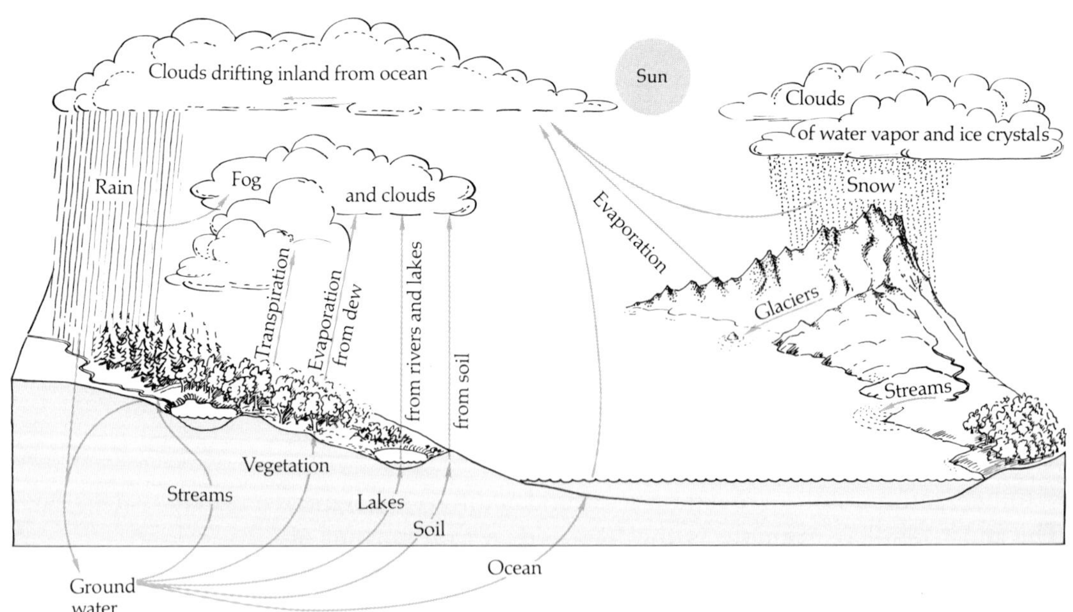

8-7 *The hydrologic cycle. (From* Principles of Geology, *3d ed., by J. Gilluly, A. C. Waters, and A. O. Woodford. W. H. Freeman and Company. Copyright © 1968.)*

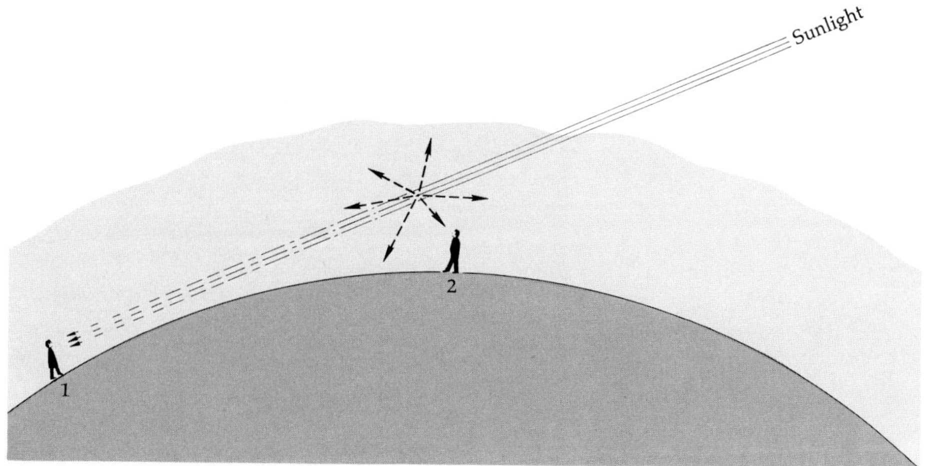

8-8 *Effects of light scattering in the earth's atmosphere. Blue light is strongly scattered, unlike the red light, and so an observer at position 1 sees a red sun at sunset, while the scattered light gives the sky its characteristic blue color as viewed, for example, from position 2.*

by a process called *Rayleigh scattering.* We recall the reference to scattering in Chapter 5; the wavelength of the light remains unchanged in a scattering process, but the direction is changed. Blue light is scattered much more strongly than red light, and, as a consequence, the red light penetrates much farther through the atmosphere. The geometrical situation is shown in Figure 8-8. As the sun sets, the amount of air that its light passes through en route to the observer is continually increasing, and thus we see a red sun. The blue light is scattered in all directions and gives rise to the blue color of the sky. Above the atmosphere, as seen by the astronauts, for example, the sky is black.

The effect of the wavelength dependence of the atmospheric penetration of light is illustrated by photographs of the same terrestrial scene taken with short (violet) and long (infrared) wavelength light as shown in Figure 8-9. The long wavelengths pass through while the short wavelengths are scattered in the atmosphere. The other part of Figure 8-9 shows a similar effect in the atmosphere of Mars.

The atmosphere is a barrier that shields us not only from the solar ultraviolet radiation, but also from meteoroids, most of which are small and burn up in the atmosphere near 100 kilometers (62 miles) altitude, due to friction with the atmospheric gases. The atmosphere also stops some of the cosmic rays, and this effect, together with the tendency of the earth's magnetic field to deflect the slower moving cosmic rays, keeps

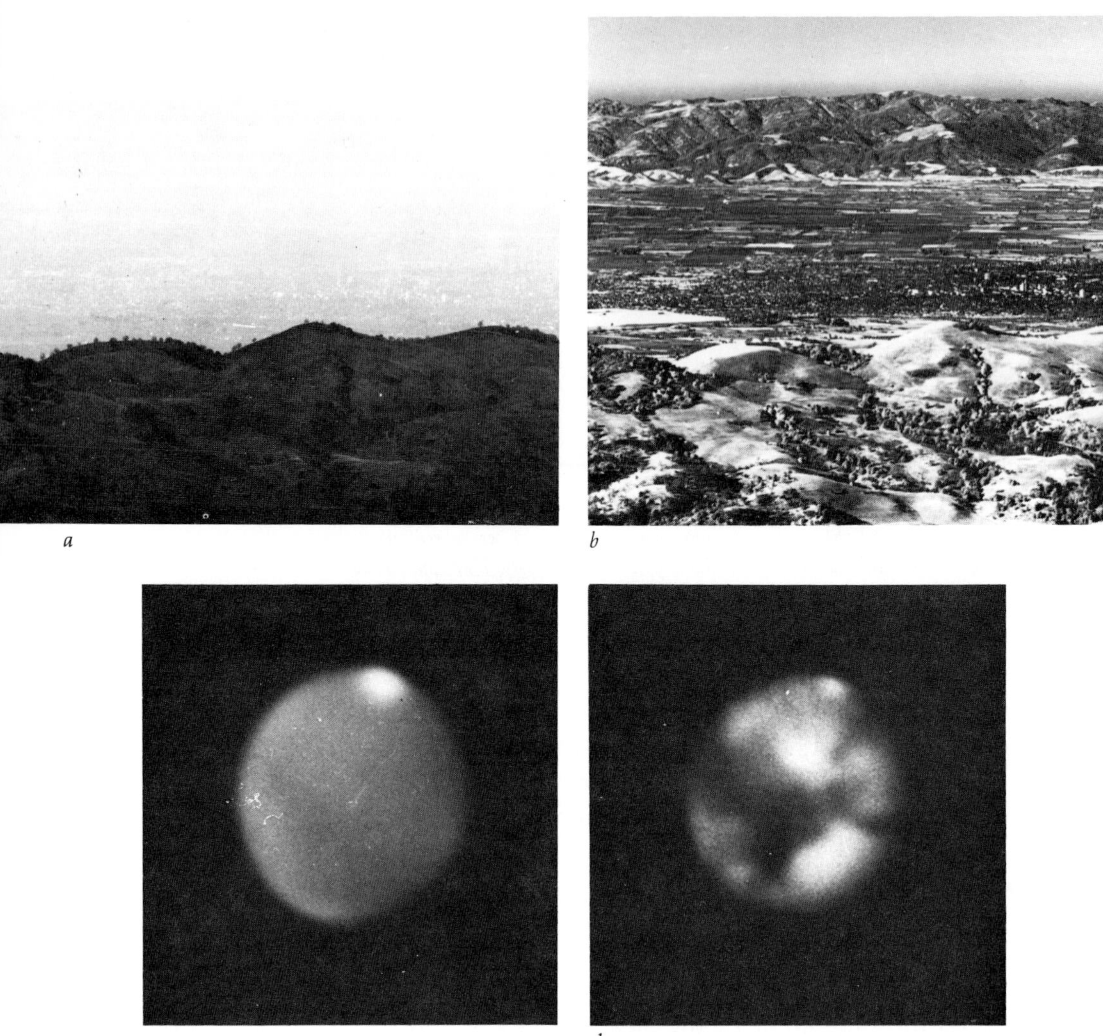

8-9 *The penetration of light through the atmospheres of Mars and the earth. The photographs are: (a) San Jose, California (13.5 miles distant) in violet light; (b) San Jose in infrared light; (c) Mars in violet light; (d) Mars in infrared light. The greater penetrating power of the longer wavelength infrared light is clearly shown in both atmospheres. (Lick Observatory photograph.)*

the genetic mutation rate of life (Chapter 3) down to a reasonably low value; biologists believe that a high mutation rate is undesirable.

Beyond the atmosphere, there is a region called the *magnetosphere,* in which electrons and positive ions are trapped in the earth's magnetic field. This region was discovered with instruments carried on the first artificial satellites and it is discussed in Chapter 15.

The surface of Venus is not directly visible; it is obscured by a thick cloudy atmosphere. Spectroscopic observations (Figure 8-10) show characteristic absorption lines (as explained in Figure 8-11) of carbon dioxide (CO_2), but for years the principal constituent of the atmosphere of Venus was assumed to be molecular nitrogen (N_2) by analogy with the earth's atmosphere (a convenient assumption, since molecular nitrogen had no strong absorption lines that could be searched for in the visible light spectrum). Radio observations of the planet indicated a temperature of about 700° K. This is surprisingly high and there was some reluctance at first to interpret it as the temperature of the surface of Venus. One alternative interpretation was that the radio emission came from a thick ionospheric layer in the atmosphere of Venus.

The question raised by the radio observations of Venus was settled in 1962 by the deep-space probe, *Mariner 2.* The intensity

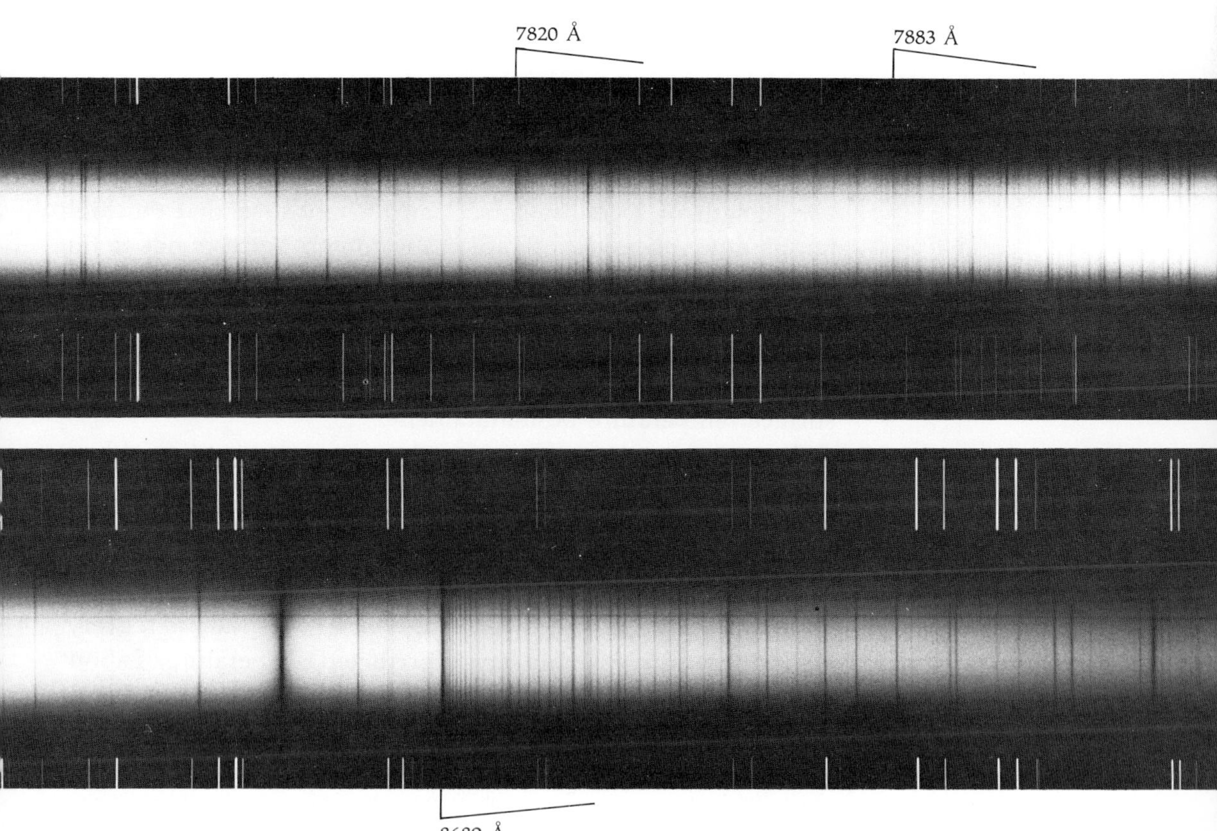

8-10 *Absorption lines of CO_2 (carbon dioxide) in the spectrum of Venus. (Lick Observatory photograph.)*

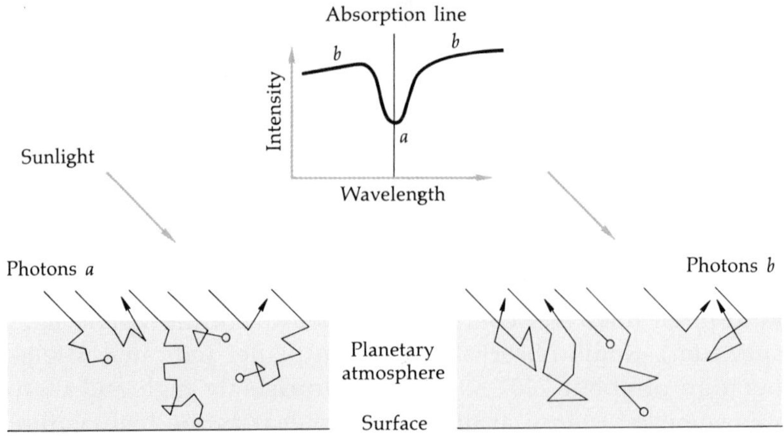

8-11 *Schematic illustration of photon diffusion and absorption line formation in a planetary atmosphere. Photons (a) corresponding to a characteristic wavelength of an atom or molecule are scattered and/or absorbed more frequently than photons (b) in the continuum where they are largely just scattered (not absorbed). Thus, fewer photons (a) than photons (b) diffuse back out of the planetary atmosphere and an absorption line is found when the spectrum of the planet is recorded.*

of radio emission would show a rather different variation across the disk of Venus, according to which of the two likely models (Figure 8-12) was correct. The *Mariner 2* observations at a wavelength of 2 centimeters clearly established that the high temperature observed by radio astronomers corresponds to the surface and not the ionosphere of Venus.* The spectroscopic results give a temperature for the cloud-tops of about 235° K. Thus, the temperature of the atmosphere of Venus decreases rapidly with height above the surface. How can the very high surface temperature be maintained?

An answer to this question involves an understanding of how planetary atmospheres are heated. As mentioned earlier, it is a relatively simple matter to calculate the temperature of a planet made of rock (and lacking an atmosphere) as a function of distance from the sun. It is determined by a balance of energy input from the sun and the rate at which rocks radiate it away. Mars, for example, would have a predicted temperature of about 250° K, a value close to the measured temperature. Similarly,

---

*This was directly confirmed on December 15, 1970, when the Soviet space probe *Venus-7* parachuted to a soft landing on the surface and measured a temperature of about 750° K at an estimated atmospheric pressure of 90 times greater than the sea level pressure on earth. The spacecraft was designed to land and function in an environment as hot as 800° K and at pressures up to 180 times those on earth. Its successful operation for 23 minutes on the night side of Venus was a major achievement.

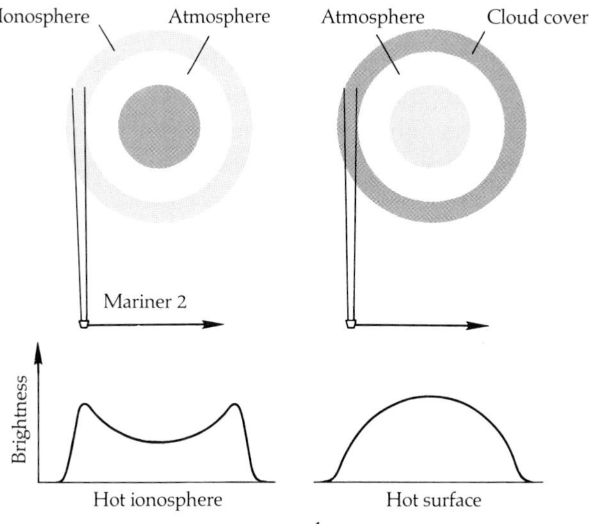

8-12 *Predicted intensities of 2-cm wavelength radio emission that would have been observed by* Mariner 2 *as it scanned past Venus if (a) the hot ionosphere theory were correct or (b) if the hot surface theory were correct. According to theory (a),* Mariner 2 *would measure maximum brightness near the edge of the planet because it would be viewing through the largest amount of hot ionosphere. According to theory (b), the maximum brightness would occur at the center, because radiation from the hot limb would be greatly reduced by the longer traverse through the relatively cool atmosphere. The* Mariner 2 *observations favored theory (b). (After Brandt and Hodge.)*

the theoretical and measured values for the earth are close to 300° K. Both Mars and the earth have relatively thin atmospheres and apparently the atmospheres have little effect. It seems that the thick atmosphere of Venus results in an efficient use of the incident solar energy with a resultant hot surface temperature.

A possible mechanism for achieving this situation is the well known *greenhouse effect.* On a hot, sunny day, anyone entering a greenhouse, or a car left with the windows rolled up, becomes conscious of this effect. As we discussed in Chapter 4, the temperature at the surface of the earth is determined by a balance of net energy received due to incident solar energy during the day and net cooling at night by radiation into space. If the loss could be reduced, the result would be a higher surface temperature; the greenhouse accomplishes just this. The visible sunlight passes readily through glass and heats the greenhouse floor, but the floor is much cooler than the sun, and instead of radiating yellow light, it radiates in the infrared. Most glass is not transparent to this infrared radiation, so it becomes trapped in the greenhouse. Thus, solar energy can get in, but the infrared radiation cannot get out. The result is a hot greenhouse.

Measurements by space probes indicate that the atmosphere of Venus is composed almost entirely of carbon dioxide; the old assumption that a substantial amount of nitrogen was present was wrong. The surface pressures found from various space probe observations are 90 to 100 times the surface pressure on earth. A thick atmosphere composed of carbon dioxide would allow the greenhouse effect to operate because, although the solar radiation does not strike the surface directly, as in the simple example of a greenhouse on earth, the atmosphere is essentially a scattering medium (changes the direction of a photon but not its energy) in visual wavelengths and thus the solar energy can diffuse down and heat the surface. The carbon dioxide is opaque (truly absorbing) to the infrared radiation from the surface, therefore the greenhouse effect and resultant high temperatures can be achieved. There may be convection in the atmosphere of Venus, which would affect the variation of temperature with height above the surface.

The thick atmosphere of carbon dioxide on Venus, and the surface temperature of more than 700° K show that it would be an extremely unpleasant place to visit. Manned exploration is not feasible in the forseeable future, and further scientific investigation will have to depend on unmanned space probes.

THE ATMOSPHERE OF MARS

Detailed studies of the atmosphere of Mars, both from space probes sent to its vicinity and from observatories on earth, have indicated a very thin atmosphere composed almost entirely of carbon dioxide. There are thin clouds and occasionally the surface is obscured by dust storms which may occur over a large area of the planet (as happened in November, 1971, when *Mariner 9* went into orbit around Mars). The surface pressure is about one per cent of the terrestrial value and the temperature is about 250° K. It has been suggested that the rate of volcanic activity on Mars is much less than on earth and this may partially explain the much smaller total mass of the Martian atmosphere. Our ideas about the surface of Mars have been radically altered by space probe results; they are discussed at length in connection with space exploration in Chapter 15.

EARTH, MARS, AND VENUS—THE KEY DIFFERENCE?

Three terrestrial planets, earth, Venus, and Mars, are similar in many ways, but their atmospheres differ markedly. The atmospheres of Mars and Venus consist primarily of carbon dioxide, whereas the earth's has little carbon dioxide but a significant amount of oxygen. The oxygen was produced mostly by vegetation, and the carbon dioxide is largely bound up in the rocks of the crust and in sea water (page 213). In 1970, the Pakistani physicist S. I. Rasool suggested that the key circumstance that led to the separate evolution of these planetary atmospheres was *the different distances of earth, Venus, and Mars from the sun.* According to Rasool's theory, most of the water on Mars would have remained in the frozen state, due to the low surface temperature that follows from Mars' distance from the sun. On the other hand, essentially all of the water on Venus would have taken the form of water vapor (steam) due to the high temperature, and ultraviolet photons from the sun would have *dissociated** these water molecules into hydrogen and oxygen. The hydrogen, being very light, would have escaped from Venus by the evaporation process (page 211). The oxygen would gradually have been absorbed into the crust of Venus through the oxidation of rocks and would not have been replenished by plants (as on earth) since the high temperature and lack of liquid water on Venus presumably prevented life from arising there. Liquid water on earth facilitates the absorption of carbon dioxide by chemical reactions in rocks of the crust, in addition to absorbing carbon dioxide directly. Thus, the absence of liquid water on Venus and Mars can account for the fact that their atmospheres are primarily composed of carbon dioxide, and the presence of liquid water on earth was crucial to the development of the present atmosphere and of life.

Rasool's theory requires careful checking, but it illustrates the point that by systematically studying the other worlds in the solar system, we can also hope to better understand our own.

ATMOSPHERES OF JUPITER AND SATURN

Intensities of infrared and short wavelength (a few centimeters) radio radiation from Jupiter and Saturn indicate atmospheric

*Ionization is the process in which the absorption of a photon by an atom or molecule gives an electron enough energy to escape from the atom or molecule. Dissociation is the process in which absorption of a photon by a molecule gives one of its constituent atoms enough energy to escape from the molecule.

temperatures of about 150° K and 120° K, respectively. Although somewhat higher than expected (page 210), these temperatures are still so low that evaporative escape of atmospheric molecules is a very slow process compared to that on the earth. Further, the gravities of Jupiter and Saturn are much greater than that of the earth or the other terrestrial planets and the combination of the dual retarding effects of low temperature and high gravity makes the escape of atmospheric molecules from Jupiter and Saturn quite negligible. Thus, these planets should have essentially the same composition as the sun and the original solar nebula.

The low temperature and solar composition of Jupiter's atmosphere are confirmed by its spectrum (Figure 8-13). The substances revealed in the spectrum are compounds of hydrogen—specifically, molecular hydrogen (H_2), methane (CH_4), and ammonia (NH_3). Helium should also be present, but it is very difficult to detect. The composition of the atmosphere of Saturn is probably the same as Jupiter's with the exception that most of the ammonia is frozen out.

Considerable winds probably occur in Jupiter's atmosphere, which is arranged in belts or bands parallel to the equator. These bands have slightly different rates of rotation, with the equatorial period of 9^h50^m being the fastest. Features such as spots occasionally are visible, as one can see in Figure 8-14. The *great red spot* is one of the most remarkable; its shape is roughly elliptical, with dimensions of 10,000 kilometers by 40,000 kilometers (6,000 miles by 24,000 miles). Although its color changes from time to time and this spot sometimes be-

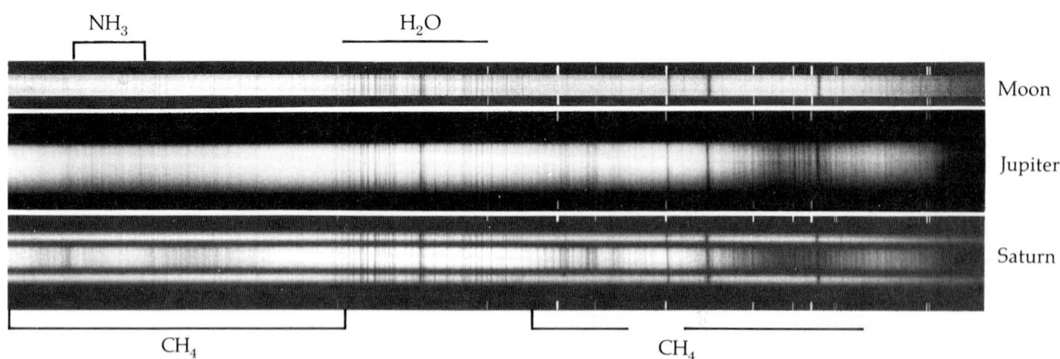

8-13 *Spectra of the moon, Jupiter, and Saturn. Lines seen in the moon's spectrum (such as those due to water vapor, H_2O) originate in the earth's atmosphere or in the solar spectrum. The spectra of Jupiter and Saturn clearly show the presence of methane (CH_4), and also ammonia (NH_3). (Lick Observatory photograph.)*

8-14 *Jupiter as photographed by the 200-inch telescope. The great red spot is clearly visible as is the satellite Ganymede and its shadow. (Courtesy of the Hale Observatories.)*

comes so pale that it disappears from view, it is usually dark red. It drifts with respect to the surrounding atmosphere and the appearance of temporary cloud features alters its motion. These observations have led some scientists to suggest that the red spot is a meteorological feature floating in the atmosphere. There are many theories of the red spot. One of the most interesting speculations* is based on the possibility that the complex molecular precursors of life (Chapter 3) might be forming now in parts of Jupiter's atmosphere. Laboratory experiments on a simulated atmosphere have produced molecules of a complex orange-red substance and it has been suggested that such molecules are responsible for the color of the great red spot.

The general appearance of Saturn (Figure 7-17), setting aside

---

*Which is far from proven!

the question of the rings, is similar to the appearance of Jupiter. The belts and bands in Saturn are, however, more diffuse in appearance and somewhat harder to observe. Both Jupiter and Saturn show considerable oblateness (caused by their rapid rotation). Likewise, the earth's equatorial bulge is caused by the earth's rotation (see page 309).

ATMOSPHERES OF URANUS AND NEPTUNE

The atmospheres of these planets are very cold as a result of their great distances from the sun; radio observations give temperatures of about 100° K. Ammonia is entirely frozen out of these atmospheres and only molecular hydrogen and methane are observed; again, helium is suspected.

Uranus seems to have very faint belts like those of Jupiter and Saturn, but features of these outer planets are very difficult to determine because of their great distance. Not much surface detail has been observed on Neptune. Both Uranus and Neptune show greenish disks.

THE MAGNETOSPHERE OF JUPITER

Intense radio signals from the planet Jupiter at wavelengths of about 10 meters (33 feet) were detected in 1955 by Bernard Burke and Kenneth Franklin at the Carnegie Institute of Washington. Their discovery is a splendid case of serendipity. They meant to observe the Crab nebula (Chapter 16) but noticed a strong source of radio emission which passed through the stationary beam of their antenna about the same time on most days. The signal was so strong that it was considered to be some kind of local interference, possibly someone going home from work every day in a vehicle with a faulty ignition. After some time a curious fact became apparent, namely, if the noise were due to a commuter, he commuted according to the sidereal clock. Recall the discussion in Chapter 4 concerning the difference between the day reckoned with respect to the sun and the day determined with respect to the stars. The sidereal day is about 4 minutes shorter than the solar day; the stars rise and set by sidereal time. The fact that the hypothetical commuter was going home 4 solar minutes earlier each successive day clearly suggested an astronomical source. Jupiter was in the right area of the sky, but at first it was not considered seriously as a possible source of the radio bursts, because only continuous thermal radiation (based on the theory given on page 129) was expected from it. Further observations proved that the bursts do in fact come from Jupiter and they are not thermal radiation.

Three fairly distinct types of Jovian radio emission have now been identified. One, occurring predominantly in the infrared and millimeter radio wavelengths, is in fact the thermal emission, as discussed on page 210. At wavelengths of roughly 10 to 100 centimeters (4 to 40 inches) *nonthermal* (that is, not resembling the predictions of the Planck Law) emission of slowly varying intensity is received from a large area, as shown in Figure 8-15. These radio waves are polarized and are typical of the radiation emitted by energetic electrons trapped in a magnetic field. Thus, we believe that Jupiter has a magnetic field, and the calculated strength at the visible "surface" of the planet is about ten times that of the magnetic field on the surface of the earth. The electrons are analogous to the particles in the Van Allen Belts (Chapter 15) of the earth's magnetosphere. The third type of Jovian emission (the kind actually discovered by Burke and Franklin) occurs at longer wavelengths of, say, 10 meters. Here the radio emission only occurs in the form of intermittent bursts. After the announcement of the discovery, the Australian radio astronomer C. Shain searched records of a radio sky survey carried out five years previously and found that Jupiter had been recorded (and unrecognized) many times.

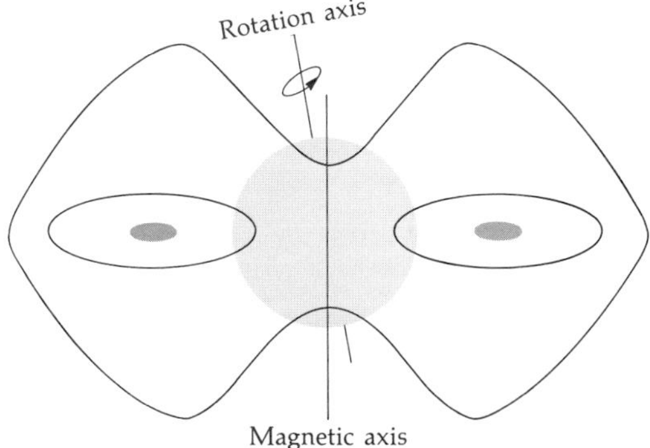

8-15 *Contours of radio emission from Jupiter as observed at a wavelength of 10 centimeters. The strongest radiation comes from the two shaded areas located about two Jovian radii on either side of the planet. The circle represents the size of Jupiter as seen in visible light.*

This reminds us of the many pre-discovery observations of Uranus (Chapter 4).

Statistical analysis of the many observations of the radio bursts from Jupiter, dating back to 1950, has led to a theory of the burst-emitting region that would have stretched the imagination of even a science fiction writer. The bursts are emitted *directionally;* they occur in three rather wide beams that rotate with the planet. However, we do *not* observe bursts every time a beam crosses the earth. The likelihood of an occurrence has been found to correlate very strongly with the orbital position of the Jovian moon Io. Apparently the satellite triggers storms of radio bursts! Their origin thus apparently involves charged particles trapped in the magnetic field, perhaps in localized concentrations, which are disturbed by the presence of a moon.

9

Our Sun

The sun, our closest star, supplies the energy that maintains life on earth. Because of its proximity it is also a unique target for astronomers; it is the only star for which we can observe the detailed features of the surface. Since serious solar study began in 1610, long before the advent of modern physics, early investigations were confined largely to observations of the positions and sizes of the solar surface features that are most conspicuous as viewed in visible light—the sunspots.

SUNSPOTS

Dark areas on the sun were noticed occasionally by observers before the advent of the telescope. The idea that these sunspots were actually planets seen against the bright background of the sun was ruled out by Galileo, who found that the spots moved across the sun in the same direction and at the same rate; in this way he discovered the rotation of the sun. Taking this solar rotation into account, we can, by analogy to the earth, define the solar equator, axis, and poles (Figure 9-1).

Watching spots move across the sun was the first of several methods used to study the sun and solar rotation (Figure 9-2).

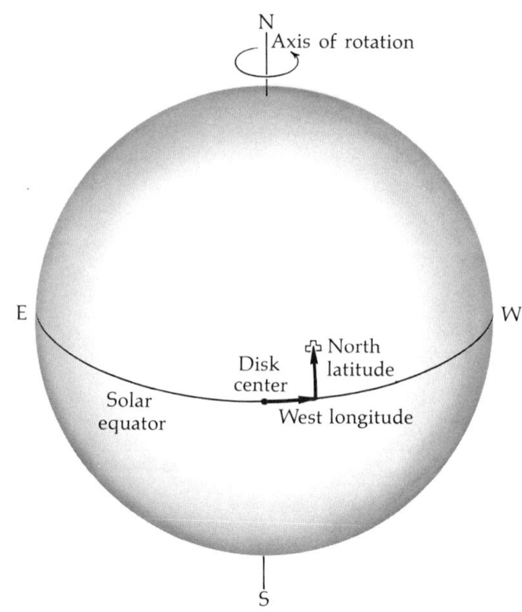

9-1 *The solar coordinate system.*

In view of the early date of discovery of sunspots and the relative ease of observation, it is surprising how slowly the basic facts were accumulated. The number of spots and spot groups changes from day to day, sometimes going slightly up, sometimes slightly down. Over longer periods of time, however (such as a few years or more), distinct trends are evident, as shown in Figure 9-3. The sunspot number increases gradually, reaches a peak, declines to a minimum value (sometimes there are no sunspots at all), and then increases again. The interval between peaks (they are called *sunspot maxima*) averages eleven years, the length of the *sunspot cycle.* By definition, a sunspot cycle begins at *sunspot minimum,* when a few new spots appear.

These basic facts were not established until 1851 (fully 240 years after the discovery of sunspots), when the work of the German amateur astronomer Heinrich Schwabe received wide attention. He had searched for a new planet between Mercury and the sun, and instead discovered a fundamental fact of solar physics—the periodic variation of the number of sunspots. (In fact, Schwabe had published his results about seven years earlier, but they had been largely ignored until 1851, when Baron von Humboldt mentioned them in his book *Kosmos.*)

Watching sunspots form (sometimes as individual spots and sometimes in groups), move across the solar disk, and gradually

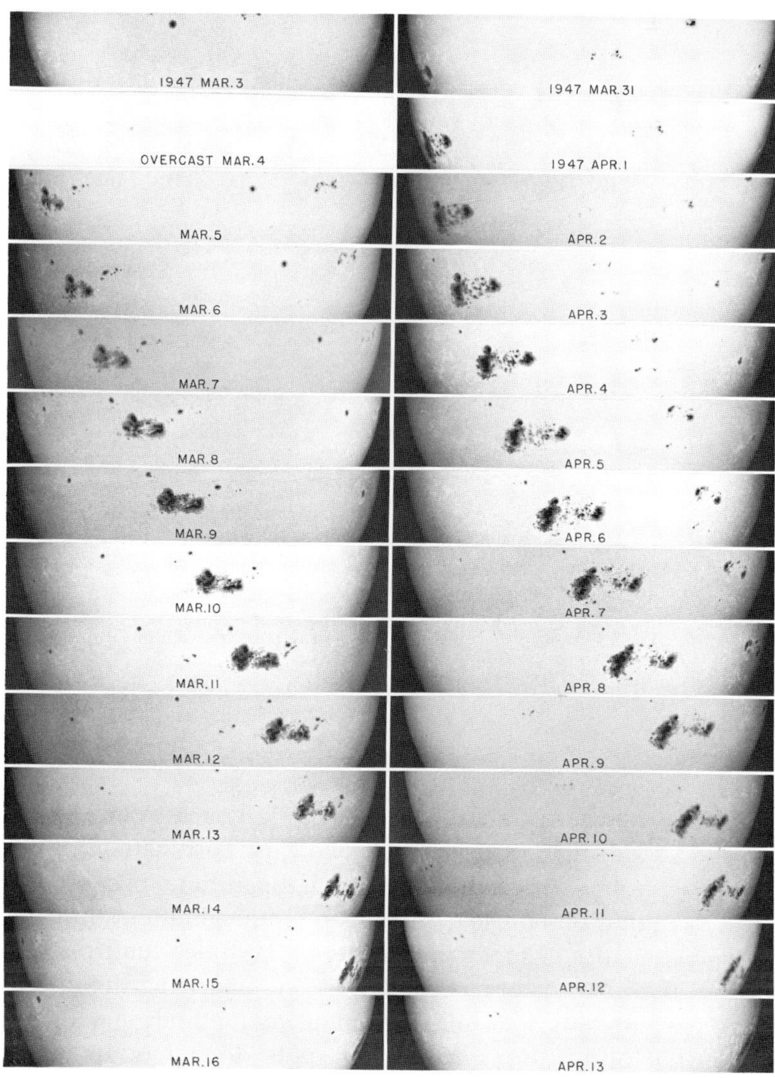

9-2 *Sunspots and solar rotation. A very large sunspot group is shown during two solar rotations in 1947. A solar rotation takes 27 days, so we see the spot group for about two weeks at a time. (Courtesy of the Hale Observatories.)*

fade away, it was natural to mark down their positions each day. These measurements have shown that the rotation is actually *not* uniform. In 1863, the English amateur astronomer Richard C. Carrington reported that the spots at low latitudes take less time for a complete rotation about the axis than do spots located closer toward the poles. This was a remarkable result. If the equivalent situation existed on the earth it would

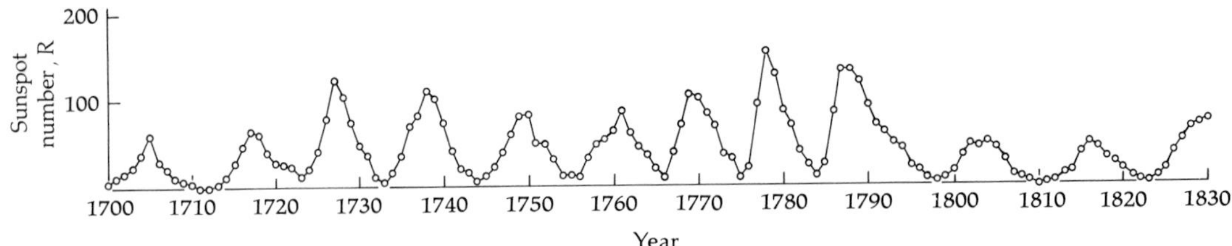

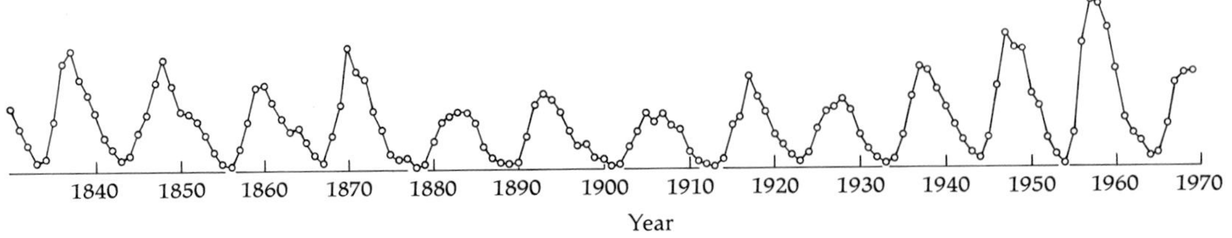

9-3 *The solar cycle during the years 1700 to 1969 as shown by the annual average sunspot number. This quantity is defined as the number of individual spots plus ten times the number of sunspot groups and is adjusted to allow for the fact that individual observers with different ability, equipment, and viewing conditions tend to count more or less spots than the average observer.*

mean that northern Canada would take longer to rotate about the earth's axis than does, say, Mexico City. Fortunately for the people who draw maps this does not happen here, because the earth is a solid body. Clearly, the sun is not solid and thus the *differential rotation* of sunspots at different latitudes confirms that the sun is a gaseous object. Table 9-1, summarizing the differential rotation, was determined from newer methods in addition to sunspot observations, since the spots do not occur at all locations on the sun.

Carrington also noted that the spots do not appear in the same locations on the solar disk throughout the sunspot cycle. Near the beginning of the cycle, the new spots form mostly between the latitudes of 20° and 30° in each hemisphere. As time passes and the spots form in larger numbers, they appear at lower latitudes, and near the end of the cycle, approaching sunspot minimum, the few spots that still form usually occur below latitude 15°. When the latitude positions of spots are marked on a graph that shows how they change during the cycle, a striking wing-shaped pattern appears (*butterfly diagram,* Figure 9-4). Individual spots from a cycle that has just ended may persist on the sun briefly while the first spots of the next cycle are forming. Thus, one occasionally sees spots at low latitudes

Table 9-1 *Differential Rotation of the Sun*

LATITUDE	ROTATION PERIOD
0° (equator)	25 days
20° N and S	26 days
40° N and S	28 days
60° N and S	31 days

from the previous cycle, and other spots near latitude 30° from the new cycle.

In addition to Carrington's two major discoveries about sunspots, he also observed the first recorded solar flare (see page 253) on September 1, 1859. These achievements boded well for Carrington's future scientific work, but unfortunately, the responsibility of running the family brewery when his father died effectively ended his scientific career.

Sunspots are found in the *photosphere* (Greek, sphere of light), the region of the solar atmosphere that produces the visible light of the sun. Although it is gaseous, the photosphere is frequently referred to as the surface or visible surface of the sun. Even ignoring the sunspots, the photosphere does not appear to be smooth and uniformly bright when examined with a telescope under conditions of good seeing. In particular, the *granulation,* a grainy pattern of small bright markings, spaced by narrower dark lanes, covers the photosphere (Figure 9-5), and motion pictures show that the individual granulation cells are temporary features that exist for only a few minutes. In fact, movies of the photosphere reveal that the whole granulation pattern is in seething motion, reminiscent of a pot of boiling water. A typical granule is only about 1,100 kilometers (700 miles) in diameter, very small in contrast to the sun itself, which has a diameter (measured at the photosphere) of about 1,400,000 kilometers (864,000 miles).

The life of a sunspot begins when two or more narrow, dark features (called *arch filaments*) form in a small region of the photosphere, and then grow larger and merge to become a recognizable spot. It is believed that the arch filaments correspond somehow to the emergence of curved lines of magnetic force from below the visible surface. This process has been observed to occur in less than one hour and such a small spot (with a diameter of only about 2,500 kilometers or 1,600 miles) is called a *pore.* The pore stage lasts from a few hours to a few days, after which most sunspots appear to fade or dissolve into

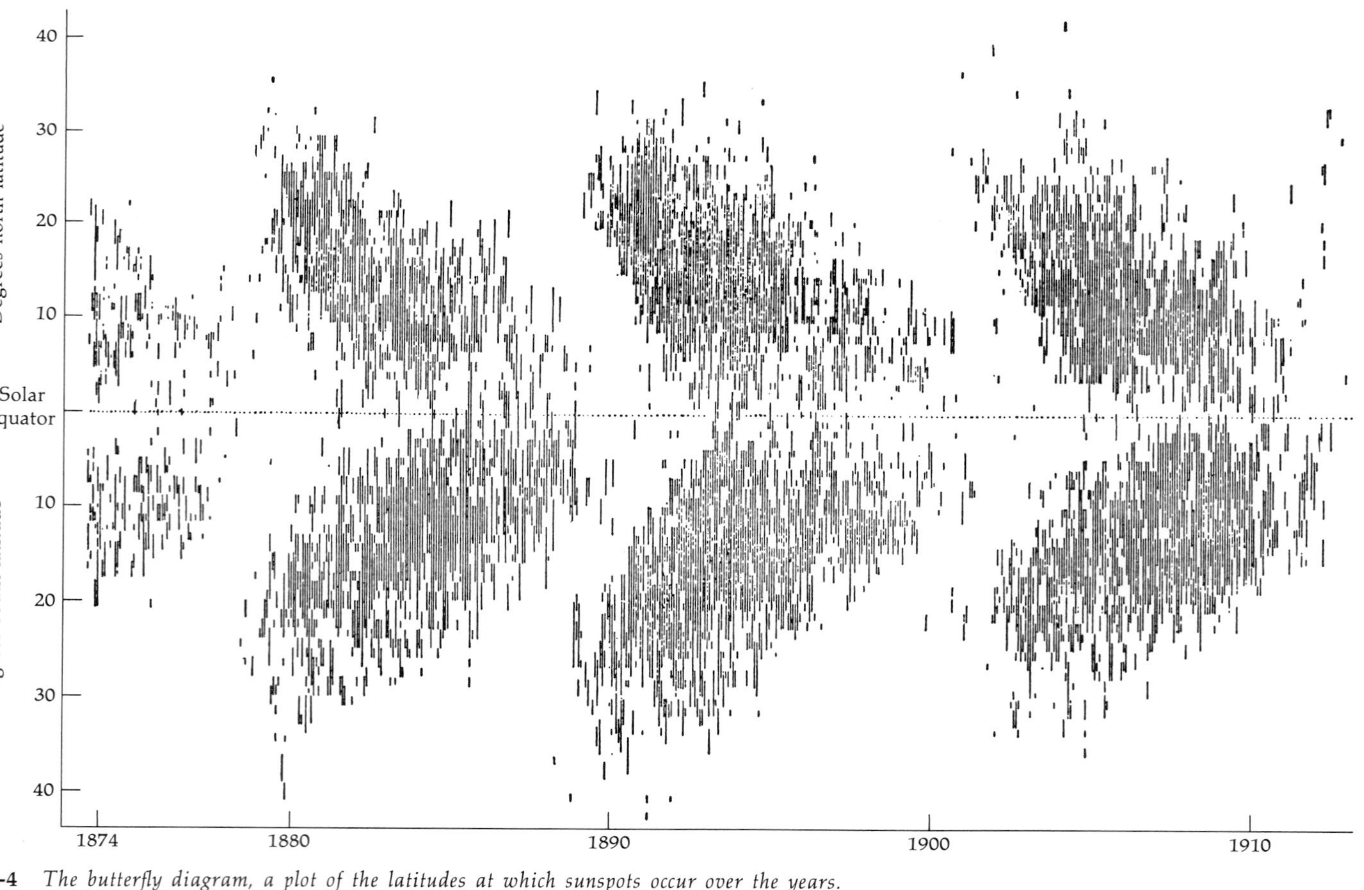

9-4 *The butterfly diagram, a plot of the latitudes at which sunspots occur over the years.* (*Courtesy of the* Monthly Notices of the Royal Astronomical Society.)

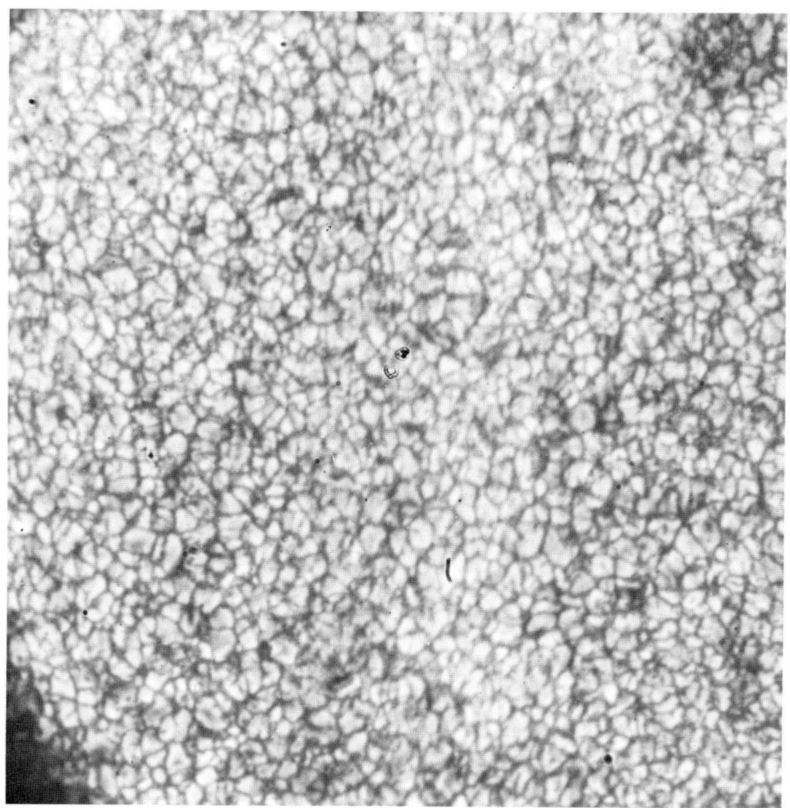

9-5 *The granules on a small part of the sun, as photographed through a telescope carried aloft by a balloon. (Courtesy Project Stratoscope of Princeton University, sponsored by the Office of Naval Research, the National Science Foundation, and the National Aeronautics and Space Administration.)*

the background pattern of photospheric granulation. A small fraction of the spots, however, grow larger and develop a surrounding, less dark region called a *penumbra;* the central dark part is now called an *umbra.* Photographs of penumbrae show that they are predominantly made up of elongated or filamentary structures that point outward from the central umbra. A photograph of a sunspot (taken with a balloon-borne camera) is given in Figure 9-6; the umbra and penumbra are clearly visible. Finally, sunspots that survive the pore stage may occur together in *sunspot groups* (Figures 9-2 and 9-7). These can achieve lengths of over 100,000 kilometers (62,000 miles).

Before discussing the physical properties and origin of sunspots and other aspects of solar activity, we will summarize the basic properties of the underlying or *quiet sun* and the source of its energy.

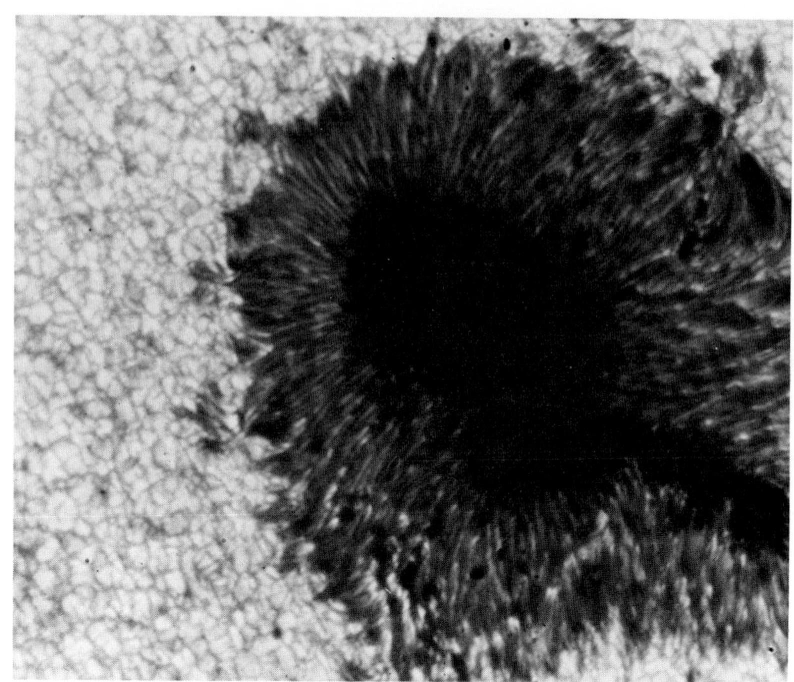

9-6 *A sunspot, also as photographed by a balloon telescope. (Courtesy Project Stratoscope of Princeton University, sponsored by the Office of Naval Research, the National Science Foundation, and the National Aeronautics and Space Administration.)*

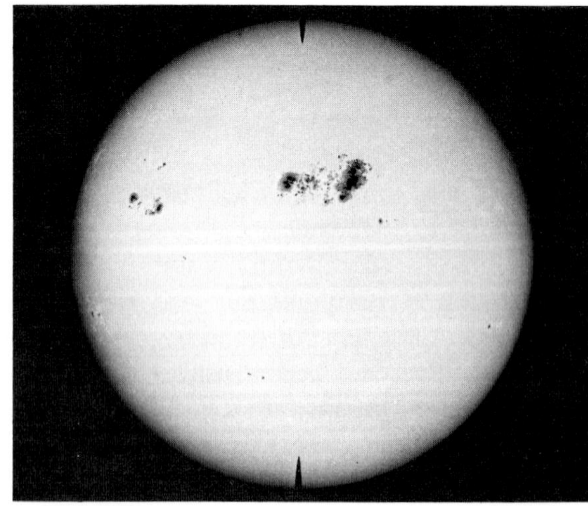

9-7 *A photograph of the sun in visible light showing sunspot groups and limb darkening. (Courtesy of the Hale Observatories.)*

THE SOLAR SPECTRUM AND THE PHOTOSPHERE

The visible light of the sun is emitted from the photosphere. By studying the spectrum of this light we can determine the physical conditions, such as temperature and pressure, under which it is formed. The most prominent features of the solar spectrum are the absorption lines that are named for the Bavarian physicist Joseph Fraunhofer (1787–1826).* Fraunhofer first noted the lines in 1817. He counted more than 500 of them (note the dark lines in Figure 9-8) in the solar spectrum and used letters to name the strongest ones. Now we know the origins of these lines; for example, the dark lines in the yellow region of the spectrum that he referred to as the D lines are produced by sodium atoms.

The solar spectrum can be given a simple interpretation on the basis of Kirchhoff's Laws (Chapter 5). Photons are supplied to the photosphere from the interior, which can be regarded as an incandescent source. The photospheric gases are at a lower density and, hence, the disk spectrum shows the characteristic absorption lines corresponding to the various elements, such as hydrogen, iron, and sodium, in the atmosphere. This interpretation can be corroborated by observing the spectrum of the next outermost layer of the solar atmosphere (*chromosphere*) on the edge of the sun, where the incandescent interior is not behind it. The chromosphere is hard to observe under ordinary conditions, but this can easily be done during a total eclipse of the sun. This spectrum (Figure 9-9) of the chromosphere consists of emission lines, as expected from Kirchhoff's Laws. It is called the *flash spectrum* because it is only observable for a brief time during the eclipse. In fact, the strong emission lines in the flash spectrum correspond to lines seen in absorption in the solar disk (photospheric) spectrum. Thus, we find a simple demonstration of Kirchhoff's Laws in the sun.

Kirchhoff's Laws are rather general statements. Looking at the process of radiation transfer in more detail, we find that photons generated in the solar interior diffuse upward through the atmosphere, being acted upon by the two different processes of pure absorption and pure scattering (discussed in Chapter 5). Pure absorption in the solar atmosphere is mostly due to H^-, the negative hydrogen ion. In this atom, the proton com-

---

*Four of the strongest Fraunhofer lines were actually discovered by the British chemist W. H. Wollaston, in 1802. Wollaston thought they represented the boundaries of different colors of sunlight.

IDENTIFYING SYMBOL		WAVELENGTH	
FRAUNHOFER	OTHER	IN ÅNGSTROMS	ELEMENT
C	Hα	6,563	Hydrogen
D		5,890	Sodium
		5,896	Sodium
	b	5,167	Magnesium
	b	5,173	Magnesium
	b	5,184	Magnesium
F	Hβ	4,861	Hydrogen
	Hγ	4,340	Hydrogen
	Hδ	4,102	Hydrogen
	Fe	4,046	Iron
H		3,968	Calcium
K		3,934	Calcium

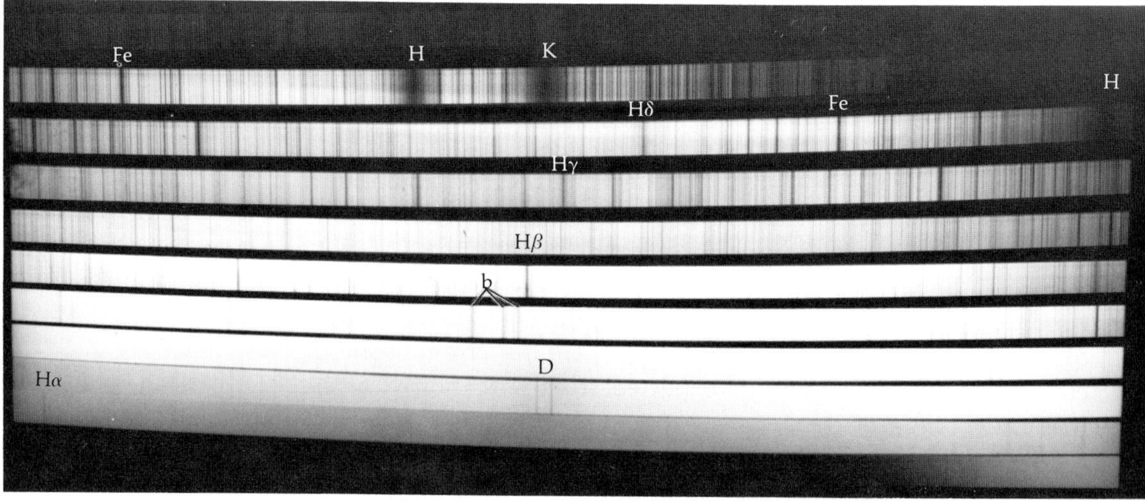

9-8 *The visible light spectrum of the sun. Some of the principal absorption lines, the elements producing them, and the wavelengths are given in the accompanying table.* See also Plate 7, following p. 260. (*Courtesy of the Sacramento Peak Observatory, Air Force Cambridge Research Laboratories.*)

pleted its outermost shell of two (Figure 9-10) and became helium-like by simply adding another electron. However, the singly charged positive proton cannot really balance the negative charge of two electrons and as a result the second electron is bound only loosely to the proton. A photon of visible light has enough energy to knock off the second electron. This effect

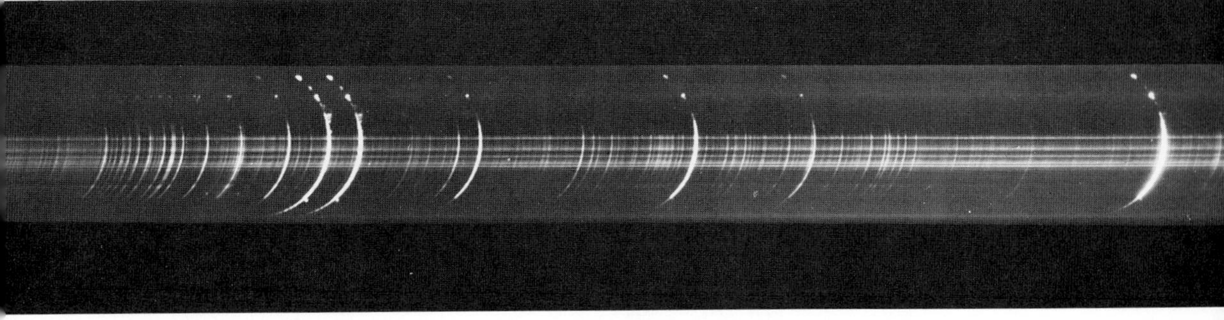

9-9 *The flash spectrum observed at the solar eclipse of January 24, 1925. The crescent shape of each individual emission line corresponds to the shape of the emitting region above the moon's limb. (Courtesy of the Hale Observatories.)*

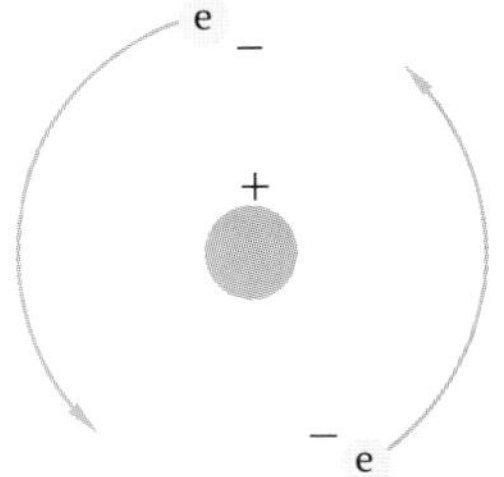

9-10 *Schematic of the H^- ion.*

provides a source of *continuous* absorption* of photons of various wavelengths and a photon of entirely different energy may be emitted later when another second electron combines with the hydrogen atom to form a new H^- ion. Thus, radiant energy migrates upward through the photosphere via photons that are continually being absorbed and re-emitted by H^- ions. If an atom (for example, sodium) that scatters photons at its characteristic wavelengths is present, then the upward migration of photons with these wavelengths is impeded and the chances of being absorbed by H^- and re-emitted at other wavelengths are increased. This process, illustrated schematically in Figure 9-11, produces the Fraunhofer lines. The model that we have just described for the formation of the photospheric absorption lines differs from the simple picture of Kirchhoff's Laws, because the incandescent (continuum producing) region and the selective absorption region overlap, a situation not covered by Kirchhoff's investigations.

---

*Continuous, because photons of different wavelengths, not just those at or near one wavelength (as in absorption line), are affected.

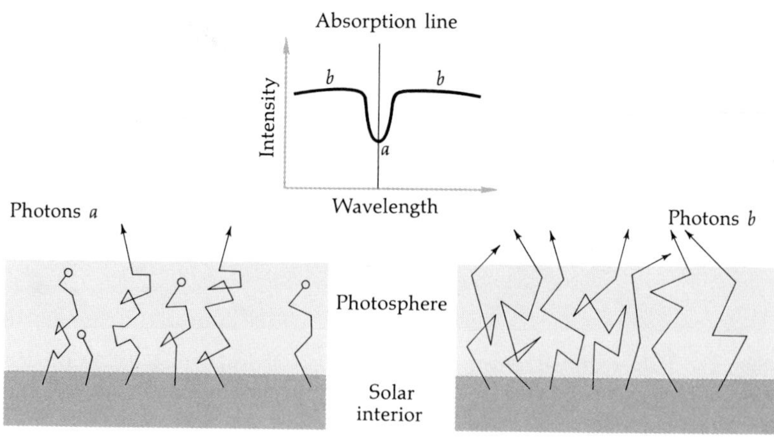

9-11 *Schematic of photon diffusion in the solar atmosphere. Photons* a *with wavelengths close to the characteristic wavelength of an atom are scattered relatively frequently, find it harder to diffuse through the photosphere, and are preferentially absorbed. Scatterings are denoted by bends and absorptions by circles. Photons* b *with wavelengths different from the characteristic atomic wavelength diffuse through the surface. The term scattering is used loosely in describing the movement of photons* b*. In fact, the process is the ionization and recombination of* H^- *(described in the text), but continuum photons* b *are involved in both the ionization and recombination. Thus, the observer outside the sun sees less of photons* a *and more of photons* b*—a situation which produces an absorption line.*

What do the Fraunhofer lines tell us about the photosphere? First, by comparing wavelengths measured for these lines in the solar spectrum with the wavelengths in the spectra of known elements, as recorded in the laboratory, we can identify elements present in the sun. Lines of one element were actually observed in the chromospheric emission spectrum in 1868, before the element was discovered on earth. It was therefore called helium (after the Greek word, *helios,* for the sun). It was subsequently recognized on the earth, in the atmosphere, in minerals, and in gas wells. Not all the elements can be detected in the solar spectrum, but it is a sensitive indicator for many of them.

The strength of a particular absorption line changes, depending on the amount of the element that occurs in the sun or other stars. The line will be darker when more of the element is present.* Thus, a careful analysis of the spectrum can show not

*The strength of an absorption line depends on other factors as well, especially on the temperature, and this is taken into consideration when a spectrum is analyzed to determine the abundances of the elements.

only what elements are present, but also their amounts. This is the basic method used to determine the compositions of the stars. When we speak of stellar abundances we refer to the compositions of the photospheres of stars, as it is those atmospheric regions that give rise to most of the light that we can observe. The interiors of stars are so hot and dense that nuclear reactions occur, such as the transformation of hydrogen to helium, and thus the abundances of the elements inside a star can differ from the atmospheric abundances. The continuous spectrum (*continuum*), namely the general dependence of light intensity with wavelength, disregarding the Fraunhofer lines, enables us to determine the temperature of the photosphere by comparing it with the Planck Law. The solar continuum is a reasonably good match for an ideal radiator curve at 6,000° K, as given by the Planck Law and as shown in Figure 5-22.

The photospheric temperature can also be inferred from the *total* radiant energy emitted by the sun (called the *solar luminosity*), the solar surface area, and the Stefan-Boltzmann Law. If we measure the *solar constant* (the rate at which solar radiation of all wavelengths is received by a surface of one cm^2 at a distance of 1 a.u. from the sun), then we can calculate the amount of radiation passing through the surface of a sphere of radius 1 a.u. Since we know the surface area of the sun we know how much energy each square centimeter must emit to produce the amount observed here at 1 a.u. Then, the temperature needed follows from the Stefan-Boltzmann Law and a value of about 6,000° K is also found.

Thus, the temperature of the solar photosphere of about 6,000° K (11,000° F) is found by two different methods. (There are other, more complex methods as well.) The agreement is not exact, nor should it be; the sun is not quite an ideal radiator as assumed in the formulae used to determine the temperature.

Does the temperature in the photosphere increase or decrease as we go into the sun? Most people would expect an increase, but can we prove this? In visible light photographs, the photospheric regions (such as granules) with the higher temperatures appear brighter, and the regions with the lower temperatures appear darker. The solar *limb darkening* as shown in Figure 9-12 provides the evidence needed. When we look across New York City from the top of a very tall building, we can see to a certain distance, depending on the amount of dust and smog in the air. Likewise, when one looks into the photosphere one sees light originating from a depth determined by the continuous absorption of the H^- ion. Because of the geometrical effect shown in Figure 9-12, one sees to a greater depth at the center

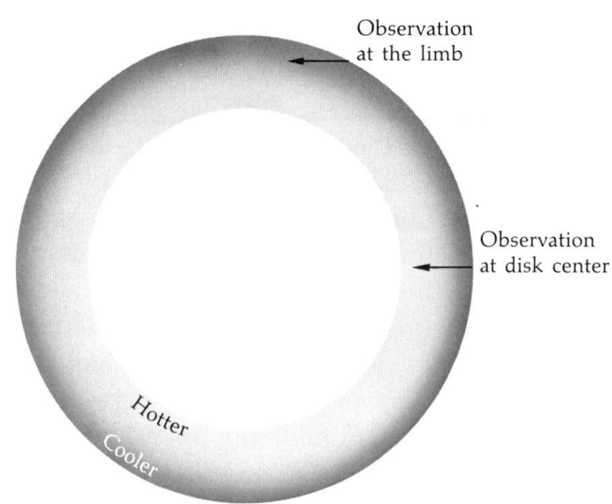

9-12 *The geometry of solar limb darkening.*

of the disk than at the limb. If the limb intensity is higher or lower than the disk center, the temperature would decrease or increase inward, respectively. The limb actually appears darker than the disk center, and hence the temperature in the solar photosphere increases inward as expected. Surprisingly, this trend reverses in the chromosphere and corona (page 250).

THE SUN'S ENERGY

The interiors of the sun and stars cannot be observed directly at present. However, we can use the total energy output of the sun, the *luminosity,* to probe the solar interior indirectly. Since the basic energy source of the sun lies in the interior, the conditions there must be capable of producing the solar luminosity.

The amount and source of energy radiated by the sun was an astronomical mystery that was solved only in the late 1930s. It was an important problem because it is the sun's energy that sustains all life on earth. The basic datum is the *solar constant,* and it is difficult to measure because it includes energy radiated over all wavelengths of the electromagnetic spectrum. Thus, one problem is that the best detector for one region of the spectrum is not necessarily the best for another wavelength region (Chapter 6). Furthermore, not all wavelengths pass through the earth's atmosphere. However, most of the radiation from the sun can reach the surface of the earth, and by making theoretical estimates for the small amount of energy that does not penetrate the atmosphere (or more recently, by observing these wavelengths from space vehicles), the solar constant has been determined to be about 2 calories per square centimeter per minute (note the unusual units; the calorie was defined in Chapter 5). The total energy radiated by the sun (*solar luminosity*) can be found by multiplying the solar constant by the number of square centimeters on the surface of a sphere of radius 1 a.u. This yields the enormous value of 9×10^{25} calories per second, or 4×10^{26} watts.

Studies of rocks and fossils suggest that the sun has been emitting roughly the same amount of energy during at least the major part of the earth's lifetime. The argument for this is rather simple: if the earth received very much more solar energy than it does now, water would exist mostly as vapor, rather than as a liquid, and if the earth received much less solar energy than it does now, water would exist primarily as ice. But the fossils of aquatic life forms, and the existence of very old sedimentary rocks formed by the deposition of eroded material in ancient seas, show that the earth has had extensive bodies of liquid water for at least several billion years. This does not rule out fairly small changes in the solar constant, which some scientists believe to have occurred and to have caused the ice ages.* Note the similarity of this argument to Rasool's theory of planetary atmospheres (Chapter 8). It is possible that a greenhouse effect occurred during the primitive atmosphere stage of the earth. In that case, a loophole exists in our argument, since the sun might have emitted somewhat less energy than it does now, and yet the greenhouse effect might have kept the earth warm. However, our argument for the rough constancy of the solar luminosity would still hold true for the time since the present atmosphere formed. Thus, an energy source must exist to have produced the solar luminosity of 4×10^{26} watts over at least the last few billion years. The road to the discovery of this energy source was a long one, and several wrong turns were taken on the way.

Attempts to understand the source of the sun's energy were made by Julius R. Mayer, a young German physician, in 1848. He was among the first to note that the sun was the ultimate source of all heat and power on earth. (Even when we burn fuel we remove from "coal storage" the sunshine energy of past geologic times.) Some of the first ideas considered by Mayer

*No existing theory of the ice ages seems convincing to most scientists. This theory is mentioned only for purposes of discussion.

can easily be discarded. If the sun were simply a big lump of hot gas radiating at the sun's present luminosity (but without any internal source for continuing to heat the gas) its temperature would drop severely in 5,000 years. If the entire mass of the sun were coal, it could sustain its present luminosity for only 4,600 years. (If the earth were a solid lump of burning coal it could emit at the rate of the solar luminosity for only five days.) Of course, the calculations ignore the problem of obtaining oxygen for the combustion and the problem of getting rid of the ashes, both of which would shorten the lifetime, but they do illustrate the magnitude of the problem.

Clearly a better source of energy was needed and Mayer suggested that the sun's atmosphere was bombarded by a continuous rain of meteoroids. By friction with the solar gas, the meteoroids would heat the gas, just as they heat their immediate surroundings when they enter the earth's atmosphere. Unfortunately, the mass of meteoroids required was huge; in particular, it was inconsistent with the known rates of meteor occurrence at the earth. Moreover, the required mass of meteoroids impacting the sun would increase the solar mass at a rate that would be detectable in the motions of the planets, since the gravitational force of the sun on each planet would be increased; this is not observed and Mayer's hypothesis must be discarded.

The next attempt was made by the German physicist Hermann von Helmholtz in 1854; the idea was also discussed by Lord Kelvin and is called the Helmholtz-Kelvin hypothesis. According to this idea, the sun is shrinking under the force of its own gravity and the solar gases are thereby compressed. We know from experience that compression heats a gas and expansion cools it (refrigerators cool their contents by expanding a gas). So the sun, if compressing itself, would produce heat. The Helmholtz-Kelvin hypothesis greatly expanded the length of time over which one could imagine the solar luminosity being maintained. The observed solar luminosity according to this theory would be produced by a shrinkage of about 60 meters (200 feet) per year in the diameter of the sun. At this rate, the sun would last some 15 million years, since its diameter is about 1.4×10^6 kilometers. (These numbers do not check exactly because the yearly rate would change as the sun shrank.)

Even a time scale of millions of years, however, is not sufficient to explain the geological evidence that requires times of billions of years. Hence, the Helmholtz-Kelvin process does not furnish the energy to support the solar luminosity over the necessary time scales. Nevertheless, we will see that such a process probably was important in the early stages of the sun's history.

The actual source of the sun's radiant energy* is now known to be nuclear energy, which is released according to Einstein's equation, $E = mc^2$. Deep within the sun a series of nuclear reactions called the *proton-proton chain* converts hydrogen (protons) into helium by merging sets of four hydrogen nuclei into one helium nucleus (alpha particle). The mass of the helium is slightly less than the sum of the four hydrogen masses (Chapter 5, page 106) and the missing mass is released as energy. Hotter stars than the sun mostly derive their energy from another chain of reactions (*carbon cycle*) in which hydrogen is also converted to helium, but carbon, nitrogen, and oxygen atoms are involved in the reactions.

The central temperature of the sun is about 15 million degrees K. At temperatures as high as this, atomic particles are highly energetic and upon collision can penetrate the nuclei of other atoms. If the sun's central temperature were too low, the collisions would not be energetic enough and nuclear reactions could not occur. In the early stage of solar formation, the Helmholtz-Kelvin type contraction played an important role in raising the central temperature of the sun above the threshhold value where the nuclear reactions could begin. At the currently estimated temperature of about 15 million degrees K, nuclear reactions can supply the sun's entire luminosity. By heating the solar interior, the reactions also raise the pressure there to the point where the difference in pressure between the interior and surface regions balances the gravitational force on the gas so that the contraction no longer occurs.

There is enough hydrogen in the sun to generate the present solar luminosity by the proton-proton chain for another 100 billion years. However, the theory of stellar evolution (Chapter 11) predicts that long before this time, changes in the internal conditions of the sun will lead to a different set of energy generating reactions.

THE SOLAR MODEL

The sun's energy is liberated by nuclear reactions in its central regions. Energy is transported outward by photons that are scattered, absorbed, and re-emitted (as discussed above) throughout the bulk of the solar interior. The gas itself must be reasonably transparent to radiation in order for this energy transport to be efficient. This condition is no longer met at a point about 85 per cent of the way from the center to the surface

---

*Based on the work of the nuclear physicist Hans Bethe in 1938.

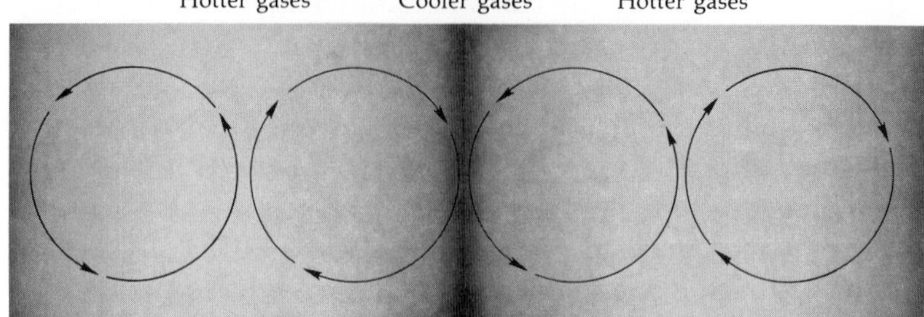

9-13 *Convection schematic (shown in vertical cross section). Hot gas masses rise and cool; cool masses sink and are warmed again. Arrows indicate direction of gas motion. As seen from above, rising masses are brighter than falling masses, since they are hotter.*

of the sun. In this region, the conditions of pressure and temperature are such that He^+ (singly ionized helium) ions form, and they can be photoionized; thus the solar material becomes relatively opaque. Since radiation becomes an ineffective method of energy transport at this point, the energy finds other ways to travel outward. From this point to just below the photosphere, the mechanism of convection becomes important for the energy transfer. In the higher parts of the convective region, the solar material is made opaque by the existence (and ionization) of, first, H (neutral hydrogen) atoms, and then H^- ions in the region just below the photosphere. Currents of solar material circulate between the hotter regions below and the cooler regions above. Convection currents can often be seen above a radiator, or a paved road warmed by the sun. This method of convective heat transfer is, in fact, often used to heat homes. Hot air or water is generated by a furnace and circulated to heat the rooms which are cooler than the area near the furnace. The convection currents are often in the form of cells with regular flow patterns, as shown in Figure 9-13. In the sun, the hotter gas is brighter than the cooler gas and the corresponding pattern influences the photosphere above, as can be seen in Figure 9-5. Thus, the photosphere shows this regular cell structure called the granulation because energy transport takes place by convection just below it.

The photosphere is a relatively thin skin 300 kilometers (about 200 miles) thick, forming the visible surface or *disk* of the sun. The temperature there is about 6,000° K. Energy travels through the photosphere by radiation, as noted on page 239. This is because the density of H^- is greatly reduced in the photosphere (compared to the upper convection zone) so that, although H^- is the major absorbing constituent in the photosphere, it is not plentiful enough to almost cut off the flow of photons as it does in the convection zone.

Photons traveling outward from the photosphere have little additional interaction with the solar atmosphere above the photosphere (because the outer atmosphere is a rarified gas, a photon collides with an absorbing atom only relatively rarely), and hence the light we measure on earth is literally from the photosphere. The solar model and energy flow are shown schematically in Figure 9-14.

Just as the atmosphere of the earth thins out with altitude above the surface, so the solar gas extends outward an enormous distance above the photosphere, growing more and more tenuous with altitude. The region just above the photosphere, reaching to heights of several thousand kilometers, is called the

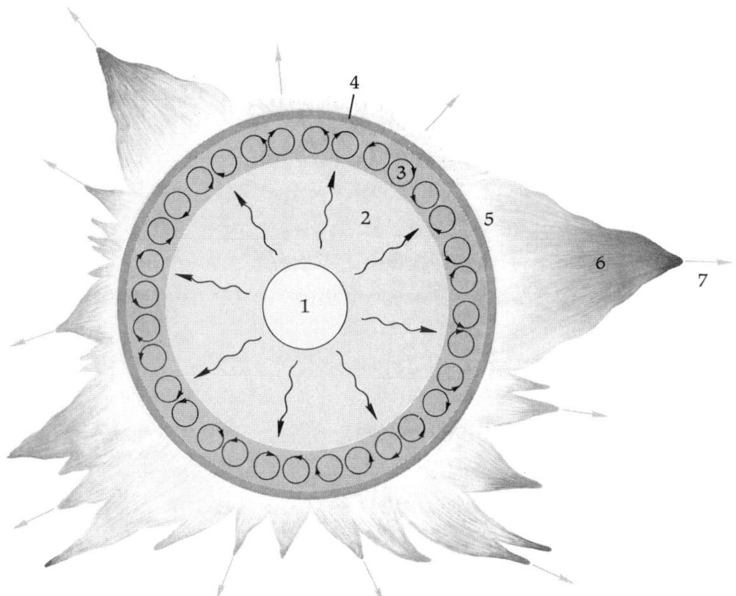

9-14 *Schematic of solar structure and energy flow. Energy is generated in the core (1) of the sun by nuclear reactions and transported through most of the interior (2) by photons. Although photons carry energy throughout the sun, convective motions (3) transport most of the energy in the region located between about 0.85 solar radii from the center and a level just below the photosphere (4) where radiative transport by photons is dominant again. Above the photosphere the sun's radiation does not interact strongly with the remaining material and light emitted from the photosphere travels outward with little hindrance. Acoustic energy generated in the convection zone (3) is deposited in the low corona (5), heating it. Throughout most of the corona (6), energy transport is by conduction. Far from the sun, the energy is still carried outward in the bulk motions of the solar wind (7).*

chromosphere ("color sphere"), named for its intense red color seen at solar eclipse. Above this is the solar *corona,* which occupies so enormous a volume of space that its size is usually expressed in units of the solar (photospheric) radius (about 700,000 kilometers). During total eclipses, astronomers on the ground have observed the coronal light out to as far as ten solar radii from the photosphere (Figure 9-15); beyond that point the coronal light is too weak to be distinguished from stray light in the earth's atmosphere, even during the eclipse. However, observations with radio telescopes show that the corona extends out even farther. Eventually, as the corona thins out with height, it must merge into the background gas of the interplanetary medium. Matter, in the form of electrons and protons, streams away from the corona and can be detected as the *solar wind* in the vicinity of the earth and the moon. (A major task of the *Apollo 11* astronauts, after making the first manned lunar landing, was to set out a solar wind collector to trap some of these particles for scientific analysis; see Figure 15-26.)

Although the mass of the solar atmosphere above the photosphere is trivial compared to the photosphere mass, and its interaction with the bulk of the solar radiation is totally negligible, the chromosphere and corona are nevertheless of unusual scientific interest. This arises from the surprising variation of temperature in the solar gas. The central region of the sun is at about 15 million degrees K and the temperature decreases continually out to the photosphere, where it is about 6,000° K. But the temperature does *not* continue to decrease above the photosphere—it actually increases to 10,000° K in the chromosphere and to 2 million degrees K in the corona. Why is this?

The convection cells beneath the photosphere are noisy and are a source of energy in the form of sound waves. Recall our discussion of longitudinal compression waves in Chapter 2. A common analogy to the generation of sound waves in the sun's convection zone is the sound generated in a boiling pot of water. You do not have to see the water boil, you can hear it. The convective motions in the boiling water are noisy. Thus, energy in the form of sound waves is generated.

Although the sun's outer atmosphere is transparent to light, sound waves are absorbed by the corona. Since the corona has a very small mass, a little acoustic energy goes a long way and the corona is heated to a temperature of 2 million degrees K. This hot gas is illuminated by photospheric radiation and this scattered light gives the corona its pearly appearance visible at the time of the total eclipse (Figure 9-15).

9-15 *The solar corona on November 12, 1966, photographed with a filter which compensates for the sharp decrease in coronal brightness with distance from the sun. The planet Venus is at upper left. (Courtesy High Altitude Observatory.)*

The high temperature of the corona and the relatively low gravity at these distances from the photosphere lead to a continual process in which the corona boils off, producing the solar wind. Thus, physical conditions in the corona cause particles to escape from the sun to interplanetary space. Equilibrium exists, however, for as the coronal material boils off, it is replaced from below.

The corona is a hot, fully ionized gas; that is, virtually every atom has lost at least one electron. (A gas in this state is called a *plasma.**) Since the solar atmosphere is mostly composed of hydrogen, the coronal plasma consists primarily of electrons and

---

*Plasma is regarded as a fourth basic state of matter (after solid, liquid, and gas) and it appears that most of the visible matter in the universe may be in this form, although it is virtually absent on the earth.

protons. As the corona expands into the space between the planets, its heat energy (corresponding to a temperature of 2 million degrees K) is converted into the energy of bulk motion of the solar wind. The solar wind temperature drops to about 100 thousand degrees K at the earth,* but the speed at which the wind moves rises from a very small value near the sun to about 500 kilometers (300 miles) per second at the earth. Typically, there are 5 to 10 solar wind particles per cubic centimeter in space near the earth.

The solar wind is very important in the physics of the solar system. It (1) shapes the earth's magnetosphere (Chapter 15), (2) pushes the ionized comet tails away from the sun (Chapter 7), and (3) strikes the unmagnetized objects in the solar system. For example, some physicists believe that the impact of solar wind particles on the lunar surface darkens the rocks and hence is possibly responsible for the relatively low fraction of sunlight that is reflected from the moon (small lunar albedo).

SOLAR ACTIVITY

Although the sunspots are the most obvious of the solar features there are many other kinds of temporary structures or disturbances on the sun, most of which seem to be associated with the sunspot cycle (Figure 9-16). These occurrences are collectively referred to as *solar activity,* and the complete description of their variations or events over the eleven-year period is called the *solar cycle.* A large sunspot with its penumbra, as described above, is just one aspect of a *center of activity.* Above the sunspot in the chromosphere is a much larger disturbed region called a *plage.* Since the photosphere is much brighter than the chromosphere in ordinary light, the latter region is studied with special filters and spectrographic instruments that isolate light of specific wavelengths that comes primarily from the chromosphere. This includes lines of Ca^+ ions (the term used for calcium atoms that have lost one electron). In pictures of the sun taken at this wavelength (they are called *calcium filtergrams* and *calcium spectroheliograms,* depending on whether they are obtained using a filter or a spectrograph to isolate the wavelengths of interest), the plages appear as bright patches. A major plage

*If the solar wind has this high temperature as it encounters the earth, why doesn't it burn us up? The answer is that it is hot, but it doesn't have much heat. The temperature is high but the number of calories is low because the total mass of hot particles colliding with the earth is very small. Furthermore, the magnetic field of the earth shields us by deflecting the charged particles of the solar wind.

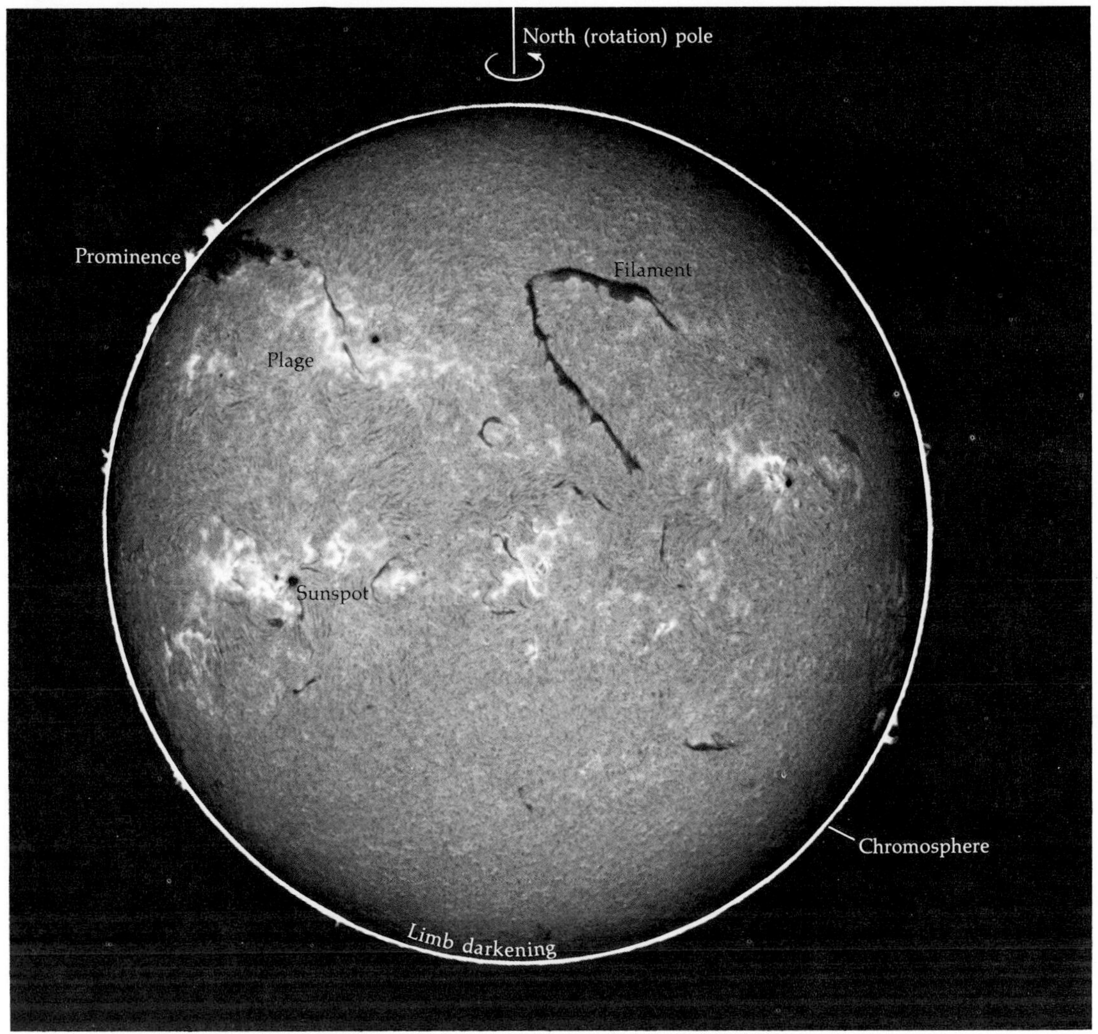

9-16 *Manifestations of solar activity as photographed on March 7, 1970, through a filter that isolates light of the hydrogen alpha line wavelength (6,563 Å). The illustration is a composite of a short exposure to show disk features and a longer exposure to show the fainter prominences and the chromosphere. (Courtesy of H. Caulk and R. Hobbs, Goddard Space Flight Center.)*

may last for several months. Occasionally, a brief, very intense burst of light, primarily from hydrogen atoms, is observed in a plage; often this *flare* (Figure 9-20) occurs above a sunspot. Above the chromosphere the center of activity extends into the corona in the form of a *condensation,* which appears to be denser and brighter than its surroundings. Straight, curved, and loop-shaped gas structures extend from the plage up into the corona. These *prominences* appear as bright objects when they are located

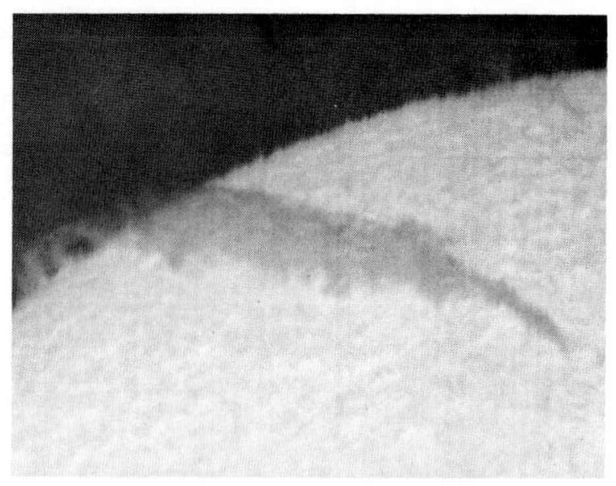

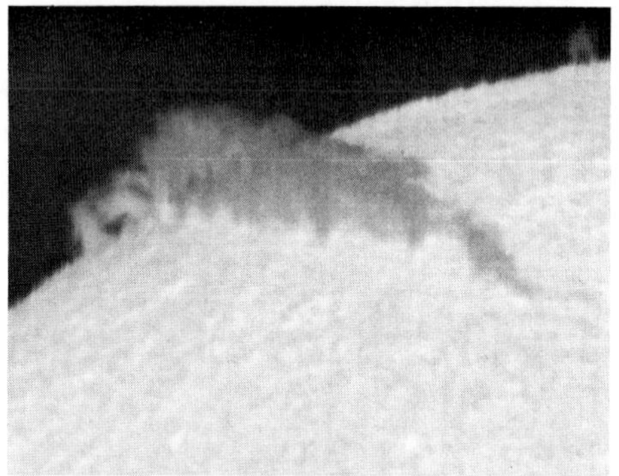

9-17 *A series of photographs made on April 9, 10, and 11, 1959 (top to bottom, respectively), showing that filaments (dark features on the solar disk) and prominences (bright features on the solar limb) are physically identical. (Courtesy Observatoire de Paris—Meudon.)*

at the limb, but are dark by contrast to the photosphere when they are seen on the disk (Figure 9-17), in which case they are called *filaments.* The fact that the filaments look dark when we actually know that they are bright prominences reminds us that bright and dark are relative terms. Actually, sunspots are bright objects, as are the dark parts of the granulation, and only look dark by contrast to their even brighter surroundings. Sunspots have a temperature of about 4,500° K and are thus some 1,500° K cooler than the surrounding photosphere.

The presence of a magnetic field can affect the emission lines produced by an atom. This was discovered in 1896 by the Dutch physicist Pieter Zeeman, who observed the change in the spectrum of a sodium flame when it was placed between the poles of a magnet. Depending on the direction of the magnetic field, an atomic line is observed to split into two or three individual lines; the separation of the lines depends on the strength of the field and the properties of the atom. The *Zeeman effect** (illustrated for a sunspot in Figure 9-18) allows detection, measurement, and mapping of the magnetic fields on the sun, using an instrument called a magnetograph. A solar *magnetogram*, recorded with such an instrument, is shown in Figure 9-19. Comparing magnetograms with filtergrams and spectroheliograms shows that sunspots, plages, and other features of the center of activity correspond to regions of strong magnetic fields that may be hundreds or thousands of times more powerful than the background magnetic field of the quiet sun. This background field is about as strong as the earth's magnetic field, which we detect by its effect on such common objects as the compass needle.

In fact, strong magnetic fields seem to be fundamental to the origin of the centers of activity, as such fields are seen to emerge on magnetograms before the sunspots appear. As the sunspots and the plages grow, so does the magnetic field, and the flares may occur in the chromosphere above the spots. The flares are accompanied by bursts of radio emission observed with radio telescopes and by *x-ray bursts* detected by satellites. The spectrum of an x-ray burst sometimes includes emission lines from iron atoms which may have lost as many as 24 of their 26 electrons due to the temperatures in the flares. These temperatures may reach 50 million degrees K (Chapter 14) and confirm that the flares involve the sudden release of enormous amounts of energy. The explanation of the great energy releases in flares is one of the outstanding unsolved problems in astronomy. Most theorists believe that the strong magnetic fields are in some way responsible for flares.

Even as the sunspots in a center of activity are shrinking from their maximum stage, the magnetic field region and the associated plage continue to grow, and above the plage the prominences (filaments) increase in size, reaching lengths greater than

---

*The principles of the Zeeman effect were applied to the spectra of sunspots by George Ellery Hale and the magnetic fields in sunspots were confirmed in 1908. Hale's contributions to astronomy are legion. He founded the Yerkes, Mount Wilson, and Palomar Observatories, and also the *Astrophysical Journal*. The 200-inch telescope on Mount Palomar was named for him and recently the Mount Wilson and Palomar Observatories have been renamed the Hale Observatories.

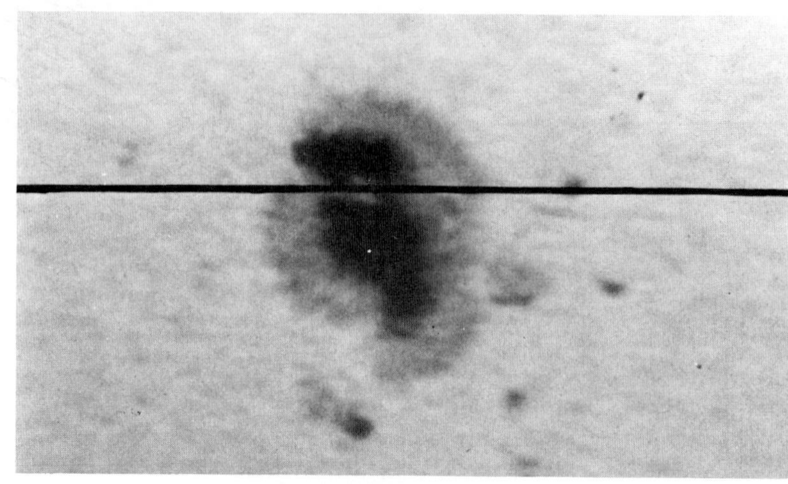

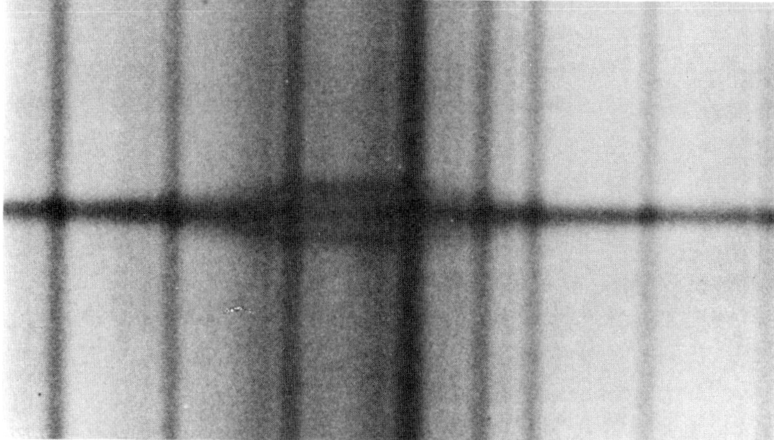

9-18 *Effects of a magnetic field on spectral lines. Top: the dark line across the photograph marks the position on the sun from which light is admitted to the spectrograph. Bottom: an absorption line in the spectrum of the admitted light. In the sunspot, strong magnetic fields split the line into three components through the Zeeman effect. (Courtesy of the Hale Observatories.)*

160,000 kilometers (100,000 miles). Although the plage eventually fades from view, prominence activity continues and long streamers may form in the corona above the weakening magnetic field region. This whole process of birth, growth, and death of a center of activity can take place over periods of ten solar rotations (270 days) or even longer.

An equally remarkable development occurs in the background magnetic field of the sun. The two solar hemispheres

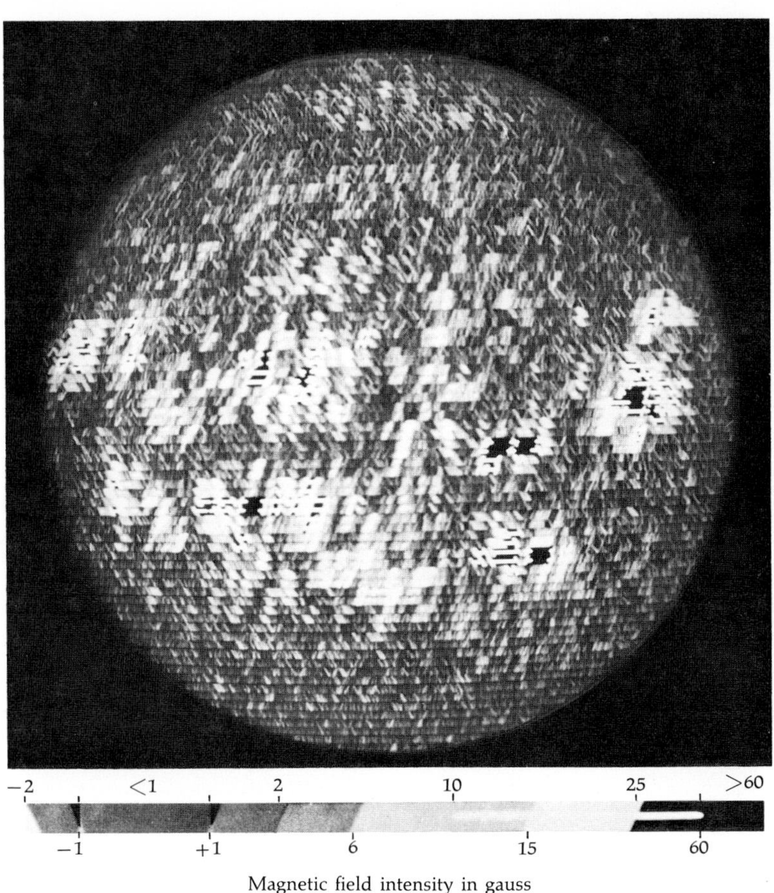

9-19 *Magnetic map of the sun for July 21, 1961. The intensity in units of the gauss is shown on the scale and the polarity by the slant to the right or left. One gauss is about three times the magnetic field strength at the equator of the earth. (Courtesy of the Hale Observatories.)*

have fields of opposite polarity (north and south), as do the earth's hemispheres. These are somehow related to the much stronger fields of the sunspots, because a sunspot group tends to evolve as two subgroups of opposite polarity, separated slightly in solar longitude. These are called the p and f spots (p for preceding, f for following; the p spots are defined as those that lead the others as the sun rotates). It is found that when the p spots in the northern hemisphere of the sun have north magnetic polarity, the p spots in the southern hemisphere have south polarity. But, in the next eleven-year solar cycle, the polarities are reversed, with the northern hemisphere p spots

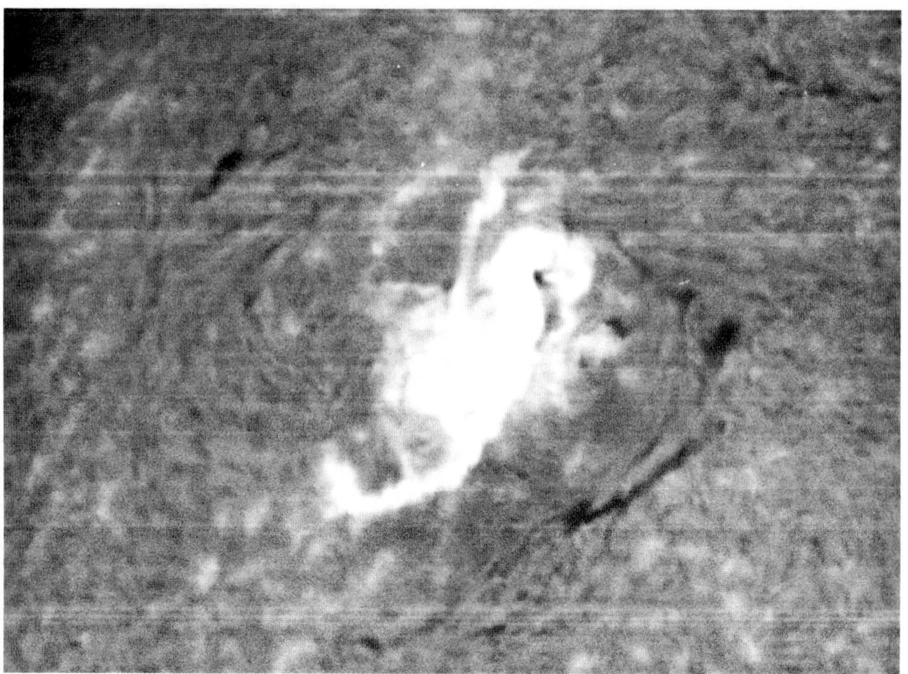

9-20 *A solar flare of July 16, 1959, photographed in the light of the hydrogen alpha line. (Courtesy of the Hale Observatories.)*

having a south magnetic polarity. This seems to be related to the fact that the background magnetic polarities of the two solar hemispheres reverse every eleven years! If this happened on the earth, one would find his north-seeking magnetized compass needle pointing toward the Southern Hemisphere.

The effect of the magnetic field on solar material is apparent in several ways. The material in the outer corona (Figure 9-15) is confined into narrow streamers. Flares (Figure 9-20) are often very directional in terms of their effect on solar features, as if the flare particles were channeled by the solar magnetic field. This confinement or inhibition of ionized material by a magnetic field stems from the effect shown in Figure 5-4; both positive and negative charged particles are deflected by a magnetic field.

Prominences are made possible by the solar magnetic fields, which support condensed coronal gas against the solar gravity. Material that condenses* from the corona to form a prominence

*By condensation, we mean a localized increase in density of the gas, not a change from the gaseous to the liquid state.

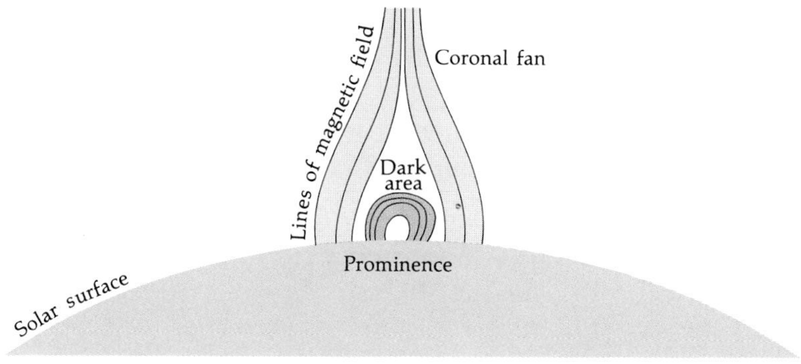

9-21 *Schematic showing the relationship of coronal structures and prominences.*

may leave a void visible as a dark place in the corona above the prominence, as illustrated in Figure 9-21. It appears that matter is continuously condensing from the corona into the prominence and "slides" down the magnetic field into the photosphere.

The strong magnetic fields found in sunspots and the fact that magnetic fields inhibit the motions of plasma particles provide an explanation for the observation that sunspots are about 1,500° K cooler than the surrounding photosphere. As we noted in our discussion of the solar model, the region just below the photosphere is heated by convection currents. This process involves the circulation of material and it is this circulation which is inhibited by the magnetic field because the plasma is constrained to move along (and not across) the lines of magnetic force. Thus, part of the energy in the convective motions must be used to overcome the resistance of the magnetic field. There is then less energy available to be carried upward to heat the photosphere, and a relatively cool region, the sunspot, is formed.

Solar activity is of interest because post-flare effects often occur on earth. A particularly large release of flare radiation by the sun can cause anomalous ionization conditions in the earth's ionosphere and disrupt radio communications; this is called a radio fade-out. In addition, coronal gas particles can be ejected as a result of a flare. One or two days later this material strikes the earth's magnetosphere, leading to the aurora as a side effect. The ability to predict flares would enable us to anticipate these events; the forecasts would be of value to astronauts in orbit or in space where they generally would be outside the protection of the earth's atmosphere and/or magnetosphere. Unfortunately, our knowledge in this area is still in an unsatisfactory state and the current flare forecasts are not very reliable.

Solar activity is also of interest because similar events must occur on other stars. Although we cannot see the surfaces of other stars, they must have "starspots" and other fascinating phenomena of their own.

Plates

1 *Spaceship Earth seen above the lunar horizon from* Apollo 8. *(National Aeronautics and Space Administration.)* See also Figure 17-1.

2 *A beam of white light strikes a diffraction grating and a spectrum is produced. (Courtesy of Bausch and Lomb.)* See also Figure 5-18.

3 *Eclipse of the sun, March 7, 1970, as observed in ultraviolet light by the Harvard College Observatory's instrument on satellite OSO-6. The different colors represent different brightnesses; observations made at the different locations on the solar disk by scanning the instrument across the sun are combined to produce the image shown here on a computer display device at the OSO Control Center. (Goddard Space Flight Center.)* See also Figure 14-5.

4 *Earth-based photograph of Mars. (Courtesy of R. B. Leighton, California Institute of Technology.)* See also Figure 15-33.

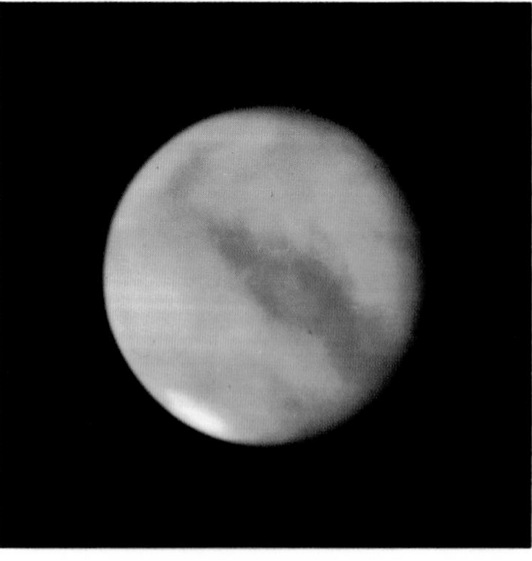

5 *Photograph of the Hyades star cluster, taken with a large prism mounted on the front end of a Schmidt telescope. The spectrum of each of the brighter stars is seen. (Courtesy of the University of Michigan.)* See also Figure 10–6.

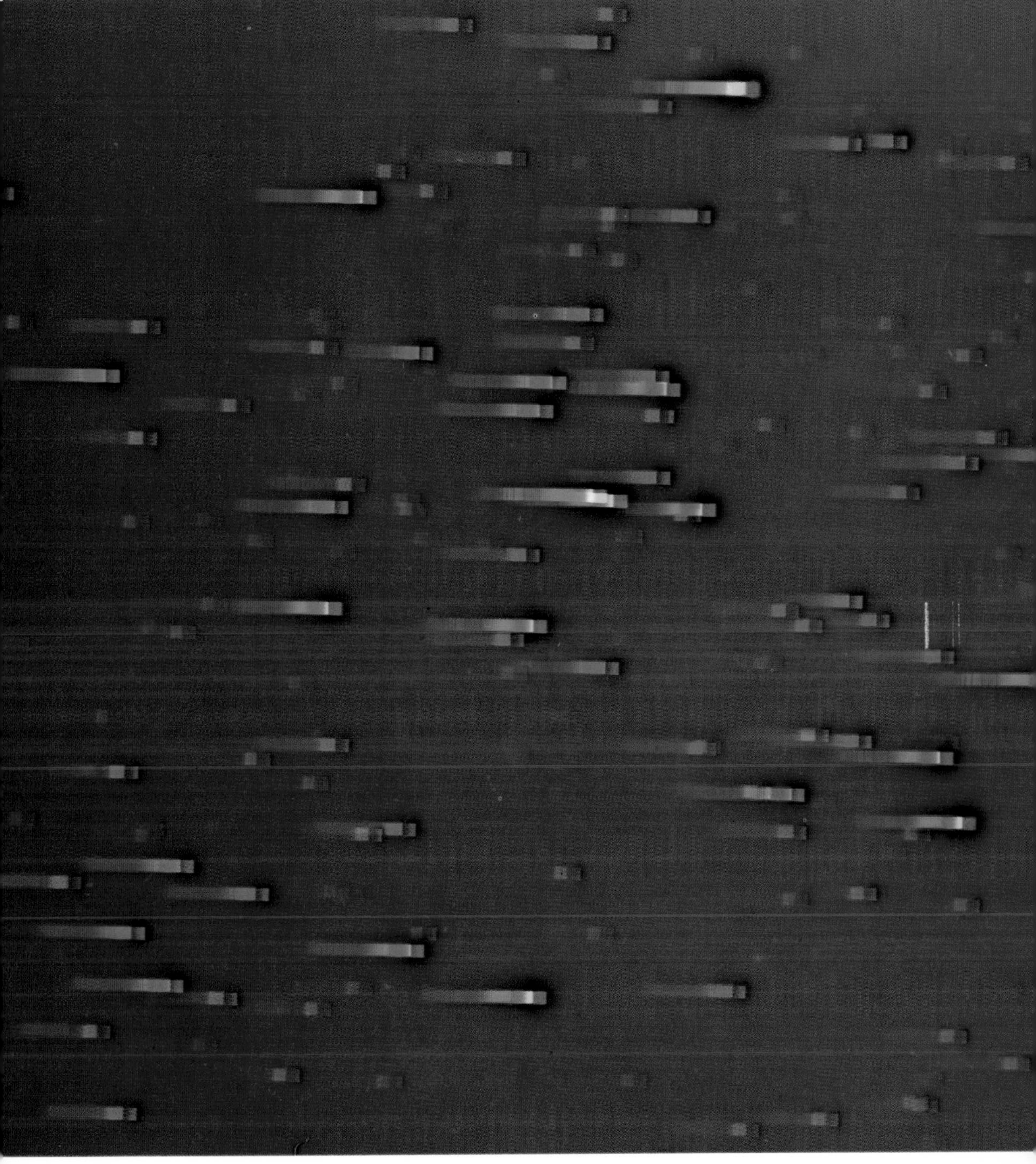

6 *Sketch illustrating Kirchhoff's laws.* See also Figure 5–20.

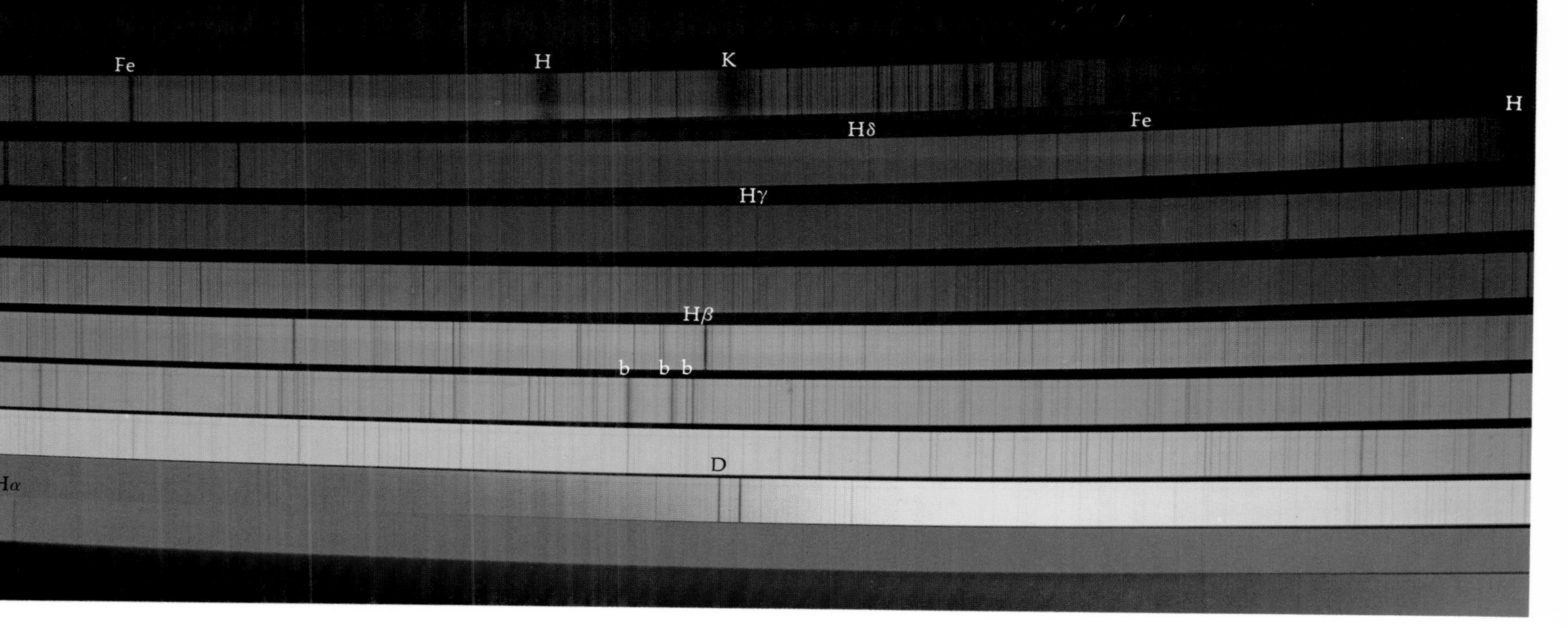

7 *The visible light spectrum of the sun. Some of the principal absorption lines, the elements producing them, and the wavelengths are given in the table. (Courtesy of the Sacramento Peak Observatory, Air Force Cambridge Research Laboratories.)* See also Figure 9–8.

Identifying symbol		Wavelength in angstroms			Element
Fraunhofer	Other				
C	$H\alpha$	6,563			Hydrogen
D		5,890	5,896		Sodium
	b	5,167	5,173	5,184	Magnesium
F	$H\beta$	4,861			Hydrogen
	$H\gamma$	4,340			Hydrogen
	$H\delta$	4,102			Hydrogen
	Fe	4,046			Iron
H		3,968			Calcium
K		3,934			Calcium

8 *The spiral galaxy in Andromeda, Messier 31. (Courtesy of the Hale Observatories. Copyright © by the California Institute of Technology, and the Carnegie Institution of Washington.)* See also Figure 13–1.

9 *The Veil nebula.* See also Figure 12–4.

10 *The Crab nebula.* See also Figure 16–4.

11 *Comet Bennett photographed on April 16, 1970. (J. C. Brandt, R. G. Roosen, and S. B. Modali, Goddard Space Flight Center.)* See also Figure 7–7.

10

The Stars—A Census

The distance of the sun is 1 a.u., while the star Proxima Centauri is about 260,000 a.u. away. The intensity of light decreases as the square of the distance increases. Thus, if the intrinsic brightness of this star happened to be exactly equal to that of the sun,* it would appear to us as 260,000 squared, or almost 7×10^{10} times fainter than the sun. Yet Proxima Centauri is the *next closest star.* We also cannot see the surface of Proxima or any other star except the sun. Despite these obstacles imposed by the great distances of the stars, we can determine some of their characteristics, including temperatures, motions, masses, sizes, chemical compositions, and luminosities. Stars are found sometimes as isolated individuals and sometimes in pairs, small groups, and clusters. We will consider both the individual characteristics and this "social behavior" of the stars, but their distances are the first item of our cosmic census.

DISTANCES–PARALLAXES

Distances are a subject of the greatest interest in astronomy. No attempt to understand the solar system, the galaxy, or the universe in terms of the forces that regulate their structure, the

*It is actually much fainter.

sources of their energy, or the processes that may have given rise to them seems possible without an understanding of the scale of each system.

In Chapter 4 we discussed the evidence that the earth moves in orbit around the sun, and mentioned that a direct consequence of this motion is that the nearby stars should show a parallax (shift back and forth during the year) with respect to the distant stars. Measurement of this effect permits a direct determination of the distance to a star. The same principle is used in more mundane applications by surveyors.

For stellar distances we measure the parallax angles as shown in Figure 10-1, and the baseline is the known diameter of the earth's orbit. The measurement is entirely straightforward in principle and the heliocentric view of the solar system requires

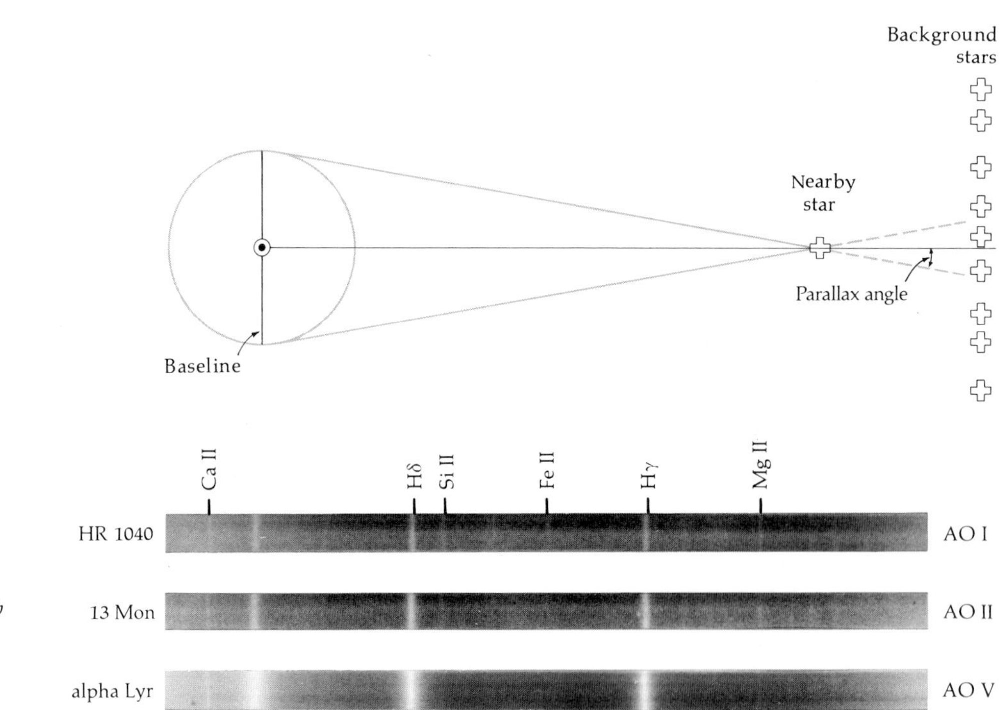

10-1 *The determination of stellar distances. a. Direct method; the geometry of stellar parallaxes. b. Indirect method; spectral lines are analyzed to determine intrinsic brightnesses of stars. The three spectra shown here were produced by stars with the same photospheric temperature (about 10,000°K). However, the supergiant (indicated by I) is much brighter than the giant (II), which is brighter than the main sequence star (V). Note that the hydrogen lines (H) are weaker in the brighter stars, but that certain ionized metal lines (Mg II, Fe II, Si II) are stronger in these stars. These effects in the giants and supergiants are related to the fact that the lines are formed in less dense and larger atmospheric regions than in the main sequence stars. (Yerkes Observatory photograph.)*

Table 10-1 *The First Parallaxes*

OBSERVER	STAR	EARLY PARALLAX MEASUREMENT (ARC SECONDS)	PARALLAX MODERN VALUE (ARC SECONDS)	DISTANCE TO STAR (PARSECS)
Henderson	alpha Centauri	1.16	0.75	1.33
Struve	Vega	0.26	0.12	8.3
Bessel	61 Cygni	0.31	0.29	3.4

that such a parallax angle indeed exists—otherwise the earth must be at rest. Unfortunately, the early attempts to measure parallax (including Bradley's discovery of aberration, page 95) were unsuccessful. This was rather perplexing because the stars were considered to be not far beyond the orbit of Saturn (recall Figure 4-14). All the principles used were correct; only the distances to the stars were vastly underestimated, and thus the expected parallax angles were overestimated.

The first certain measurements of stellar parallaxes were obtained independently at three different observatories during the 1830s. The observers were Friedrich Bessel at Konigsberg, Germany, Wilhelm Struve at Dorpat, Estonia, and Thomas Henderson at the Cape of Good Hope. Their measurements are given in Table 10-1, along with more accurate values from modern observations. The favorite unit of interstellar distance in popular science writing is the *light year,* defined as the distance that light traveling through the vacuum at 300,000 kilometers (186,000 miles) per second will traverse during one year. This distance is about 9.5×10^{12} kilometers (5.8×10^{12} miles). On the other hand, astronomers usually express distances in units of the *parsec,* which is the distance at which a star would have a parallax of one second of arc, as seen from the earth. (One parsec equals 3.26 light years, or 3.1×10^{13} kilometers.) In Table 10-1, the distance of each star, as determined from the modern parallax measurements, is given in parsecs. Note the simple relation between the last two columns: once the parallax angle (in seconds of arc) is measured, the distance is found by dividing 1 by the parallax.

The star alpha Centauri, which is listed in Table 10-1, is actually a triple star system; that is, it consists of three stars in a common orbiting system (see Binary Stars, page 281). One of the three, called Proxima Centauri, travels in an enormous orbit, with a period of several hundred thousand years. At its present location it is closer to the sun than the other two stars, and it is in fact the closest star in space to the sun. Its parallax is 0.763 seconds of arc, corresponding to a distance of 1.31 parsecs.

With the exception of Bessel's value quoted in Table 10-1, the errors in the original parallaxes are considerable. The small size of stellar parallaxes makes them difficult to measure (note that even the parallax of Proxima Centauri is less than one second of arc). As more distant stars are considered, the parallaxes get smaller. Beyond 50 parsecs, that is, for parallaxes smaller than one-fiftieth of a second of arc, the uncertainty in a parallax measurement is comparable to the measurement itself and so these parallaxes are unreliable. Such small parallaxes, although too crude to give reliable distance values, do at least show that the stars involved are far away. On the other hand, accurate (to about 10 per cent) parallaxes have been measured for about 700 stars closer than 20 parsecs. These directly measured quantities are called *trigonometric parallaxes.* Sixty stars, including the sun, lie within 5.2 parsecs of the earth.

Indirect determinations of stellar distances (necessary for stars beyond about 50 parsecs) are mostly based on techniques for estimating the *intrinsic* brightnesses of stars. How bright a star appears in the sky (*apparent brightness*) depends on (1) its intrinsic brightness, (2) the distance to the star, and (3) complicating effects that we ignore for now (such as absorption of starlight by dust in space and by the earth's atmosphere). If we somehow know the intrinsic brightness, we can figure out how distant the star must be so that it has the observed apparent brightness.

A simple example of the principle of indirect distance determination can be given for the case of a light bulb. Suppose we see a light bulb shining through the night some distance away and we wish to measure this distance. We could make direct measurements, such as by pacing off the distance, or by sighting on the bulb from two different spots with a surveyor's transit and measuring its parallax. However, we are too lazy for all that—we prefer to just point our telescope at the bulb. Now by attaching a photometer to the telescope we can measure the apparent brightness of the bulb. If we also knew its true or intrinsic brightness, we could compare this with the apparent brightness and compute the distance with the inverse square law (which says that the apparent brightness of an object decreases as the square of its distance, so that if you move the bulb twice as far away, it will seem one-fourth as bright). But how can we find the intrinsic brightness? A clever trick would be to look through the telescope and read the wattage label on the bulb. Then, knowing the true brightness, and having measured the apparent brightness with the photometer, we can calculate the distance of the bulb. There are also labels that astronomers have found in their observations of stars, and we can read these to find the intrinsic brightnesses of stars too distant to be measured by the parallax method. The astronomical labels of course are not written in English but are coded in terms of the properties of starlight. The most important example is the class of pulsating stars called *Cepheid variables* page 286). These stars are seen to get brighter and fainter and brighter again in a periodic fashion. The number of days that elapse between one time of maximum apparent brightness and the next maximum is called the star's *light period.* As described later in this chapter, the period depends on the average intrinsic brightness; in particular, the longer the interval between successive times of maximum brightness, the greater the average intrinsic brightness. Thus, looking through the telescope and counting the number of days between successive times of maximum brightness, we can determine the average intrinsic brightness of a Cepheid variable. Measuring the average apparent brightness and comparing it with the intrinsic value, we can determine the distance at which the star must be so that its known intrinsic brightness is reduced to its measured apparent brightness. The label on the star is its light period. No matter how far away the Cepheid, if we can see it, we can find its period. The period of light variation also labels the intrinsic brightness of another class of pulsators, the *RR Lyrae stars* (page 289).

By studying the absorption lines in the spectrum of a star, we can determine the physical conditions of temperature and density in the photosphere of that star (page 273). Among the many hundreds of stars for which accurate distances are known, we may find a similar spectrum, and thus similar physical conditions. It is therefore a good guess to say that the two stars have similar intrinsic brightnesses as well and, by comparing them, we can then estimate how far away the first star must be. This enables us to calibrate a method of indirect distance determination, called the *spectroscopic parallax* technique, which uses the spectrum of the star as its label. When two stars of the same photospheric temperature have different diameters, their photospheric densities, and thus their absorption line patterns are different. Therefore, the details of the absorption line pattern constitute a label we can use to estimate the intrinsic brightness and hence the distance of a star (Figure 10-1).

Astronomers are always on the alert for other ways to deter-

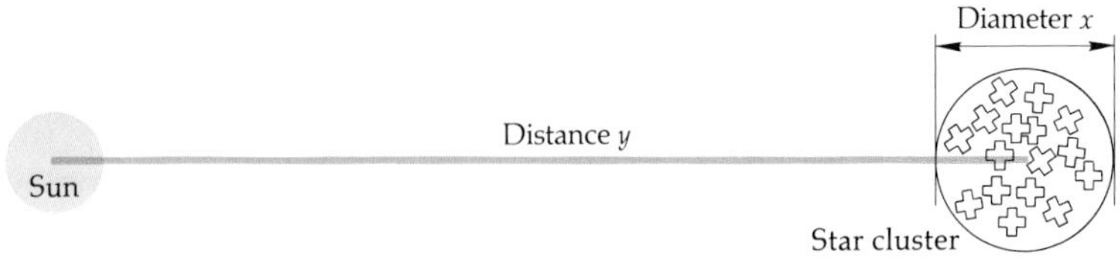

10-2 *Every star in the cluster is at distance y within an accuracy of $\frac{1}{2}x$. Distances to different stars in a cluster are all nearly the same.*

mine distance. Several methods apply to groups of stars. Stars in a cluster* are all at essentially the same distance (Figure 10-2) from the earth, and some of the group properties (Chapter 11) allow the distance to be determined. A few clusters (such as the Hyades, Figure 10-6) are close enough so that we can accurately measure their speed and direction of motion (*space velocity,* page 267). By comparing the observed motion in angular units (seconds of arc per year) with the space velocity in linear units (parsecs per year) we can determine how far away the cluster is. This technique has yielded fairly accurate distances for hundreds of stars located beyond the reach of the trigonometric parallax method.

Finally, there are the so-called *statistical parallaxes* based on a given star's proper motion (see below) across the sky. Part of the star's apparent change in position with time is due to the motion of the sun with respect to the nearby stars. As the earth travels through the galaxy with the sun, we view the stars from a constantly changing position and, hence, the nearer stars seem to continuously move with respect to the background of distant objects. Thus, the nearby stars have, on the average, larger proper motions caused by the sun's motions than those stars far away.

MOTIONS IN SPACE

The motion of a star in space with respect to the sun can be determined from knowledge of its distance, its *proper motion,* and its *radial velocity.* The radial velocity, or velocity along the line of sight, is found (in units of kilometers per second) by measuring the wavelength shifts in the spectrum caused by the Doppler effect (Chapter 5).

The velocity at right angles to the line of sight is much more difficult to determine. It can be measured as the proper motion

*The various kinds of star clusters are described on page 285.

in angular units (seconds of arc per year) as the rate of change in the position of a star with respect to background stars that, because of their much greater distance, hardly seem to move. This can be done by comparing photographs taken many years apart. The star with the largest known proper motion was discovered by the American astronomer E. E. Barnard (1857–1923); known as *Barnard's star,* it has a proper motion of 10.3 arc seconds per year. When the distance to a star is known from one of the methods discussed previously, then the observed proper motion in angular units can be converted to the *tangential velocity* in units of kilometers per second. When the radial and tangential velocities are combined by trigonometry, as shown in Figure 10–3, the *space velocity* of the star relative to the sun is known. We say "relative to the sun" because the sun has its own motion through the galaxy. At different angles to the sun's direction of motion we see an effect in the space velocities of the stars. This is analogous to the apparent velocities of other cars when we are driving on the highway; everyone may be going at roughly 70 miles per hour, but oncoming traffic seems to pass our car at 140 miles per hour. A house on the side of the road seems to be left behind at 70 miles per hour, but distant objects such as mountains on the horizon several miles away seem to move only very slowly as we cruise past them.

When the space velocities are determined for the stars near the sun, some regular patterns are observed. These can be explained in terms of the above analogy by postulating a space

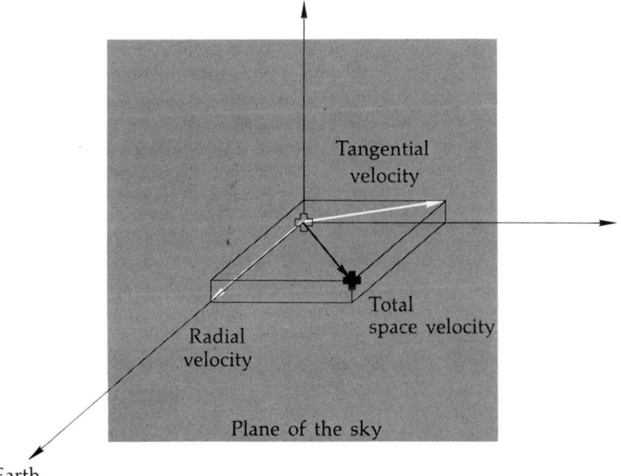

10-3 *A star's total space velocity is determined by its radial and tangential motions.*

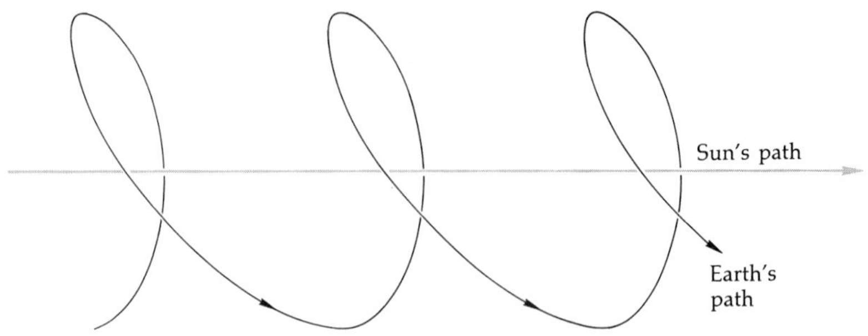

10-4 *The earth's corkscrew motion through the galaxy.*

motion of the sun of about 20 kilometers per second (12 miles per second), with respect to the nearby stars, directed toward a point in the constellation Hercules. This phenomenon shows why part of the proper motion of stars is caused by the sun's motion through the nearby stars and why proper motion depends on the star's distance from the sun. Only stars close to the sun have large proper motions; those farther away have small or unobservable proper motions.

The direction of the sun's motion relative to the nearby stars was first determined by William Herschel in 1783. The combination of the sun's motion and the earth's motion around the sun produces a corkscrew path for the earth with respect to the nearby stars (Figure 10-4).

INTRINSIC PROPERTIES OF STARS

In this section we consider some stellar properties that are independent of distance or motion. Much of this information is learned from photometry (measurements of light intensity) and spectroscopy (study of spectra) of the stars.

In photometry one usually measures the *relative brightnesses* of stars with respect to some *standard stars* whose apparent brightnesses are already known; this allows us to determine the apparent brightnesses of stars on a common scale of *magnitudes.* Magnitudes are defined (as mentioned in Chapter 4) by the statement that a star that is one magnitude brighter than another star is 2.512 times brighter than it. The filters and photomultiplier tubes used in a photometer are chosen and standardized to produce measurements in a specific range of wavelengths. One such system, the so-called UBV, gives magnitudes in the U (ultraviolet), B (blue), and V (visual or yellow) wave-

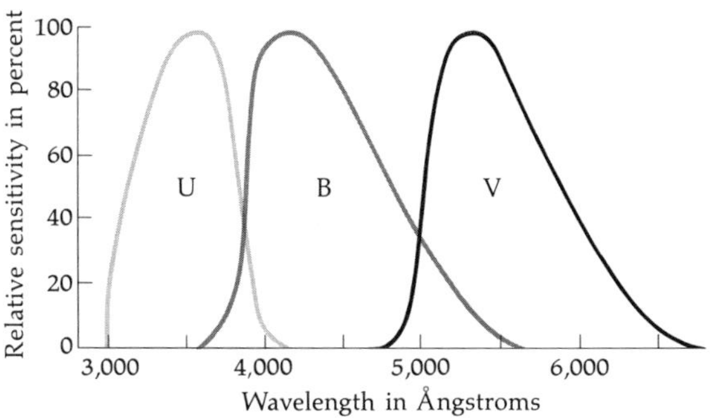

10-5 *Photometer sensitivity in the wavelength bands of the U, B, V system.*

lengths. The sensitivity of a UBV photometer to light of different wavelengths is illustrated in Figure 10-5. The zero point of the UBV scale is defined in terms of standard stars whose magnitudes have been carefully studied and agreed upon by several astronomers. Because the magnitude scale is defined so that brighter objects have smaller magnitudes, a star with B = 6.3 is 2.512 times brighter in the blue wavelength range than a star with B = 7.3.

Differences in magnitudes can be used to study the wavelength distribution of light from stars. A blue star with a surface temperature of 25,000° K produces more blue light than yellow. Thus, the B (blue) magnitude will be small (bright) and the V (yellow) magnitude will be high (faint). Similarly, an orange star with a temperature of 5,000° K emits more light in the yellow than in the blue, with a corresponding result in the magnitudes. We can subtract the V magnitude from the B value for a star, forming the quantity B–V, which indicates the photospheric temperature of the star by telling how its blue and yellow light compare in intensity. A quantity such as B–V (U–B is another example) is called a *color index,* or for short, a *color.* For example, a fairly blue star such as Alcyone (the brightest star in the Pleiades, Figure 6-17) has B–V = −0.1, a yellow star such as the sun has B–V = +0.6, and a red star such as Antares in the constellation Scorpius has B–V = +1.8. The existence of different colors in stars is shown in Figure 10-6.

All observed magnitudes must take into account the effects of absorption of light by the earth's atmosphere as well as absorption by dust in interstellar space (Chapter 12). Corrections for these effects are especially important in determining the *bolometric magnitudes* which measure the light emitted at *all*

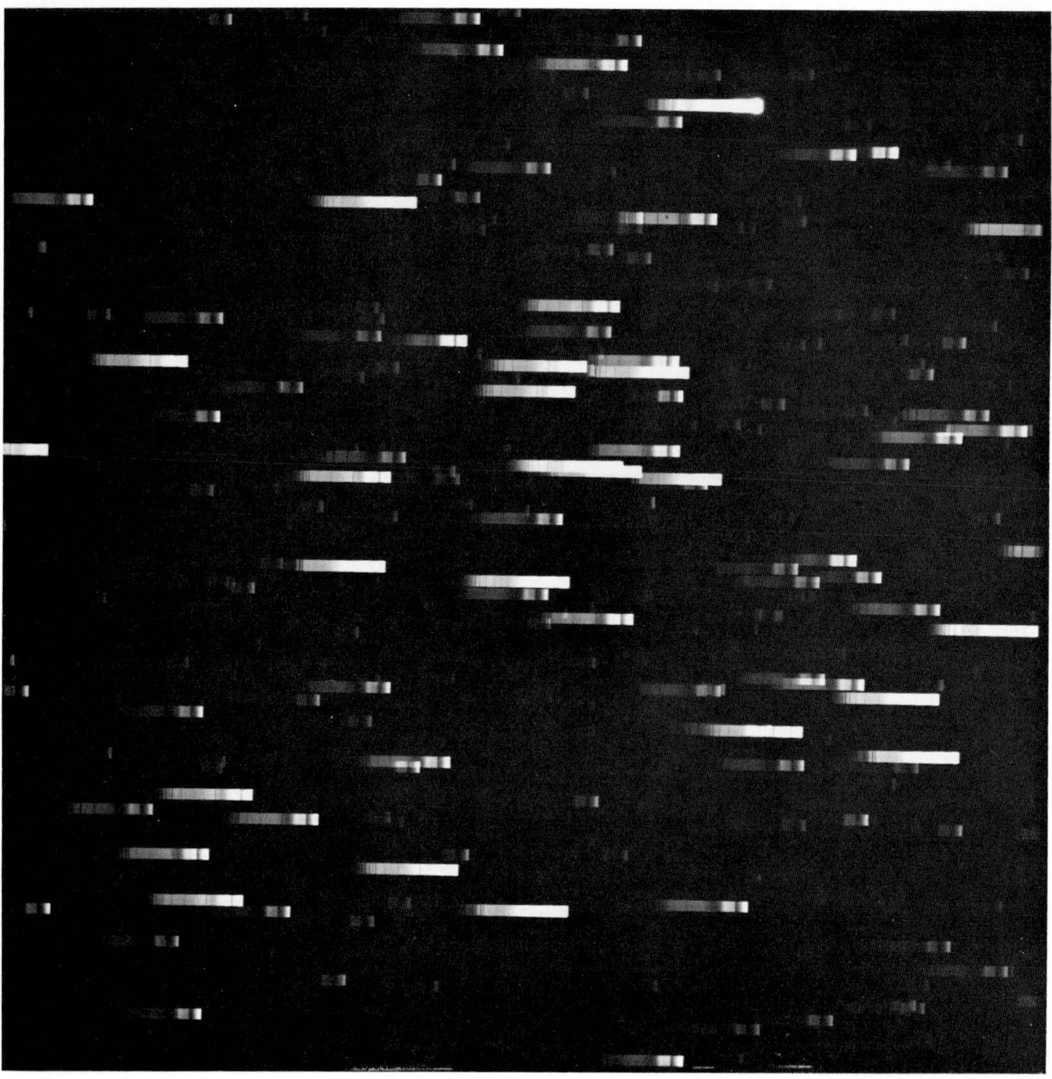

10-6 *Photograph of the Hyades star cluster, taken with a large prism mounted on the front end of a Schmidt telescope. The spectrum of each of the brighter stars is seen. (Courtesy of the University of Michigan.)* See also Plate 5 following p. 260.

wavelengths. Most of the energy radiated by the sun is carried by light of wavelengths that can pass through the earth's atmosphere. But for the hot blue stars, a large fraction of the radiation is absorbed in the air and does not reach our telescope; therefore substantial corrections must be applied to the observed magnitudes and colors of these stars.

When it is possible to determine the intrinsic brightness of a star, it is expressed in terms of an *absolute magnitude,* which is the magnitude that the star would have in a certain wavelength range* if it were at a standard distance from us and no absorption occurred in the intervening space. This distance has been arbitrarily adopted by astronomers as 10 parsecs. Absolute *bolometric* magnitudes tell us the total energy emitted per second, or the luminosity. Since the luminosity corresponds to the energy output of a star, it also tells physicists how much energy must be produced inside the star by nuclear reactions.

The photospheric temperature of a star can be estimated by comparing the wavelength distribution of its light with the Planck curves for different temperatures; the same procedure was used to determine the temperature of the solar photosphere in Chapter 9. Photometric measurements of stellar color indices are used to infer temperatures. For greater precision, allowance is made for the fact that the stars are not perfect radiators.

The temperature and luminosity can also be estimated from a detailed analysis of a star's spectrum. Spectra of stars with different temperatures are shown in Figure 10-7. (Also recall the spectra of stars with the same temperature, but different luminosities, that appear in Figure 10-1.) Stellar spectra were originally classified without the benefit of a physical theory. Similar-appearing spectra were assigned the same code letter or *spectral class* at a time when the real significance of the different spectra was not clear. Later, it was determined that stars of different spectral classes have different temperatures, and the classes into which most stars fit were arranged in a new order, according to decreasing temperature. This order is: O B A F G K M. The O stars are hot and blue (temperatures around 30,000° K), the A stars look white to the eye (temperatures around 10,000° K), and the M stars are cool and red (temperatures around 3,000° K). The sun is a G-type star. There are numerical subdivisions of these classes; for example, K3 designates an orange star, cooler than K1 or K2 and hotter than K4. The sun is G2 on this system. People who wish to memorize the sequence of spectral classes from O to M can use the mnemonic device, "**O**h **B**e **A** **F**ine **G**irl, **K**iss **M**e."

A rather reasonable idea that led the original spectrum classifiers astray was the possibility that the main differences in appearances of spectra (as in Figure 10-7) were due to the presence of different relative amounts of the elements in the various stars, so that one star, having a lot of iron, would have a spectrum dominated by iron absorption lines, whereas another, being mostly hydrogen, would just show hydrogen lines.

*For example, the U, B, or V wavelength ranges.

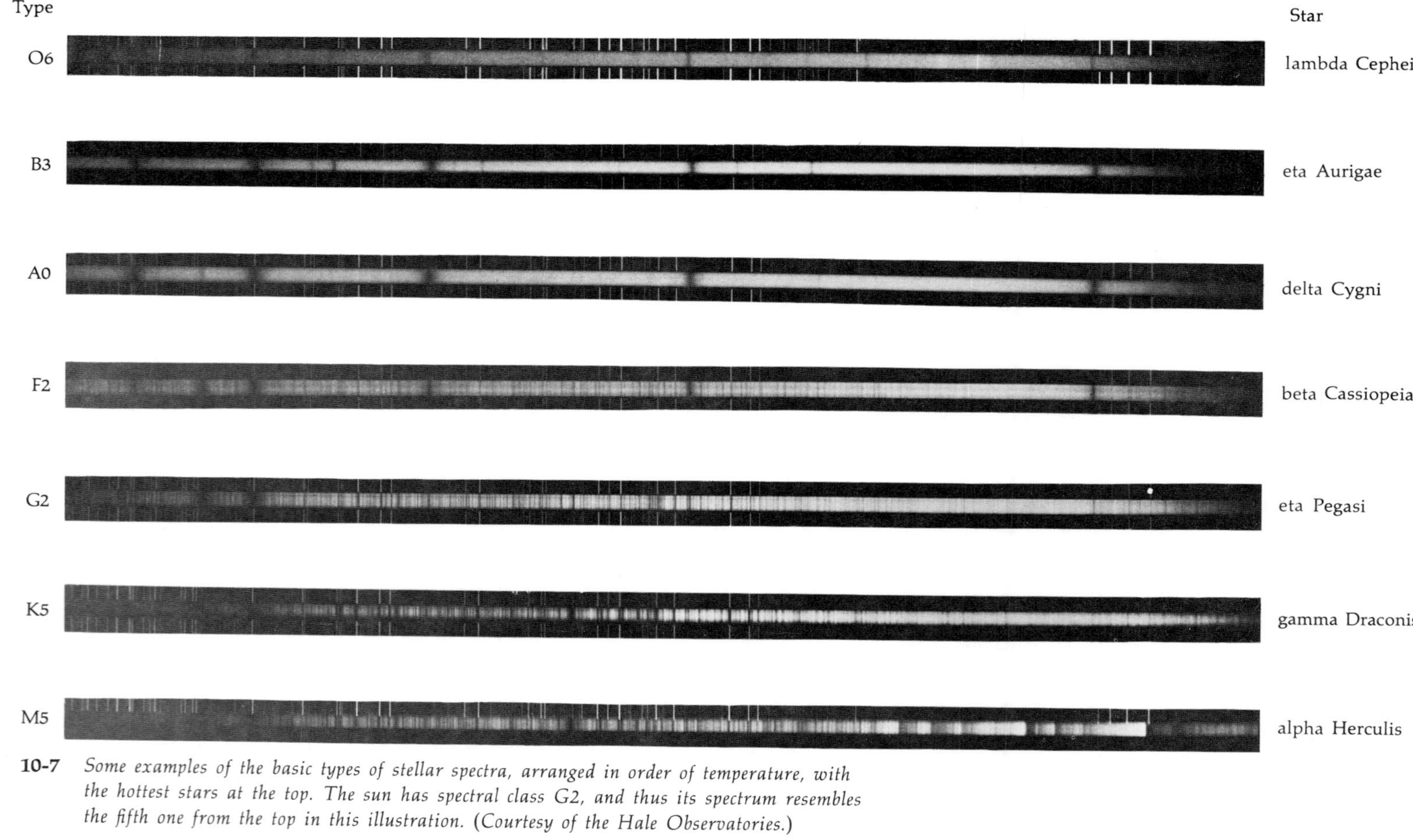

10-7 *Some examples of the basic types of stellar spectra, arranged in order of temperature, with the hottest stars at the top. The sun has spectral class G2, and thus its spectrum resembles the fifth one from the top in this illustration. (Courtesy of the Hale Observatories.)*

Although there are composition differences between stars, it turns out that they rarely have a strong effect on the overall appearance of the spectrum. The problem was solved after 1920 through the work of the Indian astrophysicist M. N. Saha, who studied the influence of temperature and pressure on the ionization of atoms. His results, when applied to the stars, show that the presence of absorption lines of (for example) neutral iron in one star's spectrum, and the absence of the same lines in a second star's spectrum does *not* mean that the latter star lacks iron. This latter star may be hot enough so that the iron is singly ionized, that is, has had one electron knocked out of the atom. An iron ion such as this absorbs light at a different set of wavelengths than does a neutral iron atom. In fact, we can calculate what fraction of the iron atoms should be neutral or ionized at various temperatures* and densities by using Saha's theory, and we indeed find that when the theory predicts (say) neutral iron, we find absorption lines of neutral iron. Extensive work of this nature has shown that the general appearance of the great majority of all stellar spectra is determined by temperature. Spectral differences due to chemical abundance variations among the stars (see Figure 10-8) or other variations (such as pressure or magnetic field) do occur but the effects of temperature usually dominate.

When we have found both the photospheric temperature and the luminosity of a star by the above methods, we are in a position to determine its size (at the photosphere). The Stefan-Boltzmann Law tells us how much energy per square centimeter is produced at a given temperature. Then we simply calculate how many such square centimeters must exist in order to produce the total luminosity of the star. In that way we learn the surface area of the stellar photosphere, and by a simple calculation (the surface area of a sphere is 4π times the square of the radius) we can find the stellar radius. As a result of such studies, we know that the more luminous stars are usually also the bigger ones.

The determination of stellar brightnesses and sizes revealed the existence of *giant stars* far larger than the sun. The bright red supergiant, Betelguese, in the constellation Orion, has a radius of about 800 times that of the sun. If placed in the sun's position, Betelguese's atmosphere would extend beyond the orbit of Mars. It is also found that giant stars have lower pressures at their photospheres than do smaller stars. The differ-

*Remember that we can estimate the temperature of a star from its color index measurements.

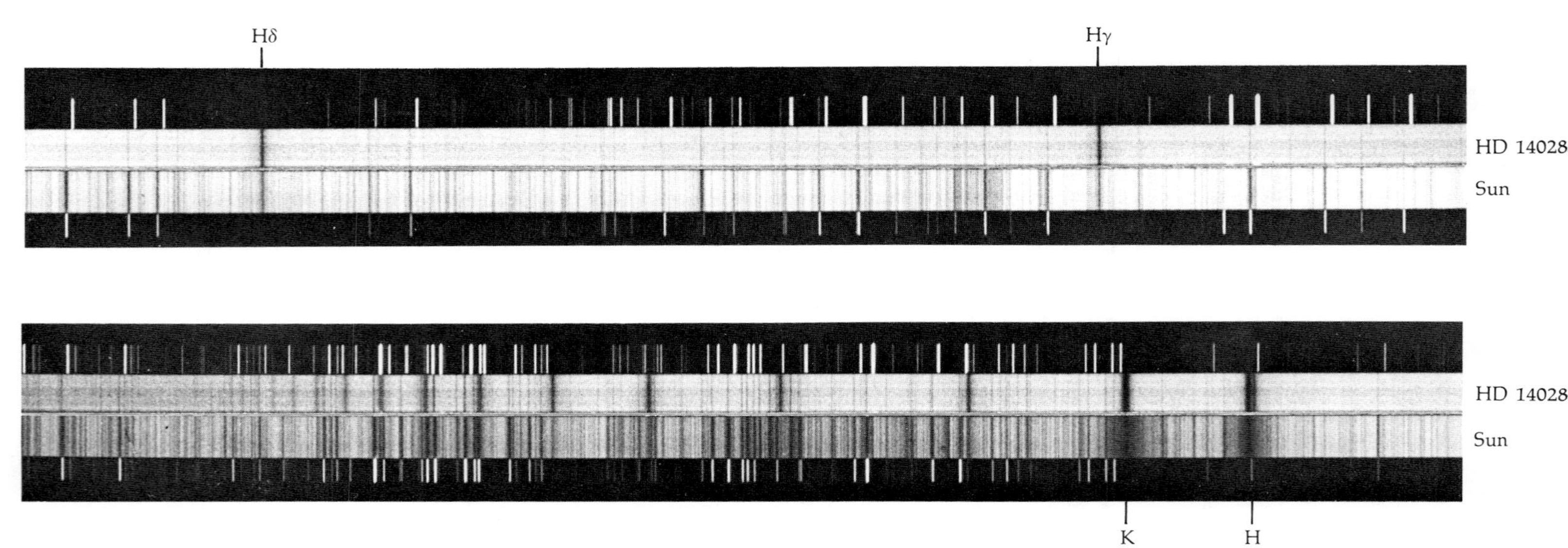

10-8 *Comparison of the same region of the spectrum of the sun and a metal-poor star (HD 140283). Notice the difference in the strengths of the lines. (Lick Observatory photograph.)*

ences in pressures cause slight differences in the spectrum. By looking for these pressure effects in a stellar spectrum, we can judge whether the star is a giant, normal size, or even smaller, and by combining this spectroscopic size estimate with the temperature judged from the spectrum, we can make a rough estimate of a star's luminosity and its intrinsic brightness in a given wavelength range. Comparing the intrinsic and apparent brightnesses, we can estimate how far away it is. This is how the *spectroscopic parallax* method of distance determinations (mentioned on page 265) works.

Finally, the basic property of a star that is the most difficult to determine is its mass. The methods used to measure mass require that the star have another object in orbit with it; fortunately such pairs or *binary stars* are fairly common. This method is analogous to the way (Chapter 4) in which the mass of the sun was found from the motions of the planets and Kepler's Third Law. Let us briefly return to a discussion of the earth's motion around the sun and consider some of the differences in the two methods of solar and stellar mass determinations. Kepler's Laws are only an approximation of the true situation. The sun, for example, is not at rest at a focal point of the planetary ellipses. Unknown to Kepler, it too performs a small elliptical motion with respect to the *center of mass* of the solar system, although this solar orbit is smaller than the sun itself. The center of mass is, in essence, the average position of the mass in a given system. For objects of equal mass this position is exactly half-way between them. If one object is more massive, the center of mass would then be closer to the more massive object (Figure 10-9), and if one object is very much more massive (as is true in the case of the sun and the earth), then the center of mass may even be within the bigger body. Thus, if the radii of the respective orbits of binary stars can be determined with respect to the center of mass of the binary system, the *ratio* of the masses of the two stars can be obtained. For example, if one orbit radius is half the size of the other, then the star with the smaller orbit is twice as massive as its partner.

In addition, Kepler's Third Law (as it was originally stated by Kepler) is not a very precise description of the way in which the orbital periods of the planets depend on the sizes of their orbits, because the period of a planet also depends slightly on its mass. The variation is such that the period of revolution of the earth around the sun (for example) depends on the *sum of the masses* of the sun and the earth. In this case, since the sun is much more massive than the earth, the error involved in neglecting the earth's mass is small (the mass of the sun is 2×10^{33} grams; the earth's is 6×10^{27} grams). For visual binary

10-9 *Illustration of the center of mass in binary star systems by considering the analogous problem with children on a see-saw.*

stars (see page 281), the period of revolution can be observed easily. If the distance is known, the sizes of the orbits can be calculated, and Kepler's Third Law allows a determination of the sum of the masses after a correction is made for the effect of the inclination of the orbital plane to our line of sight.

Once the *ratio* of the two masses of a binary star has been determined from observations of the relative sizes of the orbits around the center of mass, and the *total mass* of the two stars is found from Kepler's Third Law, we can solve for the individual masses. The study of visual binary stars has yielded about 15 fairly accurate stellar masses; these happen to be mostly of stars less massive than the sun.

Other masses can be determined from *spectroscopic binary stars.* Here, the stars are not visible separately in the telescope as they are in the case of visual binaries. The resolution is achieved not by the telescope, but by the spectrograph. If the stars are in orbit around each other, then at a given moment one star will be approaching the sun while the other recedes (Figure 10-10). Because of the Doppler effect the absorption lines from the receding star are shifted to the red and those of the approaching star are shifted to the blue. This situation can produce two sets of lines that move back and forth periodically, as shown in Figure 10-11, and then we call it a *double-lined spectroscopic binary.** The mass ratios can be inferred from the observed velocities because the more massive star will move slowly while

*Sometimes, one member of a binary system is much fainter than the other, and we can only record the spectral lines of the brighter star. However, they shift back and forth as the star orbits, and so we can identify the system as a *single-lined spectroscopic binary.*

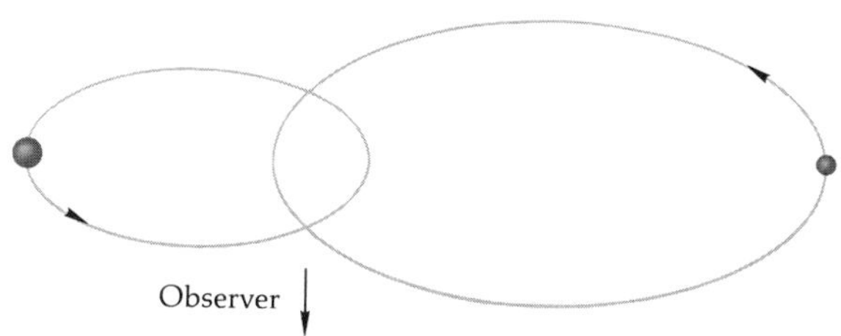

10-10 *Schematic of motions in a spectroscopic binary system. (From* Life Outside the Solar System *by Su-Shu Huang. Copyright © 1960 by Scientific American, Inc. All rights reserved.)*

10-11 *Spectra taken at two different phases of the spectroscopic binary star kappa Arietis, showing the doubling of lines (upper spectrum) due to the Doppler shifts arising from the motions of the two stars. The same lines are single in the lower spectrum. (Lick Observatory photograph.)*

the less massive star moves rapidly. The larger star moves at a slower rate than its companion because the gravitational force on each star is the same (action = reaction, Chapter 5), and a given force will produce a smaller acceleration of a more massive object than of a lesser one (Newton's Second Law, Chapter 5).

The period of a spectroscopic binary can easily be determined by observing the changes in the positions of the spectral lines, since the cycle of changes will repeat once per orbital period. The separation between the stars, however, cannot be determined when the inclination of the orbits to the line of sight is unknown. In some cases, one star eclipses the other, as seen by us. This tells us that the plane of the orbits must contain our line of sight, and so we know the inclination* is close to zero degrees. If this is the case, we can use the velocity of each star as determined from the Doppler shift of the lines in the spectrum, and multiplying the velocity by the period gives the distance traveled in one period.† If we neglect the details of ellipses and this distance is considered to be the circumference of a circle (equal to 2π times the radius), the orbit radius is determined. When the period, orbit radii, and mass ratios are known, the masses of the stars in such systems can be determined. These *eclipsing spectroscopic binaries* have provided about 25 accurate stellar masses and they happen to be mostly of stars more massive than the sun. Thus, they complement the roughly 15 masses from visual binaries, and roughly 40 stellar masses are well known—including that of the sun.

The known masses can be plotted against the stars' luminosities to determine a *mass-luminosity relation.* In this way it is found that the luminosity of normal stars (called *main sequence* stars, page 279) varies roughly with the mass of the star cubed, so that a star twice as massive as another will be about eight times brighter.

THE SUN AND STARS COMPARED

The properties of the stars are usually given in solar units (denoted by the subscript $_\odot$) for convenience, and the comparison is interesting. Most stellar masses are in the range from one-tenth to fifty times the mass of our sun (denoted by 0.1 $M_\odot$ to 50 $M_\odot$). The radii of ordinary stars range from 0.1 $R_\odot$ (one-tenth the solar radius) up to hundreds of solar radii for the giants, although two kinds of unusual, very small stars (white dwarfs, page 290; neutron stars, Chapter 11) are outside this range. Photospheric temperatures of normal stars range from 2,000° K to 30,000° K; the sun's value at 6,000° K is not near either extreme. In terms of the solar luminosity, stars that

*In this text we have defined the inclination of an orbital plane with respect to the line of sight. However, astronomers usually quote the inclination with respect to an imaginary plane at right angles to the line of sight (plane of the sky), and so you may find a different definition in other books.

†If the binary does not undergo eclipses as seen by us, the inclination is greater than zero degrees and the observed Doppler shifts only correspond to a fraction of the full orbital velocities.

are 10^4 times fainter than the sun are known, while stars 10^5 times brighter than the sun are also known. For these reasons, the sun is often described as an average star. This is really a little misleading; its properties fall midway in their ranges among normal stars, but most of these stars are smaller, cooler, and less massive than the sun. A further distinction is that the sun is a single star, whereas it appears that most of the stars in the galaxy are found in binary or multiple systems.

HERTZSPRUNG-RUSSELL DIAGRAM

In the past few pages we have discussed a variety of techniques that are used to determine masses, radii, luminosities, and atmospheric temperatures of stars. The chief differences in the appearance of stellar spectra were explained as the result of temperature differences. A convenient way of illustrating the observational data will be helpful for our further discussions of the stars and how they change with age. Such a method is a diagram of luminosity (or absolute magnitude) plotted against temperature (or color index) as shown in Figure 10-12. This type of chart was first used by the Danish astronomer Einar Hertzsprung in 1911, and independently by the American astronomer Henry Norris Russell in 1913; it is called the Hertzsprung–Russell Diagram or *H–R Diagram.* Note that the stars are not uniformly spread over Figure 10-12 but, in fact, seem to fall mostly in certain restricted locations or bands (also note—temperature increases to the left on the horizontal scale).

Many of the principal classes of stars are readily identified on the H–R Diagram. Most stars are found on the wide band (called the *main sequence*) that runs from the upper left portion of the diagram to the lower right. This main sequence was not predicted, it was simply found when the measurements of stellar luminosities and temperatures were plotted on the diagram. Note that since the stars at the left end of the main sequence are brighter (higher up in the diagram) than the ones at the right end, they must be more massive, according to the mass-luminosity relation. Since most stars are observed to fall on the main sequence, and since we find compelling evidence for evolution of stars over the H–R Diagram (that is, the position of a star in the diagram changes with time; see Chapter 11), we conclude that stars spend most of their lifetime on the main sequence. The sun, for example, is a main-sequence star. Most stars near the sun are main-sequence stars, although fainter and redder than the sun.

The *giant stars* are above and to the right of the main sequence

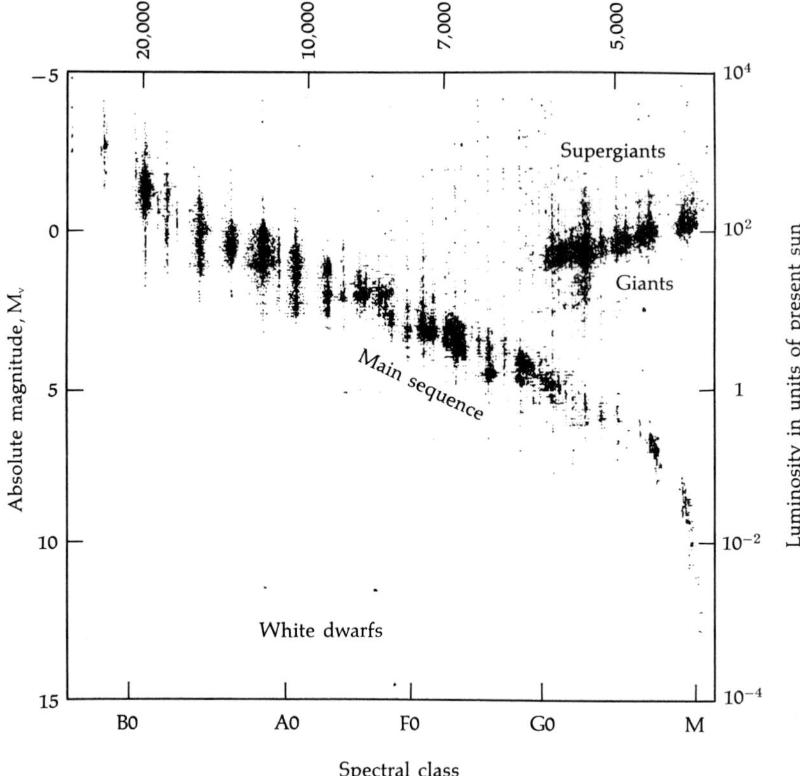

10-12 *The H-R Diagram (after W. Gyllenberg), showing both the closest and the brightest stars. Several types of stars, which fall in distinct regions of the diagram, are identified. (From* Elementary Astronomy *by O. Struve, B. Lynds, and H. Pillans. Oxford University Press. Copyright © 1959. Used by permission.)*

in the H–R Diagram. Since giants are much brighter than main sequence stars with comparable temperatures, the Stefan-Boltzmann Law tells us that they are also much larger, by factors ranging from 10 to several hundred. The biggest giant stars, such as Betelgeuse (page 273), are sometimes called *supergiants.* On the other hand, a given *white dwarf* star is much fainter than main sequence stars of the same temperature. The white dwarfs have radii about 100 times smaller than that of the sun, and are thus comparable in size to the earth.

The Hertzsprung–Russell Diagram and the other quantities discussed in this chapter provide the fundamental data for our discussion of the birth and death of stars in Chapter 11. In fact, the principal task of the science of *stellar evolution* is to explain the facts of the stellar census as illustrated by the H–R Diagram.

Studying the stars through telescopes and on photographs, it has been found that most of them occur in pairs or multiples. We have already indicated that binary stars are of great interest because studies of them can reveal the masses of stars (page 265).

In the latter part of the eighteenth century many double stars were found with telescopes. The two stars of a pair, however, were believed to be widely separated in space and seen near each other only in projection on the celestial sphere. William Herschel carefully observed many double stars in the hope of detecting their parallax, which he never found because his instruments were not precise enough. However, he observed that for some double stars the positions changed in such a way that they appeared to be revolving in orbits (they are now called *visual binaries;* an example appears in Figure 10-13). After about a quarter of a century of careful study, Herschel concluded in 1803 that many double stars "are not merely double in appearance, but must be allowed to be real binary combinations of two stars, intimately held together by the bonds of mutual attraction." Thus, Newton's laws of motion and universal gravitation had been extended to the stars.

The fact that most double stars must be physically connected had already been deduced by the Reverend John Michell. He noted that there were far too many of them to be merely chance occurrences or perspective effects. Although Michell's statistical arguments preceded Herschel's results, it was the latter's observations of orbital motions that put the matter beyond dispute.

Another demonstration of the reality of physically associated stars followed from measurements of the star Algol in 1783 by the Englishman John Goodricke. It had already been known for more than a century in Europe that the brightness of this star changes. (In fact, a story persists in the astronomical literature to the effect that the early Arabs were aware of its light variations, and therefore named it Ras al Ghul, the Demon's Head, from which we get Algol. There is no evidence for this origin of the name. In any case, there was considered to be a constellation of the demon or monster in this area of the sky in ancient times, according to the records of several civilizations.) Goodricke observed that the brightness of Algol changed by about a factor of 3 from brightest to dimmest, but his great discovery was that these changes occurred *regularly,* with a period of 69 hours. The *light curve* of Algol is shown in Figure 10-14, along with the explanation for its behavior. Algol is actually a binary system composed of a bright star and a faint

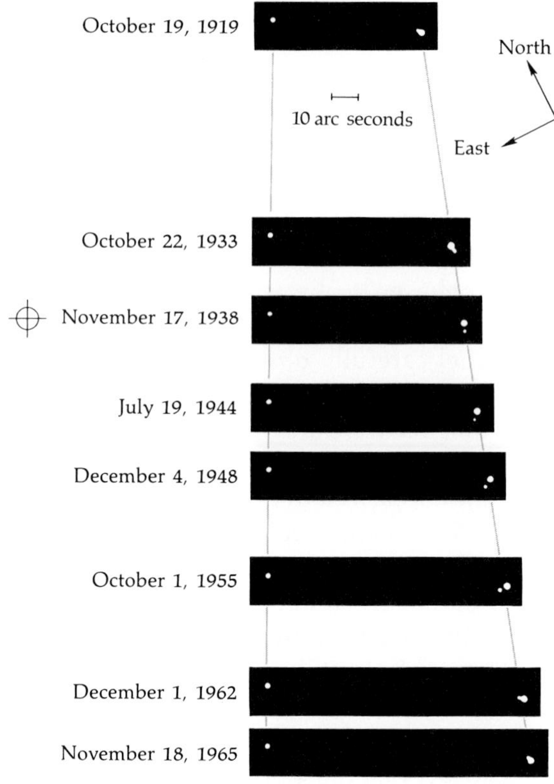

10-13 *Sequence of photographs showing the proper motion and orbital motion of the components (right) A and B of the visual binary star Kruger 60. A more distant star appears at the left. (From* Principles of Astrometry *by Peter van de Kamp. W. H. Freeman and Company. Copyright © 1967.)*

star in orbit around their center of mass, and the plane of their orbits contains our line of sight from the earth, so that it is an *eclipsing binary.* Primary eclipse (the major brightness decrease) occurs when the brighter star is obscured by the fainter star as seen from the earth. A secondary eclipse occurs when the bright star hides the faint star. The graph of brightness plotted against time is called the *light curve.* Eclipsing binaries are most valuable because the inclinations of their orbital planes to the line of sight are definitely known to be close to zero degrees (from the simple fact that we can see the eclipses) and their masses can be calculated in the way described on page 278. Eclipsing binaries are of additional value because the observations of the changes in brightness produced by one star gradually cutting off the light of the other reveal information on the

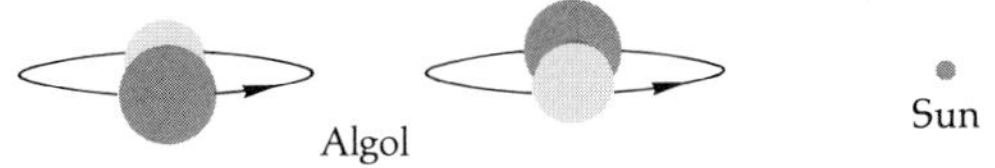

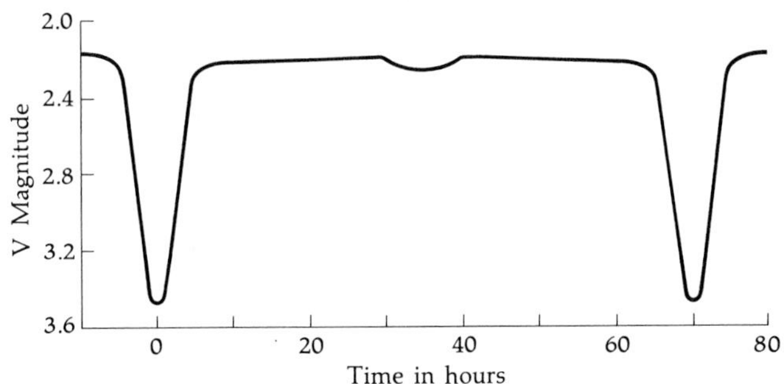

10-14 *The light curve and schematic diagram for Algol, an eclipsing binary star. The relative separation of the stars, their size, and the size of the sun are shown. (After* Matter, Earth, and Sky, *2d ed. by George Gamow. Prentice-Hall. Copyright © 1965. Used by permission.)*

sizes of the stars and even their limb darkenings (defined in Chapter 9). To estimate the sizes we (1) get the orbital velocities from Doppler shifts in the spectrum, and we (2) get the length of time it takes for one star to eclipse the second star from the light curve. Knowing the speed and how long it takes to go across the diameter of the second star, we calculate the size of the second star.

STARS AND PLANETS

A curious fact concerning binary stars is that the average separation between the members of a binary is about 20 a.u., which is roughly comparable to the distance of the Jovian planets from the sun. Could we detect a planet the size of Jupiter revolving around a star near the sun?

The answer is yes, even though Jupiter emits no visible light of its own; stars or planets with masses of less than about 0.07 $M_\odot$ have negligible luminosities because the amount of Helmholtz contraction that they undergo is not enough to raise their internal temperatures to the point where nuclear reactions can occur. Nevertheless, the existence of such an object can be detected because its gravitational attraction causes the associated visible star to execute an orbit around the common center of gravity. Thus, as seen from the earth, as the visible star moves through the galaxy, it does not follow a smooth path, but rather a wiggly one, almost a sure clue to the existence of a dark companion. If the spectrum of the bright star can be obtained, its mass can be estimated by the techniques mentioned previously; then the mass and orbital parameters of the dark body required to produce the observed wiggle can be determined. In some cases these dark bodies are comparable to the Jovian planets, as shown, for example, by results obtained by the contemporary astronomer Peter van de Kamp of Swarthmore College. He concluded that the dark companion of Barnard's star (page 267) is a large planet. If van de Kamp is correct, this is the first known case of a planet outside the solar system. He believes there is also some evidence for a second planet of Barnard's star.

The basic difference between the two interpretations of the Barnard's star system (one planet versus two planets) is in the orbital shapes. If there is only one companion, then its orbit must be a highly elongated ellipse, in order to account for the observed motion of the star, whereas if there are two planets, the orbits are nearly circular. If we can use our solar system as a basis for judgement, then the model involving nearly circular orbits is to be preferred. The parameters of the models are compared in Table 10-2, along with some data on Jupiter. (In the table, M_J represents the mass of Jupiter, 1.9×10^{30} grams.)

Barnard's star is of interest for another reason; its motion in space is carrying it closer to the sun. By the year 11,000 A.D. it will be closer than Proxima Centauri and will have a parallax of about 0.87 arc seconds and distance of 1.15 parsecs.

The existence of dark companions to two other nearby stars has also been inferred, although the evidence is not conclusive. If correct, this suggests that the occurrence of planets is not very unusual.

STAR CLUSTERS*

The social organizations of people and animals include both the lone wolves, the family groups, and the tribes or herds. Affiliation with such a group is generally related to the circum-

*The field stars and clusters that are discussed in this section, and which we can see or photograph when we observe the sky at night, are members or components of our own Milky Way galaxy, as discussed in Chapter 12. Other galaxies, like the one in Andromeda (Figure 13-1), are also seen to contain both field stars and clusters when they are observed or photographed through telescopes.

Table 10-2 *Comparison of the Planet(s) of Barnard's Star with Jupiter*

	COMPANION(S) TO BARNARD'S STAR			
	ONE COMPANION CASE	TWO COMPANION CASE		JUPITER
Orbital period	25 years	12 years	26 years	12 years
Radius of orbit	4.4 a.u.	2.8 a.u.	4.7 a.u.	5.2 a.u.
Mass	1.5 M_J	0.8 M_J	1.1 M_J	1.0 M_J

stance of birth. Photographs of the sky show that in addition to the many isolated stars that seem to follow their own paths through space (*field stars*), there are pairs of stars like the binary star Algol (page 281), triple stars like the alpha Centauri system, of which Proxima (page 263) is a member, *multiple stars,* consisting of several stars orbiting around a common center of mass, and larger groups, the *star clusters,* and *associations.* These groups include the *open clusters* like the Pleiades (Figure 6-17) and Hyades (page 270) and the *globular clusters* like the Hercules cluster (Figure 10-15). It seems reasonable to assume that the members of a binary or multiple star system share a common origin and that the members of a star cluster are likewise related by birth. A large globular cluster may have more than 100,000 member stars and a diameter of 50 parsecs or more. The term globular is derived from the round shape that is evident in Figure 10-15. The membership of open clusters ranges from less than a hundred to more than a thousand stars; their diameters range from about 3 to 20 parsecs. The members of an open cluster are not organized into the compact spherical shape of a globular cluster, and this is the origin of the term open.

Associations resemble open clusters in their lack of an obvious geometrical shape, but they are distinguished from open and globular clusters by several properties: (1) associations consist primarily of the young, hot, main sequence O and B stars, whereas clusters include stars of most spectral types; (2) the member stars of an association are not closer together on the average than other stars in the sky,* whereas the property of clusters that allows astronomers to discover them easily is that when one looks at parts of the sky, the clusters are obvious as concentrations of stars; (3) an individual association is ex-

*The existence of an association has to be checked by measuring the distances of O and B stars in a given region of the sky and seeing whether a group of them are all roughly in the same part of the galaxy.

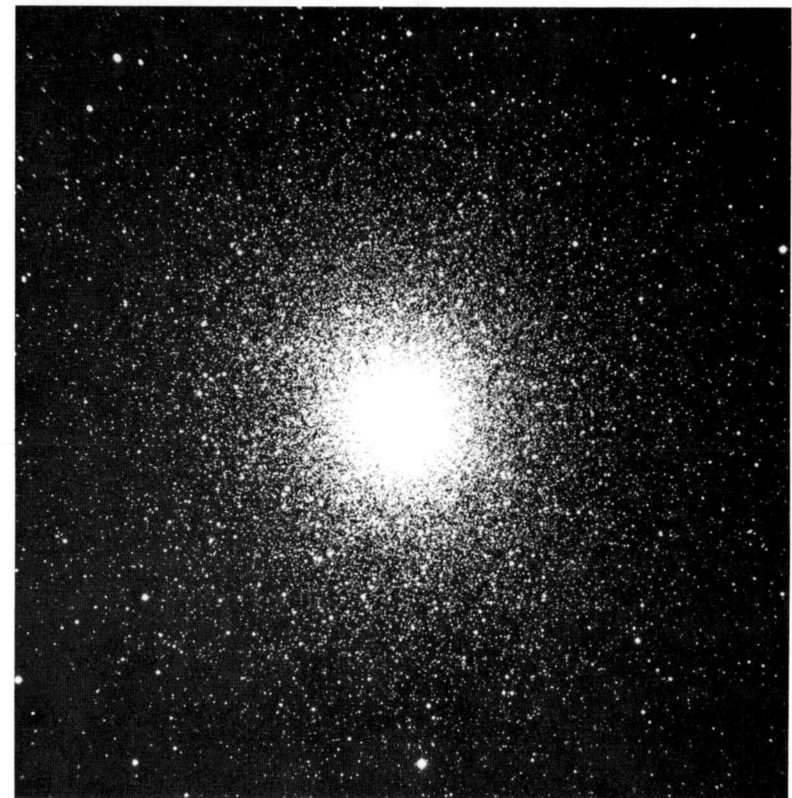

10-15 *The globular cluster M13 in the constellation Hercules. (Courtesy of the Hale Observatories.)*

panding; that is, the member stars are moving away from the center of the group. The fact that we see only young stars in the associations and that the associations themselves are expanding implies that these groups are of relatively recent origin.

PULSATING STARS

For years, light variations of all stars were considered to be caused by eclipses in binary systems. John Goodricke, the observer of Algol, also discovered another kind of variable star, named Cepheids after the first one found, delta Cephei. Hundreds of Cepheids are now known, and a typical light curve is shown in Figure 10-16. Compare this light curve with that of Algol shown in Figure 10-14. They definitely are not alike. In fact, because of the difference in shapes of the rising and

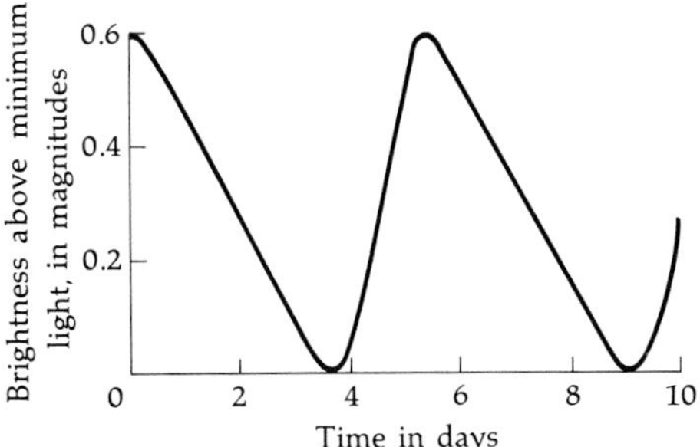

10-16 *The light curve of the pulsating variable star delta Cephei, as measured in yellow light. The period is 5.37 days.*

falling parts of the Cepheid light curve, it is hard to imagine any way that it could be produced by eclipses. Observations of Cepheid spectra showed variable Doppler shifts and the pattern of spectral lines seemed to change as well as shift as the brightness of a star changed.

The Harvard College astronomer, Harlow Shapley, investigated the Cepheids and in 1914 struck the death blow to the eclipsing hypothesis. Cepheids were clearly giant stars (because of their brightness) and he calculated that the likely dimension of a Cepheid would exceed the orbit size of the companion, as deduced from the observed Doppler shift and period according to the binary interpretation. If the companion were inside the Cepheid it would find it hard to keep orbiting and, in any case, it could hardly eclipse the Cepheid! Shapley pointed out that all the observations of Cepheids (light curves, spectra, and radial velocity changes) are consistent with the idea that these stars are pulsating. The Doppler shifts observed are literally due to the photosphere moving toward us as the star expands and away from us as the star contracts; of course, we only observe the light from the side of the star that faces the earth.

The remaining obstacles to the pulsating star hypothesis were theoretical; that is, *why* should stars pulsate? Apparently, pulsating stars are unstable and the meaning of this word is illustrated in Figure 10-17. An object in a stable configuration returns to its ordinary position or arrangement after a small disturbance occurs. An object in an unstable configuration does not necessarily return to the rest position but may collapse or oscillate back and forth around it. The latter situation applies to pulsating stars such as the Cepheids. As a star changes with

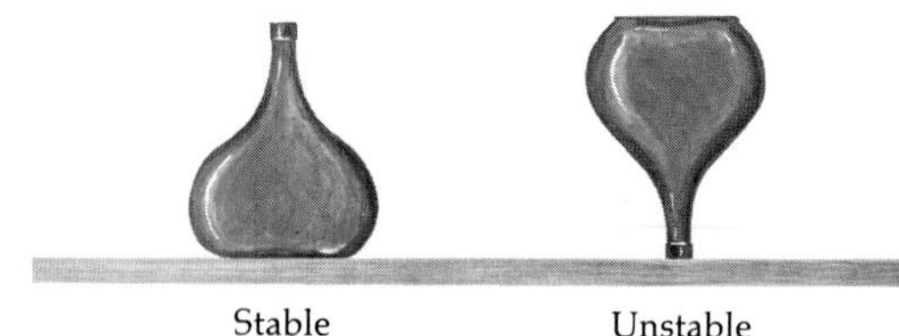

10-17 *Simple examples of stable and unstable configurations.*

time (Chapter 11) its changing internal structure may become unstable and thus start to oscillate. This problem was tackled by the English astronomer Sir Arthur Stanley Eddington (1882–1944), who established the theory of pulsating stars as a branch of astrophysics. As an example of the reasonable accuracy that was attained, the theory predicts that the increase or decrease in radius compared to the average should be less than 10 per cent, whereas the value for delta Cephei is observed to be 6 per cent. As the star increases and decreases in size, its surface temperature also changes, and this causes the changes in the pattern of lines in its spectrum.

The above discussion is just a small part of the Cepheid story. These stars are very bright and easily recognizable by means of their light variations; as such, they are good candidates for studies of distances in astronomy. The brightest Cepheids have luminosities that are 10^4 times that of the sun. They can be seen in other galaxies, such as the Magellanic Clouds (the galaxies nearest to our own, page 350). In 1912, the American astronomer Henrietta Leavitt studied the Cepheids in the Small Magellanic Cloud and noticed that the average brightnesses are related to their periods. The Cloud is so far away that all the Cepheids are effectively at the same distance from us (remember Figure 10-2). Hence, the relation between periods and apparent brightnesses was in fact a relation between periods and absolute magnitudes (although the *values* of the absolute magnitudes were uncertain, since the distance of the Cloud had to be determined). What a discovery! If the *period-luminosity relation* could be calibrated, the identification of a star as a Cepheid and the simple determination of its period of light variation would immediately give its absolute magnitude and ultimately its distance. (Such a calibration is obtained from the statistical parallaxes of the Cepheids in our galaxy; none of the Cepheids is close enough for a trigonometric parallax.) The use of Cepheids as distance indicators by this method became a prime tool in surveying our galaxy and in measuring the distances to other galaxies.

Care must be exercised to establish the correct type of variable star, as a few different types of Cepheids are known, and each has a different period-luminosity relation. Stars such as delta Cephei are called *Classical Cepheids* and another group is named after the star W Virginis. The period-luminosity relations for Classical Cepheids and W Virginis stars are shown in Figure 10-18. Thus, the classification and period determination for Cepheids enables us "to read the label on the light bulb" and calculate the distances to these stars.

The pulsating stars used as distance and population indicators are the two types of Cepheids just discussed, along with the so-called cluster variables or *RR Lyrae stars.* These stars are named for their frequent occurrence in globular clusters. However, some of them are field stars, including RR Lyrae itself. The RR Lyrae stars are rapid pulsators, with periods around one-half day. Observation of RR Lyrae stars in the same cluster shows that (unlike Cepheids) they all have about the same absolute magnitude. There is no close globular cluster, and so the calibration comes from statistical parallaxes of the few relatively nearby RR Lyrae stars found outside of clusters. These determinations can be checked by the cluster parallaxes as described in Chapter 11. The RR Lyrae stars have an absolute V (yellow) magnitude close to +0.6. Hence, if a star can be shown to be an RR Lyrae variable, its absolute magnitude (about 50 times that of the sun) is known, and comparing this quantity with its apparent magnitude we can find the star's distance.

There are still other kinds of pulsating stars, notably the *long period variables.* These are red giant stars with periods of roughly a year. A typical example is the star omicron Ceti, also called Mira (Latin, The Wonderful). It has a period of eleven months,

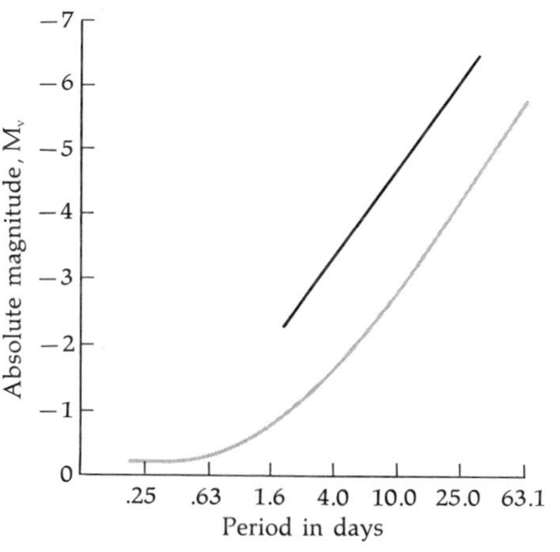

10-18 *The relationship between absolute magnitude and period for Classical Cepheids (dark line) and W Virginis stars (gray line). (From* Pulsating Stars and Cosmic Distances *by Robert P. Kraft. Copyright © 1959 by Scientific American, Inc. All rights reserved.)*

Sun

White dwarf

Neutron star

10-19 *Relative sizes of the sun, a white dwarf, the earth, and a neutron star. Neutron stars are discussed in Chapter 11. (Adapted from* X-Ray Astronomy *by Herbert Friedman. Copyright © 1964 by Scientific American, Inc. All rights reserved.)*

during which its brightness changes by a factor of 160 (about 5.5 magnitudes), so that sometimes it is conspicuous to the unaided eye, and at other times a telescope is needed to see it.

On the basis of average temperatures and luminosities of the pulsating variable stars, we find they occur in specific zones of the H–R Diagram. It is an important task of astrophysics to explain why stars with these particular temperatures and luminosities tend to be unstable.

WHITE DWARFS

Inspection of the H–R Diagram shown in Figure 10-12 reveals the existence of a group of very faint stars with typical radii only about one per cent of the sun's radius (Figure 10-19); thus, these stars are comparable in size to the earth.

The remarkable fact about white dwarfs appears when we consider their masses. A few of these stars have been observed in binary systems and therefore fairly accurate masses have been found for them. These masses range from roughly half to roughly one solar mass. But, since the white dwarfs are so much smaller than the sun, their average densities must be several hundred thousand to a million times greater than the average density of the sun (which is about equal to the density of water). The state of this material can be illustrated by recalling the Rutherford model of the atom (Chapter 5), in which the electrons orbit the nucleus and the volume of the atom is, to a large

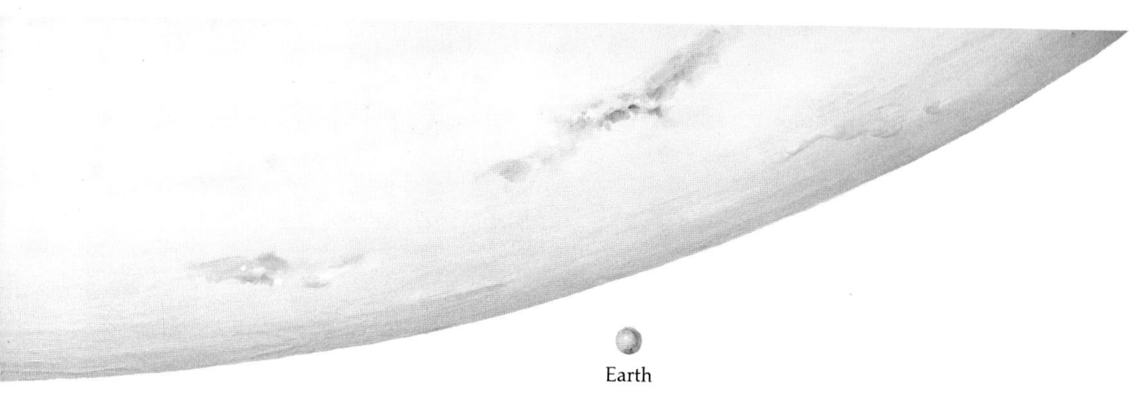

extent, empty space. The densities in the interiors of white dwarfs are so high that the electron shells around the nuclei of atoms are broken down and the nuclei and electrons are pushed closer together (this is called *electron degenerate* matter). It is believed that the white dwarfs were once normal stars, generating energy by nuclear reactions in their interiors. The pressure of this radiation helped support the gaseous matter of the star against its tendency to contract under its own gravitation. However, the nuclear fuel was eventually exhausted and the star contracted. Now it is radiating away its internal heat, slowly fading and cooling.

White dwarfs are probably a terminal phase of stellar evolution. If so, the theoretical deduction that they cannot have a mass in excess of about 1.5 $M_{\odot}$ becomes of interest because many young stars have masses above this limit. These massive stars must die in some other way, and it is thought that this happens through the supernova process (page 305). We will return to this discussion when considering stellar evolution.

EXOTIC STARS

The studies of binary stars have turned up some unusual cases. Some binaries are so close together that the surfaces of the two stars are actually in contact. These *contact binaries* are distorted by gravitational effects and there is the possibility of matter passing from one star to the other.

The variable star beta Lyrae is another kind of exotic binary system. It cannot be resolved with a telescope, but the spectrograph shows that the brighter star is a B star, and its lines shift back and forth in the way that characterizes orbital motion. The

10-20 *Schematic showing the arrangement of gas clouds in the binary system beta Lyrae. The white arrows indicate the direction of the stars' orbital motion and the black arrows indicate the direction of gas flow.*

other star is too faint to show up in the spectrum, but there are unusual spectral lines that have been ascribed to different streams of gas around the stars. Three different streams are needed to explain the observations. (1) A stream of gas flows from the more massive to the less massive star. (2) A lower density stream circles around the less massive star and returns to the source. (3) Part of the gas flow forms a spiral pattern outside of the binary system. This complex system is shown schematically in Figure 10-20.

Complicated star systems like beta Lyrae may not be very important in the scheme of things, but they remind us that there are some really intriguing problems in astronomy. The noted astrophysicist Otto Struve (1897–1963) devoted a substantial part of his professional career to studying the changing spectrum of beta Lyrae and trying to explain it in terms of models such as that shown in Figure 10-20. The problem is still not completely solved. Just a few years ago a paper entitled "A Final Model for Beta Lyrae" was read before a meeting of the American Astronomical Society; a few minutes later another speaker announced a new "Preliminary Model for Beta Lyrae."

11

Birth and Death of Stars

The title of this chapter may at first sight appear presumptuous. In Chapter 2 we established that the age of the earth was about 5×10^9 years old; men have life spans of at most a hundred years or so. Even records kept from one generation to the next do not tell us much, because recorded history covers only a few thousand years. How then can we infer that stars are born, live out their lives, and die?

The answer involves a philosophical adjustment to a difficult problem and the correct attitude toward gathering the relevant observations. The situation can be illustrated by considering the growth of trees in a forest based on one afternoon's observations. Even if you watched a few trees closely during the afternoon you would not detect their growth. But if you noticed that trees of the same species were present in different sizes, that many of the larger trees had approximaiely the same size, and that some of these larger trees had fallen over and lay dead, you might reach some useful conclusions. One reasonable hypothesis might be that trees begin small, grow to a maximum size at which (or approximately at which) they spend most of their life, and then die. A similar approach to the stellar census (as organized in the H–R Diagram) allows us to discuss the birth and death of stars.

The discussion of the various proposed energy sources for the sun (Chapter 9) centered around the sun's lifetime. From the geological evidence (page 245) we believe that the sun has had about the same luminosity for billions of years, hence, any acceptable source of energy must be capable of supporting the solar luminosity for the sun's lifetime. We found that nuclear energy was an adequate source, but it is not an unlimited one. If all the hydrogen in the sun were available for conversion into helium (that is, if it were all in the correct state of high temperature and pressure), the sun would radiate at its present rate for about 10^{11} years.

In practice we are interested in the star's main sequence lifetime, or the time that the star spends on the main sequence of the H–R Diagram. This lifetime is shorter than the hypothetical total lifetime by about a factor of 10. The reason for this stems from the fact that the burning of hydrogen to helium occurs in the hotter, denser, central region of a star. The tendency is therefore to gradually create a central core of helium with a surrounding shell where burning of hydrogen to helium still takes place, and an outer region, where nuclear reactions do not occur, composed primarily of hydrogen. The star thus becomes very inhomogeneous, as the atomic nuclei in the core have about four times the mass of the nuclei in the outer parts. The larger the core grows, the more important the inhomogeneity becomes with regard to the stability of the star. When approximately the inner 10 per cent of the star's mass of hydrogen has been burned, the star is no longer stable. Calculations show that when the star reaches this state, the outer layers rapidly expand so that the photosphere reaches ten times its original radius and thus a giant star is formed. This will happen to the sun when it reaches the age of 11×10^9 years. Hence, the sun has lived out about half of its main sequence lifetime and still has about 6×10^9 years to go before it expands to fill a substantial volume of the present solar system.

This happy situation (which gives NASA 6 billion more years to perfect interstellar travel) does not hold true for the massive, hot, and very bright stars near the upper end of the main sequence. This difference arises from the mass-luminosity relation (Chapter 10) which states that the luminosity increases very rapidly with mass. Consider a very bright (blue) O-star with a mass of 35 $M_\odot$. Its luminosity is about 10^5 times that of the sun. The main sequence age for this O star (as compared to the sun's main sequence age of 11×10^9 years) is *increased* 35 times because of the higher mass (greater fuel supply) but *reduced* by 10^5 because of the higher luminosity (or greater rate of burning). A simple arithmetical calculation then gives a main sequence time for an O star of about $11 \times 10^9 \times 35 \div 10^5$ years, or only 4×10^6 years. A bright blue star therefore exists on the main sequence for a time less than 0.1 per cent of the sun's present age. If these bright blue stars had been created at the same time as the sun, they would have long since used up their nuclear fuel. This conclusion holds true even if all their hydrogen were available instead of only the inner 10 per cent. Any bright blue stars that were born contemporaneously with the sun have already passed off the main sequence. Therefore, the O stars that we see at present must have been created no more than 10^6 to 10^7 years ago—a very short time by cosmic standards. Thus, the very existence of the bright blue stars with huge luminosities and resultant very short lifetimes shows that stars are not static creatures but are born and evolve. We will also see that there is evidence that stars die.

H–R DIAGRAMS AND THE AGES OF STAR CLUSTERS

The basic ideas of the preceding section predict certain consequences that can be tested. The H–R Diagrams of clusters are a nearly ideal aid because a cluster consists of a group of stars presumably formed from the same cloud of material at about the same time. Thus, an individual cluster represents a group of stars of similar age and chemical composition, although these properties can and do vary from one group to another.

A composite H–R Diagram is shown in Figure 11-1. Except for the adjustment of the vertical scale, which involves a determination of distance, the preparation of the H–R Diagrams for individual clusters is straightforward; a sample with the data points is shown in Figure 11-2. The distances for some clusters can be determined independently (page 266), and for others we utilize the fact that evolution is very slow and main sequence lifetimes very long for stars fainter than the sun. Thus, the lower main sequence regions of the various clusters' H–R Diagrams should be approximately the same. This idea has been used in the preparation of Figure 11-1, where the main sequences of stars fainter than the sun have been made coincident. This was achieved by sliding the individual H–R Diagrams vertically; the amount of slide determines the distance to the cluster (because it corresponds to the difference between the apparent and absolute magnitudes of the stars on the lower main sequence) and a *cluster parallax* is thus determined.

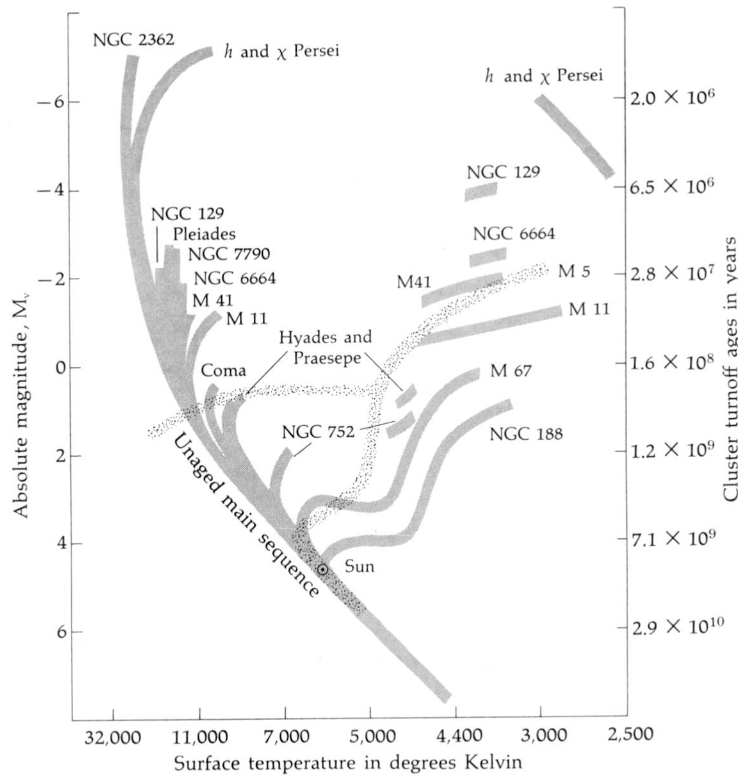

11-1 *Composite H-R Diagram, showing the absolute magnitudes and surface temperatures of stars in a variety of clusters. The cluster names or numbers are given on the diagram. The age of a cluster, defined by the position along the main sequence at which its stars turn off to the right, is marked on the vertical scale at the right. For example, the cluster NGC 188, which turns off near the position of the sun on the main sequence, has an age of about 1 × 10^10 years. (Adapted from* Pulsating Stars and Cosmic Distances *by Robert P. Kraft. Copyright © 1959 by Scientific American, Inc. All rights reserved.)*

The study of clusters and stellar evolution was actively pursued during the 1950s. The interpretation of the composite H–R Diagram (Figure 11-1) was made by the Hale Observatories astronomer, Allan Sandage. The fact that the very bright stars are still on the main sequence in the cluster NGC 2362 immediately implies that this is a very young group. Could this composite diagram, with a common main sequence at the left and segments leading off to the right, be an age sequence with the youngest cluster at the top and the oldest at the bottom?

This last idea is entirely consistent with our discussion of lifetimes in the preceding section. The most massive stars are

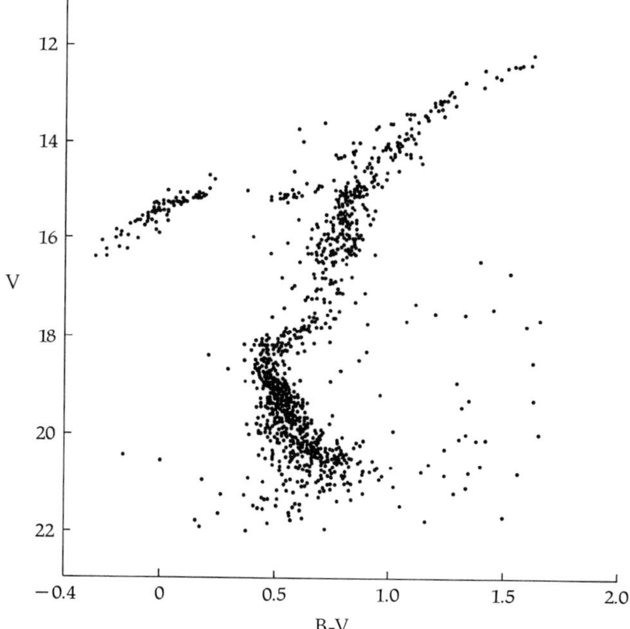

11-2 *Observational H-R Diagram, V magnitude versus B-V color, for the globular cluster M5, showing the data points that correspond to individual stars. (Courtesy H. C. Arp, Hale Observatories.)*

found at the upper left hand part of the H–R Diagram if they are still on the main sequence. The more massive the star, the shorter is its time on the main sequence; hence, as time passes, an initial main sequence would evolve schematically as shown in Figure 11-3. As the age increases, the point where the actual H–R Diagram turns off from the initial main sequence moves down the remaining main sequence to the fainter, less massive stars. Thus, the clusters can actually be dated on Figure 11-1 by the main sequence lifetime of the stars which have just left the main sequence; these *turnoff points* are converted into ages in the right-hand scale of Figure 11-1. The stars in the very old clusters, such as M5, M67, and NGC 188, have evolved off the main sequence all the way down to the sun. Their ages are about 10^{10} years or perhaps a little older, and thus they must be representative of the oldest stars in the galaxy.

The H–R Diagrams of clusters contain few stars between the main sequence segments and the giant stars to the right. Apparently stars do not spend much time in this area, that is, with temperatures and luminosities intermediate between those on the main sequence and those of red giants, or more would be seen there.

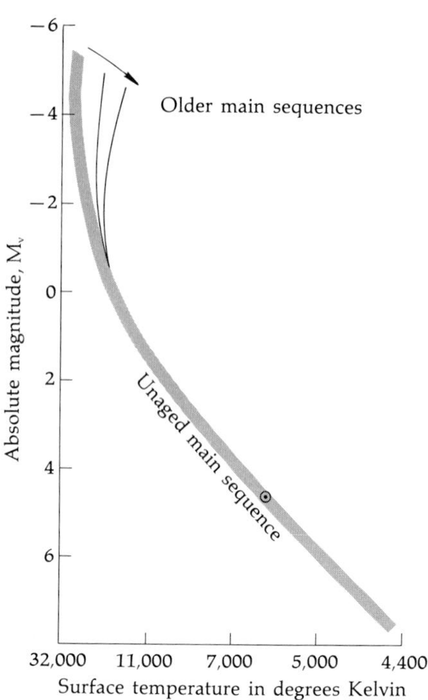

11-3 *Schematic evolution of the upper main sequences of galactic clusters. As age increases, individual stars move toward the red giant region and the older main sequences turn off from the initial or unaged main sequence.*

AN OUTLINE OF EVOLUTION

The cluster H–R Diagrams provide us with a fund of knowledge against which we can check our calculations and theories of stellar evolution. The theoretical and mathematical discussions of stellar evolution are quite complicated and are essentially too complex to be tackled by hand, by slide rule, or even by desk calculator. The development of the large, fast electronic computer enabled astronomers and physicists to carry out the calculations necessary for a study of stellar evolution. These calculations could then be compared with the observed H–R Diagrams of clusters for confirmation or improvement if necessary. In order to be valid, the calculations must place stars in the parts of the H–R Diagram where they are observed.

The question of the *birth* of stars represents the area of our greatest ignorance in stellar evolution. It also involves the origin of the solar system which we defer to the next section. The *protostars* are assumed to have formed from dust and gas clouds in interstellar space, a view consistent with our knowledge of associations. A protostar is much larger than the main sequence star that it later becomes. The luminous energy is supplied at this point by gravitational contraction in the same way as discussed on page 246 in connection with possible energy sources for the sun. Protostars are large and cool; they occupy a region up and to the right of the main sequence in the H–R Diagram. The contemporary astrophysicist A. G. W. Cameron realized a few years ago that at a certain stage the gradual contraction of the protostar would suddenly accelerate, and its rapid collapse would heat the gas to the point that the object would begin to shine as a star. A detailed theory that supports this idea was calculated in Japan by Chushiro Hayashi, who found that this was indeed the way in which a protostar evolves toward the main sequence (Figure 11-4a). If the contraction to the main sequence occurs in this way, some stars in very young clusters should be found above and to the right of the main sequence (above, because they are very large and hence fairly bright; to the right, because they are cool). This is apparently confirmed by the observations.

The contraction to the main sequence continues until the central temperature reaches about 10×10^6 degrees K and nuclear reactions can begin; the basic result of nuclear burning on the main sequence is the conversion of hydrogen into helium. The star then spends most of its life on the main sequence, a conclusion in agreement with the large number of stars found there. The time spent on the main sequence is determined by the mass and luminosity of the particular star; bigger (more massive) stars burn brighter and faster and live less long than smaller stars. When the inner 10 per cent of the star's mass has been converted from hydrogen to helium, the star moves rapidly up and to the right of the main sequence to the region of red giants. The evolution between the main sequence and giant regions is calculated to be rapid and this agrees with the fact that few stars are found there on the cluster H–R Diagrams.

The departure of a star from the main sequence occurs because of the growing helium core and the resultant inhomogeneity. The star changes structurally in two ways. Its outer part expands and we recognize the star as a giant. At the same time, the central region contracts and is heated to even higher temperatures. At temperatures of about 10^8 degrees K, it is hot enough so that a nuclear reaction that produces carbon from three helium nuclei takes place. This event in the life of a star is called the *helium flash;* it appears to stop the star (already a red giant) from moving farther upward and to the right in the H–R Diagram. Additional reactions involving helium and carbon can produce oxygen, neon, magnesium, other elements up to

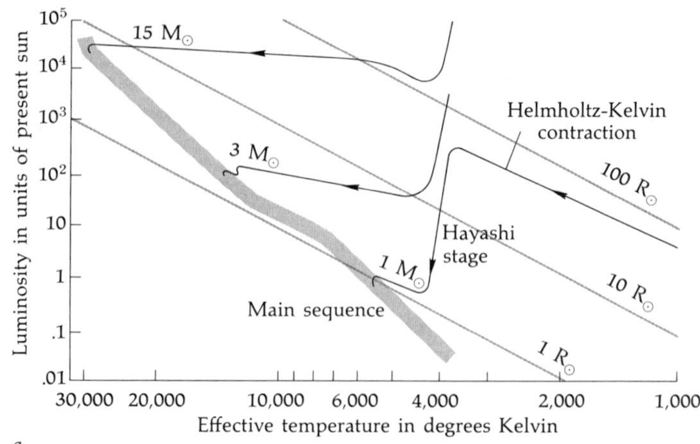

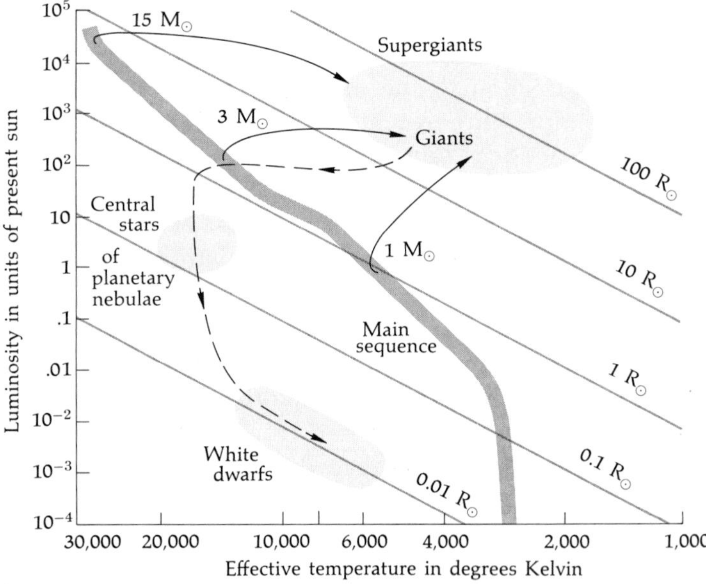

11-4 *Schematic evolution of stars on the H-R Diagram, showing stars of 1, 3, and 15 solar masses. (a) The fairly well understood contraction to the main sequence beginning with the Helmholtz-Kelvin contraction and continuing through the Hayashi stage (vertical part of track). Lines of constant radius run diagonally from upper left to lower right and are shown for sizes of 1, 10, and 100 solar radii. (b) The less understood post-main-sequence evolution beginning with movement into the giant region and continuing back toward the main sequence. The evolution probably then proceeds along a sequence of faint, hot (blue) stars, which may include the central stars of planetary nebulae, and finally to the white dwarf region. (Adapted from* The Youngest Stars *by George Herbig.*

the atomic weight of iron, and some additional energy. The net effects of the helium burning and its consequences are to reverse the direction of evolution in the H–R Diagram and to send the star back toward the main sequence.

Some observational clues to the path of a star that is leaving the giant stage are given by the H–R Diagrams of globular clusters (Figure 11-2). Notice the horizontal band of stars reaching back toward the main sequence. Perhaps the stars move along this band after the giant stage (the calculations are not particularly reliable on this subject). The evolution of the star as it moves back toward the main sequence probably takes it through one or more unstable states, as we observe that some pulsating variable stars occur in this part of the H–R Diagram.

Regardless of the path taken (still a matter of dispute), many stars reach the white dwarf stage where they are composed of electron degenerate matter (page 291). In this degenerate state the star cannot contract further and, therefore, has no energy input from gravitational contraction. It also has no nuclear energy source because all the nuclear fuel in the hot central region has been used up. Hence, the white dwarfs can only radiate away their thermal energy. The star then gets cooler and cooler, fainter and fainter, and approaches death as a *black dwarf.* This is probably a common form of terminal evolution; however, such stars have not been observed, presumably because they are too faint. The path to the white dwarf region of the H–R Diagram may be along the sequence of faint blue stars (Figure 11-4b), which includes the central stars of *planetary nebulae* (Figure 12-3).

The stellar life story discussed above is appropriate for a star of the sun's mass and is illustrated schematically in Figure 11-4. Bear in mind that the exact path taken after the red giant stage is uncertain, but that the final stage is probably among the white dwarfs.

Other terminal phases are also possible, however. We have noted (page 291) that white dwarfs with masses greater than about 1.5 solar masses probably cannot be formed. What happens to the stars with masses greater than 1.5 $M_\odot$? If they are to become white dwarfs they must lose mass—this can be accomplished in a variety of ways. A steady flow of gas (like the solar wind) might remove the mass, although the rate of loss would have to be millions of times greater than the present solar rate. There is a considerable amount of evidence for mass loss in red giant stars. For example, some of the absorption lines in the spectra of these stars seem to be produced by matter surrounding the star in a shell far outside its photospheric surface.

It has also been suggested that the planetary nebulae found around relatively small, hot stars (page 321) represent shells of gas that were somehow ejected from these stars at an earlier stage of evolution when they were red giants. However, this has not been proven. Mass loss also occurs in explosive events, as observed in the *novae* and *supernovae.*

The term nova is based on the Latin for new. When novae were first observed, they were thought to be new stars that somehow appeared and gradually faded away. Now we know that when a nova is discovered, it is possible to find the prenova (the much fainter star that exploded and produced the nova) on previously recorded photographs of the sky region where the nova is observed. After a few years, the star fades (Figure 11-5) to roughly its prenova brightness, indicating that the explosion did not disrupt the bulk of the star. However, the spectrum of a typical nova shortly after the explosion occurs does show that some matter is ejected. The presence of emission lines shows that the star is surrounded by a hot radiating gas,* and interpretation of the emission lines by means of the Doppler effect indicates that the ejected gas is expanding away from the star at speeds of typically a few hundred to over 2,000 kilometers (1,240 miles) per second. In some cases (see Figure 11-6) the nova was close enough to us so that the expanding gas cloud eventually could be resolved on telescopic photographs. Recently, such clouds have been detected with a radio telescope shortly after the novae exploded. It is believed, but not known, that a given star that becomes a nova is likely to undergo this explosive process again and again, perhaps at intervals of several thousand years, each time losing a small percentage of its mass. In recent years, studies of the old novae, the stars that have undergone the nova process, have suggested that many and perhaps all of them are actually binary stars. Theorists have speculated that the presence of a companion star somehow induces the nova explosion. One possibility among several is that matter lost by one star falls onto the other star and provides the energy to cause or trigger the explosion.

Supernovae are produced by much more catastrophic explosions than novae. A typical nova increases by a factor of 10,000 in brightness as it explodes and ejects a gas cloud, and at its

*Absorption lines with large Doppler shifts are also found in nova spectra. The shifts are toward the shorter wavelengths, indicating that the absorption lines are produced in the part of the ejected gas that is moving toward us, and is observed against the incandescent background of the star.

11-5 *Nova Herculis 1934. Photographs taken with the same exposure on March 10, 1934 (left), and May 6, 1935 (right), showing the large change in brightness as the nova faded (Lick Observatory photographs.)*

11-6 *Expanding nebulosity around Nova Persei 1901, photographed with the 200-inch telescope. (Courtesy of the Hale Observatories.)*

peak brightness must be one of the brightest stars in its galaxy.* But a supernova is sometimes brighter than *all or most of the stars in its galaxy put together* (Figure 11-7) and there may be 10^{11} stars in a large spiral galaxy. Furthermore, it is likely that much of a supernova's energy is emitted in the form of ultraviolet light,

*Both novae and supernovae have been observed in our own and other galaxies, but since the invention of the telescope we have observed supernova explosions only in other galaxies.

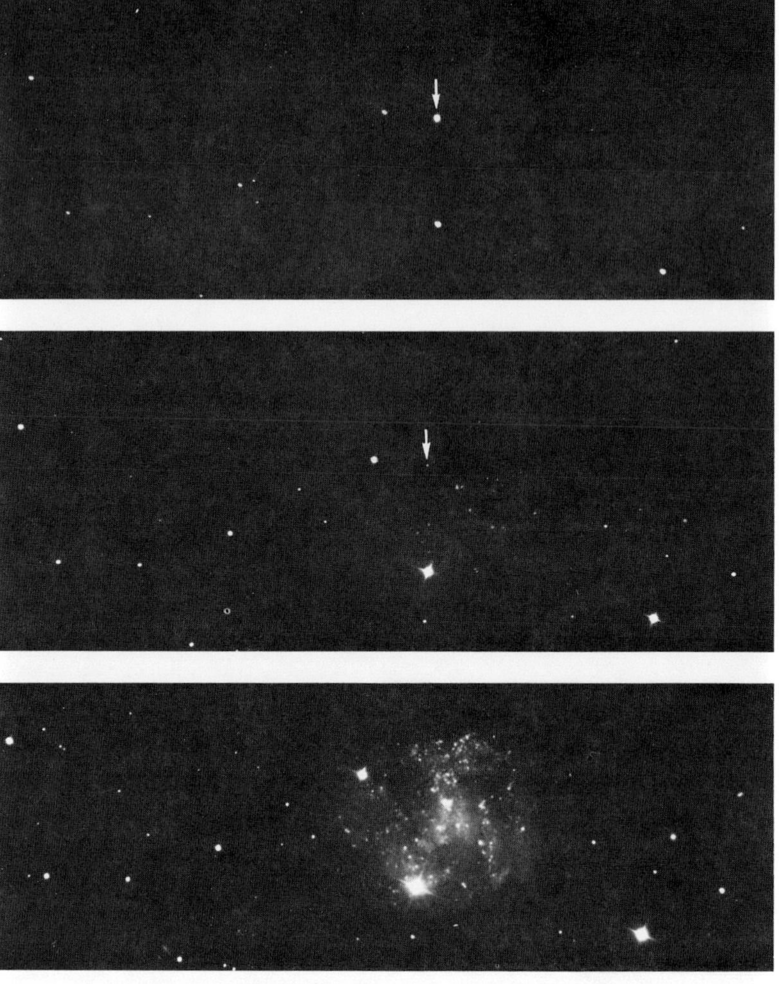

11-7 *Sequence showing brightness changes of a supernova in the galaxy IC 4182. The supernova reached maximum brightness* (top) *in August, 1937, was much fainter in November, 1938* (middle), *and was invisible by January, 1942* (bottom). *The shortest exposure* (top) *easily shows the supernova, but not the fainter stars in the galaxy. These appear in the longest exposure* (bottom). (*Courtesy of the Hale Observatories.*)

so that its bolometric luminosity is even more impressive. A substantial fraction of the mass of the star is disrupted and flung into space by the explosion, resulting in an expanding gas cloud, the *supernova remnant.* The expansion velocities of remnants range from less than 1,000 kilometers (620 miles) per second to more than 10,000 kilometers (6,200 miles) per second. The Crab nebula (Chapter 16) is a supernova remnant, as is the Veil nebula (Figure 12-4). Supernova remnants are observed to be strong sources of radio emission (about 100 of them have been found in this way), and several of them have also been found to emit x-rays.

Theoretical computations suggest that a typical supernova explosion occurs when a star of several times the sun's mass has evolved so far that all of the atomic nuclei (in its central portion) that constitute the fuel for the star's energy-generating nuclear reactions are used up. There is no longer a powerful source of radiant energy inside the star, and thus the radiation pressure is not adequate to support the outer layers. The star collapses—what we really have is an *implosion*—and this catastrophic event releases an enormous amount of energy. This takes several forms, including the radiation emitted by the supernova and the kinetic energy of the matter in the expanding remnant that is ejected as a "splash" from the implosion. The implosion compresses the star enormously, perhaps to a diameter of 10 or 20 kilometers, and the resulting highly condensed object is called a *neutron star* (Figure 10-19). Until recently, this discussion, and especially the suggested existence of neutron stars, was entirely hypothetical. Now it appears that the discovery of pulsars (Chapter 16) reveals the existence of neutron stars produced by supernova events. Supernovae probably also produce many of the cosmic ray particles that we observe at the earth, some of which, as mentioned previously, may be responsible for biological mutations.

Prior to the supernova event, nuclear reactions in the stellar interior probably produce most of the elements up to iron (atomic weight 56). In the course of the supernova implosion, the interior temperature may reach 10^9 degrees K, and nuclear reactions take place that produce the heavier elements. *The ejection of a part of the supernova into space is believed to represent the source of the heavy elements that we find on the earth, in the sun, and elsewhere in the universe.* (We recall from Chapter 1 that it is generally believed that the explosion of the primeval fireball at the birth of the universe produced hydrogen and perhaps some helium.)

We have discussed the white dwarfs, black dwarfs, and neutron stars as final states of stellar evolution. It is also believed by some physicists that some massive stars can implode beyond the neutron star point (the central density of a neutron star should be about 10^{14} to 10^{15} grams per cubic centimeter). The gravity of these hypothetical highly condensed stars would be powerful enough to prevent light waves from escaping from them. This property has led to the term *black holes.*

STELLAR EVOLUTION AND THE INTERSTELLAR MEDIUM

We know that stars are being formed at the present time and that many stars have already lived out their lives. Some of the original matter of the stars is still locked up in them in their final configuration, whatever it might be. Another part of the original stellar material, as well as heavier elements produced by nuclear reactions inside the stars, has been returned to the interstellar medium (gas and dust between the stars) by the processes of mass loss that are discussed above.

Nuclear reactions that power stars proceed in the direction whereby elements of low atomic weight are "burned" to form an "ash" consisting of elements of heavier atomic weight. The elements found in a star are conveniently divided into three categories: (1) hydrogen (by far the most common substance in the universe), (2) helium, and (3) all of the other elements, which astronomers group under the heading of heavy elements. Recall the basic nuclear reactions that power the stars. On the main sequence, hydrogen is converted into helium; in giants, helium is converted into carbon, and the combination of carbon and helium can produce elements as heavy as iron. Supernova explosions produce even heavier elements. In no case do we find evidence for the amount of heavy elements being reduced by a stellar process.

We believe that stars are formed from the gas and dust of the interstellar medium, that they convert some of this gas to heavier elements in their interiors, and that they return some of this material to the medium through mass loss. Thus the interstellar medium is a reservoir of material for star formation where heavy element abundance is continually being enriched. Hence, new stars formed from this material will begin their lives with a higher heavy element abundance than did stars formed at an earlier time. The sun is a good example of this; its central temperature is not high enough to produce much in the way of heavy elements. Even if it did, the heavy elements would probably not come to the surface. Yet, iron is observed in the spectrum of the sun and uranium is found on the earth, which we believe was formed from the same nebula as the sun. This evidence agrees with our idea that the material from which the sun and the planets were formed *had already been inside stars* where the heavy elements were produced by nuclear processes.

A competing hypothesis might be that a cosmic mixture including heavy elements was created at the beginning of the universe. But if this were true then all stars should show similar abundances of the elements. However, this is contradicted by the observation (see Figure 10-8) that some stars are deficient in heavy elements, compared, for example, to the sun. A suitable index of heavy element content is the percentage of the total mass of a star's photosphere that is made up of heavy elements, and this quantity varies over more than a factor of 10, from roughly 0.3 per cent for an old, heavy-element-poor star, to about 4 per cent for a young, heavy-element-rich star. Our sun is intermediate with a heavy element abundance of about 1.5 per cent. It is estimated that the sun is a third generation star; that is, the material from which the sun was originally formed has been inside about three previous stars. A sense of cosmic perspective may result from the knowledge that the iron in a skillet being used to prepare our food was processed from lighter atoms long ago inside some stars, perhaps even in a supernova, and so were most of the atoms in the food and indeed in our own bodies.

Stars were divided into two *populations* by Walter Baade (Chapter 13), depending on age and amount of heavy elements. Stars formed relatively recently are young in years but are made of highly processed material with a high heavy element abundance; these were called *Population I* stars. Most stars near the sun are in Population I, as are the stars in the open clusters. *Population II* stars are found especially toward the central region of the galaxy and also make up the globular clusters. These old stars were formed long ago out of relatively unprocessed material that was poor in heavy elements. The significance of the classification of stars into populations will become clear as we discuss the structure of the galaxy in Chapter 12, where a more useful breakdown into three populations, depending on location, is presented.

ORIGIN OF THE SUN AND PLANETS

The evidence that we have mentioned suggests rather strongly that the elements heavier than helium were formed in the stars. The very brightest stars now seen were formed within the last few million years, a very short time on the cosmic clock. We would now like to discuss where the birth of stars takes place, under what circumstances, and by what processes. The prevailing view is that planetary systems like our own solar system are formed at the same time and from the same material as are their central stars. Thus, the properties of our solar system are probably relevant to the study of star formation. Recall our discussion on page 199; the orbital motions of the planets and most satellites are mostly in the same direction as the rotation of the sun. In addition, most of the planetary and satellite orbits are roughly in the same plane and most of the planetary orbits are nearly circular. Our birth hypothesis must explain these properties.

At the present time, the leading theory of the origin of the sun and planets is based on the *nebular hypothesis* that was first advanced in 1755 by the German philosopher Immanuel Kant, and independently developed by the French mathematician, Pierre Laplace, in 1796. Since then it has been in and out of favor as new facts and ideas have emerged. According to the modern version of the nebular hypothesis, the solar system formed from a large, slowly turning, nearly spherical cloud of gas and dust—the nebula (Figure 11-10).

As the cloud contracted under its own gravitational force, it rotated faster and faster. This spin up was a consequence of the law of *conservation of angular momentum.* The angular momentum of a specific particle is the product of its mass, its distance from the axis of rotation, and its rotation speed (Figure 11-8).

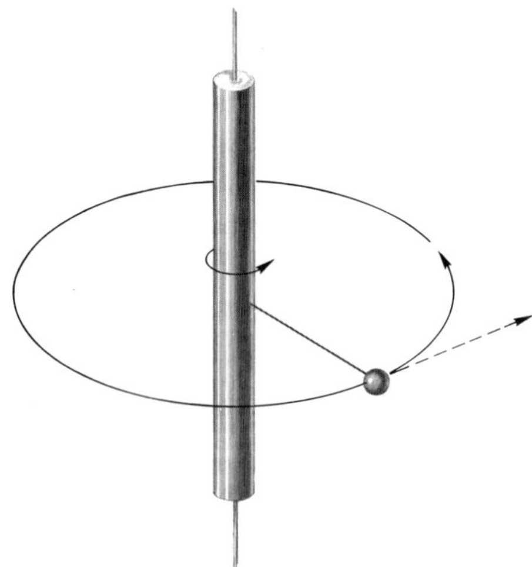

11-8 *Illustration of the definition of angular momentum through the example of a heavy steel ball fastened to a central rod with a very light rod. Turning the central rod causes the ball to whirl about it. The ball's angular momentum is the product of the distance from the rotation axis to the ball, the mass of the ball, and the velocity of the ball.*

11-9 *Illustration of the conservation of angular momentum. (From* The New College Physics *by Albert V. Baez. W. H. Freeman and Company. Copyright © 1967.)*

This definition can be generalized mathematically to describe the angular momentum of any system, whether it be a person on a spinning piano stool (Figure 11-9), the rotation of the earth on its axis, or the motion of the planets around the center of mass of the solar system. The law of conservation of angular momentum states that the angular momentum of an isolated system does not change. Thus, as the nebula contracted (Figure 11-10), the average distance of the gas particles from the axis decreased, and this conservation law required that their average speed around the axis increased.

As the nebula spun faster, it became progressively more oblate and eventually assumed the shape of a disk. (It is a well-known property of a rotating gas or fluid to flatten out in this way*; it can be regarded as an effect of centrifugal force, which is greatest at the equator and smallest near the poles of a rotating object.) The rotating nebula continued to contract, and left behind rings or blobs of matter in the disk that eventually became the planets.

There is still a great deal that we do not know. Even if the

*The planet Jupiter (Chapter 8) is oblate because of its rapid rotation. The earth is also slightly oblate—we mentioned its equatorial bulge (caused by rotation) in Chapter 2.

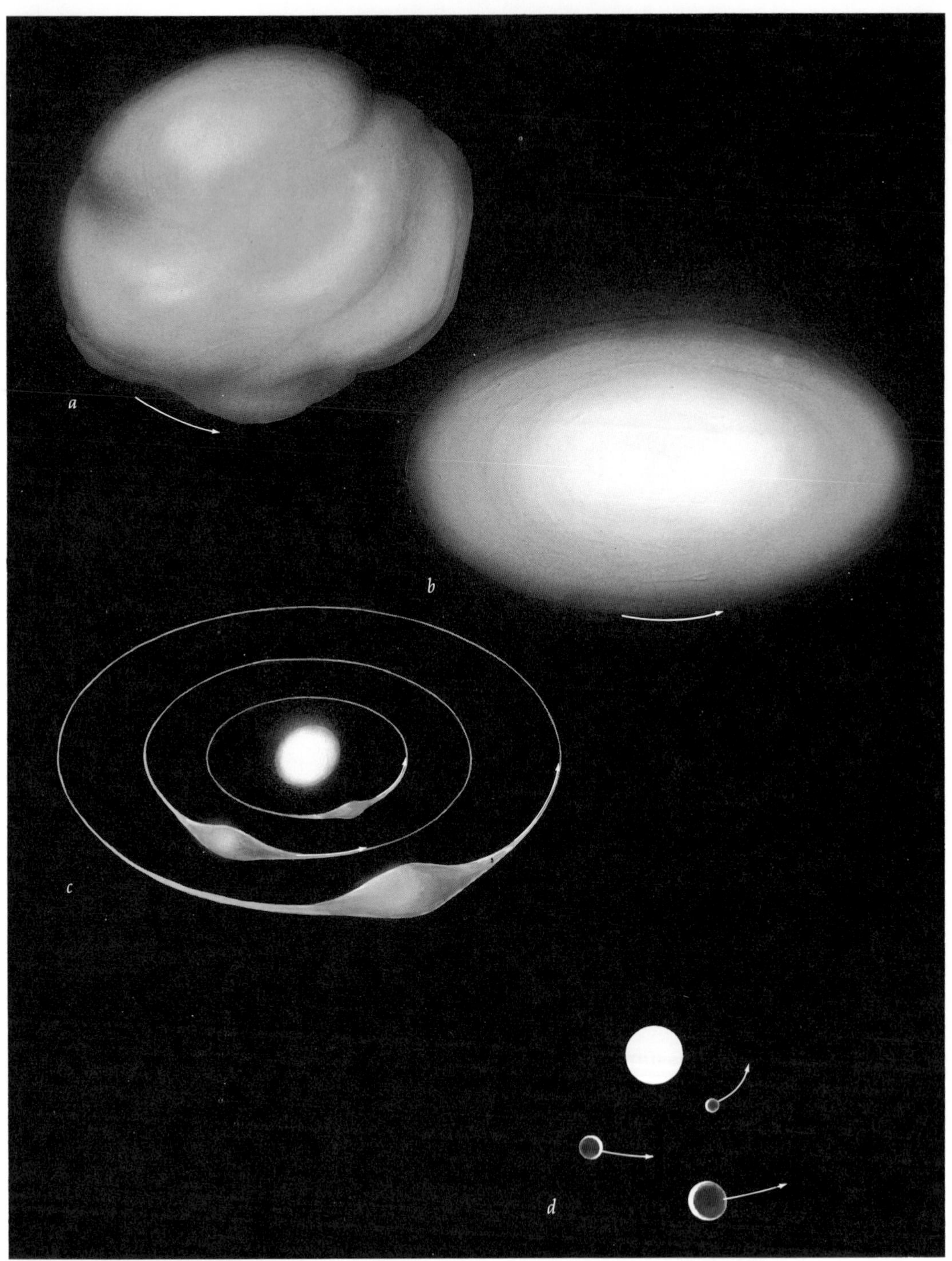

11-10 *Schematic of the nebular hypothesis. (a) The diffuse solar nebula is roughly spherical but rotates slowly and begins to contract; (b) the contraction and rotation have produced a flat, rapidly rotating disk with some central condensation; (c) as the contraction continues the protosun is formed and rings of material are left behind; (d) the material in the rings condenses into planets.*

above theory is correct in its general outline, there is very little agreement among astronomers as to exactly how the planets formed from the disk. However, it does appear that the theory is in reasonable accord with most scientists' thinking about how stars are formed. The central portion of the condensing nebula warmed as it contracted; gravitational potential energy of the matter contracting or falling toward the center was converted to heat, as we mentioned in the discussion of Helmholtz-Kelvin contraction (Chapter 9). When the contraction proceeded to the point that the central region was hot enough for nuclear reactions to occur, the sun began to shine and a new star joined the main sequence.

A serious objection that has been raised to this theory is that the calculations of the formation of the sun in this manner show that it should have most of the angular momentum in the solar system. But when we add up the angular momenta of the various objects in the system, we find that the planets, primarily Jupiter and Saturn, have the vast majority of the angular momentum, whereas the sun (which rotates once in 27 days) has only 2 per cent of the total.

One likely answer to the above objection is that the sun may have lost much of its angular momentum since it was formed. As the ionized matter of the solar wind streams away from the sun it is still attached to the sun in a sense by the lines of force of the solar magnetic field, which behave very much like elastic strings. As the material of the solar wind moves out from the sun, its distance from the axis increases, and hence (since angular momentum must be conserved) the sun spins less rapidly. The rate at which this effect presently occurs can be measured, and it turns out to be enough to slow the solar rotation rate down by a factor of 2 in about 2×10^9 years. As we believe the solar system is only 5×10^9 years old, this would not be enough to account for the fact that the sun has only 2 per cent of the angular momentum in the solar system at the present time, and so we are obliged to study the possible early history of the sun in order to search for an additional way in which mass (and thus angular momentum) could have been shed.

When we observe regions of the galaxy where clouds of gas and dust exist, so that the conditions may be right for the formation of stars by condensation, some interesting phenomena are found. In particular, bright spots (called *Herbig-Haro objects* after two contemporary astronomers) have been seen to form (Figure 11-11) in dark nebulae. These may represent the ignition of condensing protostars, or some other early

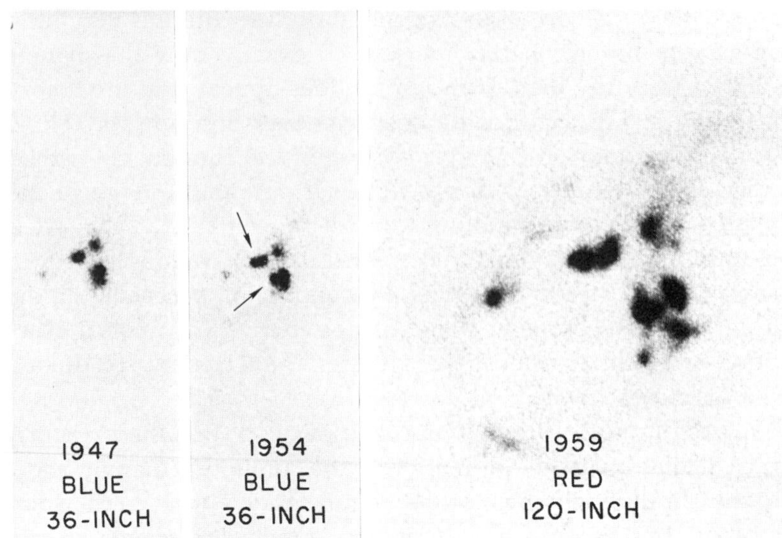

11-11 *Photographs of Herbig-Haro Object No. 2 taken in 1947, 1954, and 1959. The arrows mark the areas of striking brightness increases. These events are probably connected with the early stages of star formation; star birth itself cannot be claimed to have been observed on these photographs because some clumps decrease in brightness. (Lick Observatory photographs.)*

stage of stellar formation. Also observed in nebular regions are the *T Tauri stars,* a class of variable stars that may represent an evolutionary stage of the contracting star that follows the Herbig-Haro object stage and precedes arrival on the main sequence. The spectra of these stars show that a typical one is ejecting matter at speeds of 200 to 300 kilometers (120 to 190 miles) per second in an amount (mass per second) roughly 10^6 times that of the present solar wind. Thus, if the protosun went through a T Tauri stage, it could easily have lost enough angular momentum to explain the slow rotation of the present sun. Recent observations by Karen and Stephen Strom of the State University of New York (Stony Brook) suggest that young stars well past the T Tauri stage on the way to the main sequence are surrounded by shells of gas and dust. This fits into the general idea of the evolution of stars as summarized here.

If any significant amount of solar wind type mass loss has been going on throughout the lifetime of the sun, other stars similar to the sun which have been formed more recently should be rotating faster. Younger stars of solar type (as identified by their spectra) can be found in the clusters dated (Figure 11-1) by the turnoff method. Rotation speeds can be inferred from the widths of the line profiles. Suppose we are looking at a star at right angles to the axis of rotation. As the star rotates, one-half of the star is going away from us while the other half is coming toward us. The absorption lines due to atoms in the receding half are shifted toward the red (longer wavelengths) and the absorption lines in the approaching half are shifted toward the blue (shorter wavelengths) by the Doppler effect. The net effect on a line from the entire star is to broaden it; the faster a star rotates, the wider is a particular line, as illustrated in Figure 11-12. Thus, measurements of the line widths of stars tell us their rotation speeds. Of course, we do not observe all stars at right angles to their axes of rotation, and a statistical correction for foreshortening must be provided. The basic situation, however, remains the same. Such an investigation was carried out at the Hale Observatories in 1967 by Robert Kraft. Kraft found that solar-type stars seem to rotate more slowly as they age. Thus, the nebular or Kant-Laplace hypothesis, augmented by appreciable mass and angular momentum loss from the primary star, appears reasonable. Remember, however, that we are still ignorant of the detailed processes that produce planets.

Some other hypotheses concerning the origin of the solar system, much in vogue a few decades ago, were based on a collision of the sun with another star or, at least, a very close approach of the two. These *collisional and tidal hypotheses* do not

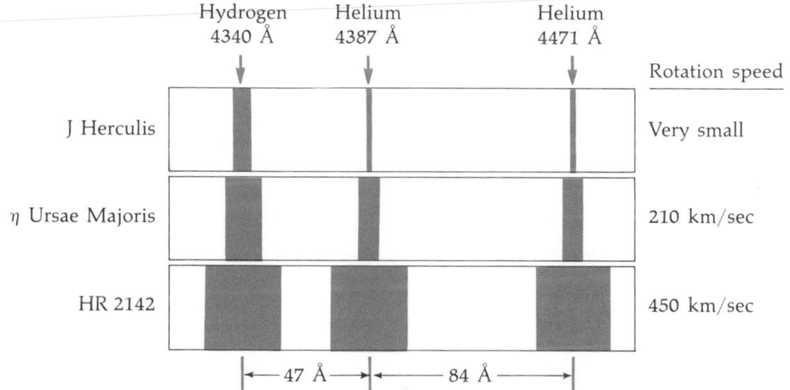

11-12 *Schematic of absorption line broadening caused by the different rotation rates of three stars. The widths of the hydrogen and helium lines in the slow rotator J Herculis are caused by processes in the star's atmosphere, but the larger line widths in eta Ursae Majoris and HR 2142 are primarily due to the rapid rotation of these two stars. (After a drawing by I. S. Shklovskii and C. Sagan based on spectra taken by O. Struve and G. A. Shajn.)*

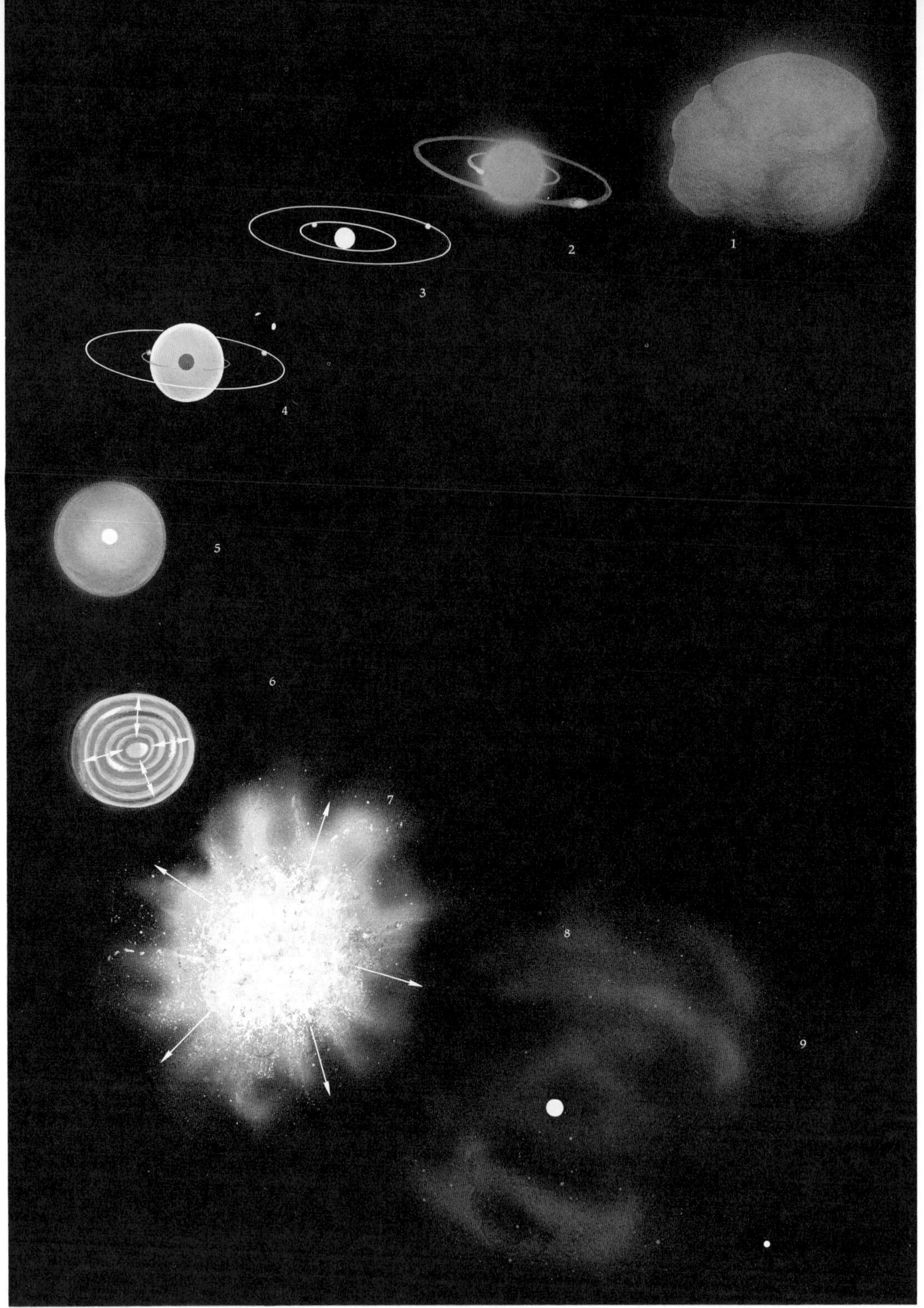

11-13 *Schematic for the evolution of a star of 1.2 solar masses. First, a cloud (1) condenses to form a star and protoplanets (2). The main sequence stage (3) ends with the star's expansion (4) to become a red giant, with the successive destruction of the planets (5). Later, the star may become a pulsating variable (6) and a nova (7). After the material thrown off in the nova outburst dissipates (8), the star finally collapses into a white dwarf (9). (After* Life Outside the Solar System *by Su-Shu Huang. Copyright © 1960 by Scientific American, Inc. All rights reserved.)*

bear on star formation, but rather only on the formation of the planets. In the collision or close approach, filaments of gas could be knocked off or pulled off the sun; the filaments then cool to form the planets. The resulting planetary orbits would be highly elongated, which is contrary to the observations. In addition, the way in which the planets would have formed is no clearer than in the nebular hypothesis. Further, another objection seems so fatal as to throw out all collision-type hypotheses: collisions between stars are extremely rare. The space in our galaxy is mostly exactly that—just empty space. Consider the dimensions involved. In units of the solar radius, the earth is 215 $R_{\odot}$ from the sun. The nearest star is about 260,000 a.u. from the sun, or roughly 5×10^7 $R_{\odot}$. If we place the sun inside a cube which has a dimension equal to the distance to the nearest star, the volume of the cube is about 10^{23} times the volume of the sun. Detailed calculations indicate that the number of collisions between the 10^{11} stars in our galaxy over the past 5 billion years would be about ten (excluding the central nucleus of the galaxy, which we cannot photograph as it is obscured by dust, and where the stars might be much closer together). Perhaps the sun suffered one of these collisions and hence the solar system is a product of a very rare event. But, we would then expect that dark planetary-like companions to stars must be extremely rare and this prediction is immediately contradicted; of the twelve nearest stars, the sun has nine planets, Barnard's star (page 267) seems to have at least one planet, and two others seem to be accompanied by dark bodies with a mass of no more than one per cent of the solar mass. The odds are therefore against the collision hypothesis and it should be discarded. The best current view holds that planets are formed as a natural by-product of star formation, roughly as outlined in the preceding paragraphs.

Figure 11-13 summarizes in very schematic fashion the discussions of this chapter as they bear on the evolution of a star like the sun.

POSSIBLE EXPLANATION OF BODE'S LAW—EVOLUTION OF THE SOLAR SYSTEM

Are the orbits of the planets the same today as they were in the early history of the solar system? Is Bode's Law (Chapter 7) a numerical curiosity or is it a predictable consequence of physical laws? Many astronomers and mathematicians have studied this problem. The orbital spacings of the inner satellites of Jupiter, Saturn, and Uranus all obey mathematical relationships similar to Bode's Law. This common pattern suggests that the distances of the planets and the asteroid belt from the sun did not occur accidentally.

Perhaps the best idea yet presented to account for Bode's Law is the *theory of dynamical relaxation.* According to this theory, the orbits of the planets were originally very different from what they are today. Remember that the orbit of Halley's comet around the sun was affected by Jupiter when it passed near that planet (Chapter 7). In the same way, in the early history of the solar system, Jupiter and the other planets may have perturbed each other, changing their orbits. This process went on until the planets were set into orbits arranged in such a way that perturbations no longer seriously disturbed them. If this were not true, the planets should still be causing major perturbations in each other's orbit. Since measurements of the planetary motions nowadays show that the orbits are changing only very slowly (in fact, the changes can be ignored for most purposes), the dynamical relaxation process must be nearly complete.

In 1969, J. G. Hills (then a graduate student at the University of Michigan) did numerical experiments to test this theoretical explanation of Bode's Law. He took eleven imaginary solar systems with different sets of planetary masses and with the sizes and shapes of the elliptical orbits picked at random, and used a computer to calculate how their orbits would change over long periods of time as the planets perturbed each other. He found that these solar systems of quite different initial properties did evolve toward stable orbital conditions, with spacings that obeyed mathematical expressions similar to Bode's Law. Furthermore, he found that this process would take place within the time available since our solar system was formed. In fact, the Jovian planets may have reached approximately their present orbits within only a million years of their formation.

12

The Milky Way Galaxy

Looking through his telescope, Galileo discovered that the diffuse band of light that is called the Milky Way consists of a vast number of stars not individually perceived by the unaided eye. Today we know that the sun is located in a large, relatively flat, and spiral-shaped system (Figure 12-1) of perhaps 100 billion stars that also contains the open and globular clusters and large amounts of gas and dust arranged in clouds and in a general medium between the stars. This system is referred to as the *Milky Way galaxy.* When we look out from the earth through the central plane of the galaxy we see the regions where most of these stars are located—the Milky Way in its original sense. On the other hand, when we look at right angles to the Milky Way in the sky, we see relatively few stars. The Milky Way galaxy is one of a great many large star systems that are now recognized as a principal type of structure in the universe. This chapter describes how, through optical and radio astronomy observations, the size, shape, and rotational motion of the Milky Way galaxy have been determined. The appearance and properties of other galaxies in the universe are discussed in the next chapter.

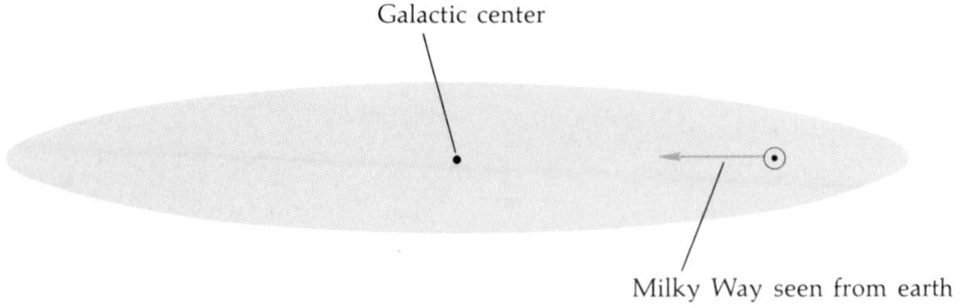

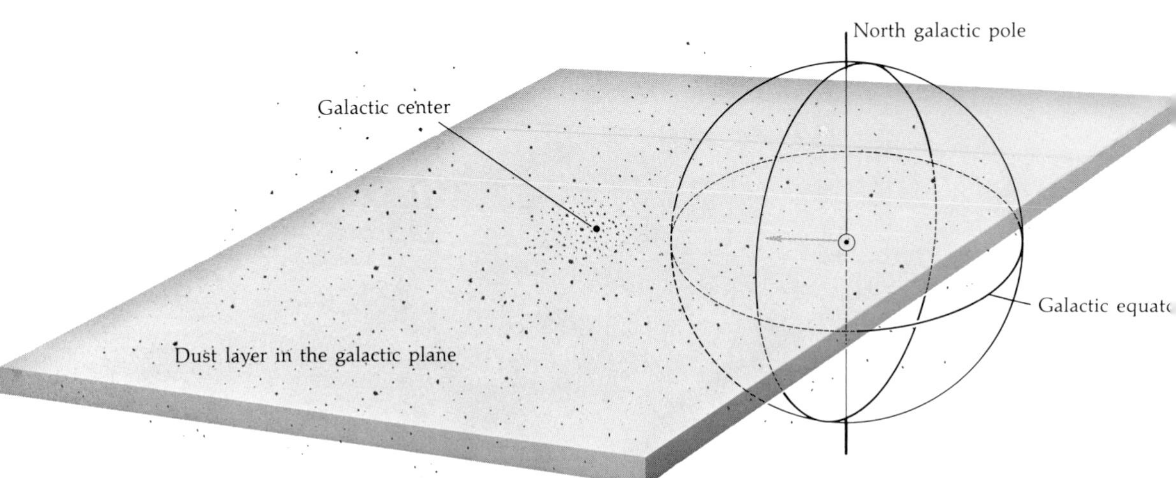

12-1 *Schematic of our galaxy. Top part shows the sun's position with respect to the stars in the galactic disk (gray area) and the geometrical effect responsible for the appearance of the Milky Way. Bottom part shows the galactic coordinate system and the sun's relation to the dust layer and the cloud of globular clusters (dots).*

It would be desirable, in trying to understand the structure of the galaxy, to be able to describe the positions of the stars in terms of their directions and distances from its center. However, finding the exact position and direction of the galactic center from the solar system is in itself a major problem. Therefore, the *galactic coordinate system* (see Figure 12-1) that we actually use is centered for convenience on the sun. The central plane (*galactic plane*) of the Milky Way intersects the celestial sphere along a circle called the *galactic equator. Galactic latitude* is measured in degrees north or south of this equator. *Galactic longitude* is measured in degrees to the east of the zero point, which is taken as the most likely direction to the center of the galaxy, as agreed upon by astronomers. The zero point lies in the constellation Sagittarius, at an unusual complex of radio and infrared emission, known as Sagittarius A.

THE SIZE OF THE GALAXY

A monumental attempt to determine the dimensions of the galactic system was undertaken early in this century by the Dutch astronomer Jacobus Kapteyn (1851–1922). He was a pioneer of modern statistical astronomy and attempted to compute the *star density* (number of stars in a unit volume) as a function of distance from the sun. The relative amounts of stars of different luminosities in the sample under study were assumed to be the same as the relative amounts among the nearby stars that can be studied directly. Knowing the relative numbers of stars with different intrinsic brightnesses, Kapteyn was able to determine the distribution that they must have in space in order to produce the distribution and apparent brightnesses of the stars as seen in the sky from the earth. His galaxy model is often called the *Kapteyn universe;* it is a small disk-shaped system with the sun at its center (Figure 12-2). On this model the star density drops to half the value near the sun in 250 parsecs at right angles to the Milky Way and in 800 parsecs in the plane of the Milky Way. A density drop to one-sixteenth the solar neighborhood value was found at 660 parsecs in the direction at right angles, and at 3,500 parsecs in the plane.

A rather different view of the galaxy was developed by

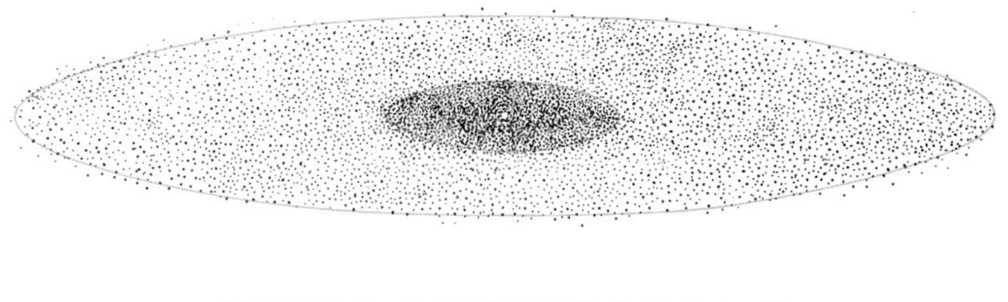

12-2 *The Kapteyn universe: a cross section diagram in the galactic plane. The inner and outer ellipses mark the distances at which the density of the stars (the dots) have decreased to, respectively, one-half and one-sixteenth the central density found near the sun.*

Harlow Shapley at the Harvard College Observatory. Distances were determined from observations of pulsating variable stars in the globular clusters (RR Lyrae stars, page 289). Shapely found (as shown in Figure 12-1) that the globular clusters are distributed equally above and below the galactic plane, although they did not seem to occur near the plane. The surprise was that most of them were found in one half of the sky and with their distribution centered about a point on the Milky Way in the direction of the constellation Sagittarius. This observation could be explained by assuming that the globular clusters were concentrated on the center of the galaxy, not on the sun. (The Kapteyn universe was centered on the sun, so if Kapteyn's model were correct, the globular clusters would be centered on a point off to the side of the galaxy.) If Shapley were correct, the sun was far out toward the edge of the galaxy, and thus the earth was not only not at the center of the solar system, but the solar system was not at the center of the galaxy. All subsequent work on the structure and dimensions of the galaxy has essentially confirmed Shapley's basic deductions, although the presently accepted distance (10,000 parsecs = 10 kiloparsecs) to the galactic center is slightly smaller than the value that Shapley deduced (14 kiloparsecs) from his globular cluster studies.

Why was Shapley right and Kapteyn wrong? Both astronomers, according to the custom of the time, had assumed (at least tacitly) that no absorbing material existed between the sun and the objects under study. If present, such material produces an additional dimming of the star besides the normal decrease in apparent brightness with distance. Distances assigned assuming no absorption are always larger than the actual distances. This discrepancy increases with the amount of absorption; that is, the error is larger for the more distant stars. We now know that a great deal of absorbing material is located in the plane of the galaxy. Kapteyn's work concerned objects that were embedded in the absorbing medium; he was limited by the circumstance that an observer in a fog sees only a very small "universe" around him. On the other hand, many globular clusters are found well above and below the galactic plane and thus the light that we receive from them is relatively free from the effects of absorption. Recall Shapley's result that globular clusters were not found near the plane of the galaxy; in fact, we now know that many of them do occur there but they are obscured by the absorbing material and were not studied by Shapley. Thus, the absorption problem was devastating to Kapteyn's studies and relatively unimportant to Shapley's results.

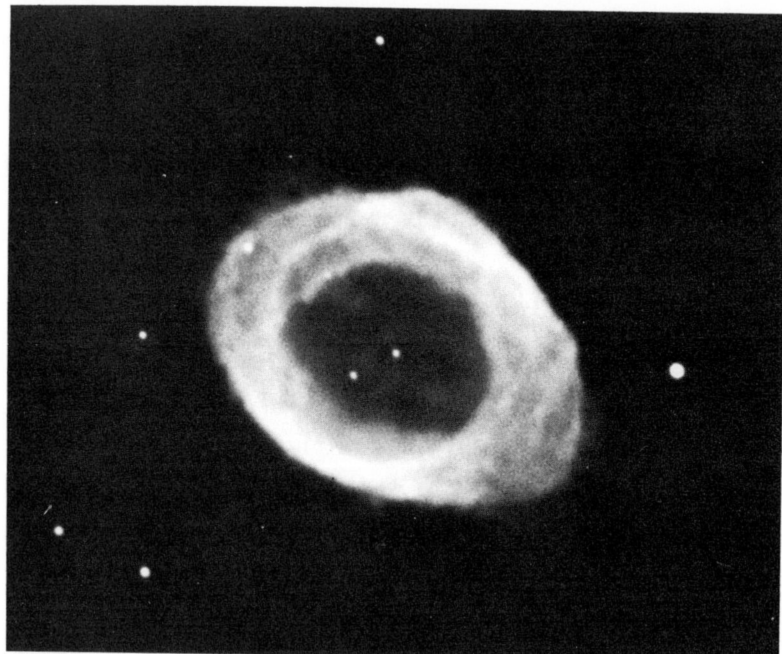

12-3 *The Ring nebula, a planetary nebula in the constellation Lyra. (Courtesy of the Hale Observatories.)*

INTERSTELLAR MEDIUM

As we saw in the last section, an understanding of the structure of the galaxy requires an appreciation of the gas and dust between the stars. Gas and dust can be thought of as coming in two varieties—obvious and devious.

The obvious examples of interstellar material have been known for years. They include the gas masses ejected from stars, such as the planetary nebulae (Figure 12-3), obvious supernova remnants (Figure 16-4) that we discussed in Chapter 11, not so obvious supernova remnants (Figure 12-4), and also the H II regions* (Figure 12-5). The term nebula is now reserved for the clouds of gas and dust that we find in interstellar space. However, in older books you will find it used also in describing galaxies outside our Milky Way system; the usage dates back to the time when the nature of the other galaxies as systems of both stars and gas, like the Milky Way, was not understood.

*So called because much of the gas in such a nebula is in the form of ionized hydrogen, which is abbreviated as H II to distinguish it from H I or neutral hydrogen.

The spectra of bright nebulae, such as the planetaries, supernova remnants, and H II regions, show emission lines as we would expect for a hot gas seen against the dark background of space. However, some bright nebulae *lack* emission lines and have spectra rather like those of stars: absorption lines and continuous light. In addition, some H II regions exhibit these continuous spectra in addition to their emission lines. These observations are explained by the presence of dust in the nebulae that scatters the light from adjacent stars.

The identification of the atoms responsible for some of the most intense emission lines in the spectra of bright nebulae proved to be quite difficult. At one time, the lines were ascribed to a hypothetical element, nebulium, just as some solar emis-

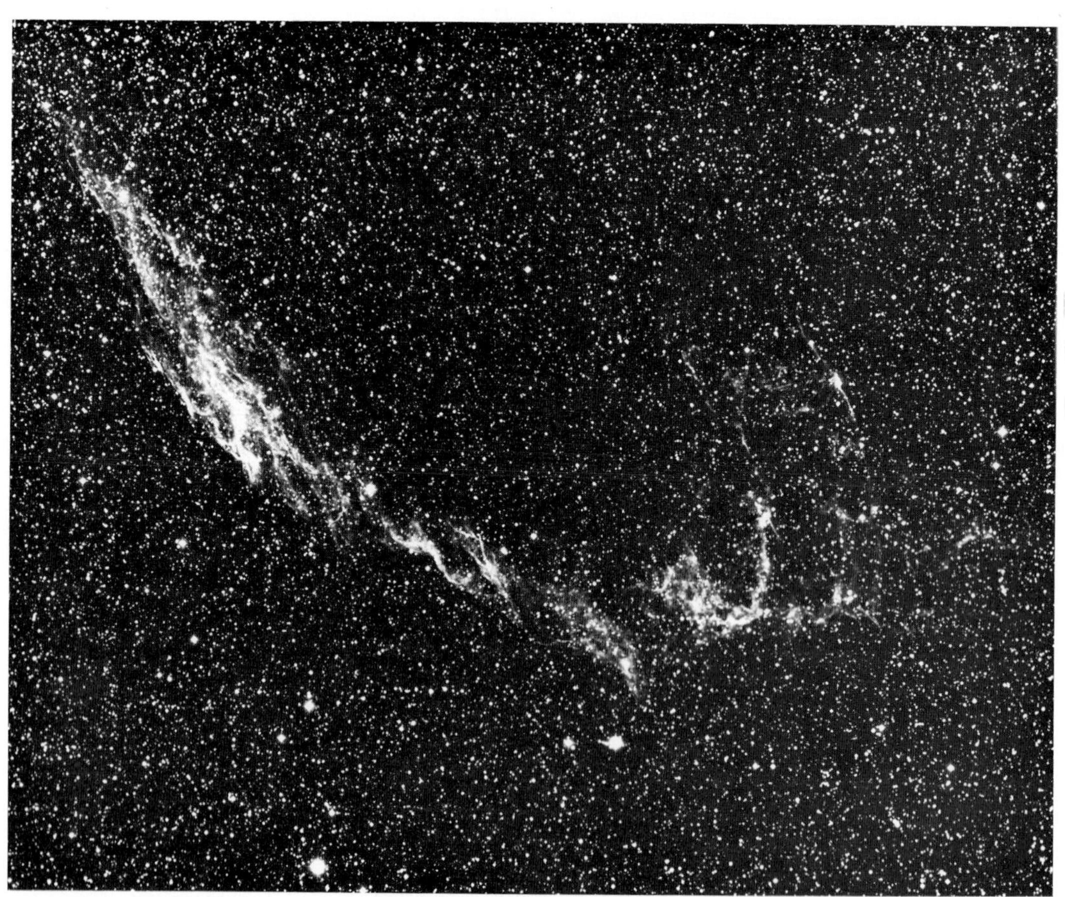

12-4 *The Veil nebula. Notice the concentration of material into narrow zones or filaments. This object is part of a large supernova remnant in the constellation Cygnus. (Courtesy of the Hale Observatories. Copyright © by the California Institute of Technology, and the Carnegie Institution of Washington.)* See also Plate 9 following p. 260.

12-5 *The Orion nebula, a bright H II region. (Lick Observatory photograph.)*

sion lines of unknown origin were ascribed to the then hypothetical helium. However, helium turned out to be a real element that had not yet been found on earth, whereas the nebulium lines have now been identified as the product of atoms of well-known elements in an unusual set of conditions. For example, Ira S. Bowen of the Hale Observatories showed that the intense nebulium lines near wavelength 5,000 Angstroms were due to doubly ionized* oxygen atoms, and that such atoms only produce light at these wavelengths when the gas density is very low (as in a nebula). Spectral lines of this type are known as *forbidden lines* because they cannot occur in normal circumstances.

The source of the energy that causes an H II region to shine is the ultraviolet light of stars that are nearby or actually located within the nebula. Photons corresponding to light of wavelengths less than 912 Angstroms have enough energy to knock the electron out of a hydrogen atom. Thus, the light of these wavelengths from a star causes the hydrogen that exists in space around the star to be ionized. The free electrons released in this way from hydrogen atoms occasionally collide with other particles, such as the doubly ionized oxygen atoms that we mentioned above. In such a collision some of the kinetic energy of the free electron is transferred to the outer electron of the oxygen ion, raising it to a higher atomic energy level. When the latter electron later drops down to a lower level, a photon is released by the ion. This is the way in which the nebulium lines are produced by *collisional excitation.* The general process by which the energy contained in an ultraviolet photon emitted by the star is eventually converted (at least partially) to light of visible wavelengths by the atoms in the nebula is called *fluorescence.* Occasionally, a free electron, produced when an ultraviolet photon ionized a hydrogen atom, meets a proton and combines with it to form a neutral hydrogen atom again. In this case, the electron may be in one of the upper energy levels of the atom, and when it drops to a lower level, one of the ordinary visible emission lines of hydrogen is emitted. This special case of fluorescence is called *recombination radiation.* In a common fluorescent light bulb, the electrical discharge (electrons) flowing through the bulb causes a vapor to emit ultraviolet light, and the ultraviolet photons strike the atoms of the bulb's inside coating, which re-radiate part of the energy in the form of visible light.

It can be seen from our discussion of the Planck Law in

---

*Oxygen atoms that have lost two of their eight electrons.

Chapter 5 that the hotter stars will produce more ultraviolet light than the cooler stars. Therefore, the hotter stars will be the most important exciters of the H II regions. In fact, the detailed theory of this process, as worked out by the Danish astronomer Bengt Strömgren, shows that only stars of the two hottest spectral classes (O and B) emit enough ultraviolet light to produce noticeable H II regions. Several stars may be involved in the excitation of a nebula like that in Orion (Figure 12-5),but an idea of the effect of a given star can be gained by considering the size of the H II region it would produce by itself in a typical part of the galactic plane where the interstellar hydrogen density is about one hydrogen atom per cubic centimeter. A main sequence O5 star with a photospheric temperature of about 56,000° K theoretically can ionize a region of about 200 parsecs in diameter, but a B1 star with a temperature of about 18,000° K will produce an H II region of less than 13 parsecs in diameter. Where the interstellar gas density is much higher (as in the central parts of the Orion nebula where the density reaches 15,000 atoms per cubic centimeter), the size of the H II region produced by a given star will be much smaller. This effect is due to the fact that at the higher gas densities, an ultraviolet photon from the star will travel a smaller distance on the average before colliding with a hydrogen atom and ionizing it. The diameter of the Orion nebula is less than 5 parsecs.

The interstellar material appears in another obvious way in the characteristic dark lanes that are seen in the Milky Way (Figure 12-6) and in *dark nebulae* such as the Horsehead nebula (Figure 12-7). In this case, the light of any stars that may be adjacent to the interstellar clouds is inadequate to cause them to shine significantly by fluorescence or even by scattering. The interstellar matter is detected by virtue of its absorption of the light from background sources, whether they be stars (as in the case of the dark lanes in the Milky Way) or a bright nebula (such as the one that lies behind the Horsehead nebula). In some cases the presence of a dark nebula is not certain, and the evidence is analyzed in terms of star counts on photographs. If the counts of the number of stars per square minute of arc are much lower in the suspect region than is statistically probable (as judged from star counts in adjacent areas), the evidence suggests that a dark nebula may be present. This can be checked, for example, by studying the spectra of the few stars seen in the direction of the suspect nebula; if they show stronger indications of interstellar absorption (as discussed below) than do similar stars outside the region, the presence of the dark nebula is confirmed.

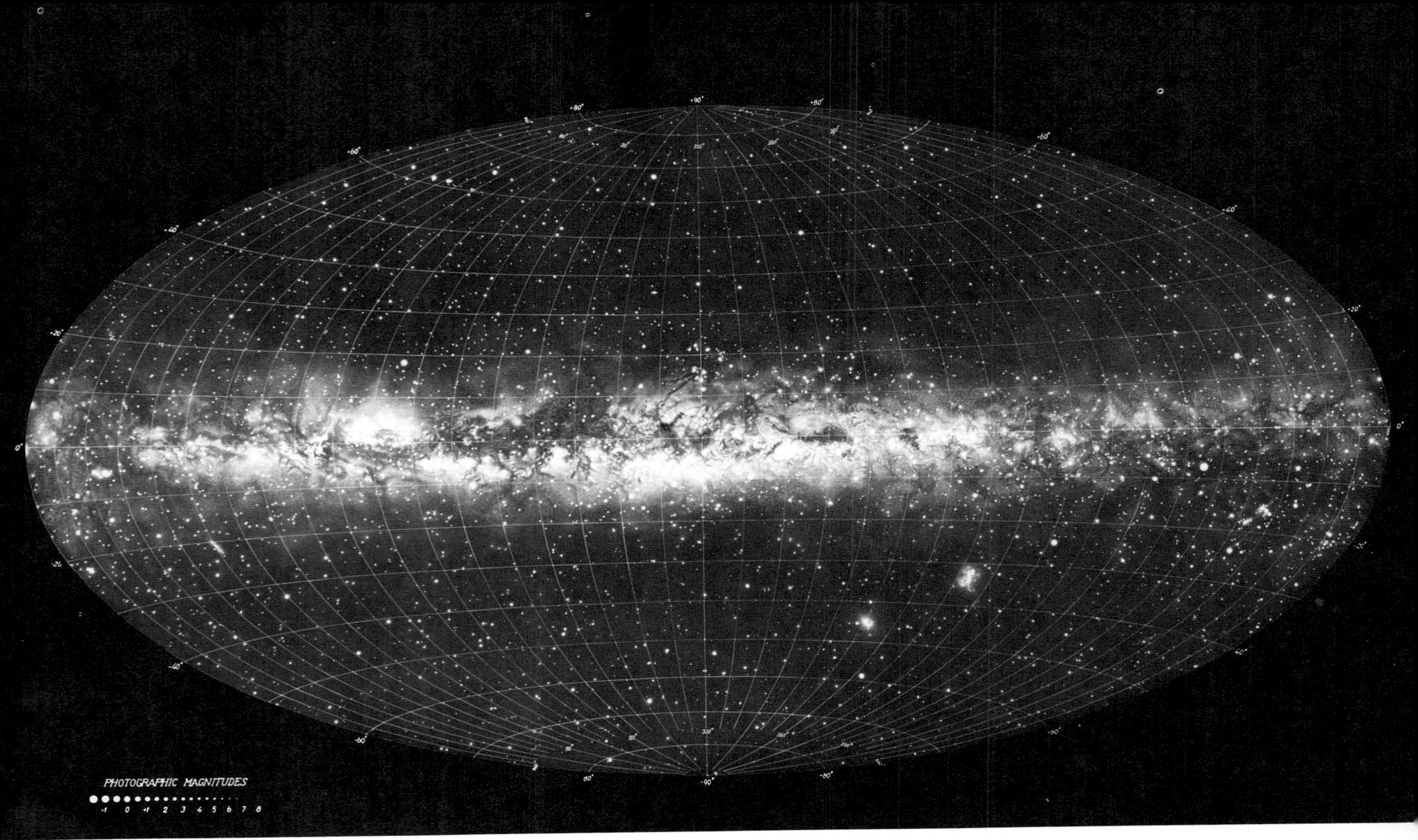

12-6 *Chart-painting of the Milky Way. (Observatorium Lund, Sweden.)*

12-7 *The Horsehead nebula, a dark cloud seen against the bright background of an H II region. (Courtesy of the Hale Observatories.)*

The "devious" form of interstellar matter that we referred to on page 321 is a thin medium of gas and dust that pervades interstellar space. Until about 40 or 50 years ago it was generally thought that no such *interstellar medium* existed. It now appears that there is an interstellar medium and that its dust component can be detected through the reddening of stars at great distances. If we study two groups of stars as nearly identical as possible, except that one group is near the sun and one is far away, we find that the colors of the stars in the distant group are redder than those of similar stars in the group near the sun. The effect is sometimes so great that an intrinsically blue star looks red to us because of the great amount of dust through which we view it. On the other hand, the absorption lines produced in the photospheric spectrum of a star are not affected by the dust, so we can use them to estimate the stellar temperatures (page 273) and hence the *intrinsic colors* of the stars. Comparison of intrinsic and apparent colors tells how much dust is present between the observer and the star.

The reddening effect of the interstellar dust is somewhat analogous to the change in the color of the sun at sunset. The molecules in the earth's atmosphere scatter blue light better than they do red light; hence, the blue light is scattered out of the line of sight while the red comes through, and the light is reddened, as illustrated by Figure 8-8. The same thing happens to starlight, but the scattering particles in this case are not air molecules but the much larger dust particles with diameters of roughly 10^{-5} centimeters. We are ignorant of the composition of the interstellar dust grains, although iron, ice, and graphite, singly and in various combinations, have all been suggested.

Not only does the interstellar dust redden the light from a star, but it also reduces the total amount of light received. The amount is different in different parts of the galaxy, but a typical dust absorption in the galactic plane is about one magnitude per kiloparsec at the wavelengths of yellow light. The existence of absorption by a general interstellar medium (as distinct from the obvious nebulae) was finally established in 1930 on the basis of detailed studies of clusters by the American astronomer Robert Trumpler. Trumpler's work can be understood in terms of two different methods for deriving the distances of open clusters:

Method A

Adopt an average diameter in parsecs for open clusters. Measure the angular size in degrees of a cluster. Use simple geometry to calculate the distance at which a cluster of a given number of parsecs in diameter will subtend an angle equal to the measured angular diameter (Figure 12-8).

Method B

This is the *cluster parallax* technique (page 295). Comparison of HR Diagrams of the cluster under study and another of known distance tells us the absolute magnitudes of the stars in the first cluster. Comparison of the absolute magnitudes and observed apparent magnitudes then gives the distance.

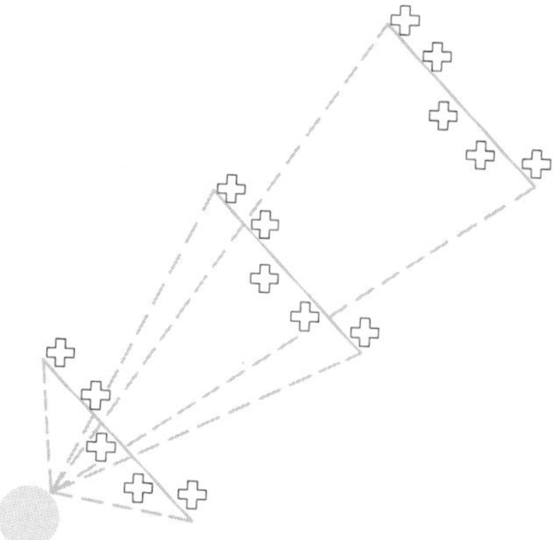

12-8 *A cluster of a given linear size (represented by the solid line) subtends smaller angles at greater distances from the earth.*

Note that Method A is not directly affected by interstellar absorption, since we measure the cluster size, not the brightness of its stars. However, Method B involves the apparent magnitudes, and therefore is affected by absorption. In particular, absorption makes the apparent magnitude fainter, and thus makes the cluster seem farther away. In fact, Method B gave larger distances for the clusters than Method A. If Method A were correct, then interstellar matter was dimming the light of the cluster stars. The other possible conclusion was that Method B gave the correct distance, but this implied that the more distant star clusters had bigger diameters than the nearby ones. It seemed more reasonable to believe that objects of similar appearance have a similar nature, including similar size, and so Trumpler concluded that interstellar absorption did indeed occur.

In 1947, W. A. Hiltner and John Hall discovered that besides dimming and absorbing the light of the stars, the interstellar dust grains also cause a slight polarization of the light.

The dust responsible for the effects described above actually constitutes only about one per cent of the interstellar material. The other 99 per cent is gas and was detected in 1904, although it was some time before the result was generally accepted. Observations of a binary star showed the usual lines which Doppler shifted to red and to blue as the stars orbited around their center of mass. However, there were other absorption lines that did *not* shift as the stars moved in their orbits. It was natural to assume that these stationary absorption lines were caused by a cloud of gas between the earth and the binary star. Subsequently, the same lines were observed in the spectra of many stars and the presence of the interstellar gas was proven. A key point was that the same lines could be found in the spectra of both hot and cool stars; the spectra of such stars should be very different, so the lines that were common to all of them were most likely due to material outside the stars. Many substances have now been identified through their interstellar absorption features in stellar spectra, including neutral and ionized calcium, neutral sodium, neutral potassium, ionized titanium, and the molecules CN and CH. Several other strong absorption features in stellar spectra are also believed to be due to interstellar matter, but the atoms or molecules that produce these absorptions have not been definitely identified.

The gas distributed throughout the galaxy can be studied with radio telescopes. Unlike visible light, radio waves can penetrate through the interstellar dust without suffering serious absorption. Thus, we can "see" farther with radio waves. Further, the physical state of the gas between the stars is such that most of the atoms emit significant energy only in the radio wavelengths.

Radio radiation from the galaxy was detected and identified in the early 1930s by Karl Jansky in Holmdel, New Jersey (Figure 12-9), while investigating radio static for the Bell Telephone Laboratories. The source of static that he discovered was

12-9 *The founder of radio astronomy, Karl Jansky, and his antenna, photographed in the 1930s. (Courtesy of the Bell Telephone Laboratories.)*

observed on a wavelength of 15 meters and crossed the meridian four minutes earlier each day, thus indicating its celestial nature (see page 228). Eventually, the position of the source was established with some certainty and found to coincide with the center of the galaxy.

This work did not attract much attention, although in retrospect it should have. However, it did interest the radio engineer Grote Reber, of Wheaton, Illinois. Reber became the first radio astronomer after Jansky, and for some years, beginning in the late 1930s, he constituted the entire radio astronomy profession. At his home in Wheaton he built his own radio telescope (Figure 12-10) for the observations. He was hampered by the general state of ignorance concerning his new field of study, as well as by the suspicion of his neighbors and the problem of children climbing on the structure. Reber was not only able to detect radio waves from the Milky Way, but he was able to map their

12-10 *Grote Reber with a radio telescope that he built and used to survey the sky. (Courtesy of National Radio Astronomy Observatory, Green Bank, West Virginia.)*

distribution on the sky at a wavelength of 1.9 meters. This radio radiation has a continuous spectrum, is polarized, and has been shown to arise from cosmic ray electrons moving through magnetic fields in the galaxy (*synchrotron radiation,* page 451). Similar emission comes from localized sources such as the Crab nebula (Chapter 16) and other supernova remnants, and also from galaxies and quasars beyond the Milky Way.

The next advance in radio astronomy, the prediction and discovery of the 21-centimeter wavelength line of neutral hydrogen, provided students of galactic structure with a powerful tool. During World War II the normal observing facilities were not available in the Netherlands and theoretical research was the order of the day. The Dutch astronomers were aware of the work by Reber and immediately saw the great advances that would be possible if a spectral *line* were observable with radio telescopes: Doppler shifts of the line, and thus gas motions in the galaxy, could be measured. In 1944, Hendrick van de Hulst predicted that such a line would indeed be produced by neutral hydrogen atoms in interstellar space. The atomic situation can be illustrated by returning to our schematic picture of the hydrogen atom as a planetary-like system with an electron orbiting a proton (page 126). In the earlier discussion we treated the proton and electron as points, but this concept is too simple. Each particle can be regarded as having a definite size and spin. Thus, there are really two ways of having the electron in the ground level or lowest energy level—the spins of the proton and electron can be in the same direction or in different directions, as illustrated in Figure 12-11. When the electron's spin flips from parallel to anti-parallel with respect to the proton, a very small change in energy results in the emission of a photon with a 21-centimeter wavelength, or frequency of 1,420 megacycles per second. At first glance, the problem with this line is that it is not emitted very often. In fact, a hydrogen atom on the average emits the 21-centimeter photon spontaneously only once in 11 million years. It may seem folly to attempt to detect photons that are emitted so infrequently, but it is not, because interstellar space is so vast and hydrogen is its most abundant constituent.

The prediction by van de Hulst was published in 1945, but partly due to war-caused disruption of normal channels of communication it was not widely known. Meanwhile, a similar prediction was made independently in the Soviet Union by I. S. Shklovskii. The efforts to observe this line in emission in the galaxy were rewarded in 1951 when it was detected by Harold Ewen and Edward Purcell at Harvard.

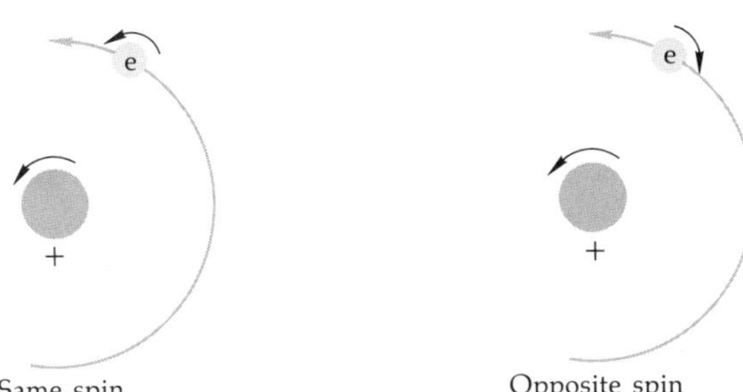

12-11 *Schematic describing the origin of the 21-cm emission line of neutral hydrogen. When the electron "flips its spin" from the same direction as the proton (left sketch) to the opposite direction (right sketch), a photon of wavelength 21 cm is emitted.*

Since 1951 a great many radio astronomers have studied the distribution and motions of the hydrogen gas in the galaxy through studies of the 21-centimeter line radiation. They have found that most of the hydrogen is confined to a relatively narrow layer (thickness about 250 parsecs) in the plane of the Milky Way, and that within this layer, the hydrogen tends to concentrate in clouds that form a pattern of spiral arms like those that we observe in other galaxies (Chapter 13). These observations are summarized in Figure 12-12; the picture of the rotational motions that has emerged from the 21-centimeter line studies is given in Figure 12-17. In regions of the galaxy, such as the Orion nebula, where the interstellar hydrogen is relatively dense, hot, and ionized, we observe not the 21-centimeter line but rather a form of continuous radio emission called *thermal radiation.* It can be distinguished from the synchrotron radiation mentioned above on the basis of a different spectral shape and the lack of polarization in thermal radio sources.

In recent years a number of other spectral lines have been found in the radio emission of the galaxy. These include radiation from the hydroxyl molecule (OH), and from water vapor, ammonia, carbon monoxide, silicon monoxide, and formaldehyde.* However, the intensities of some of these lines are unexpectedly strong and cannot be explained in simple terms such as the spin flip process in neutral hydrogen. In fact, the explanation of these molecular emission lines at radio wavelengths is one of the chief problems in radio astronomy today.

---

*Based on the results of both optical and radio studies, about two dozen molecules had been found in interstellar space by late 1971.

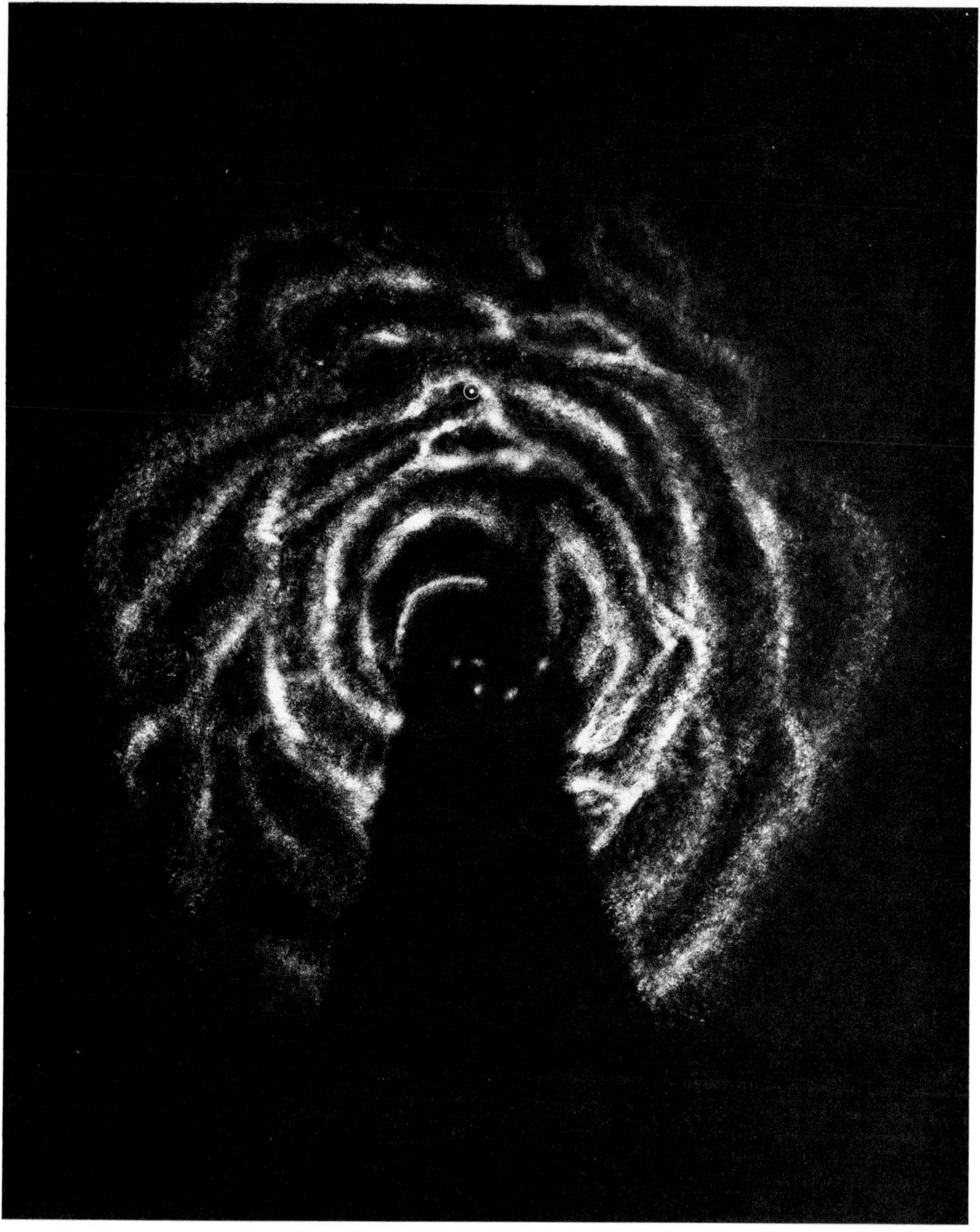

12-12 *Artist's impression of the neutral hydrogen distribution in the galaxy, based on radio telescope surveys of 21-cm wavelength radiation. The view is from the north galactic pole. The sun (some 10 kiloparsecs from the center) is marked by the symbol ⊙. (Courtesy of G. Westerhout, University of Maryland.)*

Several interesting questions are raised by the discovery of the interstellar molecules. First of all, we do not understand how many of the molecules can form and persist in space. It is especially hard to explain how several atoms can come together to form the more complex molecules that have been found in the interstellar gas, such as methyl alcohol, which contains six atoms. One would expect these molecules to be dissociated by ultraviolet light; perhaps they occur in dust clouds and are shaded by the dust. A second problem, already mentioned above, is (given that these molecules *are* present) why their radio emission is so strong. Finally, biochemists are wondering about the possibility that the interstellar molecules might somehow be related to processes that produce life. For example, the radio astronomers have found that both formaldehyde and ammonia molecules are present in several interstellar clouds, and laboratory experiments (similar to those described in Chapter 3) prove that when these two substances are present under suitable conditions, exposing them to heat or to ultraviolet radiation causes chemical reactions that yield amino acids.* This does *not* prove that living matter may somehow descend from the molecules in space, but the subject does deserve study, and such studies are underway. (Also, recall the discussion of amino acids in meteorite fragments in Chapter 7.)

STRUCTURE

Modern star count analyses similar to the statistical methods used by Kapteyn, but with full allowance for interstellar absorption, can be used to determine the distribution of stars in space. The results give a picture of a flattened disk-shaped galaxy resembling that found from the 21-centimeter line studies. However, since our observations in visible light are limited by the absorbing effect of interstellar dust, far less of the galaxy can be mapped through observations of the stars than through the 21-centimeter measurements of radiation from the interstellar gas.

The picture is complicated somewhat when the shape of the galaxy, as judged from different kinds of stars, is considered. The degree of flattening varies with the type of star that is

*The discovery of formaldehyde in the interstellar medium is interesting in itself, as this molecule is known to react under suitable conditions to form carbohydrates and subunits of the nucleic acids.

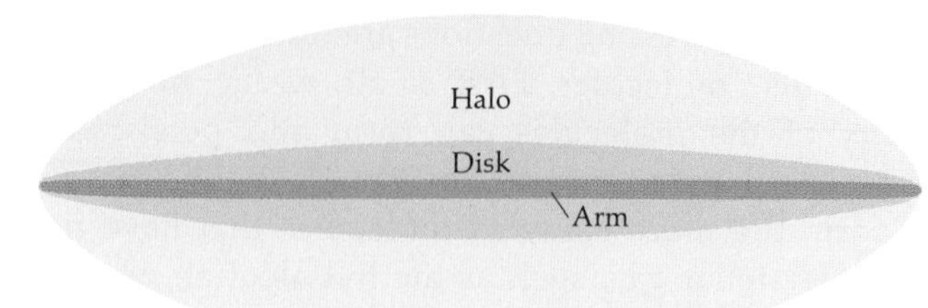

12-13 *A schematic cross section through the galaxy, showing the increasing amount of flattening as one goes from halo, to disk, to arm population.*

Table 12-1 *Kinds of Stars in the Galaxy*

	POPULATIONS		
	HALO	DISK	ARM
Typical members	Globular clusters RR Lyrae stars W Virginis stars High-Velocity stars	Bright red giants Novae Long-period variables Sun and most nearby stars	Gas and dust Supergiants T Tauri stars Classical Cepheids Open clusters
Average distance above or below the galactic plane (parsecs)	1,000–2,000	200–500	100–150
Typical speed at right angles to the galactic plane (km/sec)	75	15–20	10
Flattening (see Figure 12–13)	Slight	Fairly strong	Extreme
Distribution	Smooth and concentrated toward galactic center	Smooth and concentrated toward galactic center	Very patchy and little or no concentration toward galactic center
Fraction of heavy elements relative to the sun	0.2	0.6–1.3	2–3
Age	Greater than about 5 billion years or OLD	Between 1 and 5 billion years or INTERMEDIATE	Less than 1 billion years or YOUNG

observed, as shown in Figure 12-13 (see also Table 12-1). The very bright, young O and B stars have a very flat distribution concentrated close to the plane of the galaxy, as does the interstellar gas and dust. In an intermediate distribution (flattened but not as extreme as the stars mentioned earlier) are red giant stars and the long period variable stars. In a more nearly spherical distribution, we find the globular clusters and the *high velocity stars* (see page 341).

Looking at the flat stellar distribution in more detail we find that the stars and interstellar matter are concentrated in *spiral arms.* This kind of structure would be expected if our galaxy were similar to others such as M31 (Figure 13-1), M33 (Figure 13-3), and M81 (Figure 13-5). In our above discussion of the 21-centimeter line, we presented the models of gas distribution and galactic rotation that are derived from the radio observations (Figures 12-12 and 12-17), but we did not indicate how the distances of the hydrogen clouds are actually determined. In fact, the method involves assuming a theoretical rotation model* for the galaxy and seeing if it successfully predicts the observed pattern of Doppler shifts. The distances of individual hydrogen clouds are estimated by measuring the Doppler shifts of their 21-centimeter lines and comparing them with the model to see at what distance from the sun in a given direction the observed Doppler shift can be expected. On the other hand, the distances of the stars are determined without assuming a rotational model by using the methods described in Chapter 10, and reasonable agreement is found between the galactic structure deduced from the stars and that deduced from the interstellar gas. As shown in Figure 12-14, the sun appears to be near the inner edge of a spiral arm.

MOTIONS, ROTATION, AND MASS

The rotation of the galaxy that we observe in the 21-centimeter line might be expected on the basis of the flattened appearance of the Milky Way. Prior to the advent of radio astronomy it was possible to draw conclusions about the rotation, although the available data were much more limited. The Dutch astronomer Jan Oort showed in 1927 that the rotation of the galaxy could be detected in the radial velocities of the relatively nearby stars, *if* the stars at different distances from the galactic center

---

*i.e., a mathematical expression that represents the way in which the velocities of objects orbiting around the galactic center depend on their distances from the center.

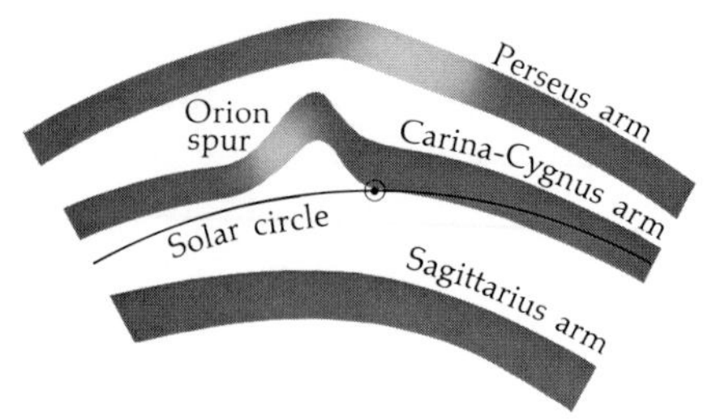

12-14 *Schematic of the spiral arms near the sun. The "solar circle" has a radius of 10 kiloparsecs.*

revolved around the galactic nucleus at different speeds. On the other hand, if the galaxy rotated as a rigid body, the relative positions of the stars would remain the same, and therefore (except for small random motions) stars would not tend to move away from each other.

This is not the case if we consider a non-rigid model. For simplicity, let the orbits of the sample stars in Figure 12-15 be circles around the galactic center. The lengths of the lines represent the different orbital speeds. If we (located at the position of the sun) look exactly toward or away from the center of the galaxy, no radial motion is seen because the motions of the sun and the stars in question are at right angles to the line of sight. We do not see any Doppler shift if we look ahead or behind in our orbit at nearby stars because these stars are moving at the same speed in their orbits as we are. Only at the positions intermediate between these examples should a Doppler shift be observed; the corresponding radial velocities should show a double-wave variation (Figure 12-16). Radial velocities are termed positive (+) when the stars are moving away from the sun and negative (−) when the stars are approaching the sun. If we look in positions (b) and (f) of Figure 12-15, the sun and stars are approaching each other and the radial velocity should be negative; in positions (d) and (h), the

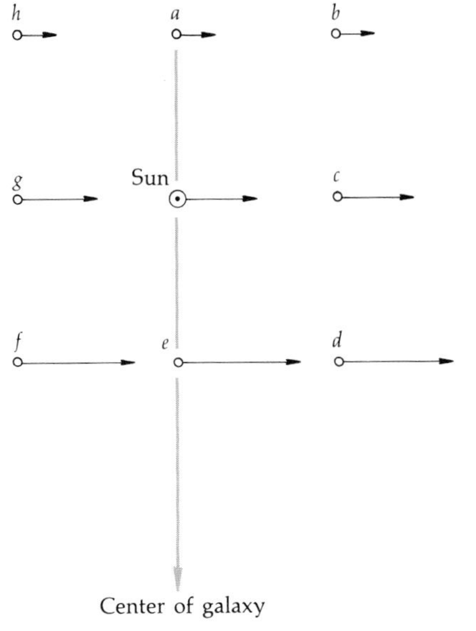

12-15 *Schematic illustration of differential rotation in the galaxy. The arrows represent the velocities around the center of the galaxy, with the velocity decreasing away from the center.*

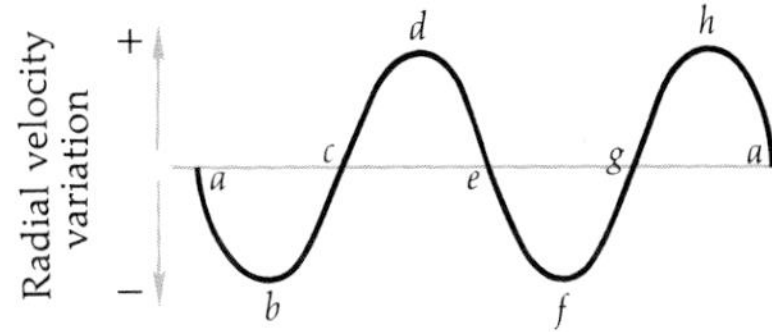

12-16 *The inferred change of radial velocities resulting from the situation shown in Figure 12-15 and described in the text. Radial velocities are termed positive (+) if the distance between the object and the sun is increasing, and negative (−) if the distance is decreasing.*

stars and the sun are moving farther apart and the radial velocity should be positive. In positions (a), (c), (e), and (g), the distance from the sun to the star is virtually constant and the radial velocity should be zero. This theory, as proposed by Oort, has been verified by detailed observations. Systematic measurements of motions, along with the known distance (about 10 kiloparsecs) to the galactic center, allow a determination of the galactic rotation velocity near the sun. This speed is about 250 kilometers (160 miles) per second.

The determination of the galactic rotational velocity near the sun can be verified by another method that does not depend on our model for the structure and size of the Milky Way. The radial velocities of other nearby galaxies beyond the Milky Way can be measured and the results inspected for a trend, as these galaxies should not be participating in the rotational motion of the Milky Way. The problem is similar to a man on a merry-go-round looking out at his surroundings to determine which way the merry-go-round is turning. Sure enough, we find that we are approaching the galaxies in one direction and receding from those in the opposite direction. This effect could be caused by linear motion of the Milky Way galaxy through space or by its rotation, or both. In fact, the direction of the galaxy's motion that we deduce from the radial velocities of other galaxies is at *right angles* to the direction of the galactic nucleus from the sun and thus it is in the right direction to be interpreted as the result of galactic rotation. There should also be an effect in the radial velocities of the nearby galaxies due to the motion of the Milky Way through space, but thus far it has not been clearly distinguished.

The internal (radial velocities of stars and hydrogen clouds) and external (radial velocities of nearby galaxies) methods agree in assigning a rotation speed for the galaxy at the sun's distance (10 kiloparsecs) from the center of about 250 kilometers per second. A circle of 10 kiloparsecs radius has a circumference of about 63 kiloparsecs; at a speed of 250 kilometers per second (or 2.5×10^{-7} kiloparsecs per year) this means that the sun takes roughly 250 million years to complete a revolution around the galactic center. Using the distance to the galactic center and this 250 million year period of revolution, we can estimate the galactic mass using Kepler's and Newton's Laws, and we find 1.5×10^{11} solar masses. This calculation is a gross simplification, but the number is correct to within a factor of 2 and thus is not a bad approximation. A more precise treatment involves determining the variation of rotation speed with distance from the galactic nucleus, which requires both the 21-centimeter measurements of galactic rotation with respect to the sun and the galactic rotation velocity of the sun. The result is an observed rotational model for the galaxy and it is represented by the *galactic rotation curve,* shown in Figure 12-17. Using the laws of motion and gravitation to interpret this curve gives an answer of about 3×10^{11} $M_{\odot}$ for the mass of the Milky Way.

The discovery and verification of galactic rotation helps us to understand a phenomenon of long standing, namely, the so-

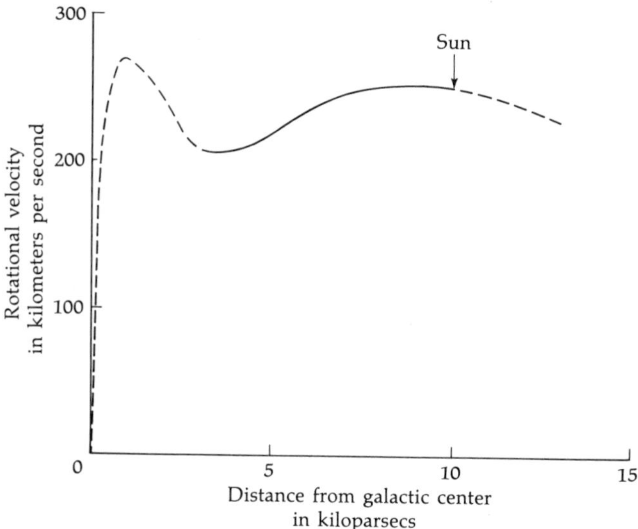

12-17 *The rotation curve of the galaxy. The solid line from 3 kiloparsecs to 10 kiloparsecs is based on 21-cm line studies and is well determined. The dashed line from 0 to 3 kiloparsecs and the dashed line for distances greater than 10 kiloparsecs represent more poorly known parts of the rotation curve.*

called *high velocity stars.* (The explanation was actually advanced by the Swedish astronomer Bertil Lindblad in 1927, before rotation of the galaxy was proven.) Most stars which have been studied near the sun have radial velocities in the range 10 to 20 kilometers (6 to 12 miles) per second. Some, however, have velocities of 65 kilometers (40 miles) per second or more with respect to the sun, and these are the high velocity stars. This situation results from the fact that most of the stars near the sun are in near-circular orbits close to the galactic plane; these have small radial velocities with respect to the sun. On the other hand, some stars are in highly elliptical orbits around the galactic center, as illustrated in Figure 12-18. These stars have low orbital velocities and when they are observed near the sun their high radial velocity actually results from the large orbital speed of the sun. This view can be checked by seeing which way these high velocity stars seem to be moving. Their apparent motions are directed back along the sun's path in the galaxy, thus verifying that the sun's motion is really responsible for their apparent high velocity. (The effect is equivalent to the apparent

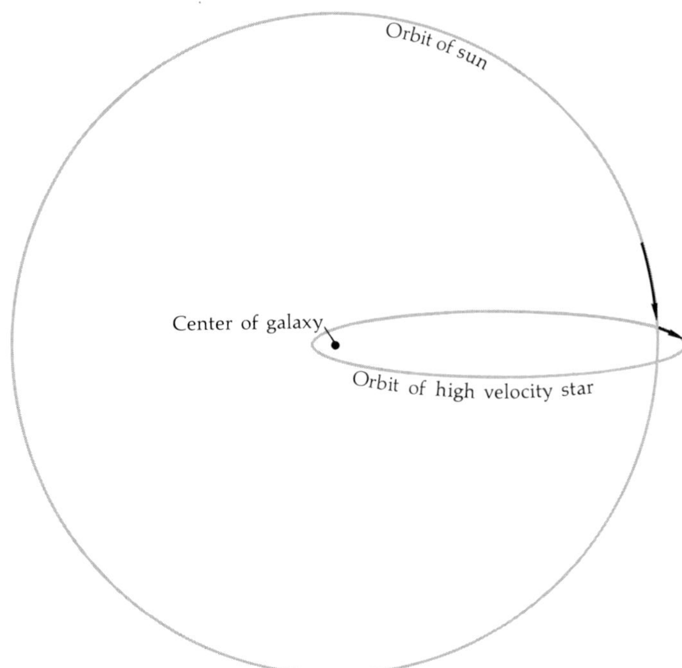

12-18 *High velocity star schematic (as viewed from the north galactic pole). The apparent high velocity is actually due to its elliptical orbit and the sun rushing by in a circular orbit.*

backward motion of slowly walking pedestrians on the side of the road as seen by an observer in a moving car.) The high velocity stars are not closely concentrated to the galactic plane, and in addition their spectra are weak in lines of the heavy elements (page 306). Therefore, they belong to a different population than do the stars in the solar neighborhood; in particular, they are *halo stars.*

THE MILKY WAY—MODEL AND EVOLUTION

Information concerning the motions, distribution in space, and spectral properties of stars can be collected and organized into a reasonable picture. The stellar populations in the galaxy require more than the two classifications introduced on page 307 and we present in Table 12-1 a simple scheme in which the stars are divided into *arm, disk,* and *halo* populations. The arm population consists of the youngest Population I objects, the disk population includes stars that we previously classified in both Population I and Population II, and the halo population consists of the older Population II stars.

Inspection of Table 12-1 suggests how the Milky Way galaxy may have evolved. The process resembles the nebular hypothesis (page 310) for the formation of the solar system. The galaxy is thought to have originated as a nearly spherical but rotating gas cloud which contracted under its own gravitational attraction. As the galactic cloud (*protogalaxy*) contracted, the denser parts of the cloud contracted at a more rapid rate to form the first stars. Thus, these stars formed from the original material, which consisted of hydrogen and a small amount (probably 25 per cent by mass at most) of helium. Stars formed near this time were poor in heavy elements and had a nearly spherical distribution, like the halo population, because the protogalaxy had not contracted very far.

As the protogalaxy contracted further and rotated faster, two processes continued simultaneously. First, as discussed in the case of the nebular hypothesis, the originally nearly spherical gas cloud was constantly flattening. Second, the gas cloud was partially replenished by mass ejected from the stars already formed within it by a variety of violent and non-violent processes, as described in Chapter 11. This matter was enriched in heavy element content due to the nuclear reactions that had occurred inside the stars. Thus, as the galactic cloud continually flattened, stars (such as the sun) formed in the region that we now call the galactic disk, and they had a greater heavy element content than the halo stars.

Much of the matter in the galaxy condensed into stars and thus the amount left in the form of interstellar gas and dust decreased. On the other hand, thanks to stellar mass loss, the relative content of the heavy elements in the interstellar medium increased. At the present time, formation of the stars with the highest heavy element content, the arm stars, is probably occurring only in isolated patches of gas and dust in the spiral arms. If the rate at which stars are formed depends on the amount of gas and dust present, this rate may have been higher in the past.

It appears that the idea that our galaxy formed from a vast cloud of hydrogen and helium, and that the stars which first formed processed these light elements in their interiors and returned heavier elements to the still condensing cloud, from which other stars have formed more recently, is consistent with the observations of the chemical compositions, ages, and locations of the stars. However, we do not know with certainty how it was that the galaxy assumed its pattern of spiral arms. A promising theory, recently advanced by C. C. Lin of the Massachusetts Institute of Technology, is based on a concept of wave-like disturbances in the distribution of the interstellar gas, with the arms corresponding to crests* of the waves.

*Regions of enhanced gas density.

13

Galaxies and the Universe

NEBULAE OR ISLAND UNIVERSES?

In the mid-nineteenth century Lord Rosse had discovered spiral structure in some of the diffuse objects that were referred to collectively as nebulae. The nature of these spiral nebulae was unclear, and by 1920 opinions on this subject had crystallized into two opposing schools of thought. In April of that year the leading American proponents of the two theories debated the matter before the National Academy of Sciences in Washington, D.C.

Harlow Shapley of Harvard took the occasion to stress his conclusion that the Milky Way galaxy was much larger than was generally suspected and that the sun was not located near its center (Chapter 12). Shapley recognized that the spirals must lie outside the galaxy, but he regarded them as much smaller than the Milky Way, and thus not very distant from it. In particular, he argued, the spirals "are truly nebulous objects," that is, clouds composed of gas and dust as is the Orion nebula, for example. Heber D. Curtis of the Lick Observatory disagreed. Although Curtis wrongly supported the concept of a rather small Kapteyn universe model for the Milky Way, he correctly argued that the spirals were at great distances, perhaps in the millions of parsecs, and that they were vast star systems like the Milky Way.

The nature of the spiral nebulae, now called *spiral galaxies,* was established largely through the efforts of Edwin Hubble at the Mount Wilson Observatory. Using the 100-inch telescope, he was able to resolve some of the brighter stars in the outer parts of M31,* the famous spiral galaxy in Andromeda (Figure 13-1). Among these stars he discovered some Cepheid variables. We have already explained (Chapter 10) how the measurement of the period between successive maximum brightnesses of a Cepheid can be used to determine its distance. Hubble applied this method and found the distance of M31. It was in accord with Curtis' ideas. A check on this result was provided by observing novae in M31 and, by assuming that their absolute brightnesses were similar to those of novae in the Milky Way, deriving their distances. M31 is indeed a large spiral galaxy, similar to the Milky Way, and in fact one of the nearest galaxies; the modern value for its distance is about 650 kiloparsecs. From the angular size of M31 and its distance, we can calculate its diameter and show that it is similar in size to the Milky Way galaxy and thus enormously larger than the Orion nebula.

THE SHAPES OF GALAXIES

The recognition of spirals as truly *extragalactic* objects was a great step forward. Studies of many other objects once termed nebulae disclosed that some of them also were distant galaxies, although they lacked spiral shape, and several types of galaxies are now known.

Spiral galaxies are basically similar to M31 (Figure 13-1) and the Milky Way. A galaxy that appears to be a spiral viewed edge on (that is, with its galactic plane parallel to our line of sight) is shown in Figure 13-2, and its similarity in appearance to the Milky Way (Figure 12-6) is obvious. By photographing the spectra of light from different parts of spiral galaxies, such as M31, and by comparing the Doppler shifts that are measured

---

*"M" stands for Charles Messier (1730–1817), French astronomer whom Louis XV called "the ferret of comets." Since nebulae were sometimes mistaken for dim comets, Messier catalogued more than 100 nebulae (some star clusters were also included, such as M13, which is pictured in Figure 10-15) to help astronomers avoid confusion in their searches for comets. Today we remember Messier much more for this catalogue than for his comet discoveries. The "nebulae" in the catalogue include both gaseous nebulae and galaxies. M31, the Andromeda galaxy, is number 31 in the catalogue.

13-1 *The spiral galaxy in Andromeda, Messier 31. (Courtesy of the Hale Observatories. Copyright © by the California Institute of Technology, and the Carnegie Institution of Washington.)* See also Plate 8 following p. 260.

13-2 *NGC 891, a galaxy seen edge on. (Courtesy of the Hale Observatories.)*

in these spectra, it has been found that these galaxies in general are rotating, just as the Milky Way does. The rotation takes place in the sense that the spiral arms "trail." A striking property of the spirals is their variety of central concentrations. M33 (Figure 13-3) shows relatively little central concentration; the spiral arms seem to dominate its appearance. NGC* 4594, on the other hand, is notable for the strong concentration of stars

*NGC stands for the New General Catalogue published by J.L.E. Dreyer of the Armagh Observatory in Ireland in 1888. Together with two later supplements (called the Index Catalogues and abbreviated I.C.) it listed over 13,000 galaxies, clusters, and nebulae. Messier 31 is also known as NGC 224.

in its central region (Figure 13-4). M31 and M81 (Figure 13-5) are intermediate between these two conditions.

Barred spiral galaxies are so named because of the long bar of stars that passes through the central region of such a galaxy (Figure 13-6). Spiral arms appear to extend from the ends of the bar. The spectroscopic observations reveal a fascinating phenomenon in the rotation of these galaxies: the bars rotate as *rigid bodies.* In other words, although a bar is composed of individual stars at different distances from the center of the galaxy, it rotates around the center as though it were a solid unit. In the solar system, the farther a planet is from the sun the slower it moves in its orbit, and as the outer planets have larger orbits, they take much longer to complete an orbit than do the inner planets. However, the stars in the outer parts of

13-3 *Messier 33, a spiral galaxy with relatively little central concentration. (Courtesy of the Hale Observatories.)*

13-4 *NGC 4594, a spiral galaxy with a conspicuous central concentration. (Courtesy of the Hale Observatories.)*

the bars in these galaxies are moving faster than the stars near the center.

The *irregular galaxies* have no special shape and they are typified by the Large and Small Magellanic Clouds (Figure 13-7), which are easily visible to the naked eye in the Southern Hemisphere. The Clouds were named after the explorer Ferdinand Magellan who saw them on his voyage through the Southern Hemisphere. (Of course, nearly everyone who lived in the Southern Hemisphere had already seen them.) Another interesting but untypical irregular galaxy is M82; photographs (Figure 13-8) taken in red light indicate that an enormous explosion in its central region has ejected gas into space around the galaxy.

Finally, we have the *elliptical galaxies,* so called because of their smooth appearance with an outline in the shape of an ellipse.

Two such galaxies, which happen to be small companions of M31, are shown in Figure 13-1; another elliptical companion to M31 is shown in Figure 13-11. Elliptical galaxies are found over a great range of size and, in fact, the galaxy of perhaps

13-5 *Messier 81, a spiral galaxy in Ursa Major. (Courtesy of the Hale Observatories.)*

the greatest known mass is the elliptical M87 (Figure 13-9). Among the smallest galaxies are the *dwarf ellipticals,* such as that in Figure 13-10.

The masses of other spiral galaxies are determined in the same basic manner we described for determining the mass of the Milky Way galaxy (page 340). The rotation curve is determined spectroscopically, and the mass can then be computed from advanced versions of Kepler's and Newton's Laws. Masses for pairs of galaxies can also be derived by a method similar in principle to the derivation of the mass of spectroscopic binary stars (page 276): the projected separation of the two galaxies is known, and their orbital velocities are found spectroscopically through the Doppler shift. Because we have only a snapshot at a given time of the galaxies in their orbits (millions of years are required to complete such an orbit), statistical methods must

13-6 *The barred spiral galaxy NGC 1300. (Courtesy of the Hale Observatories.)*

13-7 *Two irregular galaxies, the Large and Small Magellanic Clouds. Also shown is the globular cluster 47 Tucanae (lower right) which is located in our Milky Way galaxy. (Courtesy Harvard College Observatory.)*

be applied to a large number of galaxies to obtain reliable results.

The smallest known galaxies are the dwarf ellipticals with masses of about 10^6 $M_\odot$; the largest known galaxies are the giant ellipticals with masses ranging up to about 3×10^{12} $M_\odot$. The elliptical galaxy M32 (one of the companions to M31) is intermediate with a mass of 4×10^9 $M_\odot$. The masses of the Milky

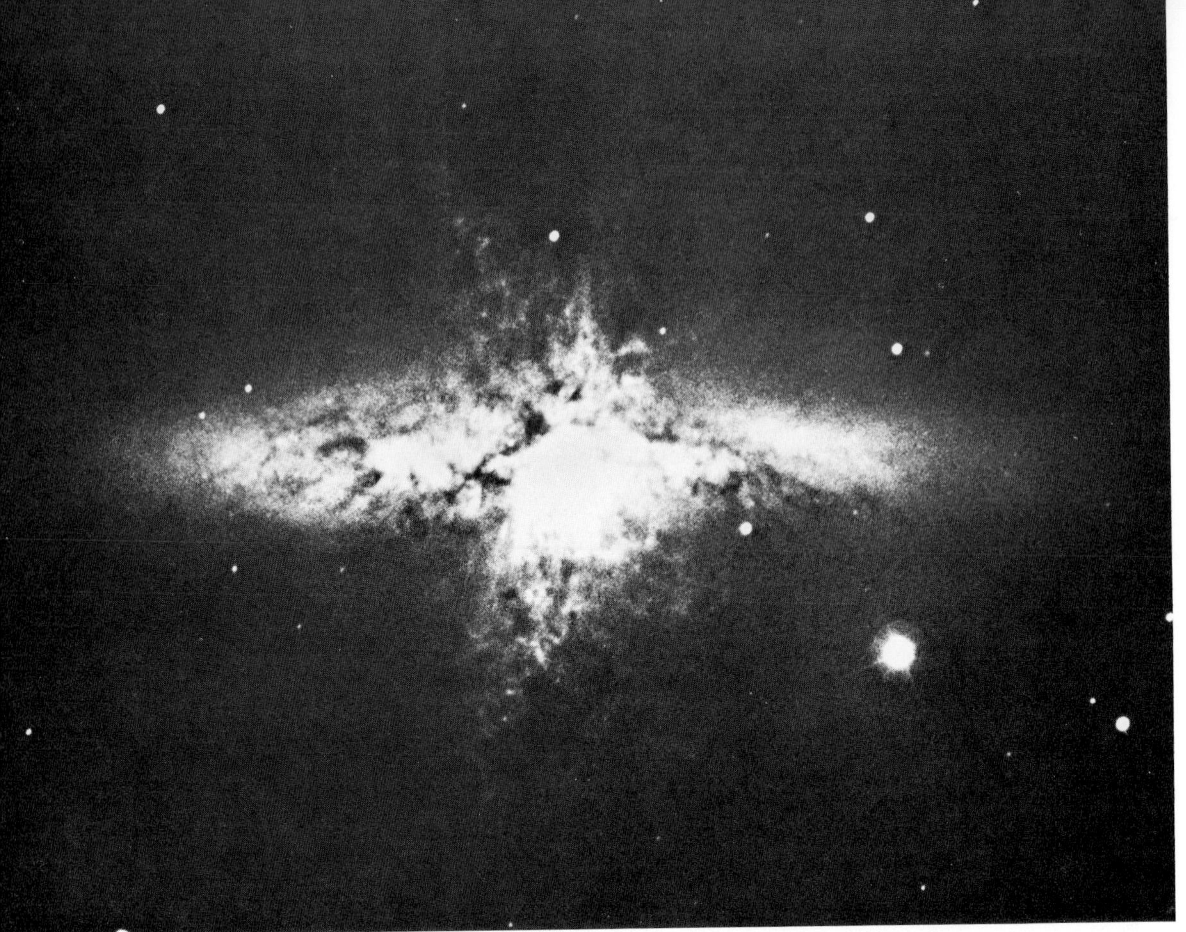

13-8 *The irregular galaxy M82, as photographed in the red light of the hydrogen alpha line. Note the filaments extending above and below the main part of the galaxy. (Courtesy of the Hale Observatories.)*

Way, M31, and M81 are all very similar, averaging about $3 \times 10^{11}\ M_{\odot}$. The spiral M33 has a mass of about $3 \times 10^{10}\ M_{\odot}$. The mass of the Large Magellanic Cloud is about $2 \times 10^{10}\ M_{\odot}$, and that of the Small Magellanic Cloud is about a factor of 10 less. Thus, galactic masses are roughly from $10^{6}\ M_{\odot}$ to $10^{12}\ M_{\odot}$, with the average mass falling in the lower part of this range.

POPULATIONS REVISITED

An understanding of galaxies as star systems requires instruments and techniques capable of resolving them into individual stars for study. The outer regions of M31 were resolved by Hubble in the 1920s; the brightest stars were found to be blue supergiants. However, the central regions of M31 were not resolved until 1944.

13-9 *The giant elliptical galaxy M87. At left is a shorter exposure which shows the central part of the galaxy and its jet. (Lick Observatory photographs.)*

This story revolves around the German-American astronomer, Walter Baade, who was on the staff of the Mount Wilson Observatory during World War II. Technically classified as an enemy alien, Baade was restricted to the Pasadena area where he made extensive observations with the 100-inch telescope. Because many of the other astronomers were assigned to wartime research, Baade had less competition for the use of the large telescope and thus had a great deal of observing time. Furthermore, Los Angeles was blacked out at night as a military precaution, so Baade was blessed with a particularly dark sky. (In fact, such good skies at Mount Wilson are approached today only when Los Angeles is heavily blanketed by smog.)

13-10 *A dwarf elliptical galaxy in the constellation Sculptor.* (*Courtesy of Paul W. Hodge.*)

Baade knew the exposure required to record images of the brightest stars in the central region of M31, assuming that they were the same kind of stars as the brightest ones resolved in the outer parts of M31. Exposures with blue-sensitive plates did not show stars at the expected exposure times. Eventually, red-sensitive plates and longer exposure times yielded images of the brightest stars in the central region of M31, as well as some elliptical galaxies near M31, such as NGC 185 (Figure 13-11).

The resolution alone was a tremendous achievement that opened up the study of individual stars in the central regions of another galaxy. The importance of this was emphasized by the circumstance that the bright stars observed in the central region of M31 were surprisingly unlike those which previously had been resolved in the outer spiral arms. This is the discovery that led to the concept of stellar populations. The brightest stars in the central region were red and somewhat fainter than the brightest stars in the spiral arms, which were blue. Baade showed that the difference could be explained (Figure 13-12) if the central region stars had an H–R Diagram like that of the globular clusters and the spiral arm stars had an H–R Diagram like that of the open clusters.

Once the concept of stellar populations is explored, many facts in our own galaxy (relating to the velocities of stars, the space distribution of stars, and the chemical composition of stars) fall into place, as discussed in Chapter 12. Spiral galaxies such as the Milky Way have both Population I and Population II stars. Ellipticals are found to contain primarily Population II stars like those in the center of M31. Irregular galaxies are largely composed of Population I stars like those in the arms

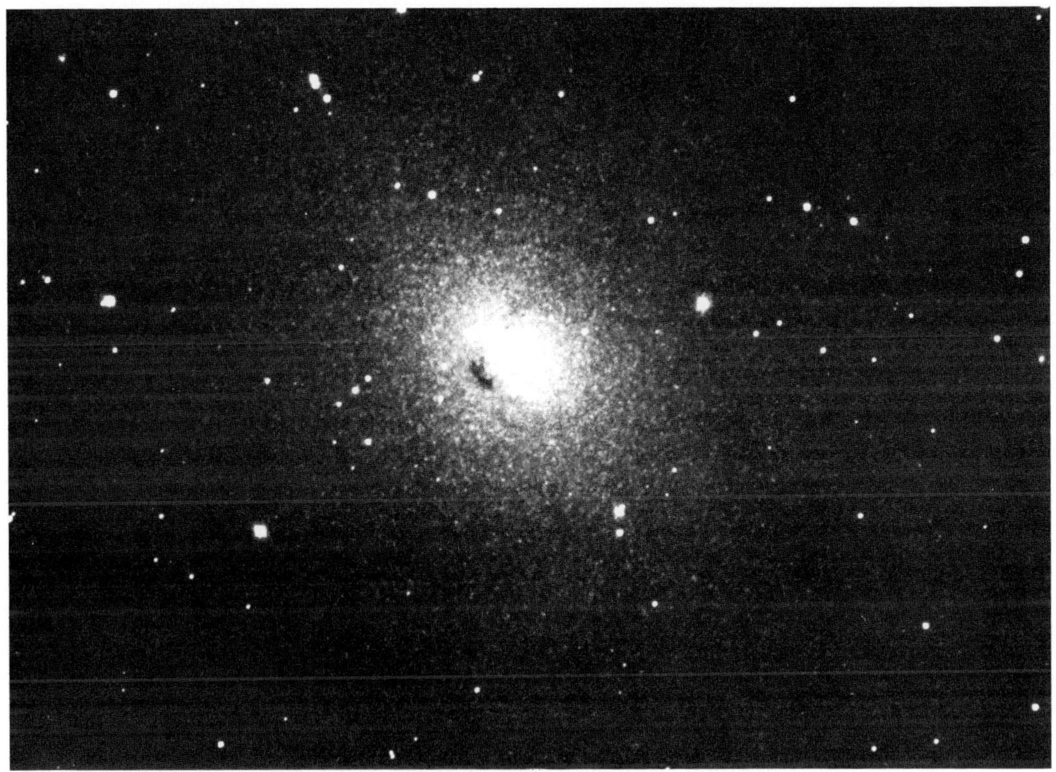

13-11 *NGC 185, a dwarf galaxy. This galaxy was among the first Population II galaxies to be resolved into stars by Walter Baade (note the stars resolved in this photo). This achievement was considered so important to astronomers that actual photographic prints of NGC 185 (instead of printed sheets) were bound into the scientific article that announced Baade's observation, in order that the reader could judge for himself. (Lick Observatory photograph.)*

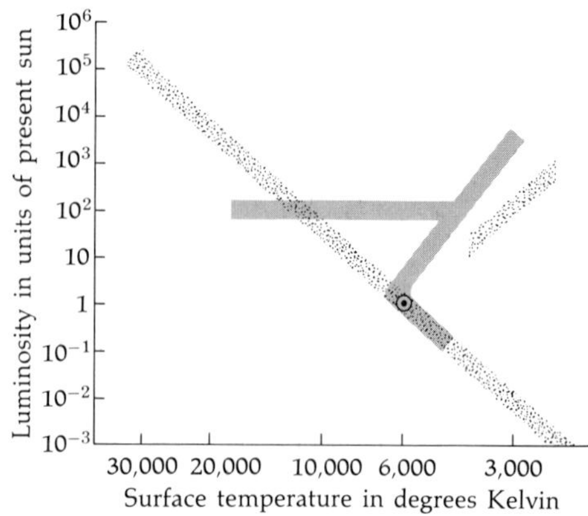

13-12 *Schematic H–R Diagram showing regions occupied by the two basic populations of stars as originally classified by Baade. Population I stars, dotted area; Population II stars, shaded area. (Adapted from* Stellar Populations *by Geoffrey and Margaret Burbidge. Copyright © 1958 by Scientific American, Inc. All rights reserved.)*

of M31. We know that galaxies must evolve, because the individual stars in them evolve, and it would be tempting to assign an evolutionary sequence to the different types of galaxies, but this cannot be done convincingly at the present time.

DISTANCES TO GALAXIES

The recognition of galaxies such as M31 as island universes basically similar to the Milky Way involved the determination of their distances, as it was necessary to demonstrate their comparable size to the Milky Way in order to show that they are similar systems. As mentioned before, the method involved the observation of Cepheid variable stars. This is a suitable technique for the few dozen closest galaxies, in which individual bright stars like the Cepheids can be resolved. (The Magellanic Clouds are the closest galaxies, at a distance of 50 kiloparsecs. The spirals M31 and M33 are each at a distance of about 650 kiloparsecs.) Individual Cepheids cannot be resolved in the more remote galaxies and thus other techniques must be employed to measure the distances.

The methods used to extend the distance scale include the

determination of the absolute magnitude of the brightest star in a galaxy of a particular type. Studies show that the absolute brightnesses of these supergiant stars in different galaxies of the same type are remarkably similar and therefore such stars can be used as *distance indicators.* This method has the advantage that the supergiants are of course the easiest stars to observe.

Observations also show that the average size of the largest H II regions in a galaxy of a particular type is nearly constant. For example, in a loosely wound spiral such as M33 (Figure 13-3), the average diameter of the five largest H II regions is about 175 parsecs. If similar nebulae are observed in a more remote galaxy of similar shape to M33, their angular diameters can be measured and the distance determined.

It appears that most galaxies occur in distinct groups called *clusters of galaxies.* A sample large cluster of galaxies, which is found in the constellation Hercules, is shown in Figure 13-13, where a variety of galaxy types can be seen. The distance-measuring techniques that involve the brightest stars and the largest H II regions can be used out as far as the Virgo cluster of galaxies at a distance of about 11 megaparsecs (millions of parsecs) and the Fornax cluster at 13 megaparsecs. However, beyond the distances of the Virgo and Fornax clusters of galaxies, individual features such as stars or bright nebulae cannot be resolved or even recognized (except for supernovae). At these distances we must use properties of the galaxy clusters themselves as distance indicators. The absolute magnitude of the brightest galaxy in a cluster can be determined in the nearby clusters (such as Virgo) and then used to calculate the distances of those more remote. The most distant known clusters are some 1,000 megaparsecs away, but some quasars (Chapter 16) are much farther. This process is simple in principle but complicated in practice. For example, the brightnesses of distant galaxies must be corrected for the Doppler effect, as they are found to have large red shifts (see next section). Thus, when we measure the V magnitude of a nearby galaxy of negligible red shift, we are measuring the light that the galaxy emitted in the wavelength range of the V filter (Chapter 10). But when we measure the light of a distant galaxy that has a large red shift with respect to the earth, the light that passes through the V filter was actually emitted at shorter wavelengths and has been red shifted. Thus, we do not get a true measure of the intrinsic V magnitude of the galaxy unless the red shift is found from the spectrum and a suitable correction is applied to the V photometer observation. The interpretation of our observations of distant galaxies is also complicated by the circumstance that

13-13 *Part of the Hercules cluster of galaxies. The "crosses" around the bright stars are due to an optical effect in the telescope. (Courtesy of the Hale Observatories.)*

we are looking backward in time. The farther away a galaxy is, the longer the light has been on its way. The light from a galaxy 1,000 megaparsecs away has been en route to us for 3.3 billion years. Hence, the observations of distant galaxies are observations of younger galaxies, and because of evolutionary changes, the stars and nebulae seen in the distant galaxies may differ systematically from those in nearby and thus "older" galaxies.

TWO BASIC PROPERTIES OF THE UNIVERSE

In this section we discuss two basic observed properties of the universe: (1) the galaxies have red shifts, and the more distant the galaxy, the greater is its red shift; (2) radio waves, with a continuous spectrum corresponding to a black body of temperature 3° K, are received by us in equal amounts from all directions in space. Any attempt to explain the nature and origin of the universe must account for these two phenomena.

The Red Shift

The distances to clusters of galaxies can be estimated by the methods described in the preceding section. Also, spectra of the galaxies can be taken and analyzed for radial motions revealed by Doppler shifts. It turns out that the absorption lines in the spectra of most galaxies are shifted to the red, as first noted at Lowell Observatory by V. M. Slipher in 1912. If these red shifts are indeed a result of motion, the galaxies must be moving away from us in a systematic fashion. Exceptions to this are a small number of relatively nearby galaxies.

In 1929 Hubble turned his attention to this problem; he studied many clusters of galaxies and found that they all exhibited systematic red shifts. Most important, the amount of the red shift depended on the distance, as shown in Figures 13-14 and 13-15. The straight-line form of the dependence of radial velocity on distance indicates that the velocity is proportional to distance. Thus, the farther the galaxy, the greater the velocity of recession. The constant of proportionality, or *Hubble constant,* is a quantity of great interest to astronomers. The value currently in use is about 100* kilometers

*Some astronomers believe the correct value may be 75 or even 50 kilometers per second per megaparsec.

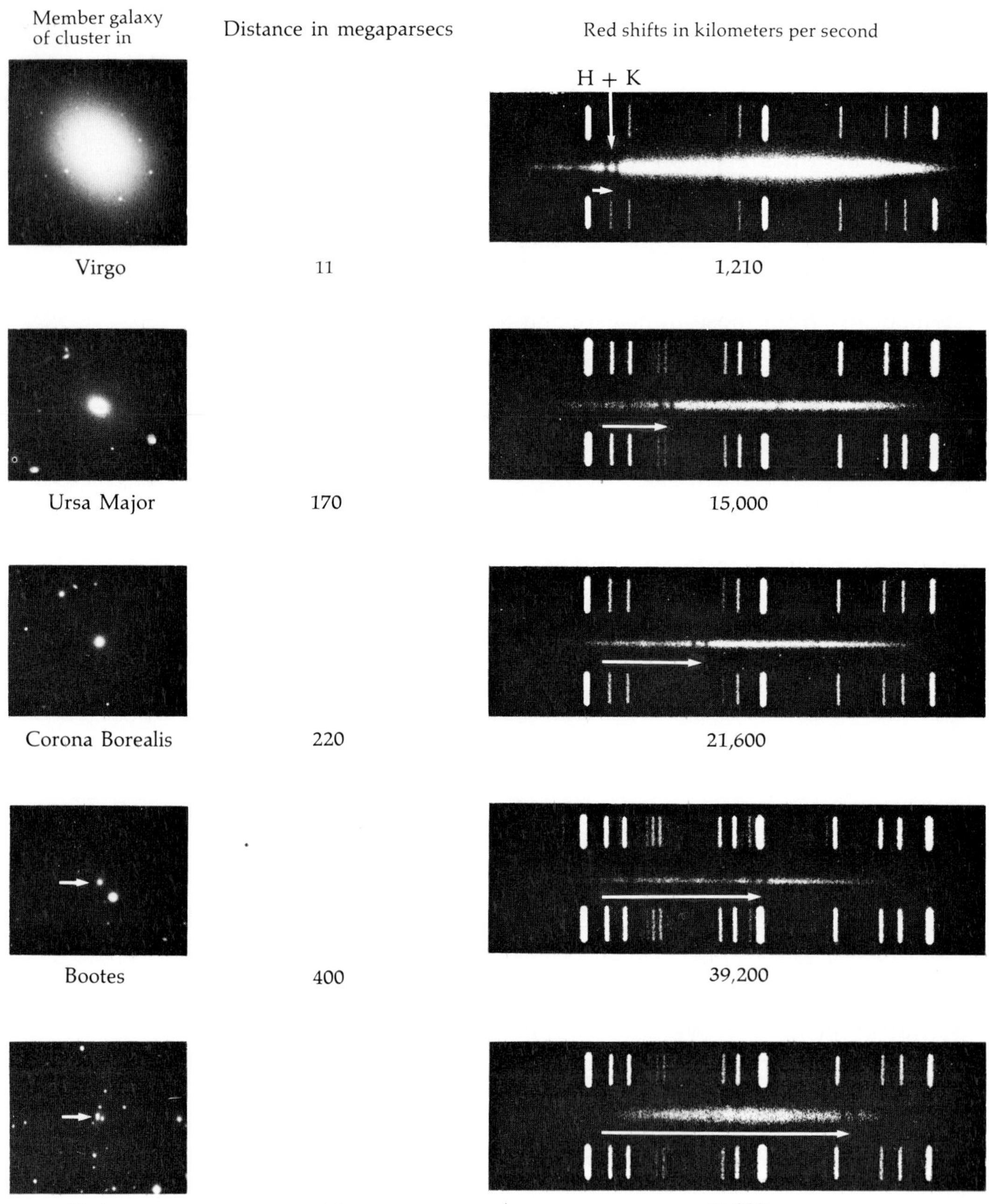

13-14 *The spectra of galaxies located in increasingly distant (top to bottom) clusters, with arrows showing the red shifts of the H and K lines of singly ionized calcium. The distances are estimated from the brightest galaxies in the clusters. The red shifts (compared to the distances) do not exactly agree with the Hubble constant of 100 kilometers per second per megaparsec because each spectrum shown corresponds to a specific galaxy in the cluster, whereas the Hubble constant is determined from the average red shifts for each of many clusters. (Adapted from an illustration supplied by courtesy of the Hale Observatories.)*

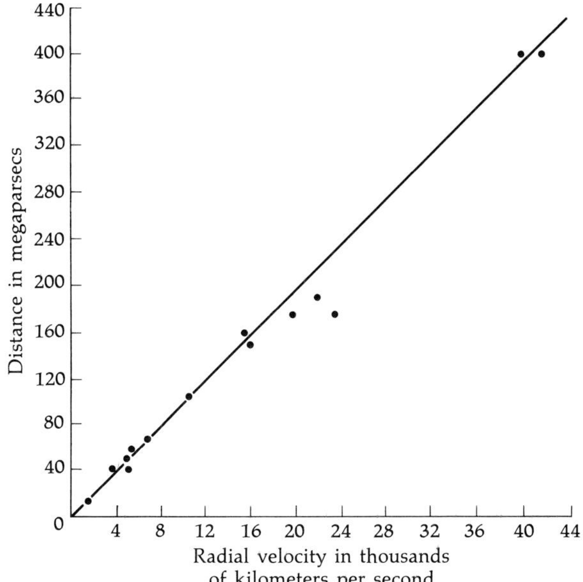

13-15 *Graph showing the distance versus radial velocity relationship for the better studied clusters of galaxies. Each plotted point represents a cluster. The straight line corresponds to a Hubble constant of 100 km per second per megaparsec.*

(62 miles) per second per megaparsec. Thus, an object 1 megaparsec distant has a recession velocity of 100 kilometers per second (62 miles per second) and an object 3 megaparsecs distant is receding at 300 kilometers per second, and the Virgo cluster galaxies have an average radial velocity of about 1,100 kilometers per second and a distance of about 11 megaparsec.

For galaxies close to the Milky Way, the velocity of recession calculated from the distance and the Hubble constant is very small, and the effect is masked by the ordinary random motions of these galaxies. This explains why some nearby galaxies are approaching us, rather than receding.

Cosmic Background Radiation

A recent and remarkable discovery relating to the general properties of the universe is that of the cosmic background radiation. The first evidence of this radiation was obtained in 1965 at the Bell Telephone Laboratories by Arno Penzias and Robert Wilson, who were investigating some unexplained receiver noise while planning a study of the radio emission from the Milky Way. In fact, the noise was actually due to radio emission from the sky, and further observations by many astronomers have established that this radiation is *isotropic:* we receive it in equal amounts from all directions. Over the wavelength range of 0.2 to 21 centimeters, where it has been well measured, the radiation approximates thermal emission (that is, follows a Planck law) at about 3° K. Studies of absorption lines due to CN molecules in the cool interstellar clouds of the Milky Way show that the molecules are heated by the background radiation, and thus confirm the existence of a 3° K radiation that apparently pervades space. This radiation was anticipated on theoretical grounds some years ago by the physicist George Gamow.

WHY IS THE SKY DARK AT NIGHT?

Olbers' Paradox is a problem first posed by the German scientist Heinrich Olbers in 1826. He asked a deceptively simple question: "Why is the sky dark at night?" The point of this question can be illustrated by considering a distribution of stars that *does not* produce a dark sky.

Suppose that the number of stars in a given volume of space is constant and that it remains constant for similar volumes at all distances from the earth. If this were true, and all stars were roughly like the sun, one could show mathematically that the entire sky should be as bright as the disk of the sun. The reasoning is illustrated in Figure 13-16. If the stars are distributed with a constant density throughout space, and if space extends without limit, it is impossible to draw a straight line from the earth in any direction without intersecting the disk of a star. The situation is somewhat akin to constructing a mosaic of the sky; as the stars at greater and greater distances are included, more and more of the sky mosaic is filled in by pieces with the same disk brightness as the sun. If the distribution of stars extends as far as one pleases, all vacant places would be filled in and the entire sky would be as bright as the solar disk.

The situation discussed in the preceding paragraph can be visualized in a different way, with a geometrical argument that involves two hypothetical regions in space, having the form of thin spherical shells (Figure 13-17). We take the simple case that (1) the two shells are of equal thickness; (2) all of the stars

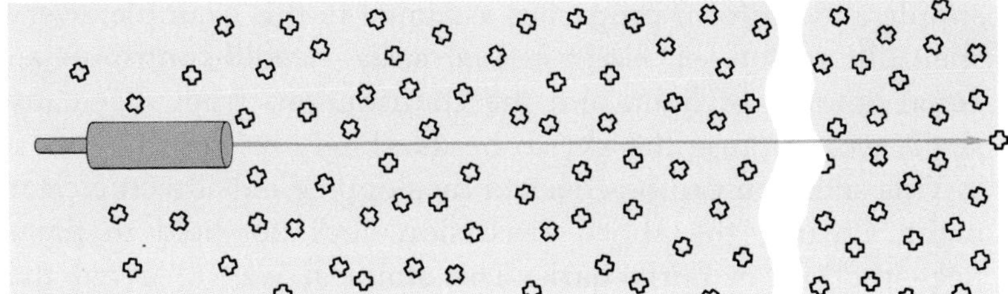

13-16 *Olber's Paradox. If the stars are uniformly distributed throughout space, a line of sight ultimately intersects the disk of a star. Even if this has not happened by a given distance from the observer, one has only to continue to ever greater distances and it will eventually occur.*

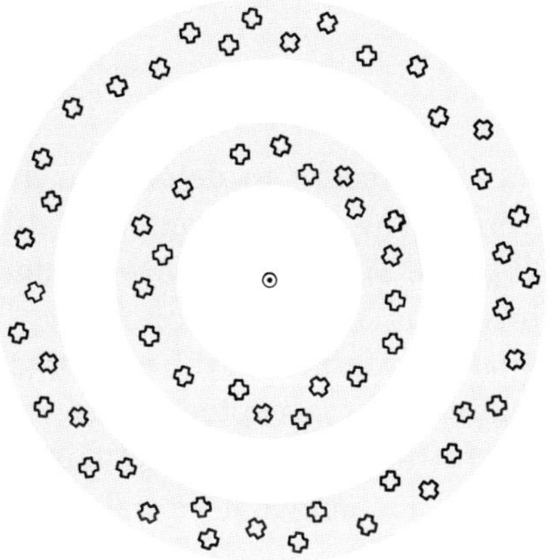

13-17 *Sketch showing that a shell of stars at a larger distance has more stars than a nearer shell with the same thickness and star density.*

in the two shells are alike (have equal intrinsic brightness); (3) the star density (number of stars per cubic parsec) is equal in the two shells; and (4) the outer shell is twice as far from the sun as the inner shell. We then ask the question: Which shell furnishes more light, as seen at the sun? Since the stars are all of equal brightness, a given star in the inner shell will appear to be four times brighter than an outer shell star as seen from the sun, due to the inverse square law. On the other hand, although the thicknesses of the two shells are equal, the volume of the outer shell is greater; in fact, it is four times greater, as one can easily calculate, and so it must contain four times as many stars. The result is that *each shell* contributes an *equal amount* of light, as seen from the sun. Thus, if space had the simple and uniform properties assumed in this example, every shell-like region—*no matter how far away*—would contribute an equal amount of light, and the contributions from the many shells would cause the sky to be dazzlingly bright.

This situation can be checked by stepping outside on a clear night. Clearly, the above discussion does not hold in practice—the sky is fairly dark. The simplest way to avoid the paradox is to postulate a limited or finite distribution of stars. The problem was regarded as solved when it was realized that the individual stars seen in the sky were, in fact, part of the Milky Way galaxy, just an isolated cloud of stars in space. However, the discovery that space is filled with galaxies reopens the question. Simply substitute galaxies for stars and Olbers' Paradox again asks: "Why is the sky dark at night?" We shall answer this question in the next section.

THE BIG BANG

Cosmology is the field of science that seeks to determine the nature and origin of the universe. The conclusions are still uncertain, but the *Big Bang* or *Primeval Fireball* theory presented here represents the current consensus among astronomers. It was first suggested by the Belgian astronomer, Abbé Georges Lemaître, in 1927, and like many other accepted ideas this theory may change or fall when further evidence is obtained.

The Big Bang idea is based on a somewhat literal interpretation of the observed red shifts in the spectra of galaxies as Doppler shifts caused by receding motions. The fact that the motions are systematically away from us reopens the problem of the center of the universe. Does this mean that the Milky Way galaxy is at the center? No. Exactly the same picture would emerge if the red shifts were measured by astronomers located in another galaxy, as illustrated in Figure 13-18. The situation is like a rubber balloon with black dots on it being blown up. All other dots on the surface are moving away from *any* given dot on the surface.

An age of the universe can be assigned by considering the past history of these motions. If the motions indicated by the red shifts were run backward, all galaxies would reach approximately the same point in space. Since objects 1 megaparsec apart are separating from each other at 100 kilometers per second, we can estimate the age of the universe (*Hubble time*) by assuming that this velocity has remained constant. A parsec

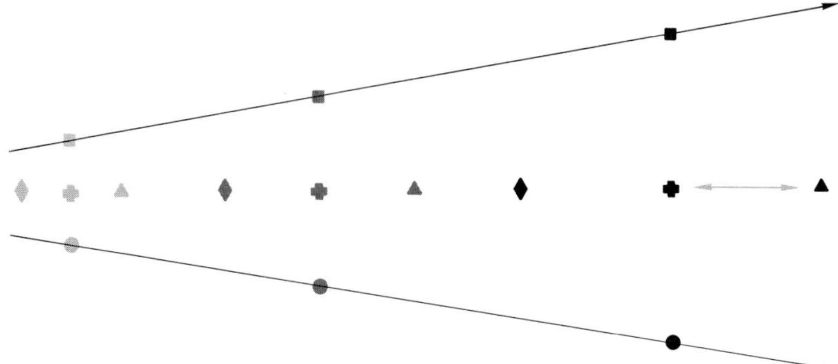

13-18 *The universal red shift does not imply that the Milky Way galaxy is at the center of the universe. A small group of galaxies in the expanding universe is shown schematically at three different times. An observer in either of the two galaxies connected by an arrow would observe the other to be red shifted. In fact, the distances between all galaxies increase on this diagram.*

is 3×10^{13} kilometers and thus a megaparsec is 3×10^{19} kilometers. The time required to traverse that distance at 100 kilometers per second is 3×10^{17} seconds, or about 10^{10} years. This value is close to the age of the oldest stars in the Milky Way. The exact age computed from the Hubble constant depends on several detailed effects, such as the possibility that the expansion rate has slowed down since the universe was born. In addition, the accepted value of the Hubble constant may be revised by further observations. Ages from 10 billion to 20 billion years are commonly quoted.

Thus, on the Big Bang scheme, the universe originated in the explosion of a single concentration of matter some 10^{10} years ago, which began the continuous expansion that we see now in the red shifts of galaxies. Some of the helium found in stars today may have originated by nuclear reactions in the dense cosmic fireball. Gamow studied this model for the origin of the universe and calculated that the explosion would have produced an enormous burst of photons. He predicted that these photons, red shifted by the expansion of the universe, would be observed in present times as radio wave photons, and this is the best current explanation of the 3° K background radiation discovered by Penzias and Wilson.

The fact that the universe is expanding at a rate that increases with distance provides the solution to Olbers' Paradox. Because the light of the more distant galaxies is red shifted by a greater amount, it is seen at the earth as less energetic than if it were not red shifted, because red photons are less energetic than blue photons (page 124). Also note that using the Hubble constant of 100 kilometers per second per megaparsec, we find that at a distance of 3,000 megaparsecs, the recession velocity would be 3×10^5 kilometers per second, the speed of light. Thus, galaxies beyond that distance (*horizon of the observable universe*) can never be seen. Figure 13–19 illustrates galaxy formation according to the Big Bang theory.

OTHER SCHEMES

The Big Bang scheme seems at present to be the most likely hypothesis for the origin of the universe. However, other theories have also been advanced. The principal one in recent years has been the *Steady State* theory proposed in 1948 by H. Bondi, T. Gold, and F. Hoyle at Cambridge University. According to this theory, there was no beginning to the universe and there will be no end. The universe has always looked more or less the way it does now and it will always look this way. Matter is continually coming into existence in the form of hydrogen atoms in space. These eventually make up new galaxies, which replace the old ones that are moving away from us in the general expansion.

On a Steady State scheme, there was no cosmic fireball and hence there should be no 3° K background radiation. If the identification of this radiation is correct, then the steady state hypothesis is wrong. However, if another viable explanation for the 3° K radiation were found, the entire subject could be reopened. During the 1960s, evidence from radio astronomy suggested that the space density (number per cubic parsec) of radio-emitting galaxies was greater in the distant past than it is now. This seemed to contradict the idea that the universe has always been the same, and seemed to rule out the Steady State theory. Now it appears, however, that more extensive radio observations do not support this conclusion.

The *oscillating universe* theory supposes that there is no beginning and no end. However, it proposes that, rather than being constant, the universe is currently in the course of an expansion that began with a Big Bang and is gradually slowing down. At some future time, gravitation will overcome the expansion effect and the universe will begin to collapse, finally reaching a point of ultimate coalescence in which the high temperature and

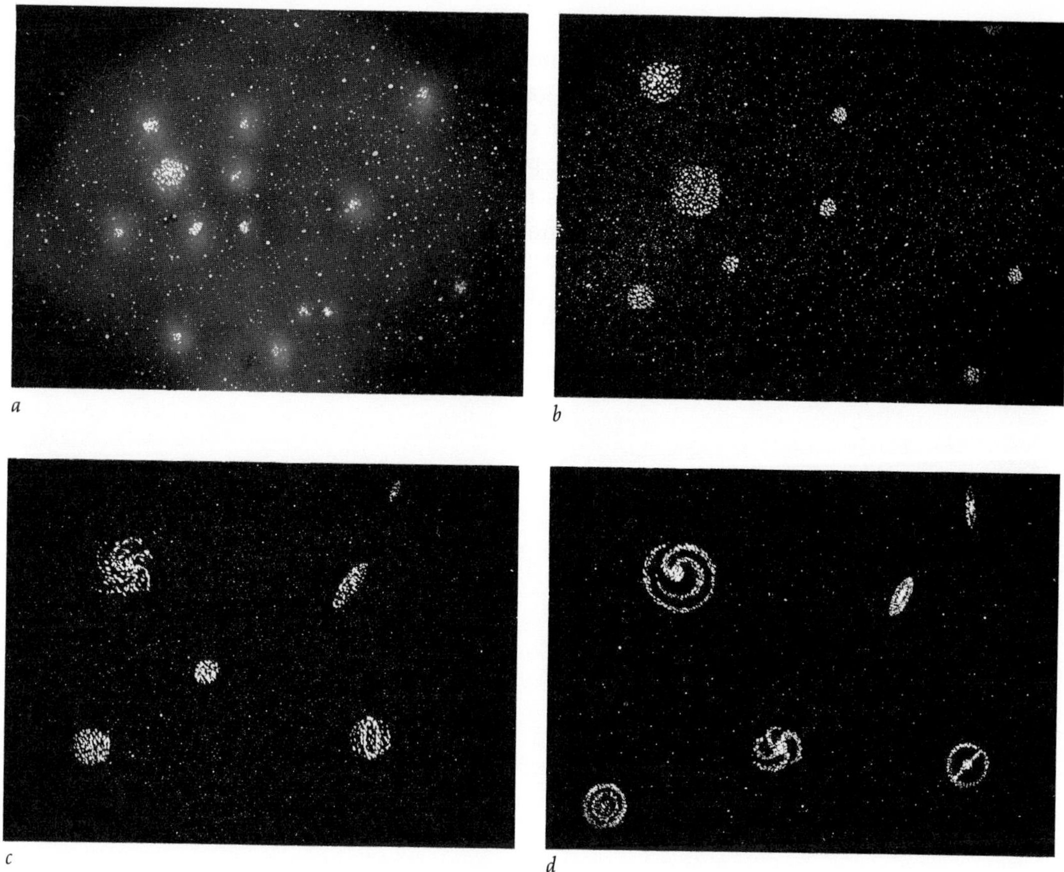

13-19 *Artist's impression of the formation of galaxies in a universe created by the Big Bang. According to the theory, the events shown here took place roughly during the period between stages two and four of the evolution of the universe, as depicted in Figure 1-2 (page 5). (a) The expanding universe was filled with ionized matter (white dots) and radiation (white glow) from the primeval fireball. It was sufficiently dense so that photons could travel only short distances before striking matter (that is, the universe was opaque). The radiation pressure kept the clouds of matter from contracting. However, the universe was cooling and when 100,000 years had elapsed, the temperature had dropped to about 3,000° K and the gas began to recombine to the neutral state. (b) The neutral gas absorbed radiation less effectively than it had done in the ionized state. Hence, photons could travel long distances without being absorbed, and so the universe was transparent. They escaped from the clouds, and the radiation pressure dropped, permitting the clouds to contract gravitationally. Thus, the protogalaxies formed (c), and evolved into galaxies (d), which we still observe today. (After* The Origin of Galaxies *by Martin Rees and Joseph Silk.*

pressure break down all matter into elementary particles; a new Big Bang occurs, and the expansion begins again.

The universe may have begun in a Big Bang, or it may be in a steady or oscillating state. In any case, it is characterized now by the process of expansion and is filled with radiation resembling that expected from a Big Bang. Further, the great distances between galaxies make it a nearly empty universe; the present density of matter in the universe averages out to about 10^{-30} grams per cubic centimeter. Expressed in other terms, this amounts to about one hydrogen atom for every 1.7 million cubic centimeters (1×10^5 cubic inches) in the universe.

14

Space Age Astronomy

ATMOSPHERIC LIMITATIONS OF GROUND-BASED OBSERVATORIES

In the discussion of observatories (Chapter 6) we touched briefly on the factors that affect the selection of an observatory location. For some astronomical research, however, there is *no suitable location on earth*—good weather and remoteness from city lights are not sufficient. This circumstance arises from four basic properties of the atmosphere: *seeing, scattering, absorption,* and *airglow.*

Seeing refers to the smeared appearance of the image of a star as seen through a telescope. It arises in turbulent air masses that bend starlight by different amounts. Thus, the atmospheric seeing effect limits the sharpness of astronomical photographs and reduces our ability to study fine details, such as the filaments and dark globules in nebulae, the narrow dark lanes between the bright granular elements in the solar photosphere, and the surface markings of the moon and planets. With the atmosphere above one's telescope it is rarely possible to photograph detail much smaller than one second of arc.

The *scattering* of sunlight by air molecules that causes the sky to appear blue (Chapter 8) also hinders our ability to study the faint solar corona. Starlight is also scattered, and this decreases the apparent brightnesses of the stars, especially at the shorter wavelengths where the effect is most pronounced.

Absorption of light by atoms and molecules in the atmosphere is a much more severe effect. Ozone absorbs ultraviolet radiation (below 3,000 Angstroms), and other gases contribute to the absorption at even shorter wavelengths so that the far ultraviolet, x-rays, and gamma rays from the sun and stars do not penetrate to the earth's surface (see Figure 14-1). Water vapor, oxygen, and carbon dioxide are among the chief absorbers at the infrared wavelengths between 1 micron (10,000 Angstroms) and 1 millimeter wavelength. However, there are a number of "windows" in the infrared spectrum corresponding to limited wavelength regions where the absorption is small. Infrared observations can thus be made in these wavelength windows, especially from mountaintops that are above much of the earth's water vapor. Beyond the infrared, in the radio portion of the spectrum, there are a few windows at millimeter wavelengths, and the atmosphere is essentially transparent between 1 centimeter and 10 meters. Beyond about 10 meters the waves are reflected back into space by the electrons in the earth's ionosphere. Thus, cosmic radiation with very long wavelengths cannot be received at the surface of the earth.

In addition to atmospheric scattering and absorption, which reduce the light that reaches us, there are the *airglow* emissions by atoms and molecules in the upper atmosphere. This dim light in the sky interferes with observations of the faintest stars,

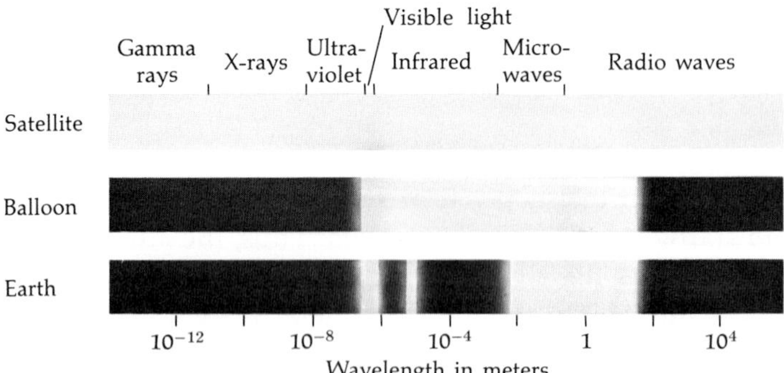

14-1 *Radiation from space is heavily absorbed by the atmosphere (dark shading) and only certain wavelengths (light shading) penetrate to the earth's surface. The situation is improved at balloon altitudes and the full electromagnetic spectrum is accessible at satellite altitudes. (From* Observations in Space *by Arthur I. Berman. Copyright © 1963 by Scientific American, Inc. All rights reserved.)*

nebulae, and galaxies. Further, since the airglow is emitted preferentially at certain wavelengths, it can obscure the spectra of astronomical objects at those wavelengths.

Space astronomy (observations from high altitude aircraft, balloons, rockets, and satellites) became necessary because of the atmospheric limitations listed above. It was principally stimulated, however, by the possibility of observing in previously inaccessible wavelength regions. *The radiation in different parts of the spectrum appears to be dominated by different physical processes, or to arise under different physical conditions, or even in different sorts of celestial objects.* There are exceptions to this statement, but generally speaking, it is valid. Take the sun, for example: in the visible region of the spectrum most of the light comes from the photosphere; in the ultraviolet most of the strong spectral lines come from the chromosphere and corona; at x-ray wavelengths most of the solar emission is due to flares and to active regions in the corona; and in the long radio wavelengths that do not penetrate to the surface of the earth, bursts of radiation are detected from particles (ejected by flares) that sometimes travel out beyond the limit of the visible light corona. Ultraviolet astronomy concentrated initially on the hot O and B stars, which emit most of their radiation below 3,000 Angstroms, and on the interstellar gas, which has its strongest absorption lines in this part of the spectrum; infrared astronomy has involved the search for new types of cool stars and for emission from dust clouds; x-ray astronomy includes the study of high temperature regions in the sun and elsewhere; gamma ray astronomy involves the search for radiation from processes in the nuclei of atoms (rather than the electron shells). Radio waves and infrared radiation are able to penetrate interstellar dust that is opaque to optical light; at these longer wavelengths it is possible, therefore, to observe previously obscured regions such as the center of the galaxy.

TRADITIONAL REMEDIES

High Altitude Observatories

For some astronomical investigations there is considerable improvement in the atmospheric limitations when one observes from a mountaintop. The decrease in scattered light is sufficient to enable limited observations of the outer solar atmosphere with coronagraphs (Figure 14-2), which has encouraged the establishment of solar observatories on mountains in the

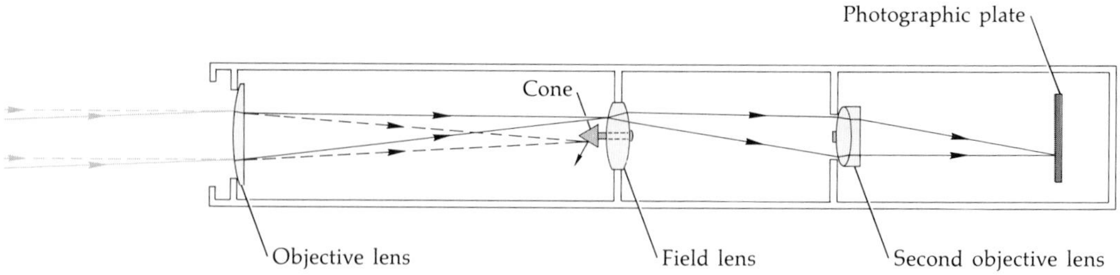

Photographic plate

Cone

Objective lens

Field lens

Second objective lens

14-2 *Optical sketch of a coronagraph showing the blocking devices, lenses, and baffles. Most of the photospheric light (dashed lines) is removed by the metal cone. Coronal light (solid lines) passes through to the plate. Stray light is caught by the baffles. A tiny mirror on the front of the second objective catches a spurious image of the sun. Finally, every optical element, but particularly the first objective, must be kept dust free, to avoid scattering of light by the dust particles. Even when the sky is clear, and the optics clean, the corona cannot be photographed to as great a distance from the solar limb as during a total eclipse—when no coronagraph is needed.*

southwestern states. The decrease in water vapor at high altitude makes a significant difference in the transmission of infrared radiation by the atmosphere. A new observatory at 4,105 meters (13,600 feet) on Mauna Kea, Hawaii, has been built for infrared astronomy. The altitude is good for astronomy, but hard on astronomers—the initial plans for the facility included a supplemental oxygen supply. Observations of the faint zodiacal light (see Figure 4-1) are especially sensitive to scattered light. This has led to expeditions to quite high mountaintops, such as Chacaltaya in Bolivia at 5,307 meters (17,630 feet). The seeing is also better on some mountains; this depends on the local weather conditions as well as the altitude. At the Pic-du-Midi (2,807 meters or 9,400 feet) in the Pyrenees, the excellent seeing enabled French astronomers to observe relatively fine details on planetary surfaces.

High Altitude Aircraft

Aircraft are especially valuable for infrared astronomy studies, as the improvement in transmission of the atmosphere at the altitudes reached by planes used for this research, namely 11,000 to 15,000 meters (35,000 to 50,000 feet), makes a considerable difference to infrared observations. One of the most remarkable contributions came from observations of the center of the galaxy in 1969. University of Arizona astronomer Frank Low, flying in a Lear jet near an altitude of 15,000 meters, used a 30-centimeter (12-inch) telescope (Figure 14-3) mounted in a port on the side of the airplane (actually in the escape hatch!) to measure infrared emission from the galactic center at wavelengths between 40 and 35 microns. This region of the galaxy cannot be seen in visible light because of the large amount of dust in the galactic plane between the earth and the center. The airplane observations show that in a region of less than 10 parsecs diameter at the galactic center, infrared energy is being released at a rate of about 8 million times the total bolometric energy emission of the sun. No really convincing explanation for this phenomenon has yet appeared. Another airplane used for this work, the Convair 990, can carry larger telescopes; flying at 12,500 meters (41,000 feet), astronomers aboard this aircraft have obtained infrared spectra of planets. Airplanes are used

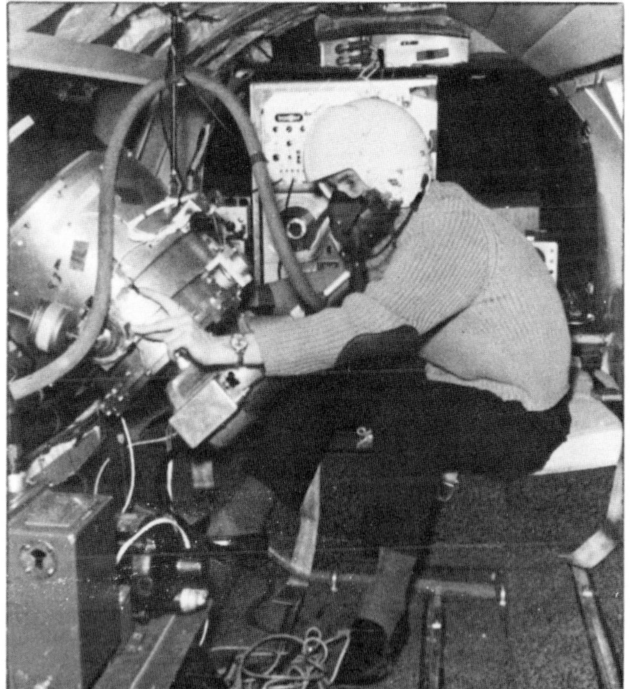

14-3 *Astronomer Frank Low at the observer's position in a Lear jet used as an infrared observatory. (National Aeronautics and Space Administration.)*

in an entirely different manner during solar eclipses. The duration of the total eclipse phase during which the corona can be seen is only a few minutes for an observer on the ground, but jets flying above the eclipse path can remain in the umbra as it tracks across the surface of the earth, and thereby multiply the duration of totality for on-board studies.

Balloons

Balloons are useful at much higher altitudes than the aircraft available to astronomers. Many balloon experiments are conducted at altitudes above 24,000 meters (80,000 feet), and some recent flights have gone up to 46,000 meters (150,000 feet). After a few attempts at manned experiments, nearly all of the astronomy in this field has been done with unmanned balloons. The telescopes can be controlled by astronomers on the ground. In one case, superb photographs of the photospheric granulation and of filamentary structure in the penumbra of sunspots (see

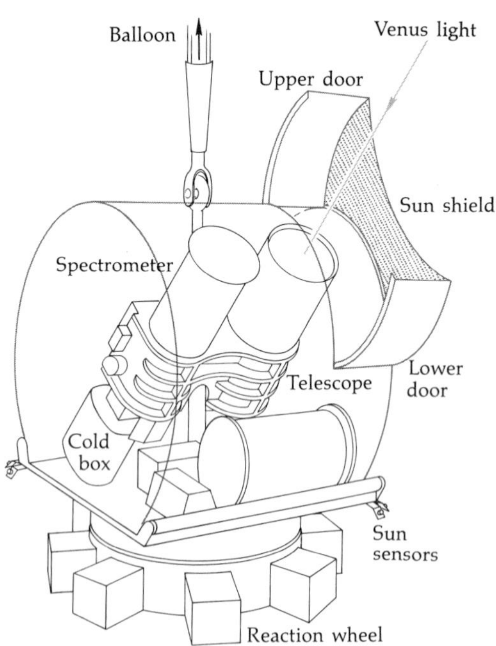

14-4 *Sketch of a ballon telescope that was used to study infrared radiation from Venus. (From* Infrared Astronomy by Balloon *by John Strong, Copyright © 1964 by Scientific American, Inc. All rights reserved.)*

Figures 9-5 and 9-6) were obtained with a 30-centimeter balloon telescope. This Project Stratoscope experiment was directed by Martin Schwarzschild of Princeton University. Another Stratoscope balloon experiment recorded an extremely compact nucleus in a peculiar galaxy—the nucleus is so small that it is smeared out by atmospheric seeing when the galaxy is observed from the ground. A sketch of a Johns Hopkins University balloon telescope used by John Strong to search for water vapor in the atmosphere of Venus is given in Figure 14-4. Since balloons can stay aloft for many hours, they were extensively used in studying the higher energy (shorter wavelength) x-rays and in attempts to detect gamma rays from beyond the solar system. The intensities expected from theoretical calculations are so weak that even the balloon flight exposure times are too short for some of this work. Observations from aircraft are tending to supplant the use of balloons in infrared astronomy because the astronomer has more direct control over his telescope. The telescope and associated instruments can be larger and heavier on the airplane than on a balloon and the equipment is recovered without damage (and hence is available for re-use). In gamma ray astronomy the trend is toward satellites, because longer observing times are needed to record the small number of gamma ray photons from a given source, and a well-designed satellite instrument can work for years in space.

Rockets

The "soft x-rays" between 2 and 100 Angstroms and the far ultraviolet waves at 100 to nearly 2,000 Angstroms are absorbed in the thin atmosphere above balloon altitudes. Therefore, in order to observe at these wavelengths, the greater heights reached by rockets and satellites are necessary. The principal contribution of rockets has, in fact, been in x-ray astronomy. Changes in the transmission of radio broadcasts through the earth's atmosphere (see page 216), which occur at the time of solar flares, led to a suspicion of flare-associated x-ray and ultraviolet emission before space experiments became possible. However, in the study of objects beyond the solar system, rocket experiments opened up a previously unsuspected field of astronomy. In fact, the discovery of the strongest galactic x-ray source, Scorpius X-1 (Sco X-1), was a happy accident in a rocket flight that failed in its intended purpose to detect x-rays from the moon.

The study of x-ray sources was at first dominated by observations made with rockets, although satellites are making an increasing contribution. More than 50 galactic and extragalactic x-ray sources have been found with rockets since the discovery of Sco X-1 by Riccardo Giacconi. Among these are the Crab nebula and several other supernova remnants. Many of them, however, have x-ray spectra that resemble Sco X-1, which has been identified with a peculiar faint blue star. The manner in which the Sco X-1 x-rays are produced has been widely debated among astrophysicists. At the present time it appears most likely that the x-rays come from a hot gas region surrounding the star, but the manner in which the gas is heated is much less certain. Some astronomers believe, from studies of shifts in visible light spectral lines of the Sco X-1 star, that the star is actually part of a binary or even a multiple system. It has been suggested that gas ejected from one member of the binary might heat up as it fell onto the other member, thereby causing the x-rays. If so, presumably the same situation would hold for other x-ray sources that resemble Sco X-1 as shown by their x-ray spectra. This idea could be tested by looking for the stars associated with the other x-ray sources and then examining the visible light spectra of these stars for possible Doppler shifts due to binary star orbital motions. Unfortunately, the x-ray instruments that have been flown on rockets were usually incapable of measuring very accurate positions for the x-ray sources. In fact, these positions are so crude that it has been extremely difficult to pick out the correct faint stars from the host of others visible on the photographs.* The x-ray sources associated with the Crab nebula and several other supernova remnants have distinctly different spectra from Sco X-1, and thus constitute a second basic class of galactic x-ray sources. A third class is known thus far from only a few examples. It is distinguished by *short lifetime.* The first of these, called Centaurus X-2 (Cen X-2), was studied with a few rocket flights, so not very much was determined about it. It is known that some time between the fall of 1965 and the spring of 1967, this strong x-ray source appeared in the sky and that it faded rapidly, becoming undetectable by September, 1967. Vela satellites have been built by the United States Atomic Energy Commission to detect atomic bomb tests by recording the x-rays generated in the blasts. On July 9, 1969, two of these satellites detected x-rays from a new source in the sky (also in or near the constellation Centaurus). It is definitely known from observations by these satellites that the source (Cen

*See the discussion of the UHURU satellite on page 385.

X-4) was not present on July 6. A month later, Japanese scientists studied Cen X-4 with a rocket launched from Kagoshima, and in late August an Australian balloon flight obtained additional data. Combining all of these measurements with the Vela satellite observations, which continued through September, it is clear that Cen X-4 rose to its maximum strength in a period of only a few days and then faded by a factor of 200 during the next few months, becoming undetectable with the available space instruments by September 24, 1969. Although some astronomers have compared Cen X-2 and Cen X-4 with the behavior of the exploding stars or novae (Chapter 11) that are well known in visible light astronomy, and have termed these two objects "x-ray novae," the plain fact of the matter is that we do not know what they are (or were).

The advantages of using a rocket for astronomical research are that one can reach altitudes of several hundred kilometers where the limitations of the atmosphere no longer apply, and that it can be done at a much lower total cost than is involved in launching a satellite into earth orbit. On the other hand, the observing time above the atmosphere is only a few minutes in the case of the sounding rockets available for this type of work. In fact, the great contributions of rocket experiments have been possible because of the very limited number of satellites available for astronomical research until quite recently. With the advent of orbiting observatory satellites, the role of the rocket experiment is increasingly becoming that of a test bed for ideas and instruments before the scarce satellite accommodations are assigned to them.

The best known rocket vehicles used by American astronomers belong to the Aerobee series. These lift typical payloads of 15 to 270 kilograms (33 to 600 pounds) to altitudes ranging from 240 to 480 kilometers (150 to 300 miles), with exposure times of two to six minutes above the level where the atmospheric effects interfere. For most work in astronomy the key component of a rocket vehicle is the guidance system. In the early history of x-ray astronomy it was not possible to point continuously at a specific target during the flight. While spinning and tumbling, the payload instruments scanned large regions of the sky. This resulted in the discovery of many celestial x-ray sources, but provided only meager data on the properties of each one. Modern rocket guidance systems allow the astronomer to select a specific target before launch and to use the full duration of the flight above the atmosphere to study that particular target.

Mountaintop observatories and telescopes flown in aircraft and balloons are still limited by the atmosphere, especially by its low transmission of stellar radiation at the very short and very long wavelengths discussed above. Rockets can carry instruments above the atmosphere, but the short duration of a rocket flight limits the opportunities for observation. The obvious solution for the space astronomer is to have a telescope permanently (or at least for an extended period of time) located above the offending region of the atmosphere. Two methods of achieving this have been widely discussed by space scientists and engineers. One method, a permanent observatory based on the moon and perhaps operated by scientist-astronauts, now seems slightly less like science fiction than it did at the time it was proposed, but the costs involved still look immense. On the other hand, the seismic (and other) instruments placed on the moon by Apollo astronauts constitute automatic observatories of a kind, telemetering information on the occurrence of moonquakes and other phenomena (Chapter 15). The other approach to a semi-permanent space telescope is that of the *orbiting observatory*, an artificial earth satellite carrying a telescope and other instruments operating via electronic commands from a control room (Figure 14-5) on the earth.

The most extensive work with orbiting telescopes thus far has been in the study of the sun. Seven OSO's (*Orbiting Solar Observatory* satellites) have performed successfully in space. A picture of one of these spacecraft appears in Figure 14-6. Instruments carried on the OSO's have recorded the occurrence and properties of the x-ray and ultraviolet radiation from the sun. The quality and range of the observations progressed as the newer satellites were launched. In the early days (the first OSO was launched in 1962) the satellite instruments simply measured the radiation from the whole sun—there was no way to map the detailed appearance of the sun in x-rays or ultraviolet. The OSO-1 observations showed that superimposed on the normal spectrum of the quiet corona there are considerable changes in the intensities of some emission lines, such as the line at wavelength 335 Angstroms, which is caused by 15 times ionized iron. Some of these changes took place gradually over a period of months; a tendency of the peaks (Figure 14-7) in the intensity of the radiation to occur at intervals of roughly 27 days suggested that much of the 335-Angstrom radiation was coming from active regions in the corona that were passing across the solar disk as the sun rotated. Bursts of x-rays

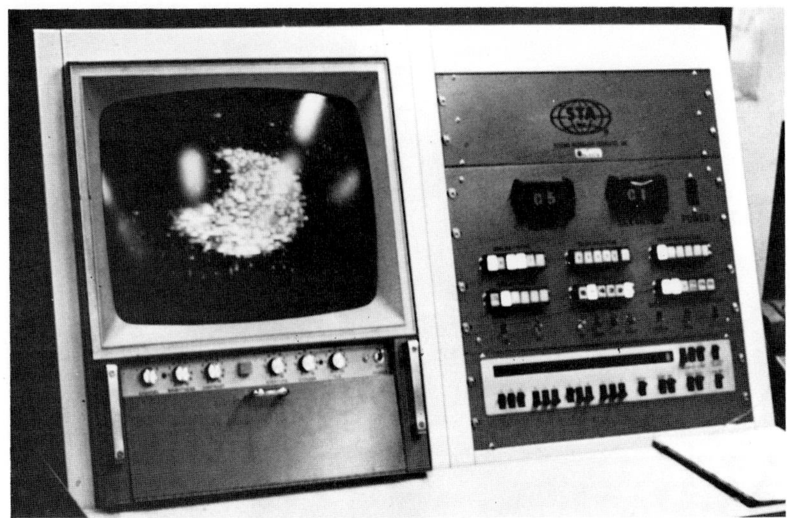

14-5 *Eclipse of the sun, March 7, 1970, as observed in ultraviolet light by the Harvard College Observatory's instrument on satellite OSO-6. The different shadings represent different brightnesses; observations made at the different locations on the solar disk by scanning the instrument across the sun are combined to produce the image shown here on a computer display device at the OSO Control Center. (Goddard Space Flight Center.)* See also Plate 3 following p. 260.

are produced by solar flares, and have been studied spectroscopically by several OSO's. Analyzing these spectra, Werner Neupert of the Goddard Space Flight Center found some surprising emission lines, such as the one at wavelength 1.8 Angstroms which is produced by iron atoms that have lost 24 of their 26 electrons. This implies that the temperature in a small region of the sun must reach 20 to 50 million degrees during such a flare. The sixth OSO, launched in 1969, was able to make maps of the whole sun in ultraviolet light every eight minutes. These maps, which had a spatial resolution of about 30 arc seconds, were telemetered to tracking stations on the earth. From the tracking stations they were relayed electronically via a computer at the OSO Control Center* to a computer printer at Harvard College Observatory in Cambridge, Massachusetts. Studying the maps as they came out of the printer, Leo Goldberg and several associates watched the development of active regions and flares. The astronomers notified the Control Center of the positions of the interesting regions on the sun. The

*At the Goddard Space Flight Center in Greenbelt, Maryland.

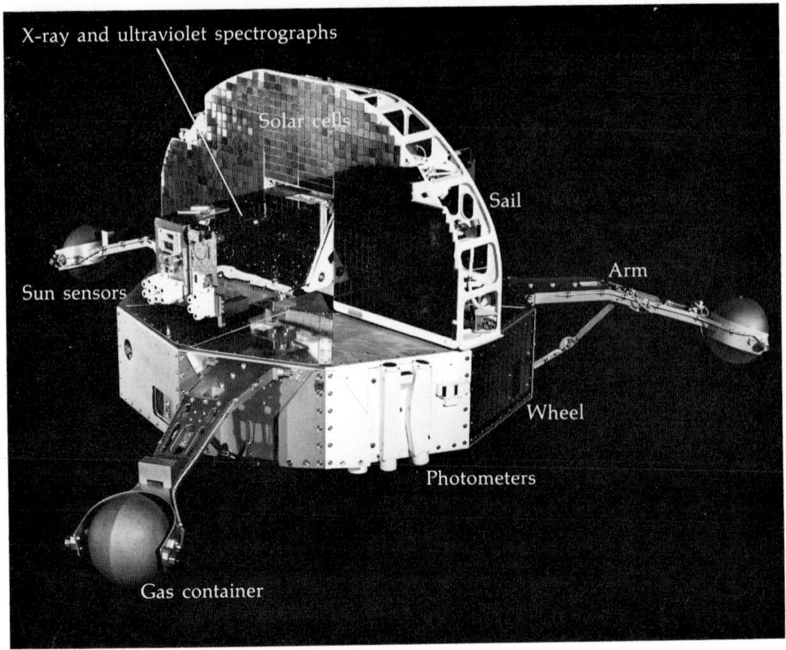

14-6 *Photograph of the OSO-5 satellite that was used to study solar flares. Principal parts are labeled; additional experiments carried in the wheel section are not indicated. The wheel rotates about 30 times per minute to stabilize the satellite as the sail points at the sun. Gas jets are used to control the rate of rotation. (Courtesy of Ball Brothers Research Corporation.)*

satellite was then commanded to point the instruments at these specific points on the sun and to record their ultraviolet spectra, or to make a series of localized maps of the interesting solar regions at a rate of one map every 30 seconds. These maps can be studied individually by the astronomers or can be assembled in the form of motion pictures that show the rapid changes in the ultraviolet light of the corona above an active region as flares occur. One movie of this type portrayed a surge of solar gas traveling rapidly outward from the limb of the sun. Measuring the position of the surge material on successive frames, the astronomers found that its velocity accelerated to a maximum of at least 160 kilometers per second (100 miles per second) and then decelerated. Similar movies of visible wavelength solar phenomena are often made at ground-based observatories. Combining ultraviolet measurements like these made by satellites with ground-based observations made in visible wavelengths, we attempt to understand the total occurrence on the sun; that is, the activity in the photosphere and chromosphere

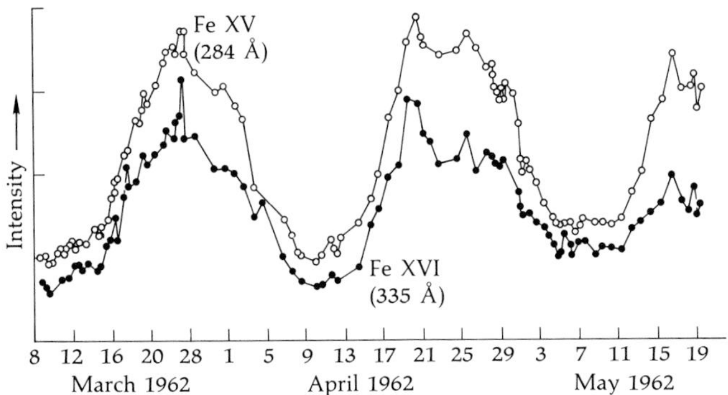

14-7 *Variable solar emissions of highly ionized iron lines as determined from the first Orbiting Solar Observatory. (Goddard Space Flight Center.)*

(seen in visible light) and the accompanying phenomena that take place higher in the chromosphere and corona (as revealed by the ultraviolet records). In addition to these phenomenological studies, the ultraviolet observations are analyzed for their physical implications. For example, it has been well known that the temperature rises very rapidly with increasing height in the chromosphere and corona (Chapter 9). But analyses of OSO-6 data by Harvard astronomers and graduate students resulted in the discovery of an unsuspected *isothermal plateau*—a narrow zone in the chromosphere where the temperature rise is much less pronounced than elsewhere.

A highly successful *Orbiting Astronomical Observatory* (OAO-2) was launched in 1968. Much larger than an OSO, the 10-foot-long OAO carried eleven instruments into orbit for ultraviolet studies of stars, galaxies, and other objects. At one end of the spacecraft (Figure 14-8) four 32-centimeter (12.5-inch) telescopes focused ultraviolet light through filters onto television cameras. This equipment (called the *Celescope* experiment) was intended to answer the basic question: What does the sky look like in the ultraviolet? Its purpose was to send back pictures of as much of the sky as possible. These comprise space astronomy's first attempts to produce an equivalent of the Palomar Sky Survey (Chapter 6) that mapped the sky at visible wavelengths. Each of the four Celescope instruments obtained its pictures in light of a different wavelength region, depending on the properties of the filters and television tubes. These wavelengths ranged from 1,050 to 3,200 Angstroms. Two Celescope pictures of the Pleiades open cluster are shown in Figure 14-9; visible-light photos were given in Figure 6-17.

14-8 *The cameras of Project Celescope. (Smithsonian Astrophysical Observatory.)*

At the other end of OAO-2 was a 40-centimeter (16-inch) telescope and six smaller instruments. Instead of taking survey pictures as in the Celescope experiment, these instruments were used to measure the ultraviolet brightnesses and spectra of individual preselected targets. This equipment was used by Arthur Code of the University of Wisconsin to study two bright comets during the early months of 1970. Observing Comet Tago-Sato-Kosaka on January 14, he made the first detection of free hydrogen atoms (that is, hydrogen atoms not contained in molecules) in a comet. This was done by observing cometary radiation in the form of an emission line at wavelength 1,216 Angstroms. This line, called *Lyman alpha,* is known to be produced by hydrogen atoms. The most exciting part of the discovery was that the hydrogen light did not come from one of the three known parts of a comet (nucleus, coma, and tail; see Figure 7-5). Instead, it came from an enormous region, slightly larger than the sun, which surrounded the comet. Thus, it appears that there is a fourth basic constituent of some

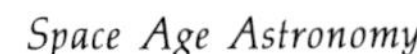

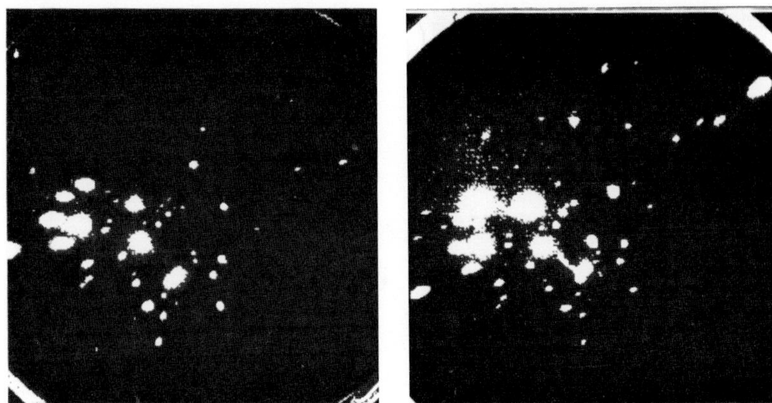

14-9 *The Pleiades as seen by the Celescope experiment. Compare with the ordinary photographs given in Figure 6-17. (Courtesy of Project Celescope, Smithsonian Astrophysical Observatory.)*

comets—the hydrogen cloud that shines in ultraviolet light. In April, 1970, Jacques Blamont of the University of Paris, using a Lyman-alpha photometer flown on another satellite, found a similar but much larger hydrogen cloud surrounding Comet Bennett. This cloud had a diameter of roughly 1.3×10^7 kilometers (8 million miles), or nearly ten times the size of the sun. Astronomers are now studying this new information to explain the hydrogen clouds. Perhaps they are produced by gas that boils off the frozen cometary nucleus; perhaps they are produced by some kind of interaction between the comet material and the solar wind. Comet Bennett was also studied by OAO-2, as were many other celestial objects. Among the many interesting findings are: (1) many galaxies are unexpectedly bright at wavelengths shorter than 2,000 Angstroms (Figure 14-10); (2) B and O stars are somewhat hotter than previously deduced from ground-based observations; (3) red supergiant stars produce strong ultraviolet emission lines in their chromospheres and may be ejecting matter at a rate as high as one solar mass per 100,000 years; and (4) the Horsehead nebula, a dark cloud as viewed in visible light (recall Figure 12-7), is *brighter* than its surroundings when observed at wavelengths below 1,600 Angstroms!

In 1970, the launch of the UHURU* satellite gave us the first orbiting celestial x-ray observatory. It was equipped to determine improved positions for many x-ray sources, leading to

*From the Swahili word for freedom; the satellite was launched from a platform off the coast of Kenya on the seventh anniversary of that country's independence.

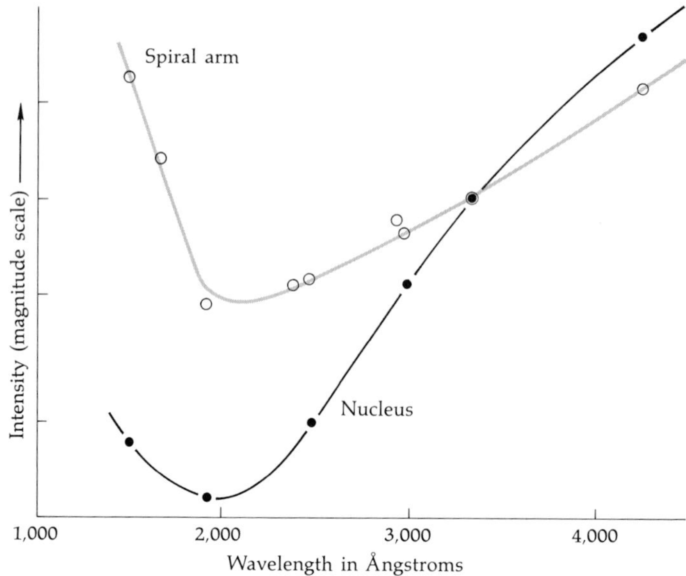

14-10 *OAO-2 observations showing intense ultraviolet radiation (wavelengths less than 2,000 Å) from Messier 31. (Courtesy of A. D. Code, University of Wisconsin.)*

better possibilities for associating them with particular optical (visible wavelength) objects. But of greatest importance was UHURU's ability to examine a given source repeatedly over a significant length of time, as distinct from a rocket observation, which lasts for only a few minutes. One of the first major discoveries through the use of UHURU was the finding that the x-ray source Cen X-3, already known from rocket observations, was varying fairly regularly, with a 5 second period. Although reminiscent of the pulsar in the Crab nebula (Chapter 16), this object appears to be a distinct class of x-ray variable star, which the theorists are trying to explain.

There are other astronomical research satellites than the few we have mentioned; they include the Radio Astronomy Explorer, which mapped the long wavelength radio waves that cannot penetrate the earth's ionosphere, and many spacecraft built or under construction by the Soviet Union and several countries of western Europe. As space astronomy advances from its initial stage of determining the basic appearance of the universe at wavelengths that we cannot see from the ground, specially designed rocket and satellite experiments are being launched to test theories based on the initial discoveries.

15

Exploration of Space

In the preceding chapters we outlined many of the techniques and results of both historical and modern astronomy. This work was carried out almost exclusively with purely *observational* data; that is, the only aspect of an astronomical object that was directly sampled was its photon radiation. Thus, unlike other scientists, astronomers who did this research did not make laboratory or field experiments. Except for an occasional meteorite, the astronomical material was beyond our reach. Rocket technology was required to send instruments or man himself to explore the moon and planets and the interplanetary medium. Although the first artificial satellite, *Sputnik 1,* was launched into earth orbit by the USSR on October 4, 1957, the development of rocketry had been underway for quite some time.

ROCKETRY

The basic principles of rocketry and satellite orbits were known to Isaac Newton. He considered the effects of firing a projectile horizontally from a high mountain. If the speed is low, the projectile travels some distance and falls to the earth's surface. If the speed is increased, the projectile travels a larger distance

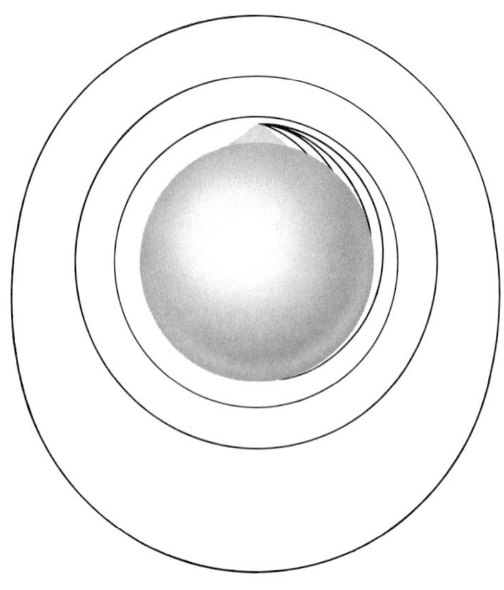

15-1 *Drawing adapted from a sketch by Newton, showing the possibility of launching artificial satellites. (From* The New College Physics *by A. V. Baez. W. H. Freeman and Company. Copyright © 1967.)*

before falling to earth. If the speed is increased to about 8 kilometers per second, the projectile does not fall to earth but travels around the earth in a low, circular orbit. The various situations for the ideal case, which ignore the effect of air resistance (friction), are shown in Figure 15-1, which is a drawing based on one by Newton, with some additional ellipses added to show the situation for higher altitude satellites. The period of a hypothetical artificial satellite was easily calculated by Newton, for the earth's circumference of 40,000 kilometers (25,000 miles) is traversed at 8 kilometers per second in 5,000 seconds or about 83 minutes. Actually, satellites are usually placed in orbits at a distance of 200 kilometers (120 miles) or more above the surface of the earth to reduce the effect of air resistance. *Sputnik 1* had an orbital period of 90 minutes.

When the speed is increased to 11 kilometers per second (7 miles per second), a rocket has reached the *escape speed* (page 212) necessary to leave a closed trajectory or orbit and pass into space beyond the earth. Sometimes one hears this situation described as "escaping from the earth's gravitational field," but this is incorrect. At escape speed the rocket is going fast enough so that the earth does not recapture it, but there is always an attractive force between earth and rocket in accordance with Newton's law of gravity. When astronauts walk on the moon under the influence of the lunar gravity, they are still clearly affected by the earth's gravity, because the moon itself is in orbit because of the gravitational force exerted by the earth. In planning a deep space probe, such as a mission to Mars, the gravitational attraction of several objects, certainly including the earth, moon, sun, and Mars, must be taken into account. The necessary computations of the orbit are complex and there are technical difficulties in launching and inserting the probe into the proper trajectory, so that mid-course corrections (firing of small rockets at times during the flight) usually are necessary to keep the probe on the desired track.

As we have seen, the theory involved in sending a rocket to Mars was known to Newton. But how can the speeds needed for orbit and escape be attained in practice? In the long run it is better to achieve the necessary speed relatively slowly than to acquire it suddenly. Too rapid accelerations tend to squash astronauts and damage the instruments. Thus, Jules Verne's (*From the Earth to the Moon,* 1865) old idea of shooting a space ship from a gun with muzzle velocity equal to the earth's escape speed is unsatisfactory. Moreover, a projectile shot from a gun, carrying no rockets, has no provision for additional maneuvers. All of these problems are solved, at least in principle, by the use of rockets. A gradual acceleration is possible and fuel reserves can be used for mid-course corrections.

The scientific principle of rocket flight is contained in Newton's Third Law, that for every action there is an equal and opposite reaction. If we produce hot gases in a rocket chamber with a nozzle at one end, the rocket and gases both push on each other with the result that the gas goes out the nozzle and the rocket goes the other way. This is the principle of the jet engine and the rocket engine. In the jet engine, air from the surroundings is combined with fuel to produce the hot gas. In the rocket all the needed substances are carried along in the form of gaseous, liquid, or solid propellants. Thus, rockets do not require air from their surroundings to function and therefore can travel through the near vacuum of space.

The application of these principles to develop useful rockets for space travel was pioneered by three men. The first was the Russian, Konstantin Tsiolkovsky (1853–1935). He was an idea man who originated many important concepts. For example, he realized that the use of a single rocket to place a payload in orbit was inefficient and that the efficiency could be greatly increased by the use of multiple stages in which the required final high speed was imparted only to a relatively small upper stage and payload. The second pioneer was Robert H. Goddard (1882–1945) of Clark University. His talent combined theory and experiment. His test flights in Massachusetts and near Roswell, New Mexico, led to fundamental innovations in rocketry, including the development of practical liquid propellant rocket motors. He was awarded over 200 patents for his technological developments. In 1959, NASA named the Goddard Space Flight Center, nucleus of its unmanned scientific satellite program, for him. The third pioneer was Hermann Oberth (1894–). Although he was associated with a number of practical and theoretical developments, Oberth's contributions were substantially through the inspiration of others by his books on rockets and space travel. Somewhat in contrast to Goddard, Oberth sought to publicize rocketry and, for example, served as a technical advisor to an early science fiction film. His efforts in both rocketry and its popularization were important factors in the rapid development of rocket technology in Germany.

The role of science fiction in these men's careers is worth noting. Tsiolkovsky, Goddard, and Oberth were all initially inspired by the descriptions of voyages to the moon by Jules Verne. Goddard also acknowledged a debt to the writings of H. G. Wells.

The Chinese in the thirteenth century are generally given credit for the use of the first (black-powder) rockets. A great deal of research and technical development, mostly in the twentieth century, has led to advanced rocket vehicles such as the *Saturn 5* shown in Figure 15-2. The rocket with payload measures about 110 meters (360 feet) in height and the weight at liftoff is about 2.9×10^6 kilograms (6.4 million pounds).

SPACE NEAR THE EARTH

Since the launch of *Sputnik 1,* a great many artificial satellites and a number of deep space probes have been used to study the properties of space around the earth. This work has resulted in a detailed, mostly unexpected picture of the remarkable phenomena that surround our planet. The thin upper atmosphere of the earth, beginning with the ionosphere (page 216), is strongly affected by the emission of ultraviolet and x-radiation by the sun. These wavelengths are absorbed in the ionosphere, and as the various events of solar activity produce different amounts of this radiation, so different amounts are absorbed by the upper atmosphere, and its density and temperature

15-2 *The lift-off of* Apollo 11 *on July 16, 1969. (National Aeronautics and Space Administration.)*

change. For example, after a major solar flare, the absorption of radiation at heights near 160 kilometers (100 miles) heats the thin air at those levels, causing it to expand. Satellites orbiting at higher altitudes encounter denser air than ordinarily, because gas from below has expanded up to their altitudes. This was discovered and analyzed by Luigi Jacchia of the Smithsonian Astrophysical Observatory and Wolfgang Priester of Bonn University from studies of the satellite motions.

Trapped Particles

Above the ionosphere there exists a region (*magnetosphere*) in which fast-moving electrons and protons are trapped in the earth's magnetic field. Spiraling along the magnetic lines of force, they bounce back and forth between *mirror points* located in opposite hemispheres. These trapped particles were discovered in 1958 by James Van Allen of the State University of Iowa, using a geiger counter carried on the first American satellite, *Explorer 1.* Two regions of trapped particles (Figure 15-3) were soon mapped and were named the *Van Allen Belts.* From additional measurements by subsequent satellites, we now recognize that the entire magnetosphere is filled with the trapped particles, and thus the idea of two distinct belts is an oversimplification. On the inner side the magnetosphere is limited by the collision of the trapped electrons and protons with atoms and molecules of the atmosphere. On the outer side it ends at a surface called the *magnetopause.* The shape of the magnetopause is shown in Figure 15-4. This surface lies at a distance of about 10 earth radii in the direction toward the sun, and it is the place where we can regard the earth's magnetic field as ending due to the pressure of the solar wind. The solar wind streams around the earth, as shown in Figure 15-4, and in the anti-solar direction a *magnetic tail* of the earth extends far out into space. The magnetic tail has been studied with many spacecraft, such as *Explorer 33* and *Explorer 35* and *Interplanetary Monitoring Platforms* (IMP's) 1 and 2, out as far as the moon, and it has even been detected out at a distance of 1,000 earth radii or roughly 6×10^6 kilometers (4 million miles) by the deep space probe *Pioneer 7,* which was launched into orbit around the sun in 1966. In the sunward direction, the *bow wave* indicated in Figure 15-4 is formed as the rapidly moving solar wind adjusts itself to stream smoothly past the obstacle of the earth's magnetosphere. It is analogous to the bow waves formed around supersonic aircraft and projectiles in the atmosphere.

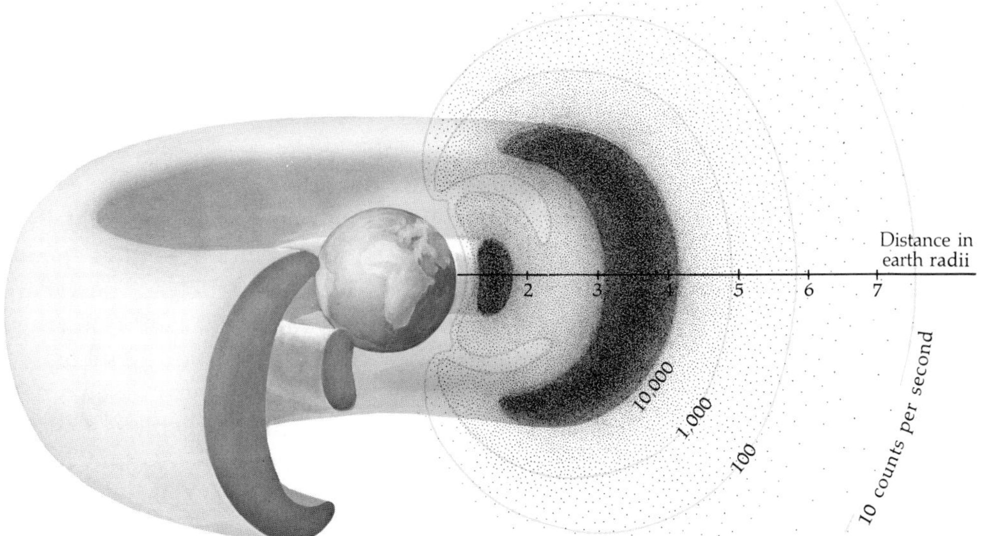

15-3 *The Van Allen Belts. The dots indicate schematically the number of trapped particles per unit volume in the earth's magnetic field. (After* Radiation Belts around the Earth *by J. A. Van Allen. Copyright © 1959 by Scientific American, Inc. All rights reserved.)*

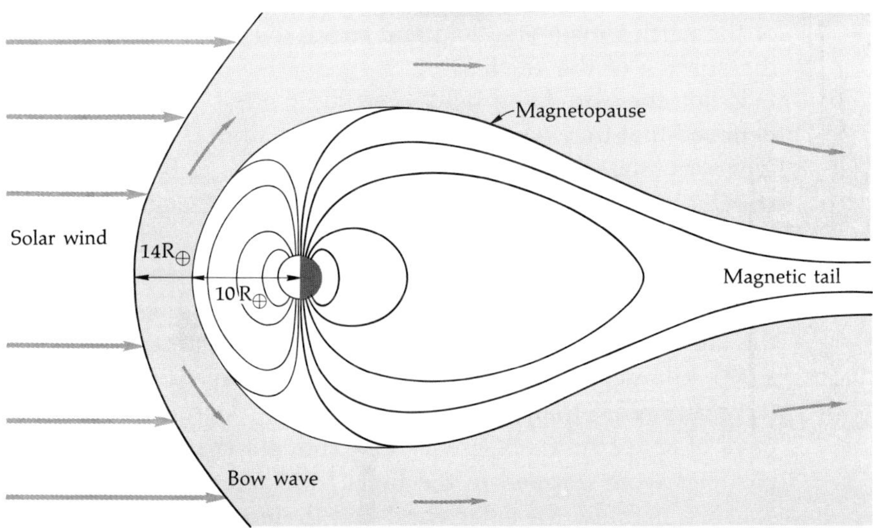

15-4 *The magnetosphere as shaped by the solar wind. The outer boundary of the magnetosphere, called the magnetopause, is not generally penetrated by the solar wind plasma. The direction of the solar wind flow is indicated by the arrows.*

For many years before the advent of space satellites, it was known that the earth has a magnetic field, with a pattern resembling that assumed by iron filings sprinkled around a bar magnet (Figure 2-15). Now from the satellite measurements, we see that although this pattern, called a *dipole field,* holds true near the earth's surface, it becomes distorted at high altitudes due to the effects of the solar wind. Sometimes a major solar flare will eject protons from the sun in small amounts but at speeds greatly exceeding those of the solar wind particles. These flare protons are deflected by the earth's magnetic field, and they strike the ionosphere above the polar caps, producing increased ionization that blacks out many radio communication channels over the poles. However, the striking auroral displays that we occasionally see in the temperate or even tropical zones of the earth after solar disturbances occur, are not caused by the flare protons. These *aurora borealis* (northern lights, Figure 15-5) and *aurora australis* (same effect when seen in the Southern Hemisphere) are produced by electrons (*auroral electrons*) that collide with the atmospheric gases and cause them to glow. (These electrons do *not* come from the solar flares either; theories of their origin are discussed below.) A third type of particle, besides trapped particles and auroral electrons, should be mentioned: the *cosmic rays* are electrons, protons, alpha particles, and nuclei of most types of atoms. They come from somewhere in our galaxy, possibly from supernovae and pulsars (Chapter 16), and probably also from beyond the galaxy, and a fraction of them have enough energy to penetrate through the outer parts of the earth's magnetic field and strike the atmosphere or even the surface of the earth.

When the Van Allen Belts were discovered, many physicists believed that they were the source of the auroral electrons. The idea was that solar activity produced changes in the solar wind which, in striking the magnetosphere, was thought to cause some of the trapped electrons to rain down on the earth and produce aurora. Unfortunately, this beautiful idea is not supported by evidence from the satellites. Among other problems, the auroral electrons have slower velocities (typically less than 2,000 kilometers per second) than the Van Allen electrons (typically more than 4,000 kilometers per second). In the early days of space research, it was also thought that the Van Allen particles were trapped in the belts from either or both of two sources, namely, the solar wind and the upward moving products of collisions between cosmic rays and atmospheric gas molecules. Once again, more detailed studies show that, in fact, these sources cannot be responsible for most of the trapped

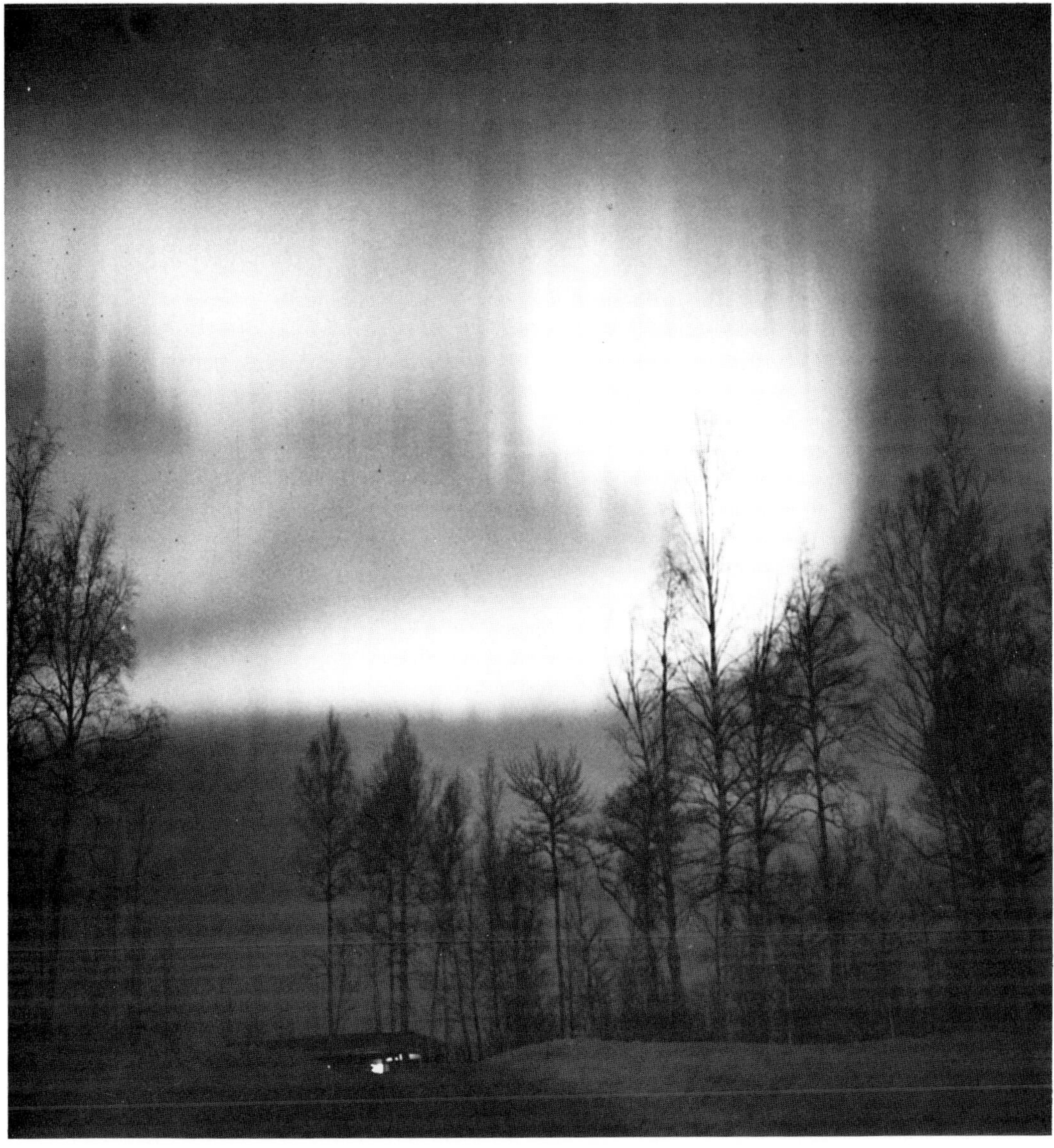

15-5 *A bright* aurora borealis *as viewed from the University of Alaska campus in the fall of 1958. (V. P. Hessler.)*

particles. Current thinking (as developed by physicists such as John H. Piddington in Australia) is that both the trapped particles and the auroral electrons may come from the region of the magnetic tail, and presumably the tail particles do originate in the solar wind, but the problems are far from solved. Thus, we know that the earth is surrounded by a region of fast-moving electrons and protons trapped in the magnetic field, and we know that aurorae are caused by somewhat slower moving electrons that strike the atmospheric atoms and molecules, but we still are not sure exactly how either the trapped particles or the auroral electrons come about.

Instruments carried on satellites and space probes have measured the strength, shape, and variations of the magnetic field surrounding the earth and have determined the distribution, composition, and velocities of the charged particles in and beyond the magnetosphere. In this latter regard they have established the properties of the trapped particles, the solar wind particles, and the cosmic rays. The properties of the magnetic field and the various sorts of particles are found to respond to variations in the solar activity. In the case of the cosmic rays (which arise outside the solar system), the solar

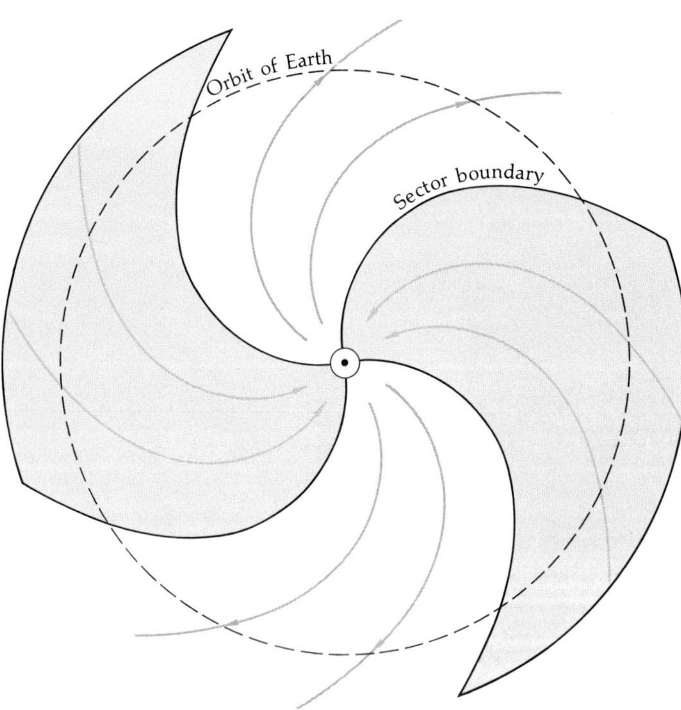

15-6 *The interplanetary magnetic field as seen in the plane of the ecliptic, showing the sector structure. The magnetic field changes from north-seeking to south-seeking in adjacent sectors as illustrated by the arrows. The magnetic field is drawn into the spiral shape shown by the combined effects of (1) the ejection of matter that carries the field out from the sun, and (2) the sun's rotation. An analogy is the spiral pattern produced in the streams of water from a rotating lawn sprinkler.*

magnetic field, as it is carried outward into interplanetary space by the solar wind, affects the numbers and energies of the cosmic particles that are able to penetrate through the solar system to the vicinity of the earth. The field in interplanetary space, arising at the sun, is weaker at sunspot minimum, and so cosmic particles are less strongly deflected by it at that time, and we then detect more of them in our part of the solar system. The measurements show that the interplanetary magnetic field does not have the simple bar magnet pattern that the earth's field assumes near the surface of our planet; rather, there is a *sector structure* that can be traced back to the general solar magnetic field away from the localized active regions on the sun. This structure, as observed in 1964–1965, is shown in Figure 15-6.

Dust in the Solar System

Satellites and space probes also carry instruments to determine the distribution of dust particles or micrometeoroids in circumterrestrial and interplanetary space. The aim is to see whether the dust particles come from the dissolution of comet tails, from debris in the asteroid belt, or from a more general medium that pervades space, and to determine whether there are concentrations of the dust at particular locations in space, such as the *libration points* shown in Figure 15-7.

In the early years of the space age, satellites detected unexpectedly large numbers of the dust particles, and some astronomers thought that the earth and moon each might have a dust cloud. Later work showed that the early experiments had given erroneous overestimates of the amount of dust particles. One of the chief techniques originally used was the method of *acoustic detection* in which vibrations caused by the impact of an interplanetary dust particle on a microphone-like device mounted on a satellite were converted to electrical signals that were measured and telemetered back to earth. Now it appears that some form of noise in the instruments masked the true impacts of dust particles. More reliable measurements were subsequently obtained by Otto Berg of the Goddard Space Flight Center, who used *time-of-flight detectors* (Figure 15-8) carried by the space probes *Pioneer 8* and *9*. The heart of such a detector is two very thin, electrically sensitive films, placed about 5 centimeters (2 inches) apart and parallel to each other. As the space probe travels through the solar system, a fast-moving interplanetary dust particle will occasionally penetrate first one

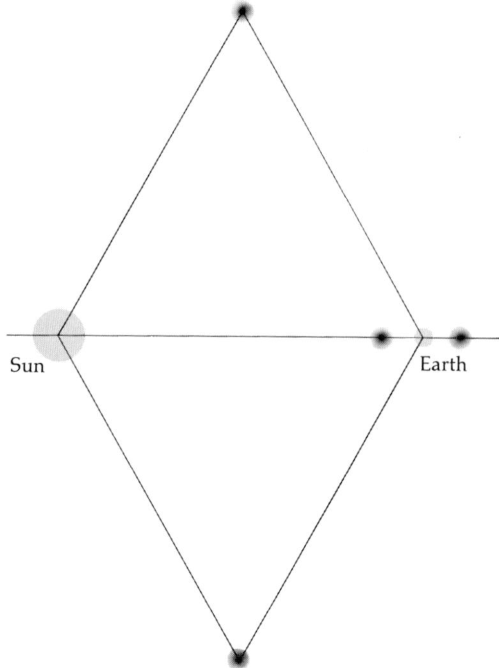

15-7 *Libration points in the sun-earth system. In theory, interplanetary dust could concentrate at the locations marked, but no large concentrations are found. Two of these points are on the earth-sun line at a distance of 0.01 a.u. on either side of the earth, and two more are located at the equilateral triangle points equidistant from the earth and the sun.*

film, and then the other. The instrument measures the position on each film and the brief time interval between the two penetrations. From this information we can deduce the speeds and directions of motion of the micrometeoroids, as well as count the rate at which their impacts occur.

The results of interplanetary dust particle measurements are still in a preliminary state, since much of the older data are now in dispute. Although the dust has been detected, there is no convincing evidence that it is concentrated around either the earth or the moon. At the average distance of the earth from the sun, and in the plane of the earth's orbit, the most numerous dust particles are those with a diameter of about 1 micron (10^{-6} meter or about 0.00004 inches), and the scarcity of smaller particles may be due to their being blown away by the pressure

15-8 *Interplanetary dust detector that was carried by* Pioneer 8. *(Courtesy of O. E. Berg and U. Gerloff, Goddard Space Flight Center.)*

of solar radiation. On the average, there are about 20 particles in the micron size class in every cubic kilometer of space (85 particles per cubic mile) in the vicinity of the earth's orbit. It further seems, based on the orbits of the dust particles, that those found in the vicinity of the earth's orbit probably come from comets. There also have been ambitious attempts to actually collect dust particles with rocket-borne traps and these experiments have brought back some particles from the upper atmosphere. Whether these are truly of interplanetary nature, or originated on the earth, is still being argued by scientists.

THE MOON

The moon's surface has long captured the imagination of laymen and amateur astronomers. A good pair of binoculars or a small telescope reveals the wealth of interesting geological or *selenological** features on the lunar surface. We begin this discussion with a summary of the ground-based investigations and then proceed to the space results. The phases of the moon were discussed in Chapter 4.

The moon is at a distance of about 60 earth radii, or about 384,000 kilometers (240,000 miles), and has a diameter of about 3,475 kilometers (2,160 miles). The mass was not so easy to

*From selene, Greek word for the moon.

determine in the days before space probes were sent to the moon. At that time the lunar mass (it is about $\frac{1}{81}$ of the earth's mass) had to be inferred from a wobble in the earth's orbit, caused by the earth executing a small orbit around the center of mass of the earth-moon system. (This wobble was detected because it produces a small monthly parallax in the apparent positions of Venus and Mars.) Using the moon's mass and diameter, we can compute its density: it is 3.3 times as dense as water—a value similar to that which characterizes the rocks in the earth's crust. The moon's gravitational attraction at its surface is about $\frac{1}{6}$ that of the surface gravity on earth; thus, an astronaut who weighs 180 pounds on earth would weigh only 30 pounds on the moon (the astronaut's mass does not change; the weight of an object is the force exerted by gravity on the object). The accuracy with which we know the mass of the moon has been improved through study of the orbits of circumlunar spacecraft such as the *Lunar Orbiter* satellites that made close-up moon photos prior to the manned *Apollo* missions. In fact, small irregularities in the moon's gravitational force were detected in this way and attributed to concentrations of mass below but near the lunar surface; they are called *mascons.* One theory of the mascons is that they are large and dense (presumably iron) meteoroids that struck the moon in the past.

Surface Features

Examining a picture of the moon (Figures 4-2 and 15-9), one notes at once the craters, the mountainous regions, and the *maria.* We now know that there is no water on the moon, but the maria (from the Latin word for seas) were so-named on the basis of their appearance as large, relatively dark areas. The pattern of these dark areas, typically several hundred kilometers in extent, and the brighter *highland areas,* produces the man-in-the-moon effect that almost everyone has noticed at some time. The heights of features on the moon's surface were originally estimated from the lengths of their shadows (note the

15-9 *A lunar landscape, the region of the Alpine Valley, as seen from* Lunar Orbiter 5. *The Valley extends through the mountainous region in the foreground to the Mare Imbrium. There are craters in the mountainous area and in the floor of the mare, and there is a rill in the valley floor. A peak is located on the mare plain in line with the Alpine Valley and another is to the right of the first; these may be volcanic in origin. "Wrinkle ridges" (probably associated with lava flows) are visible on the mare. (National Aeronautics and Space Administration.)*

15-10 *The crater Tycho as photographed by* Lunar Orbiter 5. *Notice the collapsed outer wall and the central peak (with shadow). Tycho is about 100 km (60 miles) in diameter. (National Aeronautics and Space Administration.)*

shadow of the central peak in Figure 15-10), as seen through earth-based telescopes. It was found in this way that the maria are generally low-lying and relatively flat. Radar echo timings were subsequently used to study the heights and slopes of the maria. Maria occupy about 25 per cent of the lunar surface. The other 75 per cent is covered by the relatively bright and rough mountainous regions, or highlands. The heights reached in the lunar Apennine Mountains reach about 5,500 meters (18,000 feet) above the neighboring Mare Imbrium (Sea of Storms).

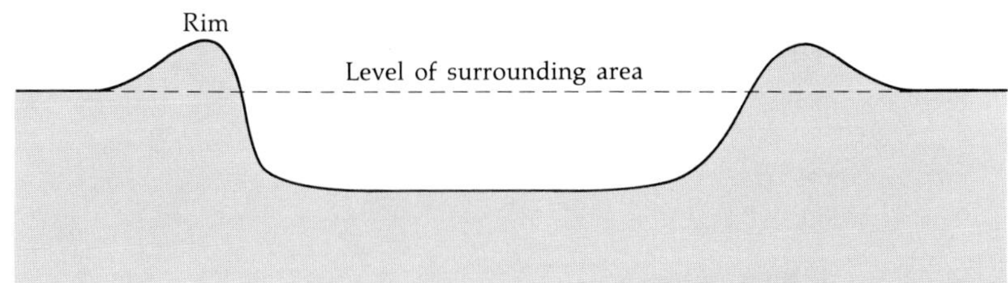

15-11 *Profile of a typical lunar crater (not to scale).*

The *craters* are found essentially all over the moon, both in the highland areas and in the maria. A cross sectional diagram is shown in Figure 15-11. The crater floor lies below the surrounding terrain, and the crater rim rises above the surroundings. Geometrical calculations based on the depths, heights, and lateral extent of crater features show that the volume of material missing from the pit is approximately the same as the volume of material piled up in the crater rim. Many craters have central mountain peaks (Figures 15-10, 15-18, 15-22, and 15-24). There are craters above 150 kilometers (95 miles) in diameter and, as the *Surveyors* (unmanned spacecraft that landed on the moon and sent back close-up photos of the surface) revealed, they range down in size to the smallest pits that can be discerned amidst the rocky soil. The ubiquity of craters cannot be overstressed. They are everywhere on the moon. There are even craters within craters, within craters. . . (Figure 15-12).

A variety of other structures decorate the surface of the moon. Presumably each type of geological feature provides a clue to the processes that have operated on the lunar surface in the past, just as the geological features of the earth reveal the erosion, glaciation, mountain building, and volcanism that have occurred here. There are linear depressions in the surface, called *rills* (Figure 15-13), and there are the conspicuous *rays,* the bright streaks that radiate from the larger craters such as Tycho (Figure 15-14). The rays extend for hundreds of kilometers and suggest that material was ejected from the crater region by an explosion or an impact. Small *domes* (Figure 15-15) are also seen on the lunar surface; their appearance is reminiscent of small mountains of volcanic origin that are found on earth.

The principal questions about the lunar surface through the years have concerned the origin of the craters and the nature of the rocks and soil. The controversy over the cause of the craters has concerned the alternative theories of *meteoroid impact* and *volcanic origin.* The fact that craters are mostly located

15-12 *Lunar craters: (above) The region of the craters Stöfler and Maurolycus as seen from earth. (Lick Observatory photograph). (below) Panorama photographed from the* Apollo 11 *Lunar Module showing numerous pits and small craters. (National Aeronautics and Space Administration.)*

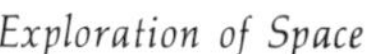

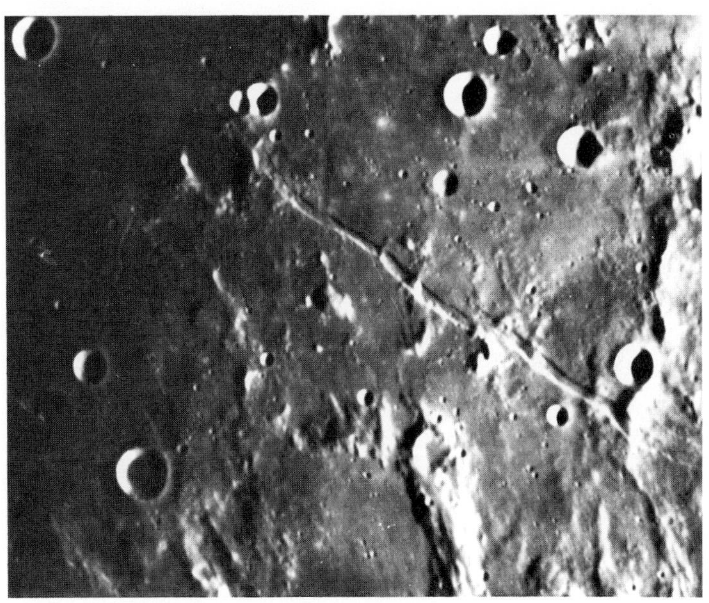

15-13 *The region of crater Julius Caesar showing the Ariadaeus Rill. (Lick Observatory photograph.)*

15-14 *The region around crater Tycho, showing the rays. (Courtesy Lunar and Planetary Laboratory, University of Arizona.)*

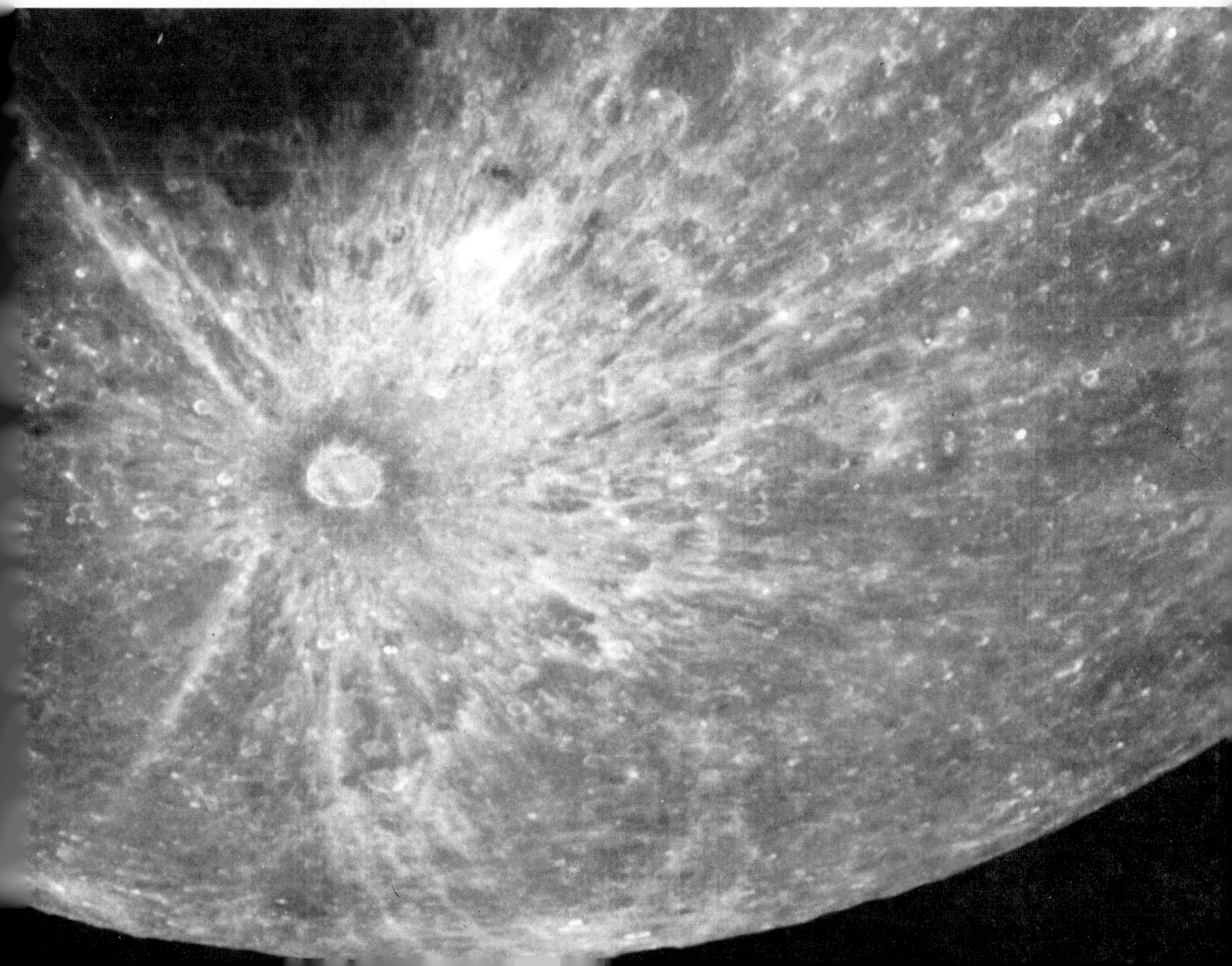

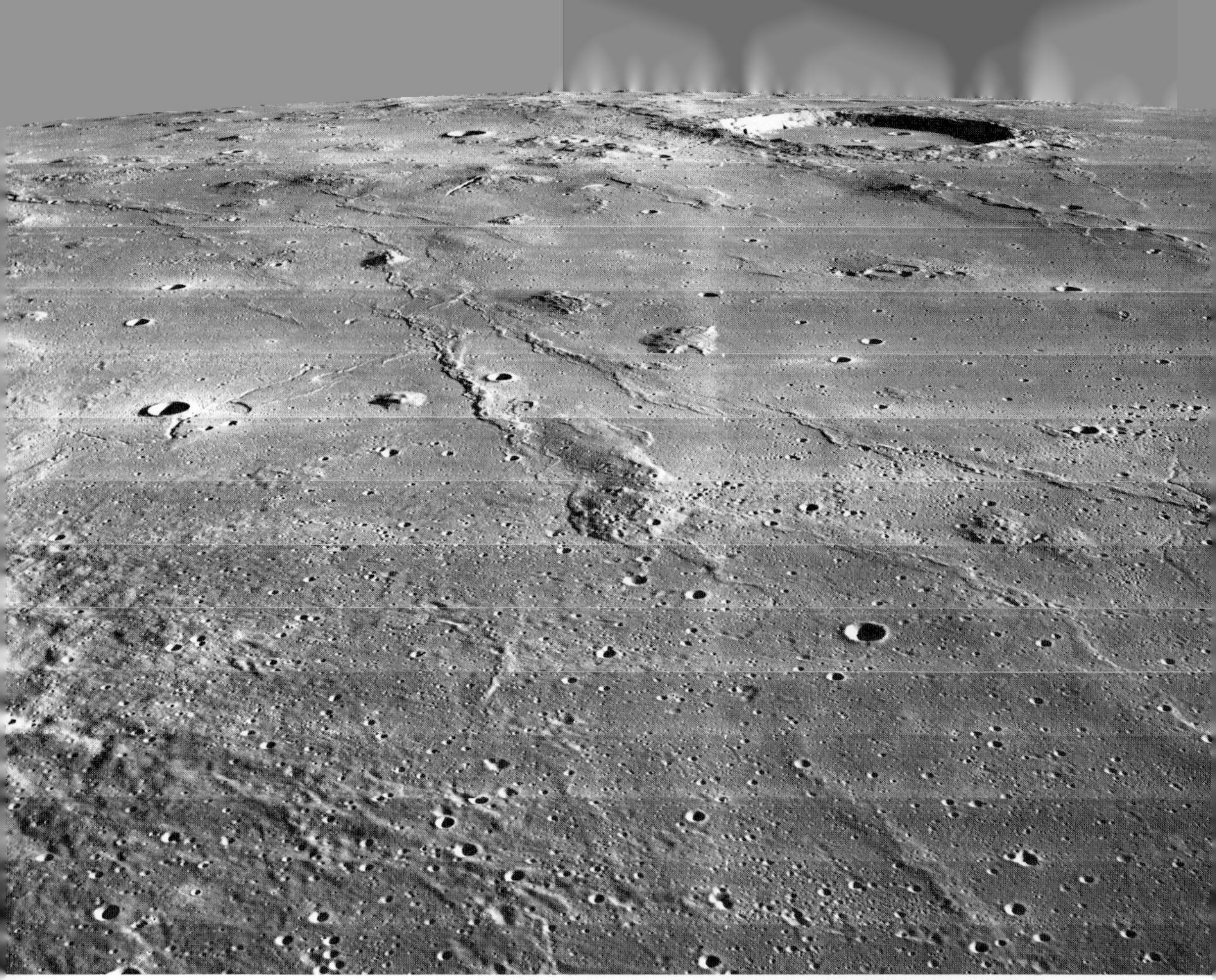

15-15 *Lunar domes. (above) The region of the Marius Hills, showing several dome features thought to be volcanic in origin; also note the many wrinkle ridges. The crater Marius is in the background. (below) Close-up view of the region north of crater Gruithuisen. The large symmetrical dome on the left is about 20 km (12 miles) in diameter. (National Aeronautics and Space Administration.)*

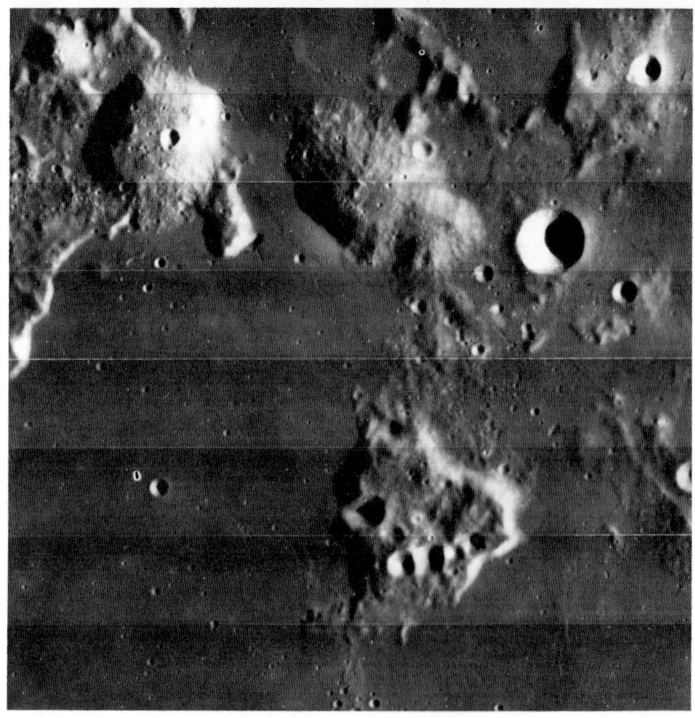

at random on the moon supports the impact hypothesis, since volcanoes on earth usually occur in chains, as found, for example, in the Aleutian Islands.

The appearance of the rays suggests an impact to most astronomers, although a volcanic origin for them might be possible. The approximate equality in volume of the material in the rim and the material missing from the pit strongly suggests an origin by meteoroid impact, at least when one compares volcanic and meteor craters on earth. Since we know that erosive processes on the moon are much less effective than on earth (because there is no water and virtually no air), at least *some* of the craters that we see must have resulted from impact by the smaller bodies in the solar system. In fact, the prevailing body of evidence suggests that most of the lunar craters originated by impact.

On the other hand, there certainly is evidence for some volcanism on the moon. The domes mentioned above are most likely volcanic; they hardly could be due to impact. Some craters are not randomly distributed but found in *crater chains,* which could be interpreted as volcanic (Figure 15-16). In addition, the phenomenon of filled-in or *drowned craters* (Figure 15-17) argues

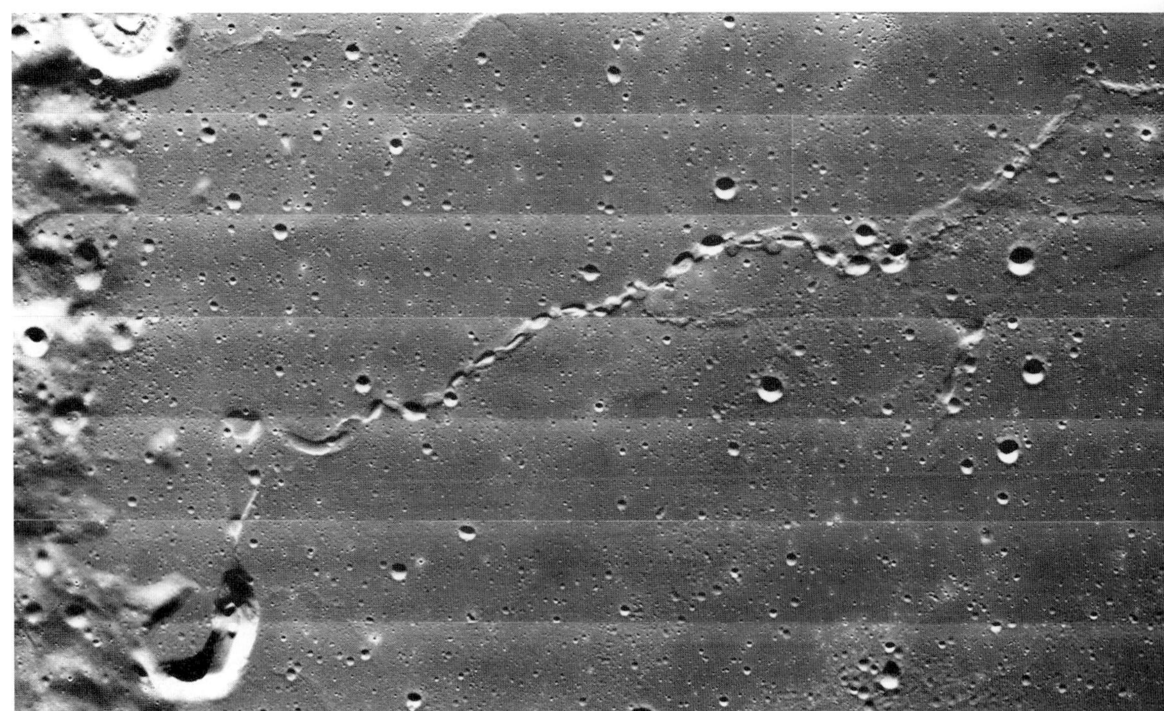

15-16 *A lunar crater chain (about 35 km or 22 miles in length) as observed by* Lunar Orbiter 5. *(National Aeronautics and Space Administration.)*

15-17 *The drowned crater Archimedes (right) is seen in this view which also shows the craters Aristillus (bottom left) and Autolycus (top left). (Lick Observatory photograph.)*

for volcanism in the form of lava flows. Finally, there have been occasional reports of gas clouds or reddish glows observed on the lunar surface. These have been discounted many times in the past, but an early record (1783) of such an occurrence was made by a truly expert observer, William Herschel. Several astronomers, notably N. A. Kozyrev in the Soviet Union, have reported seeing gas clouds in the crater Aristarchus (Figure 15-18). These observations, if correct, could be the result of local subsurface heating.

Rocks and Soil

Now what about the surface material? This can be studied in a variety of ways. The total light reflected from the lunar surface is relatively low. The ratio of the amount of light that is reflected in all directions to the total light that falls on a moon

15-18 *The crater Aristarchus as recorded by* Lunar Orbiter 5. (*National Aeronautics and Space Administration.*)

or planet is called the *albedo* and this fraction is only 0.073 for the moon; individual regions do reflect much more (highlands) or much less (maria). The albedo may indicate a dark rock-like basalt or a lighter rock that has been darkened by solar wind bombardment, a process that has been checked and simulated in the laboratory. The brightness changes dramatically with the relative angles of the earth, moon, and sun, as explained in Figure 15-19; this implies a rough lunar surface. When we observe the full moon, we are looking with the sun behind us and we see no areas in shadow. At other lunar phases many areas are shadowed by local height irregularities, from mountain ranges to tiny craterlets and rock particles, and so the lunar surface seems less bright. Observing the way in which radio waves of different wavelengths, beamed to the moon by powerful radar transmitters, are reflected by the lunar surface, we can determine that the surface is very rough, on the scale of 1 centimeter (0.4 inch) and less. Well before men walked on the

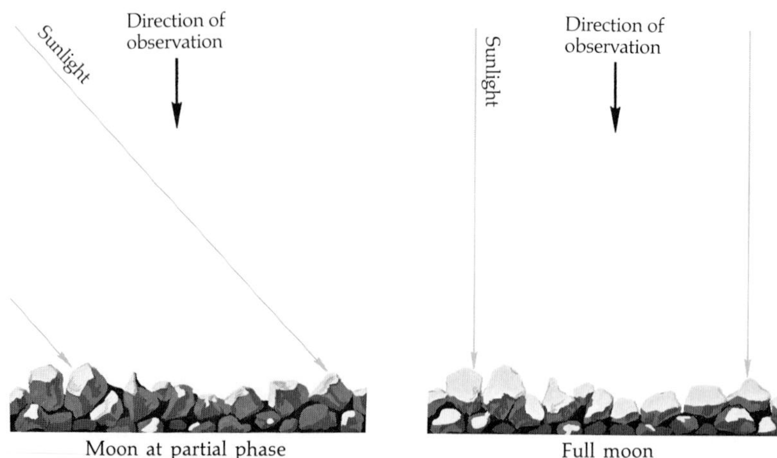

15-19 *Schematic diagram showing the increase of brightness of the lunar surface at full moon. At partial phase, the observer sees* both *illuminated surface and surface in shadow. At full moon,* only *illuminated surface is visible.*

moon the radar echoes showed that it must be largely covered with small rock particles and other irregularities (perhaps tiny crater pits) of these sizes.

Visible light from the illuminated lunar surface is, of course, reflected sunlight. But at longer wavelengths, electromagnetic radiation received from the moon is dominated by the thermal emission of the surface. Thus, further information about the lunar surface material came from measurements of its infrared and short-wavelength (millimeter and centimeter waves) radio emissions. Measurements of this radiation give the temperature of the emitting region, regarded as approximately an ideal radiator. The different wavelengths are emitted at different average depths,* so we can probe the first 30 centimeters (12 inches) or so of the lunar surface. These studies show that the surface temperature changes from about 373° K† to 123° K as the lunar day goes from noon to shortly after sunset. On the other hand, at depths of several centimeters, the change in temperature is already much less, and amounts to only a few degrees K below 30 centimeters. This provided strong evidence that the moon was covered with a very effective insulating layer, composed of some kind of rock dust. However, it left several questions begging: How deep is the dust layer? How compact

*The longer the wavelength, the greater the depth.

†Recall that the boiling and freezing points of water at sea level are 373° K and 273° K, respectively (page 112).

is the dust? Could a spacecraft land on it safely or would it settle in unstable sands, shifting due to the impact of landing?

The advent of the space age enabled scientists to test deductions about the moon, which were based on studying its visible, infrared, and radio properties, by directly sampling the surface. Further, we have seen the surface close up (Figure 15-20) and from photographs of the hidden side of the moon have found that although both sides have the same type of features, there are noticeably more highlands and fewer maria on the far side. A Soviet spacecraft, the *Automatic Interplanetary Station,* took the first pictures of the far side in October, 1959; Figures 15-21 and 15-22 show portions of the lunar surface not visible from earth, as later photographed by the *Lunar Orbiter* spacecraft.

The *Ranger* spacecraft (in 1964 and 1965) transmitted photographs of the lunar surface en route to hard landings on the surface. Then, in February, 1966, a Soviet probe, *Luna 9,* achieved the first soft landing on the moon and sent clear views

15-20 *Impressions on the lunar surface. (left) Astronaut's footprint from the* Apollo 11 *mission; (right) the padprint of the* Surveyor *spacecraft which landed on April 20, 1967 as photographed by the* Apollo 12 *astronauts on November 20, 1969. (National Aeronautics and Space Administration.)*

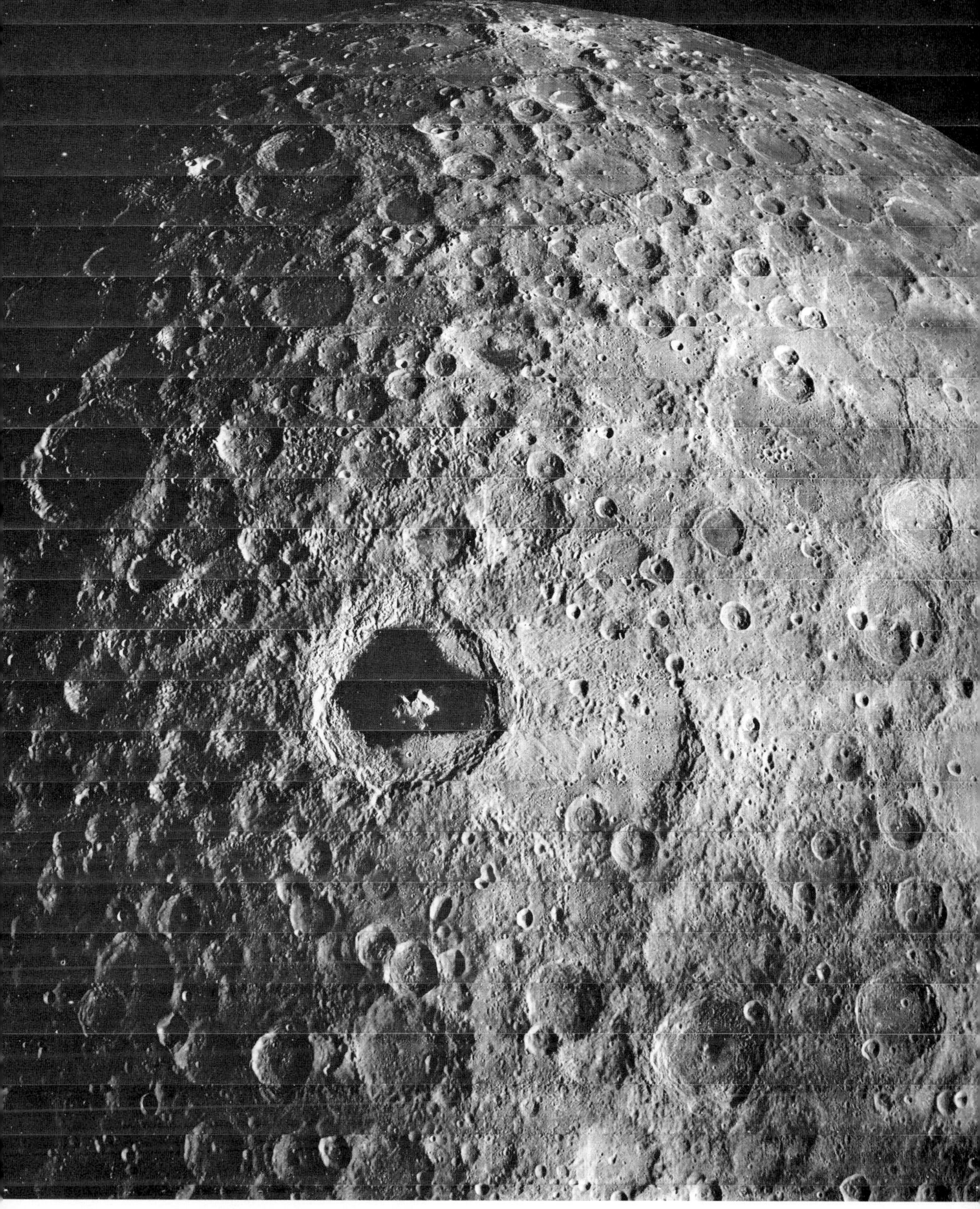

15-22 *View of the moon's far side showing the crater Tsiolkovsky. (National Aeronautics and Space Administration.)*

5-21 *The Oriental Basin as seen from lunar orbit. This region is on the moon's western limb and is not clearly visible from earth. (National Aeronautics and Space Administration.)*

of its surroundings back to earth. This showed that the dust coating of the lunar surface did not prevent it from providing a stable platform for a spacecraft. The first of the *Surveyors* had a soft landing in June, 1966; they obtained many photos (Figure 15-23). Later, several spacecraft were put into lunar orbit.

What did we learn from these activities? The soft landings of several space probes established that the lunar surface can support the weight of these craft, an important consideration in determining the nature of the material and the prospect that it would be safe for manned space vehicles. Close-up views of the landing pads and their impressions (or padprints) showed that the lunar surface material near the spacecraft was soil-like and was compactible. The panoramic views telemetered back

15-23 *Rubble-strewn lunar landscape as photographed by* Surveyor 7. *The rock in the foreground is about 0.6 meters (2 feet) across and cast a 1.2-meter (4 foot) long shadow at the time when the picture was taken. (National Aeronautics and Space Administration.)*

15-24 *Oblique view of crater Theophilus—about 100 km (60 miles) in diameter and about 6.5 km (4 miles) deep—as seen by* Lunar Orbiter 3. (*National Aeronautics and Space Administration.*)

to earth were of rock-strewn, cratered landscapes, and established that the presence of lunar craters extends down to pits of very small dimensions (Figure 15-23). The photographs from orbiters showed oblique views (Figures 15-24, 15-25) and vistas that increased our comprehension of the lunar surface. It was clearly very rough, and manned landing sites would have to be carefully chosen.

Manned exploration of the moon was achieved in July, 1969, by the *Apollo 11* crew of Neil Armstrong, Edwin Aldrin, and Michael Collins.* Armstrong and Aldrin gathered rocks from the surface and took core tube samples of the soil. They used a foil to collect solar wind particles (Figure 15-26), set up a reflecting device that was used for laser beam measurements by earth-based astronomers, and installed a lunar seismic station (Figure 15-27) to study moonquakes. The solar wind foil was rolled up and brought back to earth along with the rock and soil samples. They also took close-up stereo photographs of the appearance of the lunar soil.

The *Apollo 11* astronauts found that there is indeed dust on the moon; it seemed to be everywhere and to cling to everything,

---

*Collins, the Command Module Pilot, remained in lunar orbit while the other two astronauts explored the region of Tranquility Base.

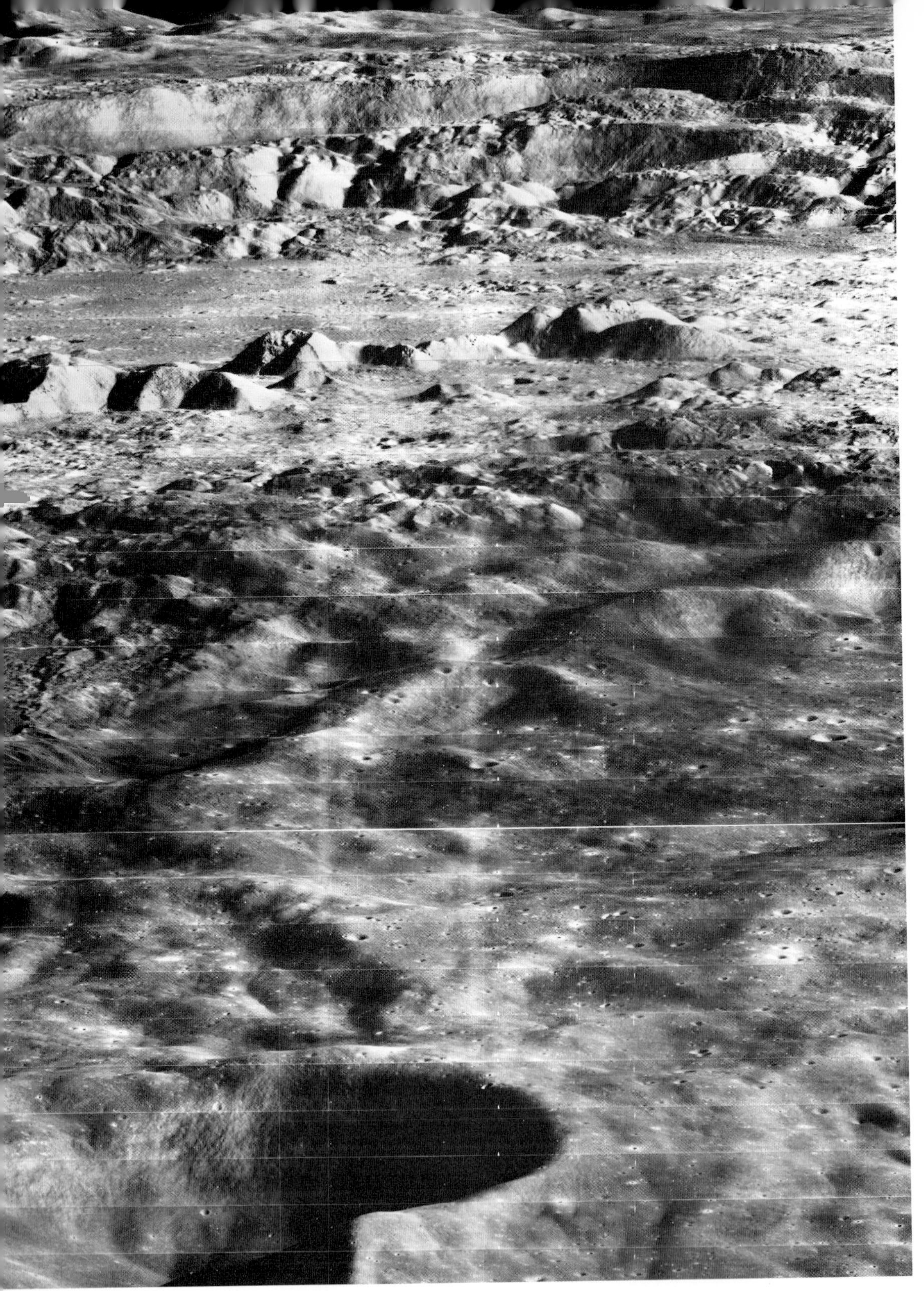

15-25 *Panoramic view of crater Copernicus. (National Aeronautics and Space Administration.)*

15-26 *The solar wind collecting foil placed on the lunar surface by the* Apollo 11 *astronauts. (National Aeronautics and Space Administration.)*

15-27 *The seismometer placed on the moon by the* Apollo 11 *astronauts. (National Aeronautics and Space Administration.)*

including their spacesuits and equipment. From the analysis of the rock and soil samples, and by study of visual and photographic observations on the moon, we know that the lunar surface in the vicinity of the landing spot in the Mare Tranquillitatis consists of a few centimeters of sandy soil lying atop a layer (the *regolith*) of larger fragments of broken rock that is several meters deep. Beneath the regolith is the *lunar bedrock,* which is exposed in the bottoms of craters. Many rocks lie on the surface, and are of two basic types: *igneous rocks* (including basalt) that were probably formed by the cooling of lava on or close below the lunar surface, and *breccia,* or rocks formed by the pressing together of small fragments of older rocks. In addition, there are iron meteoroid fragments and glass particles in the soil.

The igneous rocks, soil, and breccia were dated by radioactive methods similar to those described in Chapter 2. The soils and breccia gave ages up to about 4.5 billion years.* The igneous rocks gave smaller ages, ranging up to 4.0 billion years. These values, especially for the soils and breccia, are subject to further study and might be revised.

The shapes of the lunar rocks are especially interesting; exposed surfaces of rocks lying on the moon tend to have more rounded contours than the unexposed bottom sides. This suggests that the tops have been eroded or weathered by the impacts of tiny meteorites. Some of the rocks are partially covered by glassy crusts, and Armstrong discovered that in several craters of about a meter in diameter, there were droplet-shaped blobs of glass (larger than the tiny glass particles found in the soil samples). Similar droplets and glass crusts appear in the centers of smaller craterlets that were photographed with the stereo camera.

Several theories have already been proposed to explain the lunar glass. Thomas Gold ascribed the glass that is specifically found in small craters to melting of rock by radiation from extremely large solar flares. The point is that reflection from the crater walls would have concentrated the heating effect at the crater bottoms, where Armstrong found this glass. However, there is no proof that such giant solar outbursts actually occur. Another idea is that the glass was formed by rock melting due to the escape of hot gases from the interior of the moon. A third suggestion, which seems most reasonable to the authors, is that the various kinds of glass, including the soil particles and the rock crusts, originated when meteoroids fell on the moon,

*The oldest rocks found on earth have an age of about 3.5 billion years (Chapter 2).

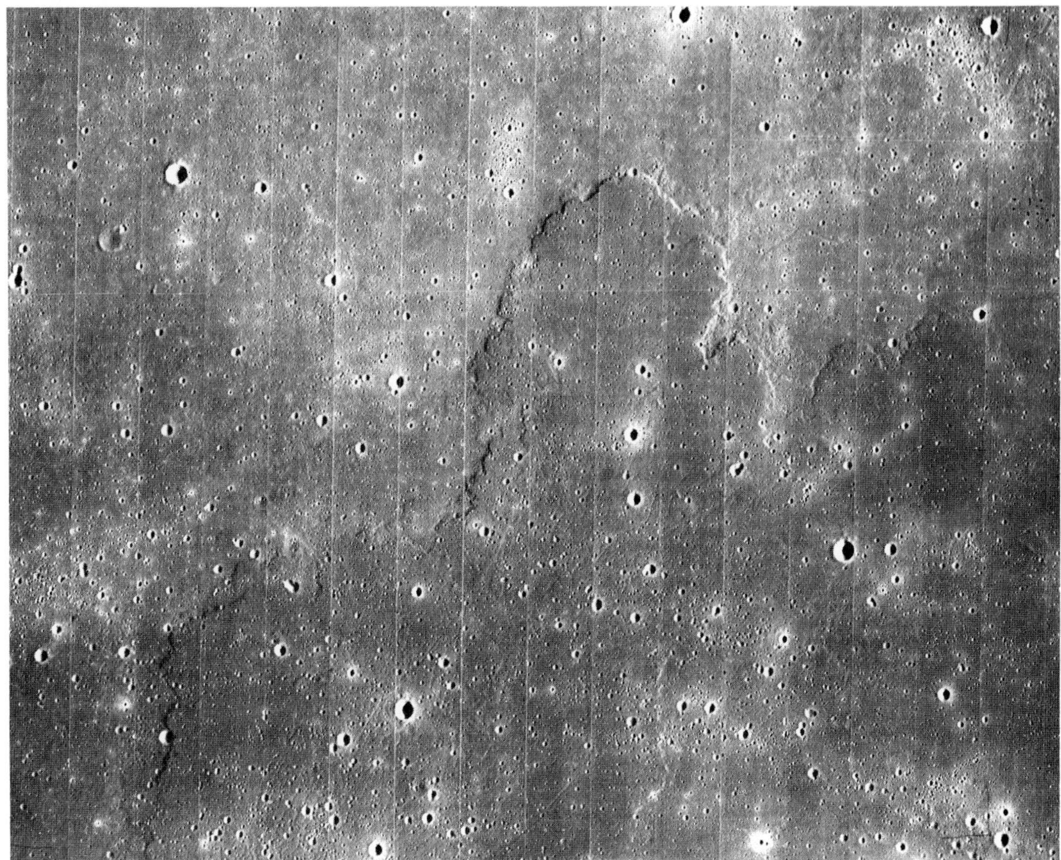

15-28 Lunar Orbiter 5 *view of Mare Imbrium. The area in the upper part of this photograph is at a lower elevation than the region in the lower part, which is apparently covered by a lava flow.* (*National Aeronautics and Space Administration.*)

heating the material they struck, and splattering the molten rock over the surrounding areas.

The rocks have the same bulk composition as the soil, consisting of minerals similar to those formed from magma within the earth. There are some differences from earth rocks; for example, the titanium content is higher in the lunar rocks and they have many glassy inclusions. The lunar lavas (Figure 15-28), as compared to terrestrial lavas, are found to have been very hot and very dry (no minerals with water bonds).* Such lavas would flow more easily than terrestrial ones; this may explain the flat appearance of the maria. There was no evidence of living, dead, or fossilized organisms in the lunar samples.

*Recall the discussion of molecular bonds in Chapter 5.

Placing seismometers on the lunar surface allows us to probe the interior of the moon, just as earthquake measurements are used to study the interior of the earth (Chapter 2). The lunar seismometers were much more sensitive than the usual terrestrial ones, which have sensitivity to weak quakes effectively limited by background noise caused by large storms on the oceans and by human activities. Compared to the situation on the earth, there was a remarkably low rate of quakes observed by the Apollo* seismometers, despite the great sensitivity of these instruments (they were able to detect motions of the surface as small as 1 Angstrom). However, several kinds of seismic events were observed: first, the shocks of meteoroid and spacecraft impacts; second, moonquakes which usually occurred when the moon was nearest or farthest from the earth. The latter type are thought to be triggered by the earth's tidal force, therefore they also involve an external influence. Since quakes on the earth are associated with geological processes driven by convective motions in the mantle (Chapter 2), the scarcity of moonquakes (other than those attributed to meteoroid impacts and the earth's tidal force) suggests that the interior of the moon is solid and thus relatively cool. Very recently, a third kind of moonquake, which occurs repeatedly at intervals of several hours during a period of a few days, was discovered. These *moonquake swarms* are not yet understood.

After the astronauts have returned to the command module, which has been circling the moon, the lunar module ascent stage can be sent crashing onto the lunar surface as a test. The *Apollo 12* ascent stage struck about 75 kilometers (47 miles) away from the *Apollo 12* seismometer and caused the moon to ring, vibrating at a rate of one cycle per second for about 40 minutes. Such behavior is rarely found on earth, and only after an exceptionally strong quake. (However, similar activity was recorded after the *Apollo 14* lunar module ascent stage impact and also after many of the natural events.) Several suggestions to explain the lunar vibrations were considered and rejected. The spacecraft impact could have kicked up lunar material into high trajectories, from which it rained down on the surface over many minutes of time. Or perhaps the impact started one or more landslides down crater rim slopes, which persisted for the observed length of time of the ringing. On closer examination, neither of these alternatives was tenable. A more plausible explanation that has been advanced for this

*Here we summarize some results from the *Apollo 11, 12, 14,* and *15* seismometers.

phenomenon interprets it in terms of inhomogeneous material or *rubble.* As we recall from Chapter 2, seismic waves are reflected or bent by changes in the medium through which they are propagating, and this circumstance allowed geophysicists to discover the various layers and discontinuities within the earth. If the lunar material were very inhomogeneous, some waves would propagate more or less directly through the medium and reach the seismometer quickly, while other waves would get reflected many times and take a long time to reach the seismometer. Thus, multiple reflections of seismic waves and a thick, inhomogeneous layer of rubble may explain the seismic events on the moon that seem to last so long.

What is the evidence on the temperature of the lunar interior? As mentioned above, the seismic data favor a relatively cold interior. Furthermore, the magnetic field carried by the solar wind* has been observed by space probes to *pass through the moon* with relatively little change. For reasonable possible compositions of the lunar interior, this would not be possible *unless* the central temperature were low (below about 1,000° K). A cold lunar interior also is indicated by the *shape* of the moon. The moon has three unequal radii, as measured (1) along its rotational axis, (2) in the direction of the earth, and (3) in the direction along the moon's orbit. The radii directed toward the earth and along the moon's orbit are, respectively, 1.1 kilometers (0.7 mile) and 0.7 kilometer (0.4 mile) longer than the polar radius. These differences imply large stresses in the interior that could not be supported if it were hot. In other words, a fluid or near-fluid would have adjusted itself to a spherical shape. Thus, we have several different lines of evidence for a cold lunar interior.

However, minerals in the rocks found on the lunar surface include some that must have crystallized at temperatures in the range of 1,300° K to 1,500° K. Thus, although we have arguments for a cold interior (less than 1,000° K), it appears that there must have been a hot surface layer (greater than 1,300° K) at one time. The evidence on the interior agrees with the *cold moon* theory of lunar origin, proposed by the chemist Harold

*The magnetometer placed on the moon by the *Apollo 12* astronauts detected a very weak magnetic field, about 0.13 per cent of the field at the equator of the earth. *Apollo 14* astronauts Alan Shepard and Edgar Mitchell used a portable magnetometer to measure the lunar field at two different locations, and found values of 0.14 per cent and 3.4 per cent, respectively, of the earth's field. These observations do *not* indicate an overall lunar magnetic field similar to the earth's and weaker, because if the moon had such a field, it would have been detected by the two magnetometers on the lunar orbiting satellite *Explorer 35,* which obtained negative results. Instead, the magnetic fields found near the *Apollo* landing sites are probably due to magnetized structures (such as large iron meteoroids) in their vicinities.

Urey. According to him, the moon formed by accretion of asteroidal-type matter. There does not seem to be a persuasive theory yet on the origin of the once hot outer layer. Its heating has been ascribed to the effects of impact, to a possible higher abundance of radioactive isotopes in the outer layer, and even to electrical effects. Thus, the thermal history of the moon is still an open question in several respects.

Further Exploration

The *Apollo 12* astronauts installed a number of instruments on the moon, including devices to study moonquakes, the solar wind, and magnetic fields. They landed on the Oceanus Procellarum in November, 1969, close by the location where *Surveyor 3* had soft-landed in April 1967 (Figure 15-29). There were differences in the regolith and rocks at the *Apollo 12* site, compared to those found at Tranquillity Base. For example, most of the rocks brought back by the *Apollo 12* astronauts, Charles

15-29 *The* Surveyor 3 *soft-lander being inspected by an astronaut; the* Apollo 12 *Lunar Module is in the background. (National Aeronautics and Space Administration.)*

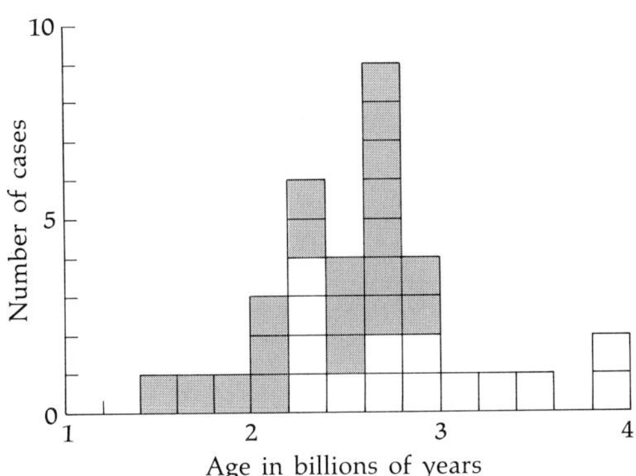

15-30 *The ages of 34 of the lunar rocks brought back by* Apollo 11 *(unshaded) and* Apollo 12 *(shaded) astronauts, as measured by the potassium-argon radioactive dating method. Each square represents one rock. The chart, a histogram, shows that ages between 2 billion and 3 billion years were most common in this rock sample.*

Conrad and Alan Bean, were about a billion years younger (Figure 15-30) than those retrieved by Armstrong and Aldrin, and the regolith near the *Apollo 12* site was thinner than that at Tranquillity Base. The *Apollo 14* astronauts made a geological traverse up the boulder-strewn slope of Cone Crater, found that breccia greatly outnumbered igneous rocks at their landing site in the Fra Mauro highlands (this was not true at the *Apollo 11* and *12* sites in the maria), and operated a thumping device to produce artificial moonquakes. Analysis of seismometer records of the P waves (Chapter 2) from these events showed that the Fra Mauro regolith was 8.5 meters (28 feet) thick and underlain by another layer of about 50 meters (160 feet) thickness. The *Apollo 15* astronauts rode their lunar rover to the edge of Hadley Rill and observed what appeared to be layers, tilted far from the horizontal, on the side of Mount Hadley.

It will be many years before the results of the various *Apollo* landings are fully studied. In addition to these manned expeditions, there has been a Soviet spacecraft that obtained a small soil sample and returned with it to earth, and also the *Lunokhod 1* robot that studied soil properties, photographed surface features, and survived on the lunar surface for 10.5 months after its November, 1970, landing. It is clear that lunar studies now fall in the geologists' domain and that future explorations should employ both automated devices and scientifically trained astronauts.

There are three distinct traditional ideas on the origin of the moon. (1) It could have been formed elsewhere and captured by the protoearth long ago. (2) It could have formed near the earth as a by-product of the same process that formed the earth. (3) It could have once been a part of the earth and separated from it by fission. Each of these ideas is open to objection when examined in detail, and recently this led to a fourth idea—that the moon was formed by the conglomeration of a considerable number of smaller satellites that were previously captured by the earth. This process may have been somewhat piecemeal, and several of the smaller satellites may have remained in orbit long after the moon was formed, subsequently colliding with it to produce some of the maria and the larger craters. Although this concept is persuasive, the origin of the moon is still a subject of considerable debate. Urey's cold moon theory (page 421) would seem to be compatible with either the first or the fourth of the theories listed here. As the evidence from direct exploration of the moon accumulates, we should be able to clarify our ideas about its origin.

The earth-moon system is changing. This fact is intimately tied up with the tides, as shown in Figure 15-31. The moon's gravitational attraction pulls the water on the side toward the moon a little away from the earth and the earth a little away from the water on the side opposite the moon. The general phenomena of tides are well known to many people, but are complicated by the local topography so that tidal effects are more dramatic in some locations than in others. The sun is also a tide producer. When the sun and the moon are on the same

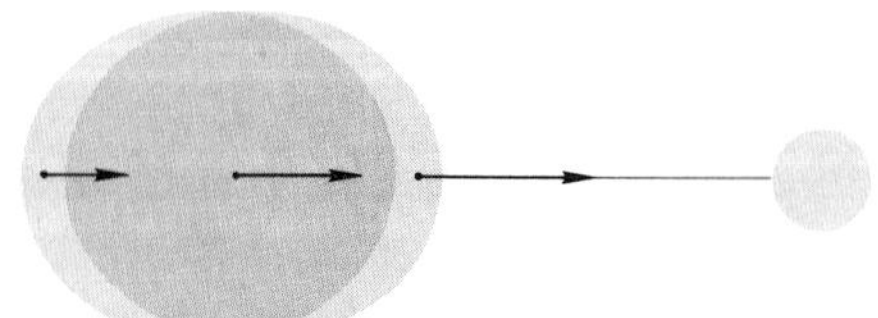

15-31 *Schematic diagram of the moon's tide-producing forces (shown by the heavy arrows). The differences in the forces exerted at different points in the earth produce the tides on both sides of the earth. View is from the plane of the moon's orbit around the earth. (Not drawn to scale.)*

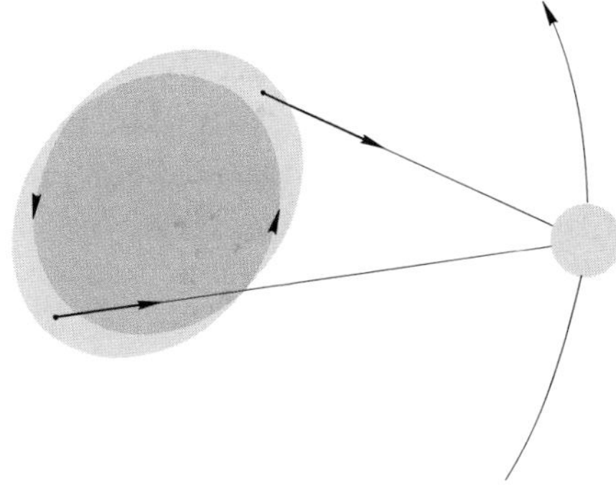

15-32 *Schematic diagram showing the earth's tidal bulges. The tidal forces are shown by the long arrows. The force on the part of the bulge nearer the moon is larger than the force on the more distant bulge and, hence, the earth's rotation is slowed. The view is drawn as seen from above the north pole of the earth. (Not drawn to scale.)*

or opposite sides of the earth, the highest tides (called *spring tides*) are produced. When the moon and the sun are 90 degrees apart, as seen from the earth, the lowest (*neap*) tides occur. One generally thinks of tides in the oceans, since those are the ones that we notice, but the same forces also cause them in the solid part of the earth.

If the earth were not rotating on its axis, its lunar tidal bulges would occur on the line directed toward the moon. However, it is rotating and this causes the earth's tidal bulge to precede* (in the direction of the earth's rotation) the earth-moon line (Figure 15-32). The pull of the moon on this tidal bulge is slowing down the earth's rotation at the rate of about 0.002 seconds per century. This means that the day was shorter in the past. In Devonian times (some 350 million years ago) the day must have been about 22 hours long, and this value can actually be checked. Some reef corals show yearly bands which are composed of about 365 finer rings, which biologists have shown correspond to a daily growth. Fossil corals of a similar type from Devonian time show 400 fine rings per yearly band.

*This happens because it takes a finite amount of time for the tidal bulge to form or to recede, and during this time, the earth's rotation carries the bulge away from the earth-moon line.

Since there is no evidence for (or reason to suspect) any real change in the length of the year, each day must have been shorter, and we can calculate the length of the day at that time as about 22 hours.

Previously, the earth probably exerted a tidal force on a faster rotating moon, causing it to slow down to its present rotation rate of once per month. Tides produced in the moon by the earth are the reason that the moon keeps the same face toward the earth.

The earth also has an *equatorial* bulge (Chapter 2), which is a feature of its shape and not a changing effect of tidal forces. The pull of the moon on this bulge is largely responsible for the phenomenon of precession. The earth's rotation axis is tilted by 23.5 degrees from the direction at right angles to the plane of the ecliptic. Hence, the equatorial bulge is not in this plane, although the pull of the moon and other solar system bodies tries to bring the bulge into it. The bulge is not pulled into the plane of the ecliptic any more than the earth's pull brings the moon down from orbit. Instead, the result is the precession of the earth's rotation axis around an axis perpendicular to the plane of the ecliptic with a period of 26,000 years (Figure 4-10; see the discussion on page 74).

The slowing down of the earth as revealed by the coral studies leads to an interesting situation. As mentioned on page 308, the quantity *angular momentum* is conserved in any isolated system. The angular momentum can correspond to an object's rotation on its axis, or its revolution in orbit, or both. If the earth's rotation is being braked by the tides, its angular momentum is decreasing and must reappear somewhere else. Presently, it is taken up in the moon's orbit around the earth: the moon's orbital angular momentum is being gradually increased and the moon is slowly spiraling outward, away from the earth. As it moves away, the orbital period increases in length according to Kepler's Laws. Thus, the day and the month (in its original sense) are both lengthening, but the day at a faster rate. Some 5 to 10 billion years from now, the day and the month will be equal at approximately 43 of our present 24-hour days. At that time, the earth will keep the same face toward the moon and the lunar tides will no longer change. Beyond this point, the further evolution of the earth-moon system is a much more speculative subject. Even when the moon's tidal influence on the earth stops, tides will still be produced by the sun. This will tend to slow the earth's rotation even more and make the day longer than the month. When this happens, the lunar effects must start up again but in the opposite direction, shortening the day and bringing the moon closer to the earth. The situation is complex because the solar-produced tides in the earth slow down the rotation while the lunar-produced tides speed it up. The detailed calculations indicate that the moon will spiral in toward the earth, continually coming closer. Eventually, it will get so close that the difference in the earth's gravitational pulls on the near and far halves of the moon will literally tear the moon apart. When this happens the lunar remains might take the form, at least in part, of a ring of particles around the earth, very much like the rings of Saturn.

THE NEW MARS

The *canals* of Mars, discovered by Giovanni Schiaperelli at Milan in 1877, were up until very recently the subject of great controversy. The Italian term used by Schiaperelli, *canali,* means "channels," and thus admits of a natural explanation, but it was generally mistranslated as "canals," meaning artificial waterways. Percival Lowell in Arizona became the greatest advocate of an artificial origin. Although a few of the largest canals appear as dark, fuzzy streaks on photographs, in general the results of photographic exposures* were not very clear (Figure 15-33) and most observers made sketches of the reddish martian surface while viewing it "live" through their telescopes (Figure 15-34). This task required experience, visual acuity, and perhaps some imagination. The observers agreed about the larger dark and light areas on the planet, but their maps of the long, narrow canals were very different (indeed, some people did not see canals). Some astronomers reported that groups of canals tended to radiate from (or meet at) certain spots, which were compared to oases in a desert. Lowell drew an almost rectilinear system of canals that resembled nothing less than a vast engineering project which, some believed, the enterprising Martians had constructed to bring water down from the polar caps to irrigate the temperate and equatorial regions.

Mars has seasons for the same reason that the earth has them (Chapter 4); its axis is tilted about 24 degrees to the plane of its orbit. As the seasons change so do the polar caps, as the early observers noted. When it is summer in the northern hemisphere of Mars, the north polar cap is small, but in the southern hemisphere, winter prevails and the south polar cap is large. Soon, as spring comes to the southern hemisphere, the

*Full disk views taken by Mariner spacecraft are shown in Figure 15-35.

15-33 *Earth-based photograph of Mars. (Courtesy of R. B. Leighton, California Institute of Technology.)* See also Plate 4 following p. 260.

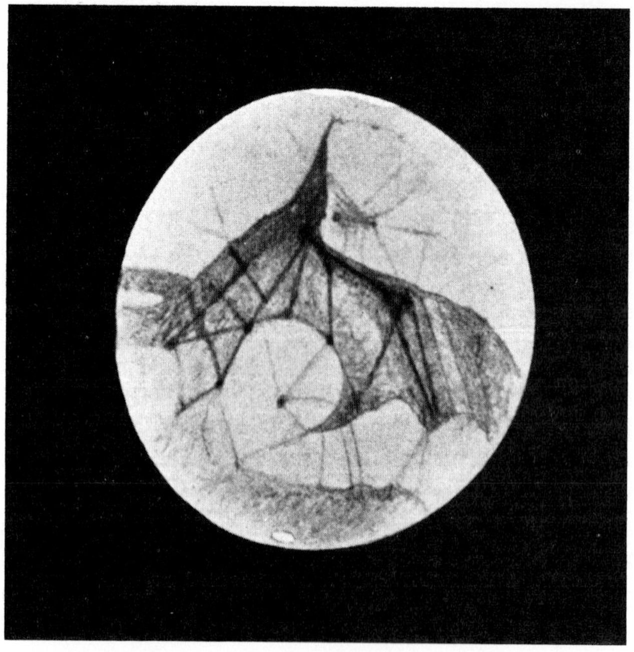

15-34 *A drawing of Mars by Percival Lowell showing canals. (Lowell Observatory photograph.)*

a

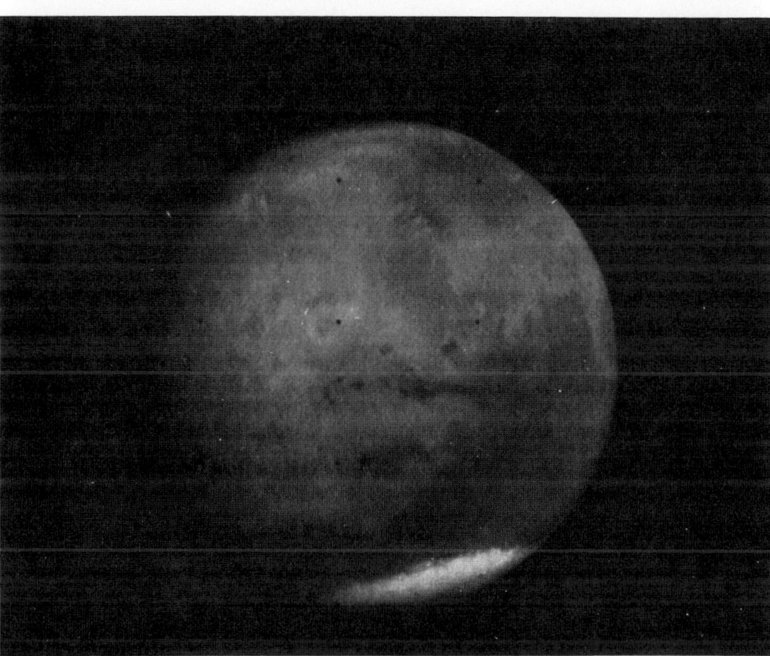

b

15-35 *Full disk views of Mars obtained by a* Mariner 4 *television camera on August 4, 1969. (a) As seen from a range of 711,000 km, and (b) as seen from a range of 535,000 km. (Courtesy National Aeronautics and Space Administration, Jet Propulsion Laboratory.)*

south polar cap is seen to decrease in size, retreating toward the south pole. At the same time, however, the appearance of the southern hemisphere changes; the dark areas in the region just outside the polar cap get darker and the effect seems to spread gradually toward the equator. This progressive enhancement of the contrast of the dark areas was called the *wave of darkening.* (The same phenomenon happens half an orbital period later, in the Martian northern hemisphere, when spring occurs there.)

The observation that the wave of darkening proceeded from polar latitudes toward the equator, and that it coincided with the decrease in the size of the polar cap led to the idea that the two effects were causally related. It was natural to believe that the polar caps were formed in whole or in part of frozen water, and that as the cap receded in the spring, water vapor was carried equatorward by the wind, or liquid water spread down over the planet, thanks to the convenient network of canals. The water might cause the darkening by producing physical or chemical changes in the rock and soil of the Martian surface, or it might be triggering a much more interesting phenomenon. Perhaps the wave of darkening indicated the presence of life on Mars; that is, perhaps the water was rejuvenating vegetation that had lain dormant or dead throughout the winter season. A difficulty was that the spectra of Mars showed little evidence for water; some astronomers believed the polar caps might actually be dry ice, frozen carbon dioxide.

Thus, before the advent of the deep space probe, astronomers studying Mars were faced with three basic questions: Was there life (at least vegetation) on Mars? What were the canals—natural rock channels, artificial canals produced by an extraterrestrial civilization, or figments of the zealous observer's imagination? And what were the polar caps? Now, as a result of several closeup studies of Mars by *Mariner* spacecraft, ideas about the planet have narrowed and centered about new evidence. The polar caps are indeed thin layers of frozen carbon dioxide; the planet is pock-marked with craters, much like the moon; only a very few of the largest canals correspond to actual surface features—these are elongated dark areas of still unknown nature. Certainly there is no planet-wide array of canals visible on the pictures telemetered back to earth from the *Mariners.* Most of the canals that were reported by the visual observers were apparently nothing more than chance arrangements of small dark patches, seen with inadequate resolution, that the observers interpreted as line-like features.

The Martian craters were discovered on pictures obtained by Robert Leighton of the California Institute of Technology, with a television camera mounted on *Mariner 4*, the first successful Mars probe. *Mariner 4* approached within 9,900 kilometers (6,200 miles) of the planet after launch in November, 1964. More detailed pictures were obtained by Leighton and his colleagues with instruments carried aboard the *Mariner 6* and *Mariner 7* spacecraft (Figure 15-36) in the summer of 1969. Figure 15-37, a *Mariner 6* close-up, shows how much some parts of Mars resemble the moon. Some of the craters in the upper portion of a *Mariner 7* picture (Figure 15-38) occupy the region of Mars where earth-based astronomers had sketched a canal named Cantubras. A mosaic of wide-angle pictures taken by *Mariner 7* (Figure 15-39) shows a portion of the south polar cap, and proves that the polar ice is not very deep as we can clearly see crater outlines. A close-up view of several of the craters just outside the polar cap edge near the left side of Figure 15-39 is shown in Figure 15-40; "snow" or "ice" is clearly seen within the craters. Measurements of the intensity and spectrum of

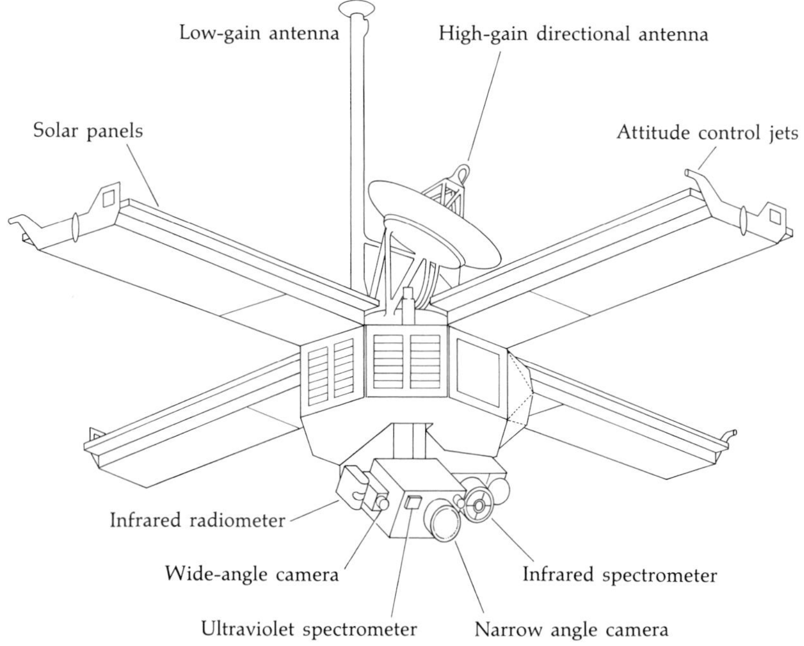

15-36 *Sketch of the Mariner spacecraft used to obtain close-up pictures of Mars. (From* The Surface of Mars *by R. B. Leighton.*

15-37 *A cratered portion of the Martian surface (103 km wide by 77 km high) as transmitted by* Mariner 6. *(Courtesy of the National Aeronautics and Space Administration, Jet Propulsion Laboratory.)*

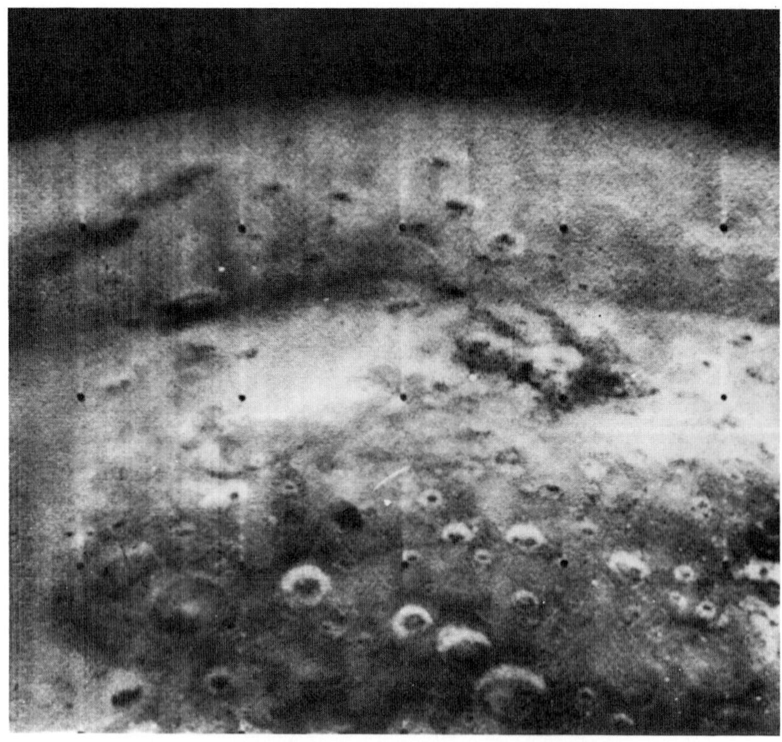

15-38 Mariner 7 *picture of a region on Mars. (National Aeronautics and Space Administration.)*

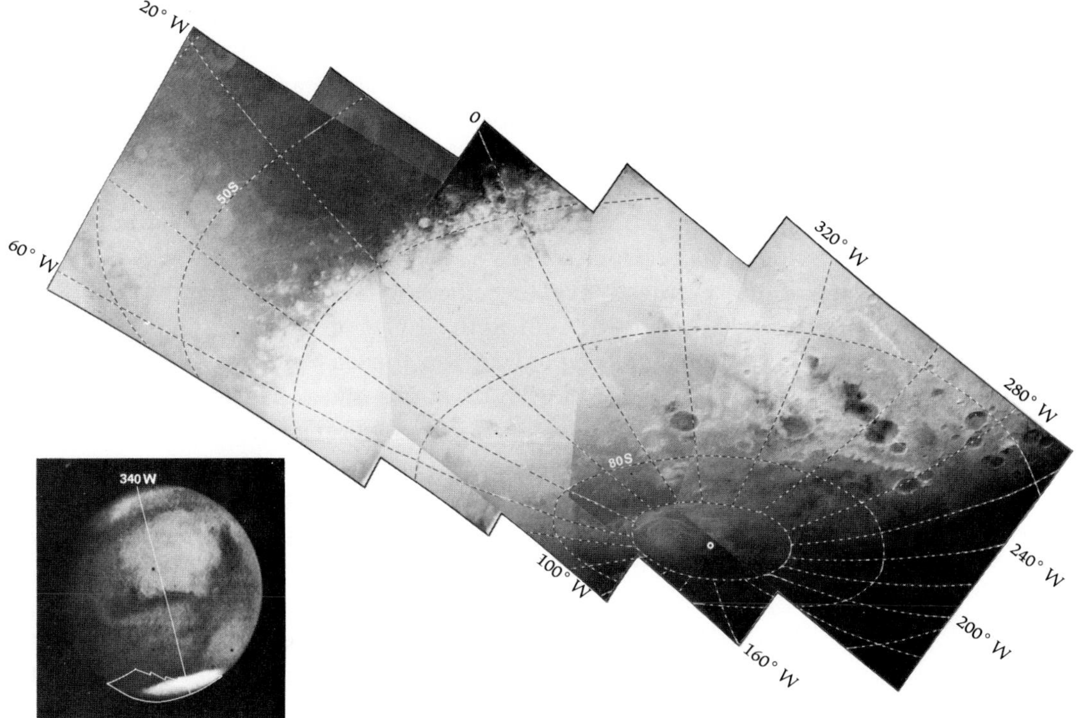

15-39 *Mosaic of Mariner pictures of the Martian south polar cap, showing craters and "snow"; Martian latitude and longitude are marked. (National Aeronautics and Space Administration.)*

infrared radiation from polar regions were made by instruments carried on *Mariner 6* and *Mariner 7*. These show that the white polar cap material is frozen carbon dioxide.

The quantity and composition of the thin Martian atmosphere can be estimated in several different ways: (1) spectrograms taken with ground-based telescopes show that it is largely CO_2 with a trace of water vapor; (2) Mars is fairly bright in the violet due to Rayleigh scattering in the atmosphere* (recall that Rayleigh scattering is more efficient at the shorter wavelengths, Chapter 8). Photographs of Mars in the violet show regions in the atmosphere while photographs in the infrared show the Martian surface. Visible light pictures show both surface and atmospheric phenomena. The amount of atmospheric gas required to produce the observed violet brightness can be calculated. (3) The atmosphere can be probed by radio waves. Item (3) is the *occultation* technique, in which a radio telescope on

*These first two observations were made by conventional methods with earth-based telescopes.

earth records the changes produced in the radio transmissions of a *Mariner* spacecraft as it passes behind the atmosphere of Mars and finally is eclipsed by the planetary disk. The observed effects on the radio waves (they are refracted) can be used to determine the atmospheric structure of Mars. (4) Molecules in the upper atmosphere of Mars emit a faint airglow (just as they do no earth, see page 372), including ultraviolet emission lines. A spectrometer on *Mariner 6* measured the wavelengths of these lines, revealing the presence of carbon monoxide, oxygen, hydrogen, and ionized carbon dioxide.

The surface temperature on the day side of Mars is about 250° K and the density of atmospheric gas is very low; the surface pressure is only about one per cent of the earth's atmospheric pressure at sea level. However, strong winds must occur on Mars because extensive dust storms have been observed. The density of the Martian atmosphere may also vary with the change in the seasons.

In November, 1971, *Mariner 9* went into orbit around Mars. The first data showed that Mars has an equatorial bulge, that both of its moons are oblong in shape and cratered, and that some of the Martian craters resemble volcanic craters on earth.

Despite the thin atmosphere, the dearth of water and oxygen, the exposure (thanks to the lack of atmosphere) to the ultraviolet radiation, cosmic rays, and meteorites that our own atmosphere

15-40 *Close-up view of "snow"-filled martian craters. (National Aeronautics and Space Administration.)*

largely shields us from, one still wonders: Is there life on Mars? The wave of darkening phenomenon shows that changes occur of a type that vegetation growth might explain. So far as advanced or intelligent life is concerned, we do not see any evidence of civilization (roads or cities) in any of the roughly 200 close-up television photographs taken by *Mariners* 4, 6, and 7. On the other hand, a study of weather satellite photographs of the earth showed that one had to examine thousands of pictures in order to find the slightest evidence of our civilization, and these satellite pictures had a resolution better than that obtained with *Mariner* 4, although somewhat poorer than that of the best *Mariner* 6 and 7 photos. In the early 1960s, several scientists believed that they had identified lines of various compounds of nitrogen and oxygen in spectra of Mars taken with earth-based telescopes, and that these compounds were present in amounts sufficient to prevent the existence of life as we know it. Their spectral observations were subsequently reinterpreted in terms of other chemical substances, and further, as the American astronomer Carl Sagan pointed out, there is a greater concentration of one of these compounds, nitrogen dioxide (NO_2), over Los Angeles than had been claimed for Mars. Sagan and his colleagues concluded: "Life in Los Angeles may be difficult, but it is not yet impossible. The same conclusion applies to Mars." At the present time we would have to agree with this—conditions are hostile on Mars, but it is not hard to imagine that at least primitive life forms might exist there.

SPACE TRAVEL AND LIFE IN THE UNIVERSE

The possibility of life elsewhere in the universe has become more relevant as our capability for space travel has developed. Our nonunique situation has been emphasized throughout this book. We have every reason to expect habitable planets around other suns in this galaxy and throughout the universe.

Just what are the conditions that make a planet habitable? The crucial parameter appears to be the temperature. *Our life chemistry depends on the production of complex carbon compounds and the presence of liquid water as a solvent.* The complex carbon compounds required generally do not form at temperatures much above the boiling point of water. Active life as we know it is then truly restricted to approximately the range of liquid water. On this basis we expect that the temperatures on Mercury and Venus rule them out as abodes of life. Water itself is probably absent on the moon and on the asteroids; thus, life as we know it should not arise on these bodies. The Jovian planets also appear unattractive because of their low temperatures. We are on less certain ground at the lower temperatures because other substances, such as ammonia, could play the role of the liquid solvent. Also, there is a need for caution because there is always the possibility of an entirely different life chemistry where, for example, silicon replaces carbon as the basic biochemical constituent. On the other hand, physicist Robert Jastrow has argued for basic similarities in the evolution of extraterrestrial life forms. As he pointed out, "You can't have your eyes on your tail; you have to see where you're going." Mars is somewhat on the cold side, but is the prime possibility for life in the solar system outside the earth. This fact is a strong motivation for space research. Obviously, the discovery even of primitive plant life on Mars would be an event of immense scientific interest.

If, as we have argued, planets are common and the building blocks of life can originate by natural processes in the oceans and atmosphere of a newly developing planet, life should be widespread throughout the universe. Some civilizations should be more advanced than we are and some less. This opens the possibility of interstellar communication with and visits to or from other civilizations. Surely a discovery in this area would be the most impressive event in human history. Some persons claim to be already carrying out their own private communications efforts with other civilizations; we refer to the reports of *Unidentified Flying Objects* (UFO's) and contacts with extraterrestrials traveling in them. Most of these reports are easily explained away and some appear to have been outright frauds. However, a small number cannot be dismissed (this is *not* the same as saying that they are proven!) and we should keep an open mind on the subject. As we shall see below, interstellar travel may be a long-term requirement for the survival of advanced civilizations.

The problem is simply this. In about 6×10^9 years the sun will evolve into a red giant star and the earth will be uninhabitable. The requirement for liquid water, as opposed to ice or steam, on a planet like the earth demands a solar constant between 0.5 and 4 calories per square centimeter per minute. When the sun becomes a giant star, its luminosity will increase by 10^2 to 10^3 times. As the process begins, the sun's surface temperature will decrease and its surface will appear redder than now and its radius will increase. On earth the oceans will boil away and eventually the top layer of the atmosphere, the *exosphere,* will be so hot that all atmospheric gas molecules will escape in a relatively short time. The probable length of the sun's red giant stage is thought to be several hundred million years. At its greatest extent the sun's surface will contain the orbits of Mercury and Venus and may even reach the orbit of earth. Our temperature requirements at this time could be met on the satellites of Saturn, Uranus, and Neptune. Colonies could be established in relative comfort to ride out the sun's giant phase.* Even if the earth remained afterward, there would be little point in returning to the charred cinder that was our birthplace because the sun would be on the road to becoming a white dwarf, a star far too dim to supply our energy requirements.

Thus, the onset of the sun's red giant phase will be, in effect, man's eviction notice from this solar system. At that time interstellar travel and the colonization of planets circling other stars will move from the realm of science fiction into a necessity for the survival of the species, if we are intelligent enough to last until that time.

THE COST OF ASTRONOMY AND SPACE RESEARCH

Since we have discussed the exploration of space in this chapter, and even the eventual need to move the population of earth to some planet of another star, it seems appropriate to discuss briefly the cost of space activities. The monies appropriated for science and space exploration are often discussed in connection with our many social and ecological problems. It would be surprising if the reader did not question the cost of the space program described above.

Let us begin by recalling the situation in the latter part of the nineteenth century. Some support of astronomy was undertaken by the government, such as the operation of the U.S. Naval Observatory in Washington, D.C., but much of astronomy was supported by private funds. In this matter the United States carried on the English tradition. Private donors supplied the cash to found Lick Observatory on Mount Hamilton near San Jose, California, in 1888. Similarly, funds were given for the Yerkes Observatory in Williams Bay, Wisconsin, in 1897. Since the prestige of the donor was involved, the quality of the telescope generally increased with each succeeding observatory.

In the twentieth century the Mount Wilson Observatory (near Pasadena, California), the Mount Palomar Observatory (north of San Diego, California), and the McDonald Observatory (near Fort Davis, Texas) were founded with private funds.

*Not for everyone. Even very optimistic calculations suggest that the rate at which people could be launched into space is less than the rate at which they are presently being born on the earth.

The situation began to change in the 1950s when the creation of two major observatories was made possible by funds from the National Science Foundation. The Kitt Peak National Observatory with optical stellar and solar telescopes was built near Tucson, Arizona, and the National Radio Astronomy Observatory was established at Green Bank, West Virginia. These two installations provide facilities for use by all qualified astronomers, including graduate students. More recently, the Kitt Peak facility acquired a sister installation for observing southern skies, the Cerro Tololo Inter-American Observatory near La Serena, Chile.

Other astronomical research was sponsored by branches of the military, such as the Office of Naval Research and the Air Force Cambridge Research Laboratories. In particular, the Air Force developed the Sacramento Peak Observatory, Sunspot, New Mexico, whose work is wholly in the domain of unclassified basic research on the sun.

So far as astronomy is concerned, the transition from largely privately supported "little science" to largely federally-supported "big science" was very much the result of the beginning of the space age in 1957. The launch of *Sputnik 1* and the realization that the United States was not the leader in some vital aspects of space science came as a shock and, to many people, a challenge. Money was appropriated to develop programs to remedy the situation. The principal recipient of this money was the National Aeronautics and Space Administration (NASA), founded in 1958. Federal budgets are organized on the basis of a fiscal year, meaning the year ending on June 30. For example, the fiscal year 1960 ended on June 30, 1960. The NASA budget began in fiscal 1959 at 330 million dollars and rapidly increased to a peak of 5.25 billion dollars in fiscal 1965. Since then the NASA budget has decreased each year; the request for fiscal 1971 was down to 3.40 billion dollars, and the appropriation was a smaller sum. Obviously, most space science is supported through the NASA budget, but NASA is also the principal source of funds for astronomy as a whole.

What relationship does the NASA budget bear to the entire federal budget? The complexities of the federal budget cannot be underestimated, but for each year a summary is prepared by the Executive Office of the President, Bureau of the Budget (now reorganized as the Office of Management and Budget), entitled *The Budget in Brief.** Figure 15-41 (from this publication for fiscal 1971) shows the estimated spending by function. The

*Available from the Superintendent of Documents, U.S. Government Printing Office, Washington, D.C. 20492.

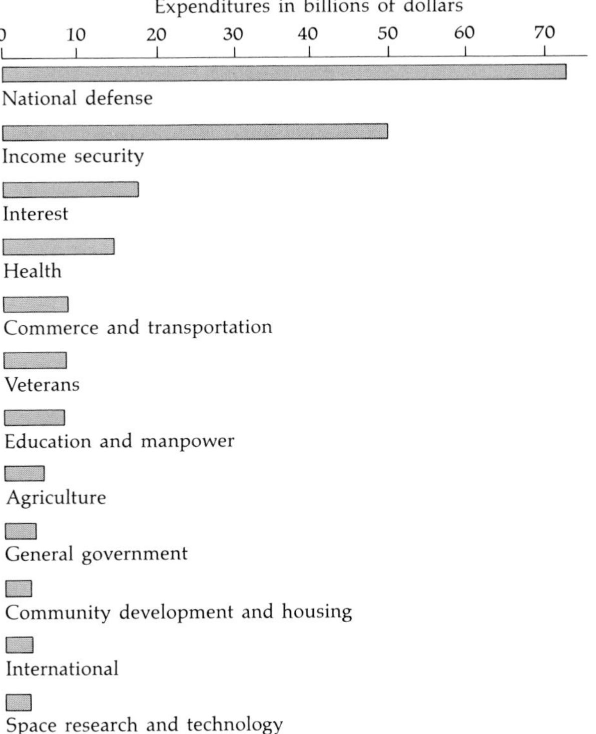

15-41 *Items in the United States Federal Budget as proposed for fiscal year 1971, by function.*

total budget comes to about 200 billion dollars. Most of the items are fairly straightforward, but some clarification is desirable. The total budget includes substantial amounts of trust fund expenditures, such as social security and medicare payments. Here the Government acts essentially as an insurance company; these are not tax revenues. Total trust funds in the fiscal 1971 budget come to approximately 55 billion dollars.

A very large slice of the budget goes for current national defense activities and the obligations due to past conflicts, as represented in the outlays for veterans' benefits and the interest on the national debt, which results largely from deficit spending during war years. Viewed this way, 70 per cent of the tax dollar goes to pay for present and past defense costs. A discussion of the arguments pro and con concerning the various expenditures in the Federal Budget is beyond the scope of this book. Nevertheless, the figures presented in Figure 15-41 will serve as the beginning of any meaningful discussion in this area.

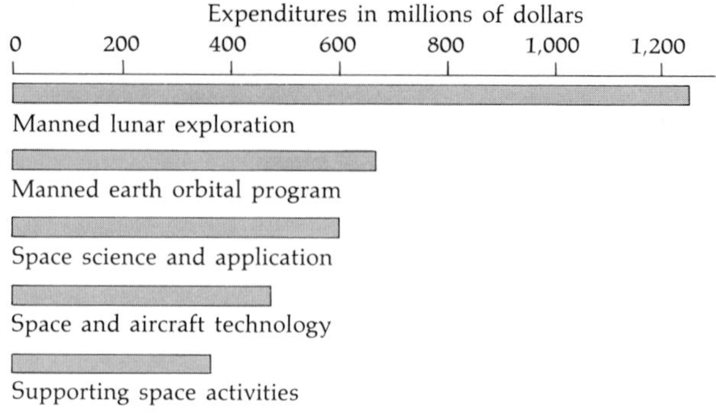

15-42 *Breakdown of items for Space Research and Technology in the U.S. Federal Budget, as proposed for fiscal year 1971.*

The breakdown of expenditures for Space Research and Technology, as it was proposed by NASA for fiscal 1971, is given in Figure 15-42. The largest expenditures are in the category of "Manned Lunar Exploration"; that is, the Apollo Program. The category "Space Science and Application" at 612 million dollars pays for unmanned astronomical satellites, such as OSO and OAO (Chapter 14) and for space science and astronomy. By contrast, the other major non-military supporter of physical science is the National Science Foundation (NSF) with a total estimated budget of 528 million dollars in fiscal 1971. The NSF sponsors the national observatories mentioned on page 438 and a good deal of research at university observatories, but its funds must go to support many other scientific disciplines besides astronomy.

The opinions expressed here are our own, but the authors cannot claim to be unbiased. (We have taught in universities and have both worked for a national observatory and for NASA.) We believe that research in astronomy and the exploration of space are related, desirable, and important activities that transcend national boundaries and express much of the best in man. They challenge thought and open new horizons for us today as they did for men in Galileo's time. It has become a cliche, and yet we feel bound to repeat it, that the *Apollo 11* moon landing proved that men can commit themselves to a seemingly impossible task and accomplish it. There is a great deal that needs to be done here on earth, and there are many competing priorities, human and environmental; nevertheless, we believe that there is an important place for astronomy and the exploration of space.

16

Problems in Modern Astronomy

We know a fair amount today about our local portion of the universe, the solar system. And we have some idea about the nature of stars and the makeup of our own galaxy. Some of the questions that dominated astronomical and philosophical thinking for a long time have apparently been answered; for example, we know roughly how old the earth is and that its location is certainly not central to the scheme of things. We have good reason to believe that we exist in an expanding universe. On the other hand, many great problems remain, and these include questions that in some cases could not even have been asked before the development of modern astronomy, physics, and mathematics. Some of them have been singled out for presentation in this chapter; they are among the "hot topics" that have been vigorously debated at astronomers' conferences in the past few years. Although far from settled, and perhaps not even very well understood, these problems are included here to give you a taste of what lies beyond the elementary survey presented thus far—what is happening in astronomy today.

THE GENERAL THEORY OF RELATIVITY

Astronomy, as we have seen, attempts to describe the matter and radiation that exist in space and to explain the physical processes that occur among them. A related and more fundamental problem, one that is more easily raised than solved here, is: What is space? To Newton and many other scientists and philosophers, we existed in a three-dimensional universe; an important property of space was that light traveling through the vacuum moved in a straight line. In the twentieth century Albert Einstein revolutionized our conception of space by describing it with the aid of an additional dimension—time—and by suggesting that space has curvature due to the presence of matter. In his theory, a light wave traveling past a large concentration of matter, such as a star, would move in a curved path due to the gravity of the star. In most cases this would be an extremely small effect, so that very precise measurements would be necessary to distinguish the curved light paths from straight lines. Einstein's *General Theory of Relativity* represented an attempt to describe the nature of space-time, and it related the properties of space to the presence of matter.

Several types of astronomical observations have been made in order to distinguish between the pre-Einstein or Newtonian space and the theory of general relativity. Chief among them are measurements of the deflection of starlight near the sun and of the properties of Mercury's orbit.

Gravitational Deflection of Light

If Einstein were right, the path of a light wave would be slightly curved as, coming from a distant star, it passed by the sun. Ordinarily we cannot see stars near the sun, due to the great brightness of scattered sunlight in the atmosphere. However, stars *are* visible during a total eclipse. Since the light rays passing close by the sun would be bent (Figure 16-1), this would result in an apparent change in the position of the star, as compared with its position observed when the sun was not in the picture. The amount of the shift predicted by Einstein's theory was 1.75 seconds of arc for a star located at the solar limb. Observations of this effect have shown that it does occur, and that it agrees with the numerical prediction within an accuracy of about 10 per cent. Since the relevant pre-Einstein laws of physics (called *Newtonian theory* for brevity) do not account for such a shift, the starlight deflection observations indicate that general relativity is a better theory.

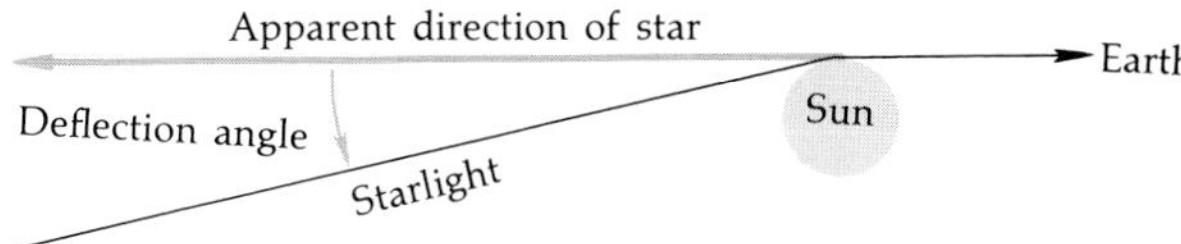

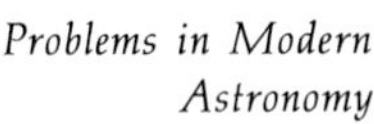

16-1 *Schematic diagram showing the deflection of starlight near the sun.*

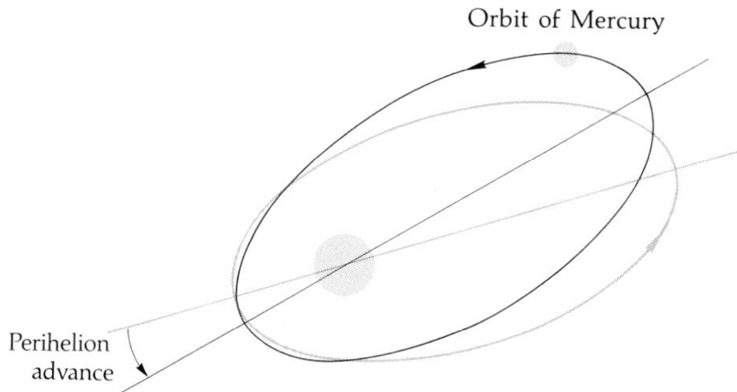

16-2 *Diagram illustrating the advance of the perihelion of Mercury.*

Perihelion Advance of Mercury

As seen from the earth, the orientation of Mercury's orbit slowly changes, as shown in Figure 16-2. A convenient way of describing the amount of change is in terms of the motion of the perihelion point about the sun; this is called the perihelion advance of Mercury. It amounts to 1°33′20″ per century. Most of this quantity is an apparent effect, due to the precession of the earth (Chapter 4), and about 10 per cent of it is due to gravitational forces exerted on Mercury by the other planets. However, the observed perihelion motion of 1°33′20″ per century is 43″ per century greater than can be accounted for by these two effects that are known from Newtonian theory.

Several possible causes for the excess perihelion motion of 43″ per century were suggested. They included gravitational force due to (1) a hypothetical planet* located between Mercury and the sun, or (2) dust and gas in the solar system. Another possibility was that the sun is not a perfect sphere but is fatter at the equator than across the poles (that is, oblate), in which

*Called Vulcan.

case its gravitational effect on Mercury would also produce a perihelion advance. In fact, careful searches have failed to show any evidence of a planet inside the orbit of Mercury. Further, observations of the interplanetary gas and dust show that there is not enough of it (in terms of total mass) to produce the excess perihelion advance of Mercury. On the other hand, an additional perihelion motion of Mercury *is predicted* when the gravitational effects are calculated according to the general theory of relativity, and the calculated value amounts to 43.03″ per century in the correct direction! This fine agreement with observation has been regarded for some time as the best evidence favoring Einstein's work over Newtonian theory.*

Is General Relativity Sacred?

Careful readings of Einstein's works suggest that in some respects he formulated general relativity with substantial reliance on philosophical ideas about the nature of space, time, and gravitation, rather than on strict mathematical derivations from established facts. Nevertheless, Einstein was able to account for the excess perihelion motion of Mercury, and he was able to predict a previously unknown effect, the gravitational deflection of light, which has since been observed. To the average scientist, struggling to take the next step forward in a narrow field of specialization, the triumph of Einstein's theorizing, based on broad philosophical considerations, seems to be one of the most profound achievements of the human mind.

Despite what we just said, successful verification of Einstein's predictions does not prove the theory of general relativity. What it does is to establish general relativity as the best available description of space, time, and gravitation that has been tested by observation. Since Einstein, other theories have been proposed. Chief among them is the gravitational theory of Carl Brans and Robert H. Dicke. According to Einstein (as indeed, according to Newton), the gravitational force exerted on one object by another one only depends on the properties and separation of the two objects themselves. In the Brans-Dicke theory there is an additional contribution to the gravitational force apparently exerted on one object by another, and this contribution arises from matter throughout the universe. According to Brans-Dicke, for example, the weight of a 200-pound

---

*Note that Newton's theory was not wrong, but that general relativity represents an improvement in our knowledge of physics that allows us to explain phenomena that were unknown in Newton's day.

man on earth would consist of about 188 pounds due to the earth's gravitational force (attraction), and about 12 pounds due to the effects of distant matter in the universe. According to Newton and Einstein, the entire 200-pound weight would arise from the properties of the earth and the man.

Furthermore, since the Brans-Dicke theory states that the apparent weight of an object on earth depends on the distribution of distant matter in the universe, and since the expansion of the universe draws the distant matter farther and farther away, the theory also predicts a very slight decrease of the weight with time.

How can we determine whether the Brans-Dicke theory is better than general relativity? Calculations based on these theories predict slightly different values for the deflection of light and for the motion of Mercury's perihelion. The difference between the two predictions for light deflection is less than the 10 per cent uncertainty in the solar eclipse measurements of light deflection. Therefore, the eclipse results are not sufficiently precise to discriminate between the two theories. However, the Brans-Dicke theory would predict a Mercury perihelion advance of about 39″ or 40″ per century. Since the observations of this quantity yield a value of 43.1″ per century with an uncertainty of 0.5″, it would seem at first sight that general relativity is in better agreement with the perihelion motion of Mercury than is the Brans-Dicke theory. However, Dicke pointed out that of the three pre-Einstein suggestions for causing the excess perihelion motion, only two (the possible planet Vulcan, and a large amount of interplanetary matter) have been ruled out. Suppose the sun did have an oblate shape—it would then be responsible for a portion of the 43″ per century motion, and the remaining motion would be smaller than predicted by Einstein. In particular, this remainder might agree with the value predicted by the Brans-Dicke theory.

At Princeton, Dicke and H. Mark Goldenberg set out to measure the sun's shape with a precision instrument of their own design (Figure 16-3). They found that the sun is indeed oblate, that is, slightly flattened at the poles. The amount is small, with the diameter at the equator being 0.005 per cent greater than the diameter through the poles, but if the sun has this slightly flattened shape, then it produces a Mercury perihelion advance of 3.5″ per century. Subtracting 3.5″ per century from the observed motion of 43.1″ per century, we get 39.6″, in agreement with the Brans-Dicke theory.

As we have seen, there is an intriguing piece of evidence for the Brans-Dicke theory of gravitation. However, the whole subject is very much in dispute. No one has yet repeated the

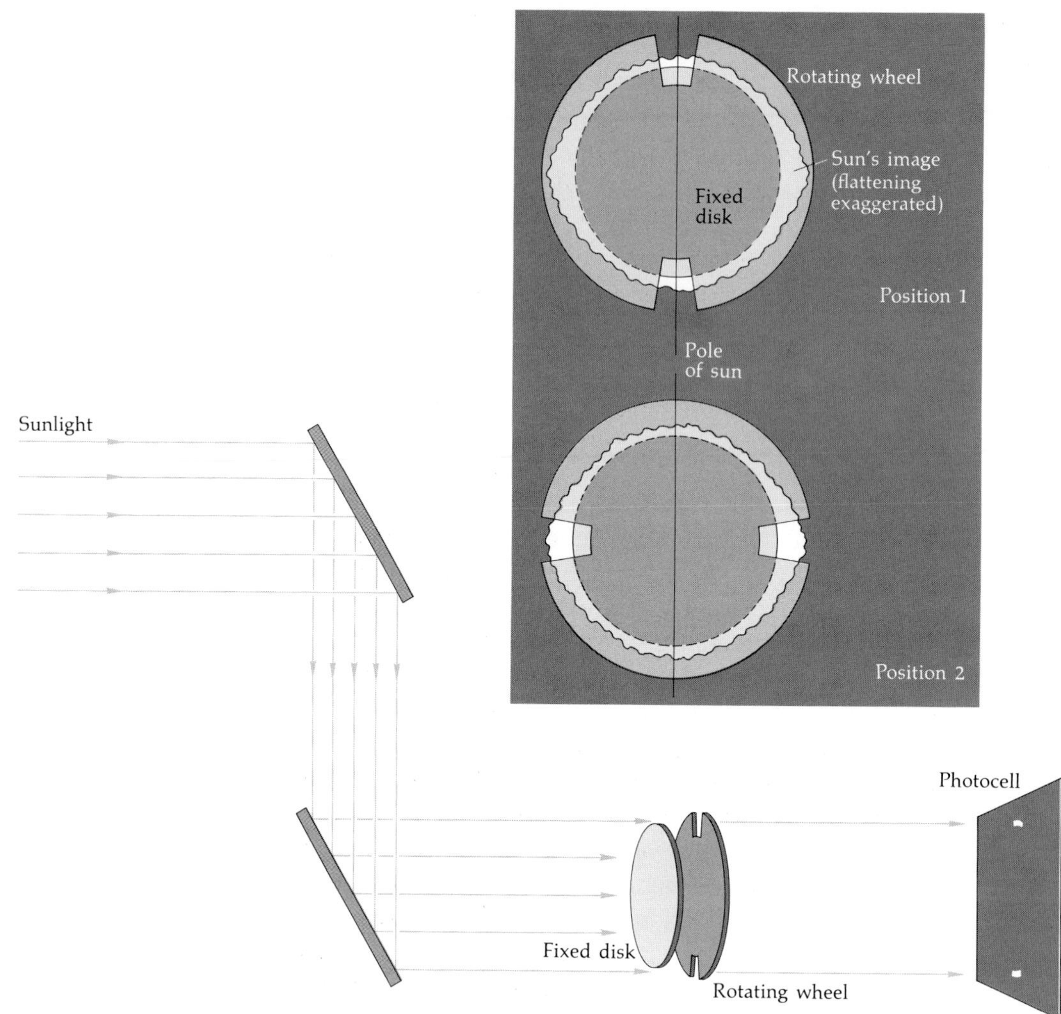

16-3 *Schematic of the Dicke-Goldenberg device for measuring the sun's shape. The solar image is reflected by two mirrors and passed into the analyzing device (below). The fixed disk blocks off all but the limb of the sun and the rotating notched wheel allows only the selected part of the limb image to pass on to the photocell. If the sun is oblate (flattened at the poles and bulged at the equator) the situation is illustrated by the upper sketches. When the notches are lined up with the sun's poles (position 1), less light reaches the photocell than when the notches are lined up with the sun's equator (position 2). Hence, an oblate solar image registers a varying brightness on the photocell. The amount of oblateness that was actually found by this method is extremely small. Details of the imaging optics are omitted here.*

measurements of the tiny solar oblateness, and the fact that these measurements which favor the Brans-Dicke theory were supervised by one of its authors who must have had at least an unconscious bias should not be wholly ignored. Besides the desirability of having additional observations made by an independent research group, there is also a considerable dispute in the pages of scientific journals about the meaning of the oblateness if it is real. Some astronomers think the oblateness is at most a skin effect, confined to the upper part of the photosphere. In this case, it would not reflect the overall distribution of matter inside the sun, and if that is true, then the Mercury perihelion advance would not be affected, and general relativity would still be in good shape. On the other hand, if the sun really is an oblate spheroid, then the flattening must be a distortion caused by rotation. However, it is a simple matter to calculate the amount of rotation required to flatten the sun to the observed value, and the computed spin rate is much faster than the observed solar rotation of once per 27 days. In fact, Dicke suggested that inside the sun there is a core rotating once every day or two! That would certainly produce the observed oblateness and would obviously change many of our ideas about the interior of the sun. It might also shed some light on the curious circumstance that many stars rotate faster than the solar photosphere value of once per 27 days.

Dicke's idea that the solar core is rapidly spinning at the present time, despite the observed slow rotation of the photosphere, has stimulated a lively controversy. A good number of physicists, mathematicians, and astronomers disagree with Dicke, and a small body of scientific literature has built up, consisting of their detailed mathematical proofs that (for a variety of reasons) the inside of the sun cannot be spinning faster than the outside. However, there are also very authoritative, convincing articles by Dicke and others, showing that this virtually unobservable solar interior *can* indeed rotate faster than the photosphere. The behavior of tea leaves stirred in a cup has even been put forth as evidence in the controversy, although the various authors disagree about whether the tea leaf motions are quite pertinent or totally irrelevant to the inside of the sun.

As this book was being completed, a substantial number of new observations and experiments were underway to discriminate between the predictions of general relativity and Brans-Dicke gravitation. They included measurements of the effect of the sun's gravity on the time required (*propagation time*) for radar pulses to make return trips from the earth to Mercury or to Venus and similar measurements of the propagation times of radio signals sent to the Mariner deep space probes as they traveled around the sun after passing close by Mars. Other experiments included observations of the perihelion motion of the asteroid Icarus, measurements of laser light echos from an optical reflector placed on the moon by the *Apollo 11* astronauts, studies with an advanced type of coronagraph that allows astronomers to measure the bending of starlight near the sun without an eclipse, and observations of the bending of radio waves from quasars by the sun.

There are indications from several of these new experiments that general relativity makes better predictions than does the Brans-Dicke theory. In particular, the Mariner signal propagation times are in agreement with relativity. Thus, when one thinks of the literature controversy that was stimulated by the report of the solar oblateness, one might very well be tempted to compare this learned debate on the rotation of the invisible solar interior with the medieval disputes about the number of angels that can dance on the head of a pin. But as the reader will have recognized, the answer to this question is related to not only the properties and evolution of stars like the sun but also to the most profound conjectures about the nature of space, time, and matter in the universe.

PULSARS AND THE CRAB NEBULA

Astronomers, like most people, have a great interest in something new. Indeed, it is the prospect of discovering a new comet, asteroid, or other object that inspired some of us to become astronomers. Others are more interested in investigating physical laws to derive a better understanding of the celestial phenomena recorded by the observers. Rarely, however, have scientists been fortunate enough (or clever enough!) to *predict* an entirely new *class* of objects that was subsequently actually found. Such, however, was the case in the matter of the neutron stars.

The story of neutron stars and pulsars begins in China in the year 1054 A.D. when Yang Wei-Tek recorded the presence of a brilliant "guest star" in the heavens. The object—a bright star where none had been seen before—was so luminous that it was visible in broad daylight for more than three weeks. Gradually, however, it faded, but it continued to be visible for some 650 days, after which it was too faint to be seen by the unaided eye. Today we identify the phenomenon observed by Yang as a *supernova,* the death of a star by an especially violent and rare means (see Chapter 11).

Centuries went by, and we have no information on the supernova in this interval since it could not be seen, until finally the telescope was invented. Another century passed, and in 1731, an English physician, John Bevis, finally turned a telescope to the direction in which (unknown to him) the guest star had been observed, and discovered a dim, fuzzy object. (Because of its appearance, it was later named the Crab nebula by the Earl of Rosse, who discovered its filaments; Figure 16-4.) During the next 235 years observational evidence about the nebula accumulated. By 1966 three of its main properties could be summarized as follows: (1) It consists of an irregular cloud of gas expanding at a rate of about 1,000 kilometers per second (presumably due to the original explosion). (2) There are luminous wisps in various parts of the nebula that are moving at velocities much higher than that of the nebular expansion. (3) It is the source of optical, infrared, radio and x-ray radiation of great intensity. Further, it had become clear that only one proposed theory could explain the continuous spectrum of the Crab nebula, that is, the relative intensities of radiation in the different wavelengths. According to this model, the light, radio waves, and x-rays were produced as *synchrotron radiation* by electrons moving at great speeds (approaching the speed of light) through a magnetic field in the nebula.

Strange light at General Electric

At this point in the story of the Crab nebula, we will digress briefly from astronomy and talk about atomic accelerator machines. In 1944–1945, a number of frontier investigations in experimental physics centered on the bombardment of atoms with streams of energetic particles produced by such accelerator devices as the cyclotron. Just as astronomers keep hoping and asking for bigger and better telescopes, the physicists, then as now, always seem to need ever more powerful accelerators to advance the study of nuclear structure. At the time in question, the aim was to develop machines in the hundred MeV (million electron volt*) range and eventually break through to 1 BeV (billion electron volts). There is a basic limitation in the cyclotron that prevents it from reaching these high energies. More advanced instruments, the betatrons, were in existence, but the cost involved in pushing them to higher energies became prohibitive.

*The electron volt or eV is a unit of kinetic energy often used by physicists. We say that temperature of the atoms in the solar photosphere is 6,000° K; a physicist could equally well say that their average energy was about 0.5 eV.

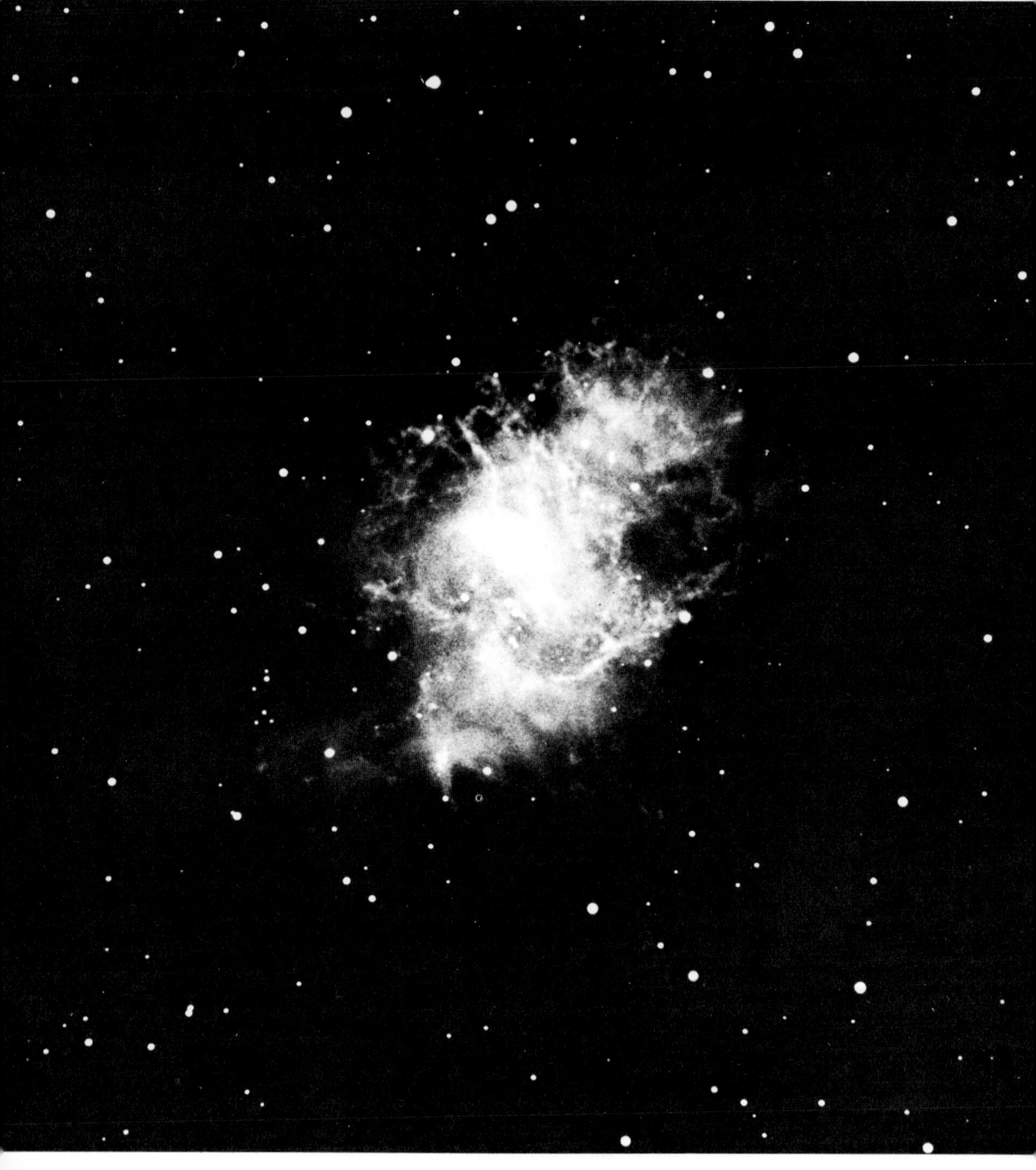

16-4 *The Crab nebula. (Courtesy of the Hale Observatories. Copyright © by the California Institute of Technology and the Carnegie Institution of Washington.)* See also Plate 10 following p. 260.

The time was ripe for a basic improvement in accelerator design and the design parameters for such a machine, the *synchrotron,* were described by Edwin M. McMillan of the University of California in a brief paper in the *Physical Review.* McMillan discussed a synchrotron that he planned to build at Berkeley. However, shortly after this paper was published, a group of scientists at the General Electric Company in Schenectady began work on a synchrotron. The specific purpose of their machine was to produce energetic electrons. When the synchrotron was built, the General Electric workers observed a fascinating effect through a glass window in its side: as the electrons moved along they emitted an intense light in the direction of motion. Appearing dull red in color when the synchrotron was idling at 30 MeV, the light brightened as the machine was gradually turned up; at 80 MeV, slightly above the design goal of the synchrotron, the radiation was a brilliant bluish-white. There was another interesting effect; viewing the light through a Polaroid filter, the researchers found that although easily visible in one orientation of the Polaroid, the light became imperceptible as the filter was rotated through a 90 degree angle. Thus, it obviously was polarized.

The possibility of light emission by electrons moving at high speed in a powerful magnetic field, such as existed in the synchrotron, had been discussed half a century before by the physicist A. Lienard, and it had also occurred to McMillan as a possible side effect of his proposed machine. However, the strange light found at General Electric became known as *synchrotron radiation.* We mentioned in Chapter 5 that the path of a charged particle, such as an electron, will be deflected or bent by a magnetic field. In fact, in the case of the Crab nebula, the electrons actually spiral around the lines of magnetic force, and the synchrotron radiation that they emit is polarized in the direction perpendicular to the magnetic field lines (Figure 16-5). Synchrotron radiation may be emitted at any wavelength, depending on the velocity of the high-energy electron and the strength of the magnetic field.

The idea that the Crab nebula radiation was due to synchrotron emission was put forward by I. S. Shklovskii, and was soon confirmed when the nebula was observed with a polaroid filter.

Shklovskii's theory solved a problem—the radiation was produced by the synchrotron mechanism—but it eventually raised another one when x-rays from the Crab nebula were observed. The theory allowed one to calculate how long an electron of given energy would continue to emit radiation. In

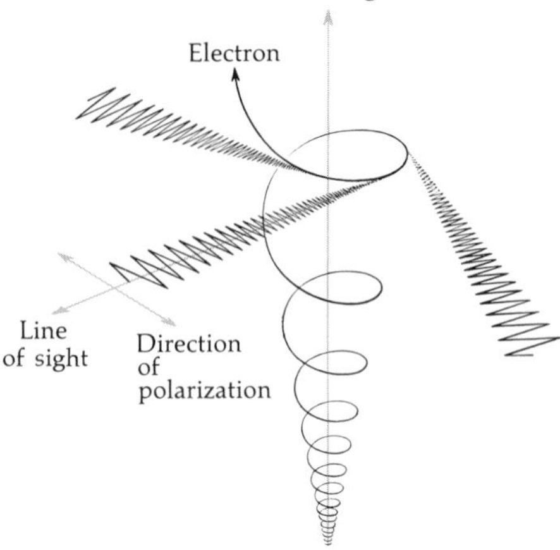

16-5 *Schematic showing production of polarized synchrotron radiation by electrons spiraling in a magnetic field. (Adapted from* Exploding Galaxies *by A. R. Sandage. Copyright © 1964 by Scientific American, Inc. All rights reserved.)*

the case of x-ray emission, this *radiative lifetime* was only about a year. But the supernova explosion that produced the Crab was observed over 900 years ago. How then could electrons of radiative lifetime one year still be present to emit the x-rays? Presumably, some unknown energy source capable of producing fast electrons was still present in the nebula.

Neutron stars

In the 1930s the Soviet physicist, Lev Landau (and later the American, J. Robert Oppenheimer), proposed a new state of matter that might exist in a previously unsuspected form: the neutron star. The matter in such a star would be compressed to the enormous densities characteristic of the nuclear core of an atom. At the center of the neutron star, the density may be 10^{14} or 10^{15} times greater than that of water. As a result, all of the mass of an ordinary star like the sun, compressed to this enormously high density, would occupy a sphere with a diameter of only some 20 kilometers (12 miles). Under certain circumstances such a star might be formed (Chapter 11) in an *implosion* of the central region of a supernova, while an *expanding* gaseous remnant, like the Crab nebula, would also be generated. Many astronomers did not believe in this process, however, and calculations indicated that the neutron stars would be very hard to observe (due to their small size), so they were generally regarded as a physicist's dream. Although theoretical analyses of the properties of the hypothetical neutron stars were pursued by a few true believers (such as A. G. W. Cameron of Yeshiva University), the lack of observational evidence for the existence of these objects was discouraging, and this is where the problem of the neutron stars stood in 1967.

Pulsars

In July, 1967, radio astronomers at Cambridge University, led by Antony Hewish, discovered a highly unusual object. Most previously known celestial sources of radio waves were relatively steady emitters, although a few (like the sun) produced bursts of radio waves at erratic times. However, the Cambridge source seemed to be emitting bursts (pulses) at perfectly spaced intervals (Figure 16-6) like the ticks of a clock. Although the strength of the pulses varied, the observations indicated that the new radio source, called a *pulsar,* was keeping better time than most man-made clocks! This incredible discovery led to a great effort to discover more pulsars, and to analyze their properties in an attempt to explain them. Among the first 60 pulsars to be found, the pulse-spacings (pulse periodicities) ranged from one-thirtieth of a second to over three seconds.

The extreme regularity of the pulse-spacings required an explanation. The three likely *clock mechanisms* that were generally considered were: (1) *pulsation* of a star, which might

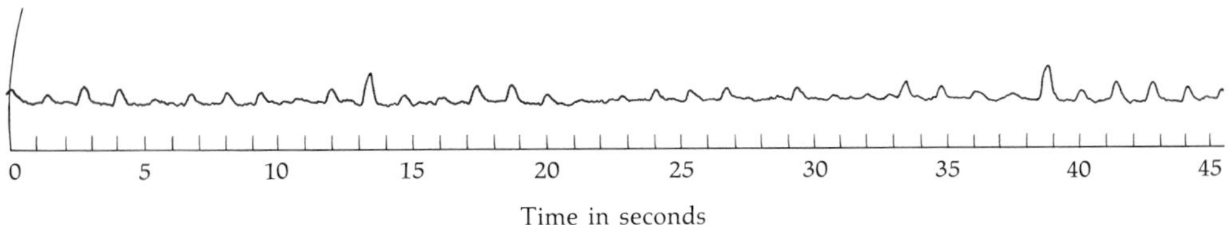

16-6 *The observed pulses from CP 1919, the first pulsar, showing pulses spaced by about 1.3 sec. Note that they are not of equal strength.* (From Pulsars *by Antony Hewish.*

Table 16-1 *Typical Parameters of Compact Stars*

	WHITE DWARF	NEUTRON STAR
Diameter	0.5 to 2 earth diameters	20 to 40 km
Mass	0.4 to 1 solar mass	0.5 to 4 solar masses
Central density*	10^5 to 10^9 gram/cm^3	10^{14} to 10^{15} gram/cm^3

*Density of ordinary water = 1 gram/cm^3.

resemble the pulsations of a Cepheid variable star, but much faster; (2) *orbital motions* of a binary star; (3) *rotation* of a star. Two types of clock stars were considered for each of these theories: white dwarfs and neutron stars. These were selected because the shortness of the pulsar periods implied that small, dense objects were involved. (Typical parameters of white dwarfs and neutron stars are given in Table 16-1; also see Figure 10-19.) For example, an ordinary star cannot pulsate as fast as the slowest pulsar; two stars like the sun cannot orbit with a pulsar period because from Newtonian mechanics we know that in order to orbit so rapidly their separation would have to be less than their own diameters; finally, an ordinary star cannot perform one rotation in a time as short as a pulsar period because, at such speeds, matter would spin off at the equator, disrupting the star.

During the first few months after the pulsar discovery announcement, there was considerable debate over what both the clock mechanism and clock stars might be. However, three additional discoveries convinced astronomers that the pulsars were rotating neutron stars:

(1) two pulsars were found to be associated with supernova remnants;
(2) the period of one of these two pulsars, namely, the object NP 0532* *in the Crab nebula,* was extremely short: $\frac{1}{30}$ of a second;
(3) comparison of measurements over several months showed that many of the pulsar periods were not constant but were gradually lengthening.

In connection with discovery (1) we recall that the hypothetical origin of a neutron star took place in a supernova event.

*NP 0532 stands for "National Radio Astronomy Observatory Pulsar near Right Ascension 05 hours, 32 minutes." NP 0532 was discovered by David Staelin and Edward Reifenstein at the N.R.A.O.

Discovery (2) ruled out white dwarfs, since the size and the distribution of matter in the interior of a white dwarf are incompatible with rotation, orbital motion, or vibration in so short a period as $\frac{1}{30}$ of a second. Discovery (3) ruled out orbital motion or vibrations of neutron stars, because calculations showed that the periods would be expected to remain constant or to gradually *decrease.* The remaining possibility, which has survived all observational tests so far, is rotation of a neutron star. A typical neutron star, with properties as given in Table 16-1, could easily rotate 30 times per second without breaking up. Theorists are studying the ways in which the radiation might be beamed. Apparently a very strong magnetic field, perhaps even 10^{13} times that of the earth, may be involved. As the stars rotate, the beams sweep around like those of searchlights (Figure 16-7) and each time such a beam sweeps past the earth we observe a pulse.

The pulsar NP 0532 in the Crab nebula is thus apparently a neutron star that is spinning at a rate of 30 rotations per second. It has also been observed at optical wavelengths (Figure 16-8). The observations of its pulses show that the rotation period is slowly increasing. This slow-down of the spinning neutron star represents a loss of energy that must reappear in some other form, according to the law of conservation of energy. (For example, when an automobile is stopped, some of its kinetic energy appears in the brakes in the form of heat.) In fact, it has been calculated that the energy lost by the rotating neutron star in this way is somewhat larger than the total energy emitted by the Crab nebula in the form of electromagnetic radiation.

If the identification of pulsars as neutron stars is correct, then the evaluation of the energy loss involved in the slowing down of pulsar NP 0532 has solved the basic problem of the Crab nebula. The energy source of the nebula, which enables it to shine in the x-ray portion of the spectrum nine centuries after the supernova explosion, is the gradual slowing of a dense, rapidly rotating object, the neutron star. The spinning neutron star is losing its rotational energy, and the presence of a strong magnetic field allows the energy to reappear in the form of the kinetic energy of fast-moving electrons (precisely how this happens is also still being debated). The electrons encounter the magnetic field of the nebula, and produce synchrotron radiation at all wavelengths from the x-rays to the radio waves. Some of the fast particles produced by the neutron star may escape from the nebula and thus pulsars like NP 0532 may represent a source of cosmic rays.

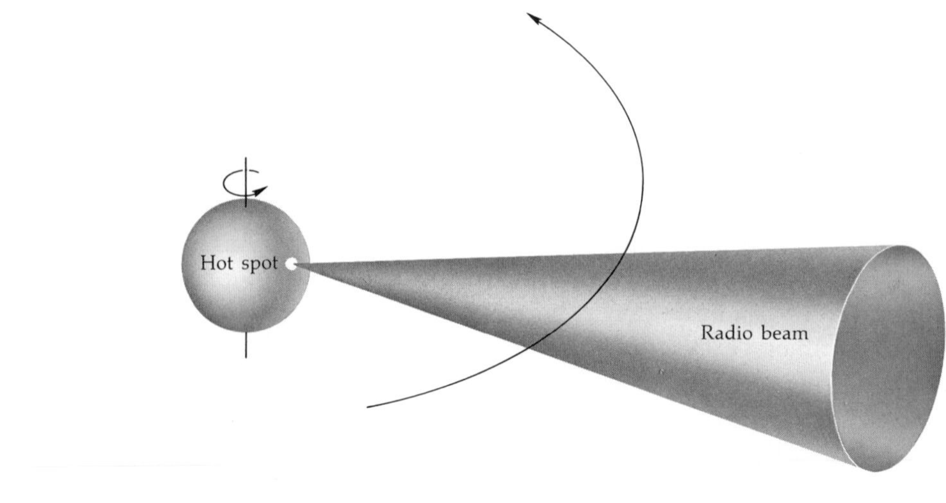

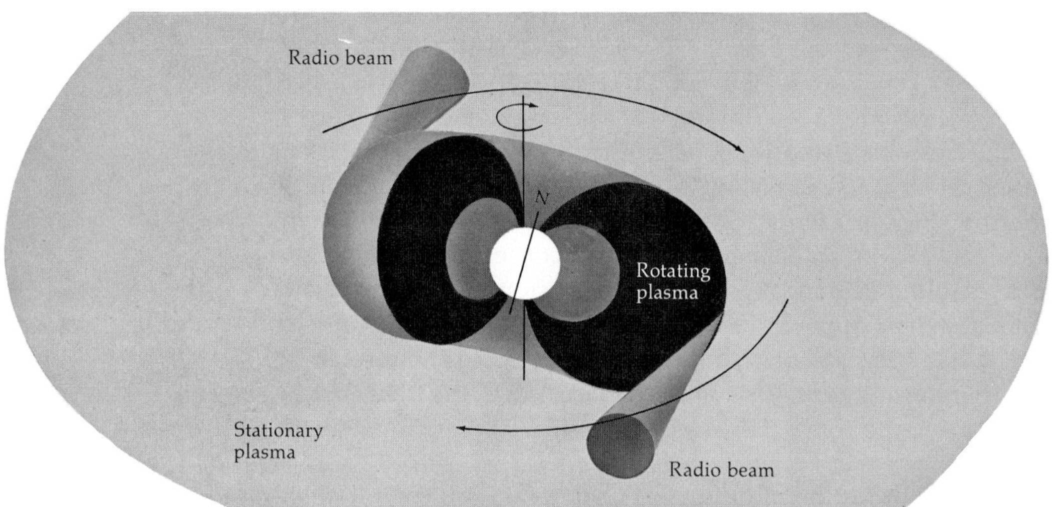

16-7 *Artist's impression of two beaming models for pulsars. In the upper diagram, the pulsar radiation originates at a "hot spot" on the surface of the star. This might be the magnetic pole, and in the case drawn here, the magnetic pole would be at the star's equator. In the lower figure the pulsar emission originates well out in the surroundings. In either case, as the rotating beam sweeps past the earth, we detect a "pulse."* (*After* Pulsars *by Antony Hewish. Copyright © 1968 by Scientific American, Inc. All rights reserved.*)

QUASARS

As radio astronomers mapped the sky (recall Figure 6-25), they provided the celestial coordinates—Right Ascension and Declination—of their newly discovered radio sources to the optical astronomers, who searched the regions of the radio sources to see what kind of stars, nebulae, or galaxies might be responsible

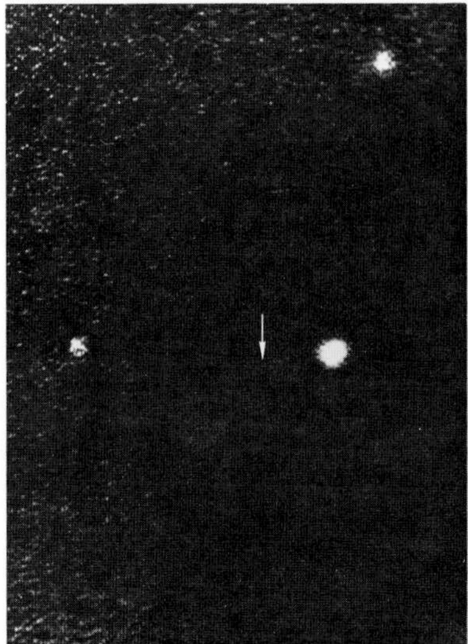
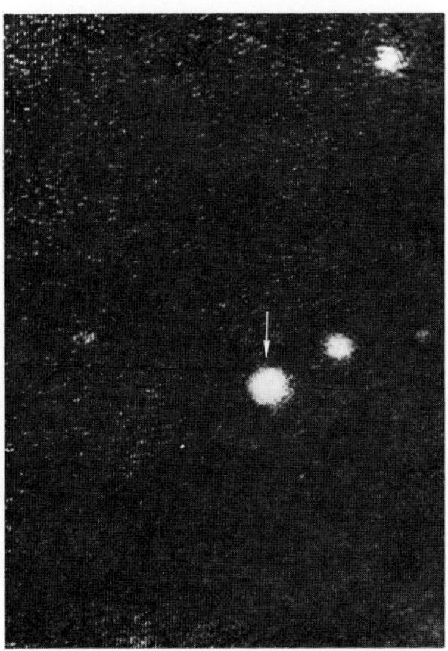

16-8 *The Crab nebula pulsar (NP 0532) in optical wavelengths. Right: during the pulses; left: between the pulses. (Courtesy J. S. Miller and E. J. Wampler, Lick Observatory.)*

for the radio emission. Some of the radio sources turned out to be well-known objects, such as the Crab nebula, the Orion nebula, and the M31 galaxy. Surprisingly, some very distant and thus optically faint galaxies turned out to be among the strongest radio sources. Even stranger, some of the radio sources did not seem to have optical counterparts, or else the uncertainties in the radio telescope measurements of the source positions were so large that it was impossible to decide which of the many stars and other objects in a given region was the radio source. In the early 1960s a few radio sources, named 3C 48,* 3C 196, and 3C 286, were identified with faint stars (Figure 16-9), but when the spectra of the stars were photographed, they showed emission lines rather than absorption lines and the patterns of the lines were unfamiliar. In fact, it was impossible to decide what the atoms were that produced these emission lines.

The breakthrough came in 1963 when the radio astronomer Cyril Hazard in Australia observed the moon to occult (that is, eclipse) a hitherto unidentified radio source, 3C 273. Hazard's radio telescope was not ordinarily capable of making really precise position measurements, but it enabled him to record

*This is astronomical shorthand for the Third Cambridge Catalogue of Radio Sources, as mentioned in Chapter 6.

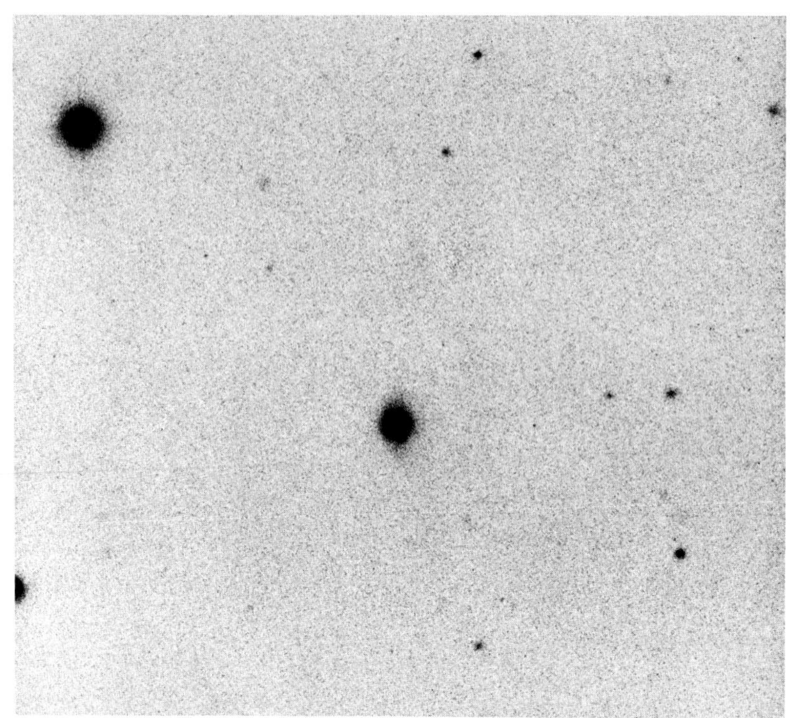

16-9 *Quasar 3C 48 resembles a dim star; this is a negative print, so the stars appear as black dots. (Courtesy of the Hale Observatories.)*

the exact times when the moon occulted 3C 273 and when 3C 273 emerged from behind the moon. Since the orbit of the moon is known with great precision, Hazard was able to calculate an accurate position of 3C 273 from his timings of the occultation. At Palomar, Maarten Schmidt photographed the spectrum of a faint blue star at this position (Figure 16-10). The emission line pattern of this star was recognizable. It was a well-known set of hydrogen lines, but it was remarkable. The lines had a very large red shift, amounting to 45,000 kilometers per second, or 15 per cent of the velocity of light, as one would find in a galaxy at a distance of 450 megaparsecs. Thus, 3C 273 was apparently *not* a star in the Milky Way galaxy. With this clue in mind the puzzle of the spectral line patterns of 3C 48, 3C 196, and 3C 286 was soon cleared up; they had been unrecognized because they had equally unexpected and even larger red shifts. From the circumstance that these objects at first had seemed to be stars, they were named *quasi-stellar radio sources* or, more popularly, *quasars.*

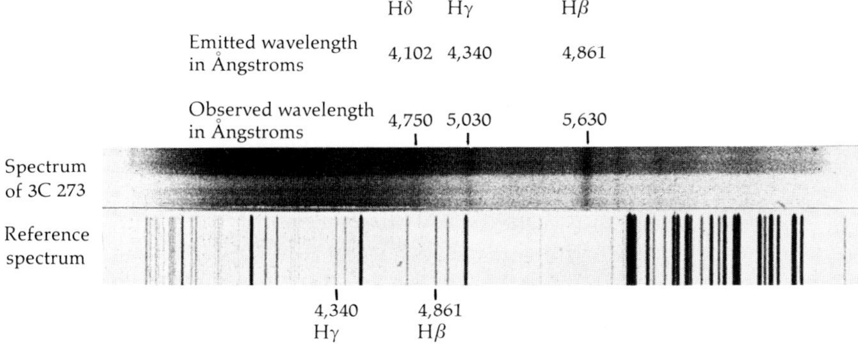

16-10 *Laboratory reference spectrum (below) is compared with the spectrum of the closest known quasar, 3C 273 (above). The upper half of the quasar spectrum was given a longer exposure than the lower half and thus appears darker on this negative reproduction. Note that although the laboratory spectrum shows the hydrogen line "Hβ" to be at wavelength 4,861 Angstroms, this line appears at 5,630 Angstroms in the spectrum of 3C 273, due to the red shift. (From* The Evolution of Quasars *by Maarten Schmidt and Francis Bello. Copyright © 1971 by Scientific American, Inc. Spectrogram courtesy of Maarten Schmidt.)*

Today, more than 200 quasars are known, defined in the strict sense as radio sources identified with stellar-appearing objects whose optical spectra show large red shifts. Also known are radio-quiet quasars or *blue stellar objects* that appear to be virtually identical to quasars in their optical properties but that are not strong radio sources. The celestial object of greatest observed red shift is the quasar 4C 05.34, which appears to be receding at a speed close to 88 per cent of the velocity of light, as deduced from spectrograms obtained by Roger Lynds and D. Wills at the Kitt Peak National Observatory.

The great red shifts of quasars imply that they are at great distances. But the quasars are also strong radio emitters in terms of *apparent* radio brightness as observed at the earth. These two circumstances taken together imply that the *absolute* radio brightnesses of quasars are immense. Indeed, the radio luminosity of a typical quasar is more than a million times that of a normal spiral galaxy like the Milky Way.

Improvements in radio astronomy techniques since Hazard's observations were made have allowed very high spatial resolution to be obtained in simultaneous measurements of the same quasar by radio telescopes in opposite hemispheres of the earth. These measurements have shown that the angular diameter of the emitting region of a quasar is usually very small. Combining these angular diameters with the distances computed from the red shifts and the Hubble constant, it is found that most of the radio emission of a typical quasar occurs in a region of less than a parsec in size. A spiral galaxy like the Milky Way has a diameter of about 30 kiloparsecs. Thus, a quasar somehow produces a million times more radio energy than a normal spiral galaxy in a region of 30,000 times smaller diameter than the galaxy! In fact, the optical radiation of a typical quasar is also enormous, and in the case of the most thoroughly studied quasar (3C 273), infrared emission has been detected, which exceeds its optical and radio energy by far.

A great challenge to theoretical physicists has been to account for this enormous energy emission of a quasar, concentrated in such a small region. Some tried to get out of the problem by saying that the quasars were not at the great distances calculated from their red shifts and the Hubble constant. Instead, they suggested that the quasars were much closer, and that their red shifts were not due to the expansion of the universe but due to the *gravitational red shift.* (Einstein had shown that the gravity of a star would produce a slight red shift in its light; the *gravitational red shift theory* of quasars required that they have extremely strong gravitational fields in order to produce the observed high red shifts.) On the other hand, the *local theory of quasars,* extensively discussed by James Terrell of the Los Alamos Scientific Laboratory, suggested that the quasars were relatively close to the Milky Way. According to this theory, their huge red shifts were caused by ejection of the quasars at high velocity from an explosive event in the Milky Way or a nearby galaxy.

The arguments over the nature of the quasar red shifts have continued, but most astronomers believe that the *cosmological theory of quasars* is correct. This is the simple idea that the red shifts of quasars are due to the same effect as those of galaxies, namely the expansion of the universe. One way of verifying this hypothesis would be to find quasars located in clusters of galaxies. At the California Institute of Technology, John Bahcall and James Gunn explored this possibility. They found several quasars and quasar-like objects that do appear to be associated with clusters of galaxies, on the basis of their positions in the sky and similar red shifts. The objects studied by Bahcall and Gunn are relatively close as quasars go, so they are not strictly typical of the class, but these findings do encourage us to think that if we accept the distances of galaxies that are estimated from their red shifts, we should also accept the cosmological theory of quasars.

The spectrum and polarization of the radio emission from many quasars have been measured, and it seems clear that the radio waves are produced by the synchrotron process, as in the Crab nebula (page 451). However, the question of the *source* of the enormous radio energy of a quasar is still unanswered. The problem is even greater than originally suspected because Frank Low at the University of Arizona has measured infrared radiation from quasars and found that in some cases it is much stronger than the radio emission. A great many speculative ideas have been put forth and we mention only a few here to give the reader a feeling for the kind of thinking about quasars that is going on today.

The Swedish physicist Hannes Alfvén has revived the old idea of the possible existence of large amounts of *anti-matter* in connection with the energy source of quasars. Anti-matter consists of atoms resembling those found on earth, but composed of particles with opposite charges, that is, having negatively charged nuclei and positively charged electrons. If a cloud of anti-matter encountered a cloud of ordinary matter, the result would be the annihilation of the atoms and the release of an enormous amount of energy.

Collisions between stars have also been widely discussed as an energy source for quasars. According to these theories, a quasar may consist of a large number of stars, packed much more closely together than in a globular cluster. Thomas Gold of Cornell University has calculated that collisions between these stars could supply the energy of a quasar. However, Stirling Colgate of the New Mexico Institute of Mining and Technology suggests that the most important effect of the collisions would be to cause smaller stars to coalesce into larger ones, which evolve faster and become supernovae, according to the general ideas presented in Chapter 11. Thus, a large number of supernova explosions might account for the energy of a quasar. The *gravitational collapse* theory was proposed by Fred Hoyle (Cambridge University) and William Fowler (California Institute of Technology). According to their ideas, hypothetical superstars with masses millions of times greater than that of the sun, contract under the influence of their own gravitation, attaining great densities (as in a neutron star), and energy is released as the matter falls together. Most recently, physicists have drawn analogies between quasars and pulsars. One suggestion is that a quasar derives its energy from the slowing down of a large rotating mass, and another idea is that many individual rotating magnetized objects are found within a quasar and provide its energy.

Quasars, once thought to be stars in the Milky Way, are now regarded as among the most distant objects in the universe. They are also the most energetic phenomena known. They represent a challenge to the physicist who must account for their enormous energies, and to the cosmologist who must explain their significance in terms (perhaps) of a special stage in the evolution of galaxies or of the universe. As distant sources of intense radiation, they offer the possibility of studying *intergalactic matter* in terms of its effects on the transmission of quasar radiation, just as interstellar matter was studied through its reddening, polarizing, and line absorption effects on the light of stars.

17

Epilogue

The preceding chapters have outlined the continuing story of man's probing and exploration of the universe. At every turn we have encountered new horizons. Here, we review these explorations and our changing world view.

THE EARTH MOVES IN MANY WAYS

The history of astronomy has involved a continuing realization that the earth is not the center of the universe, the earth is not fixed, and the heavens do not revolve around it. Just how many motions of the earth are there?

First of all, the earth rotates on its axis, thereby producing day and night and the apparent rising and setting of the stars. The axis of rotation performs a conical motion, tracing out a complete circle in the sky every 26,000 years, the phenomenon of precession.*

The center of mass of the earth-moon system travels around the sun in an elliptical orbit; the earth also performs a small

*There is a slight wobble, called *nutation,* in this precession motion.

orbital motion around this center of mass. The net result is that the earth's path around the sun is an ellipse with wiggles in it.

The earth also partakes of the sun's motion in the galaxy. The sun, traveling in a roughly circular orbit around the center of the Milky Way, is now moving toward a point in the constellation Hercules. Thus, the earth never returns to the same point in space; it does not trace out the same ellipse repeatedly but partakes in a corkscrew motion as seen from this point of view (Figure 10-4).

The earth and sun both partake of the motion of our galaxy, which is in orbit about the center of mass of a local group of galaxies, which includes the Andromeda galaxy (Figure 13-1) and the Magellanic Clouds (Figure 13-7). Thus, together with the local group, the earth participates in the general expansion of the universe as revealed in the red shifts in the spectra of distant galaxies.

HOW LONG AGO?*

Many aspects of our understanding of the universe can be summarized in a time sequence of events. It begins with the explosion of the primeval fireball. In this expanding medium the galaxy and the oldest stars formed about 10 billion years ago. The sun condensed from a cloud of gas and dust after the galaxy had flattened into a disk-like configuration. About 4.5 billion years ago, the solid bodies of the solar system, including the earth, were formed. The oldest known rocks on earth formed about 3.5 billion years ago, or at least a half billion years after the oldest known lunar rocks. The present atmosphere of the earth began to appear through exhalation from the crust at about the same time, and the oceans were forming by about 4.2 billion years ago and had developed by about 3 billion years ago. The primary broth arose over 4 billion years ago; life developed and important amounts of molecular oxygen (O_2) were injected into the atmosphere by the green plants. The atmospheric composition as we presently know it (pollutants excepted) was reached by 1 billion years ago.

Although life developed earlier, the fossil record and our generally accepted knowledge of geological time extend back only 600 million years. About 250 million years ago the sun

---

*Review Figure 1-2 in connection with this section.

began the trip around the center of the Milky Way galaxy that it is now completing. The dinosaurs were developing at this time and they were the dominant animals on earth for about 100 million years. The Rocky Mountains were formed about 80 million years ago. Nuclear reactions within the youngest stars now observed on the main sequence were ignited less than 3 million years ago. The development of man was underway at this time, and Advanced Australopithecus appeared about 1 million years ago.

Early *Homo sapiens* appeared about 250,000 years ago. By 35,000 years ago the Cro-Magnon, virtually modern man, had arisen. About 26,000 years ago the earth's rotation axis pointed to alpha Ursae Minoris; since then it has swept around a small circle in the sky and is now pointing roughly at that same star again. Man's written historical record dates from about 5,000 years ago, when the earth's axis pointed to the star Thuban (alpha Draconis), and Thuban was thus the North Star.

About 930 years ago Yang Wei-Tek at Khaifêng, China, observed a brilliant "guest star" in full daylight. Some 248 years ago Pluto was in the same location as now in its orbit around the sun; 22 years ago the sunspots were arranged on the sun roughly as they are now; 11 years ago the spots were also roughly as numerous and roughly at the same latitudes, but their magnetic polarities were opposite to the present situation. One year ago the earth was at the same position in its orbit around the sun. One month ago the moon was at about the same phase.

Light arriving at the earth from the sun left there about 8 minutes ago, while light from the moon left about 1.3 seconds ago. During the time that this light traveled from the moon to the earth, the neutron star at the heart of the Crab nebula, produced by the supernova event recorded by Yang, spun around 39 times.

THE FUTURE

Where do we go from here? The earth is the only place we know where an atmosphere exists that Man can breathe, although we have probed the universe with telescopes and are now beginning the direct exploration of the nearest astronomical bodies. The footprints of men are on the moon and Mars may well be next. Hopefully, this exploration can continue in the name of all men—men who have come to view themselves as integral parts of nature, every atom in their bodies having once

17-1 *Spaceship Earth seen above the lunar horizon from* Apollo 8. (*National Aeronautics and Space Administration.*) See also Plate 1 following p. 260.

been among or inside the stars, every substance they use having come from some spot in the total ecology of the globe and likewise destined or doomed to return to the environment of Spaceship Earth (Figure 17-1).

Table 17-1 *The Earth Moves in Many Ways*

MOTION	PERIOD
Rotation on its axis	1 day
Precession of its axis	26,000 years
Revolution about the earth-moon center of mass	1 month
Revolution about the sun	1 year
Sun's motion in the galaxy	250 million years
Galaxy's motion in the expanding universe	More than 10 billion years (if periodic*)

*Periodic if oscillating universe theory is correct.

Table 17-2 *Events in History*

EVENT	APPROXIMATE TIME (YEARS AGO)	(SECONDS AGO)
Explosion of the cosmic fireball	13 billion	4×10^{17}
Oldest stars in our galaxy formed	10 billion	3×10^{17}
Formation of the sun	5 billion	1.5×10^{17}
Formation of the earth with present mass	4.5 billion	1.4×10^{17}
Primary broth arises as oceans begin to form	4.2 billion	1.3×10^{17}
Formation of the oldest known rocks on earth	3.5 billion	1.1×10^{17}
Formation of exhaled atmosphere completed	3.5 billion	1.1×10^{17}
Oldest fossils formed—bacteria and blue-green algae	3.4 billion	1.1×10^{17}
Formation of oceans (at approximately their present volume) completed	3.0 billion	1.0×10^{17}
Plants began oxygen production	2.0 billion	6×10^{16}
Atmosphere formed as now known	1.0 billion	3×10^{16}

Table 17-2 *(Continued)*

Production of abundant fossil record	600 million	1.8×10^{16}
Sun began last revolution about galactic center	250 million	6×10^{15}
Dinosaurs were dominant life form	100 million	3×10^{15}
Rocky Mountains formed	80 million	2.5×10^{15}
Brightest stars on the main sequence began nuclear reactions	3 million	9×10^{13}
Man developed to Advanced Australopithecus	1 million	3×10^{13}
Early *homo sapiens* appeared	250 thousand	8×10^{12}
Modern man appears (Cro-magnon)	35 thousand	1.1×10^{12}
North celestial pole in same direction as now	26 thousand	8×10^{11}
Man's historical record began	5 thousand	1.5×10^{11}
Pluto began last revolution around the sun	248	8×10^{9}
Sun at same point in sunspot cycle	22*	7×10^{8}
Earth at same point in orbit around the sun	1	3×10^{7}
	TIME (DAYS AGO)	
Moon at same phase	29.5	2.6×10^{6}
Earth at same rotation position with respect to the sun	1	8.64×10^{4}
	TIME (MINUTES AGO)	
Sunlight arriving at earth now left sun	8	5×10^{2}
Moonlight arriving at earth now left moon	—	1.3
While the light traveled from moon to earth, the central star of the Crab nebula spun around 39 times.	—	—

*Not a misprint for 11; review page 257.

Appendixes

APPENDIX 1

Powers of Ten Notation

There is an enormous range in the scale of objects studied in astronomy. For example, the wavelength of the hydrogen line used by radio astronomers to map the spiral pattern of the galaxy (page 334) is 21 centimeters or 8.3 inches; the radius of the earth is 6,400 kilometers or 640,000,000 centimeters (250,000,000 inches); the distance to the sun, one astronomical unit (page 181), is 15,000,000,000,000 centimeters (5,900,000, 000,000 inches); the distance to the nearest star beyond the solar system is 1.31 parsecs (page 264) or 4,300,000,000,000,000,000 centimeters (1,600,000,000,000,000,000 inches); and the distance to the Virgo cluster of galaxies (page 359) is about 11,000 kiloparsecs or 34,000,000,000,000,000,000,000,000 centimeters (13,000,000,000,000,000,000,000,000 inches). Very small quantities also are inconvenient to write out in inches. For example, the radius of the helium atom is about 1 Angstrom or 0.00000001 centimeter (0.000000004 inch).

The range in sizes described above is so great that it is very inconvenient to use any one unit, such as the centimeter, for every kind of length—there are just too many cases in which an annoying number of zeros has to be written down. On the other hand, it is also troublesome to use so many different units (centimeter, kilometer, astronomical unit, or parsec) and have to keep referring to Appendix 2 for the respective values of these units. A technique that allows us to minimize the number of zeros involved in using a common unit over a large range is the "scientific notation" or "powers of 10" method of representing numbers. It also happens to be very handy for doing simple calculations.

The power of ten simply locates the decimal point as shown in the following listing:

10^{10} = 10,000,000,000
10^{9} = 1,000,000,000
10^{8} = 100,000,000
10^{7} = 10,000,000
10^{6} = 1,000,000
10^{5} = 100,000
10^{4} = 10,000
10^{3} = 1,000
10^{2} = 100
10^{1} = 10
10^{0} = 1
10^{-1} = 0.1
10^{-2} = 0.01
10^{-3} = 0.001
10^{-4} = 0.0001
10^{-5} = 0.00001
10^{-6} = 0.000001
10^{-7} = 0.0000001
10^{-8} = 0.00000001
10^{-9} = 0.000000001
10^{-10} = 0.0000000001

Note that one thousand is simply $10 \times 10 \times 10$ or 10^3. One hundredth is $1 \div (10 \times 10)$ or 10^{-2}. Obviously, the listing could be extended to as large or as small powers of ten as desired. A number which does not happen to be an even multiple of ten is also easily represented by this method. For example, 365 is simply 3.65×100 or $3.65 \times 10 \times 10$ or 3.65×10^2.

This technique allows us to conveniently rewrite some of the cumbersome numbers quoted at the beginning of this appendix. The distance to the Virgo cluster of galaxies can be written as 3.4×10^{25} cm, while the radius of the helium atom can be written as 1×10^{-8} cm.

The use of the powers of ten notation allows easy multiplication and division using the rules that *the powers add when multiplying* and *the powers subtract when dividing*. For example,

Multiplication: $(200) \times (3{,}000) = (2 \times 10^2) \times (3 \times 10^3)$
$= (2 \times 3) \times 10^{2+3} = 6 \times 10^5$

Division: $\dfrac{5{,}000}{200} = \dfrac{5 \times 10^3}{2 \times 10^2} = \dfrac{5}{2} \times 10^{3-2} = 2.5 \times 10^1 = 25$

Division: $\dfrac{200}{4{,}000} = \dfrac{2 \times 10^2}{4 \times 10^3} = \dfrac{2}{4} \times 10^{2-3} = \dfrac{1}{2} \times 10^{-1}$
$= 5 \times 10^{-1} \times 10^{-1} = 5 \times 10^{-2} = 0.05$

Division: $\dfrac{600}{20{,}000} = \dfrac{6 \times 10^2}{2 \times 10^4} = 3 \times 10^{2-4} = 3 \times 10^{-2} = 0.03$

A little practice with these ideas will enable the reader to make power of ten calculations with ease. Even if you do not become a scientist, powers of ten may help you to understand other things of interest; for example, the total revenue of the U.S. Government from personal income taxes in 1968 was 7.82×10^{10} dollars, and the population was about 2×10^8. Thus the average tax per person was

$$\frac{7.82 \times 10^{10}}{2 \times 10^8} = \text{about } 3.9 \times 10^2 \text{ dollars.}$$

APPENDIX 2

Units

LENGTH, AREA, AND VOLUME

The scientific units of length are often confusing to citizens of the United States and other English-speaking countries. However, we are a minority in this respect and in fact the "scientific" or metric system of units is used in everyday affairs in a great many countries. It is rather difficult to defend our own system in which the thickness of a thumb once determined the inch, the length of someone's foot determined the unit "foot," and the distance from the nose to the end of the thumb of Henry I of England was the basis of the yard. Some of these origins may be legendary, but the units are certainly not fundamental nor is conversion from one unit to another convenient.

The *metric system* is based on the *meter,* which was defined as one-millionth of the distance from the North Pole to the equator along the arc shown in Figure A-1. The distance from Barcelona, Spain, to Dunkirk, France, was surveyed between 1792 and 1799, and the length of the entire arc calculated from astronomical observations. One meter is about 1.1 yards. On the metric system convenient units are generated by changing the power of ten. Common examples are:

kilometer = 10^3 meters (0.6 miles)
centimeter = 10^{-2} meters (0.39 inches)

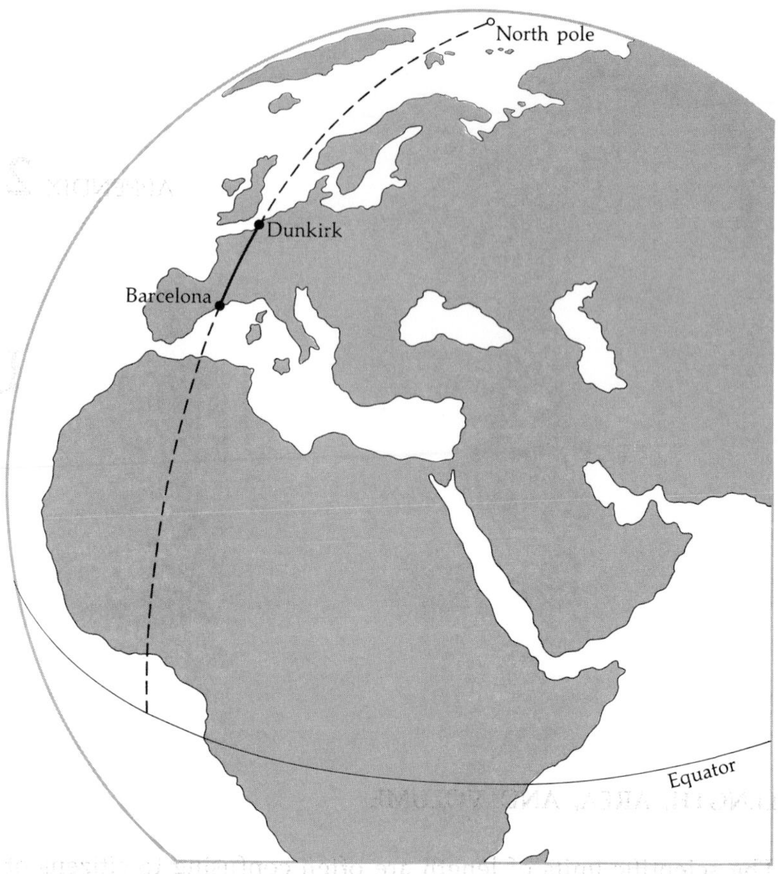

A-1 *The meter was based on the quadrant of the earth's meridian passing through Dunkirk and Barcelona, as shown. (From* Standards of Measurement *by A. V. Astin. Copyright © 1968 by Scientific American, Inc. All rights reserved.)*

millimeter = 10^{-3} meters
Angstrom = 10^{-10} meters or 10^{-8} cm.

The centimeter is the most frequently used as the unit of distance in scientific work. Other units of length (usually quite large) are used in astronomy, such as:

earth's radius = 6.4×10^{8} cm
sun's radius = 7.0×10^{10} cm
astronomical unit = 1.5×10^{13} cm
light year = 9.5×10^{17} cm
parsec = 3.1×10^{18} cm
kiloparsec = 3.1×10^{21} cm
megaparsec = 3.1×10^{24} cm.

Currently, the meter is defined as 1,650,763.73 wavelengths of a specific atomic line of an isotope of the gas krypton emitting light under standard conditions. (Emission lines produced by gases are discussed in Chapter 5.)

Surface area is usually given in square centimeters or square meters:

1 square centimeter = 0.16 square inches
1 square meter = 10.8 square feet.

Volume is usually given in cubic centimeters or cubic meters:

1 cubic centimeter = 0.061 cubic inches
1 cubic meter = 35 cubic feet.

A specific version of the metric system that one frequently encounters in physics books is the *c.g.s.* system, in which the basic units of length, mass, and time are the centimeter, gram, and second. In the *m.k.s.* system, these basic units are the meter, kilogram, and second.

MASS

For practical purposes, the basic unit of mass, the gram, can be defined as the mass of 1 cubic centimeter of water, as measured at a temperature of 4 degrees centigrade. The kilogram (10^3 grams) is another common unit. A mass of 1 kilogram weighs about 2.2 pounds on the earth's surface.

TIME

The basic unit of time is the second. It can be defined as a certain fraction of the rotation period of the earth. Greater precision comes through the use of *atomic clocks,* devices that provide a time source based on the frequency of an optical or radio emission line. There are about 3.16×10^7 seconds per year.

ANGLE

Almost everyone knows that the circle is divided into 360 degrees. This convention may possibly be traced back to an

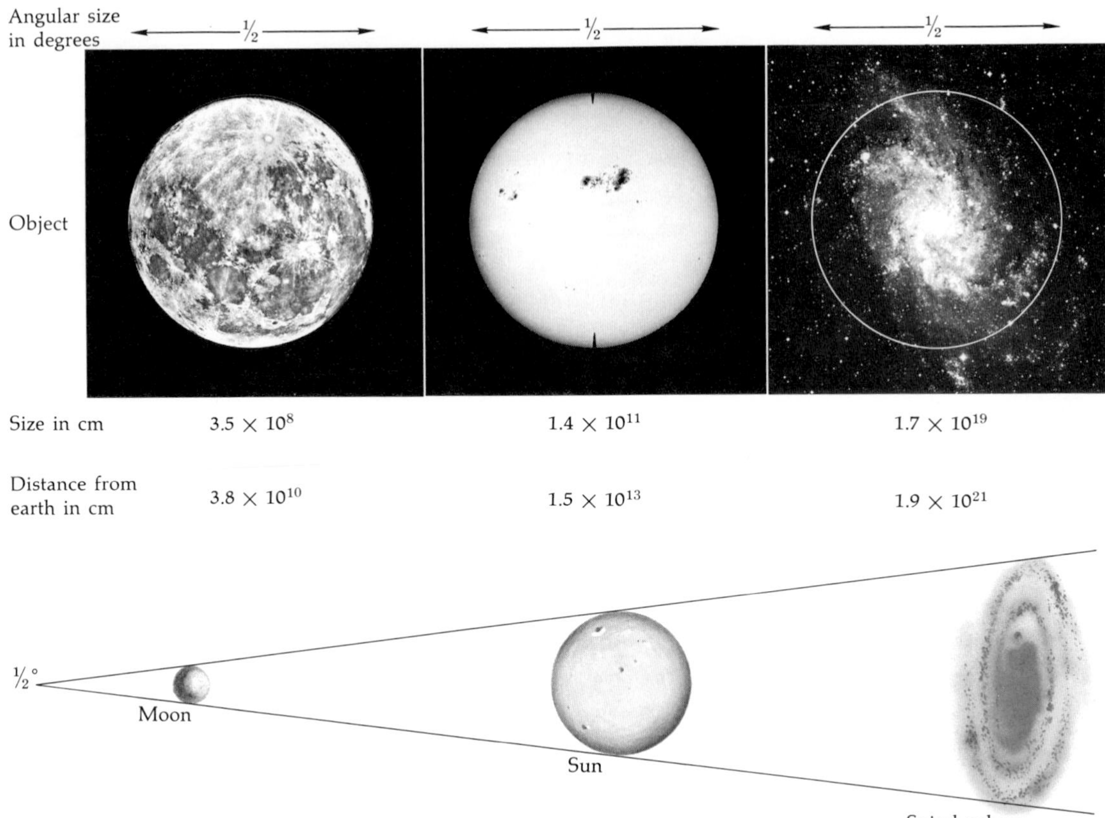

A-2 *The concept of angular size. Bottom—If the angle in the example were about one-half degree, the moon, sun, and the galaxy M33 (which have vastly different linear sizes) would fit into it as shown. Top—Photographs printed to the same angular scale show the effect cited. (Courtesy of the Lick Observatory, Hale Observatories, and Hale Observatories, respectively.)*

early, inaccurate determination of the number of days in a year; if there were 360 days per year, and if the earth's orbit were a circle, the earth would move around the orbit at the rate of one degree per day. The degree is further subdivided into 60 minutes of arc (60′) and the minute into 60 seconds of arc (60″). Thus, there are $60 \times 60 = 3{,}600$ seconds of arc in one degree.

The concept of angular size of an object is used throughout astronomy, because it constitutes an apparent size that we can measure. On the other hand, in order to know the true (or "linear") size, as measured in length units (for example, kilometers) we have to know how far away the object is; note that objects of entirely different true sizes can have the same *angular size* as seen in the sky; this effect is shown in Figure A-2.

APPENDIX 3

Additional Reading

An older book written with the same general philosophy as ours is

Through Space and Time, by Sir James Jeans, London: MacMillan, 1934.

Contemporary books with the same general philosophy and on about the same level are

Knowledge and Wonder, by Victor F. Weisskopf, New York: Doubleday, 1966.

Red Giants and White Dwarfs, by Robert Jastrow, New York: Harper and Row, 1969.

On a somewhat more technical level, we recommend

Matter, Earth and Sky, by George Gamow, Englewood Cliffs, N.J.: Prentice-Hall, 1965.

Intelligent Life in the Universe, by I. S. Shklovskii and Carl Sagan, San Francisco: Holden-Day, 1966.

Introduction to Natural Science, Part 1, the Physical Sciences, by V. Lawrence Parsegian, Alan S. Meltzer, Abraham S. Luchins, and K. Scott Kinerson, New York: Academic Press, 1968.

Introduction to Natural Science, Part 2, The Life Sciences, by V. Lawrence Parsegian, Paul R. Shilling, Floyd V. Monaghan, and Abraham S. Luchins, New York: Academic Press, 1970.

Two good standard astronomy texts to refer to are

Exploration of the Universe, by George Abell, New York: Holt, Rinehart, and Winston, 1964; Brief Edition, 1969.

Principles of Astronomy, by Stanley P. Wyatt (Second Edition), Boston: Allyn and Bacon, 1971.

To learn or refresh your knowledge of basic mathematical concepts, read

Mathematics, A Human Endeavor, by Harold R. Jacobs, San Francisco: W. H. Freeman and Company, 1970.

We think you will also enjoy

The Immense Journey, by Loren Eiseley, New York: Random House, 1957.

Well-illustrated accounts of the latest developments in astronomy and space research appear regularly in *Scientific American* and *Sky and Telescope,* and occasionally in *Natural History* and *Smithsonian* magazines. In book form, some of the most exciting developments in modern astronomy are described in

Violent Universe by Nigel Calder, New York: Viking Press, 1969.

For the history of great advances in the astronomy of the past, told in the words of the masters themselves, browse through

A Source Book in Astronomy, edited by Harlow Shapley and Helen E. Howarth, New York: McGraw-Hill, 1929.

Source Book in Astronomy 1900–1950, edited by Harlow Shapley, Cambridge, Mass.: Harvard University Press, 1960.

APPENDIX 4

Basic Astronomical Equations

Here are the simple mathematical expressions that are relevant to the content of this book—for readers who wish to know and use them.

KEPLER'S THIRD LAW

The average distance of a planet from the sun, d (in astronomical units), and its period of revolution, p (in years), are related by

$$p^2 = d^3$$

This equation was used in constructing Figure 4-20, page 87.

NEWTON'S SECOND LAW

The force, F, the mass, m, and the acceleration, a, are related by

$$F = ma.$$

The acceleration is in the same direction as the force. Recall that an acceleration is a change in velocity and that no acceleration means a constant velocity. Hence, Newton's First Law of Motion (page 91) is really a special case of his Second Law. In the c.g.s. system, the units are dynes (F), grams (m), and centimeters per second per second (a). The dyne is defined as the amount of force required to accelerate a mass of one gram by one centimeter per second per second.

UNIVERSAL GRAVITATION

The gravitational force F between two masses m and M that are separated by a distance r is

$$F = \frac{Mm\,G}{r^2}.$$

The units of force, mass, and distance are dynes, grams, and centimeters, respectively, in the c.g.s. system, and G is the constant of gravitation which has the numerical value of 6.7×10^{-8}. See page 92.

EINSTEIN'S EQUATION FOR MASS-ENERGY EQUIVALENCE

The energy equivalent E of a mass m is given by

$$E = mc^2,$$

where the units of E and m are ergs and grams, respectively (c.g.s. system), and c, the speed of light, is 3×10^{10} centimeters per second. See page 117.

RADIATION LAWS

The total energy E emitted by the surface of a black body or ideal radiator is given by the area under the Planck curve and this is called the Stefan-Boltzmann Law,

$$E = \sigma T^4.$$

In the c.g.s. system, the constant σ is about 5.67×10^{-5} erg/cm^2 sec deg^4 and T is in degrees Kelvin.

The *wavelength* (λ) at which the maximum energy is emitted by an ideal radiator is given by Wien's Law,

$$\lambda = \frac{0.29}{T}$$

where the units of λ are centimeters and T is the temperature in degrees Kelvin.

PARALLAX EQUATION

A star's parallax p (in seconds of arc) is related to its distance d (in parsecs) by

$$d = \frac{1}{p}.$$

HUBBLE'S LAW

The distance D of a galaxy is related to its red shift velocity V by

$$D = \frac{V}{H}.$$

Here H is Hubble's constant, about 100 kilometers per second per megaparsec, and the units of D and V are megaparsecs and kilometers per second, respectively. See page 361.

THE MAGNITUDE EQUATION

If two stars have brightnesses b_1 and b_2, their magnitudes m_1 and m_2 are related by

$$m_1 - m_2 = 2.5 \log_{10}\left(\frac{b_2}{b_1}\right).$$

Thus, if star 2 is 100 times brighter than star 1, $\log_{10} b_2/b_1$ equals 2, and the magnitude difference $m_1 - m_2$ is 5, as required by the definition of magnitude on page 76. This equation can be applied to the brightnesses and magnitudes in a specific wavelength interval, such as that of the B or blue magnitudes, page 268.

APPENDIX 5

Summary Tables

Properties of the Sun

Mass	2×10^{33} grams, or 3×10^{5} earth masses
Diameter of the photosphere	1.4×10^{6} km, or 109 earth diameters
Temperature of the photosphere	5,750° K
Temperature of the corona	2×10^{6} °K
Rate at which mass streams away in the solar wind	2×10^{12} grams/sec
Rate at which solar energy is released in photons (luminosity of sun)	4×10^{26} watts, or 4×10^{33} ergs/sec
Rate at which mass is converted into energy by nuclear reactions in the core of the sun	5×10^{12} grams/sec
Rate at which solar energy is received by the earth (solar constant)	2 calories/cm^{2}-min or 1.4×10^{6} ergs/cm^{2}-sec
Estimated age	5×10^{9} years
Distance from the center of the galaxy	10,000 parsecs
Speed of motion through the galaxy	250 km/sec

Properties of the Planets

PROPERTY	MERCURY	VENUS	EARTH	MARS	JUPITER	SATURN	URANUS	NEPTUNE	PLUTO
Average distance from sun (km)	0.58×10^8	1.08×10^8	1.50×10^8	2.28×10^8	7.78×10^8	14.3×10^8	28.7×10^8	45.0×10^8	59.0×10^8
Average distance from sun (a.u.)	0.39	0.72	1.00	1.52	5.20	9.54	19.2	30.1	39.4
Average orbital speed (km/sec)	48	35	30	24	13	9.6	6.8	5.4	4.7
Orbital period (sidereal)	88 days	225 days	365 days	1.88 years	11.9 years	29.5 years	84 years	165 years	248 years
Radius (km)	2,439	6,050	6,378	3,394	71,880*	60,400	23,540	24,600	7,000 ?†
Mass (grams)	3.3×10^{26}	4.9×10^{27}	6.0×10^{27}	6.4×10^{26}	1.9×10^{30}	5.7×10^{29}	8.8×10^{28}	1.0×10^{29}	1×10^{27} ?†
Mass (earth masses)	0.056	0.81	1.00	0.11	318	95	15	17	0.18 ?†
Escape velocity (km/sec)	4.2	10.3	11.2	5.0	61	37	22	23	4?†
Average axial rotation period (sidereal)	58.7 days	243 days	$23^h\ 56^m$	$24^h\ 37^m$	$9^h\ 55^m$	$10^h\ 38^m$	$10^h\ 49^m$	14^h	6.4 days?†
Tilt of equatorial plane to orbital plane	7°	6°	23.5°	24°	3°	27°	98°	29°	?†

*Radius measured at height in atmosphere where there are 10^{14} molecules per cubic centimeter.
†Question marks indicate the many uncertainties about Pluto.

*Satellites of the Planets**

SATELLITE	AVERAGE DISTANCE FROM PLANET (km)	ORBITAL PERIOD (SIDEREAL DAYS)	DIRECTION OF ORBIT (WITH RESPECT TO PLANET'S ROTATION)	TILT OF ORBIT TO PLANET'S EQUATOR	RADIUS (km, WHERE ACCURATELY KNOWN)	MASS (gm, WHERE KNOWN)
Earth						
Moon	384,000	27.3	direct	$23\frac{1}{2}° \pm 5°$‡	1,738	7.3×10^{25}
Mars						
Phobos	9,000	0.32	direct	1°	13§	-
Deimos	23,000	1.26	direct	2°	7§	-
Jupiter						
I Io	420,000	1.77	direct	0°	1,830	7.3×10^{25}
II Europa	670,000	3.55	direct	0°	1,460	4.8×10^{25}
III Ganymede	1,100,000	7.15	direct	0°	2,550	1.5×10^{26}
IV Callisto	1,900,000	16.69	direct	0°	2,360	9.5×10^{25}
V	180,000	0.50	direct	0°	-	-
VI	11,500,000	251	direct	28°	-	-
VII	11,700,000	260	direct	26°	-	-
VIII	23,500,000	737	retrograde	33°	-	-
IX	23,700,000	758	retrograde	25°	-	-
X	11,700,000	253	direct	28°	-	-
XI	22,600,000	692	retrograde	16°	-	-
XII	21,200,000	631	retrograde	33°	-	-
Saturn						
I Mimas	190,000	0.94	direct	2°	-	4×10^{22}
II Enceladus	240,000	1.37	direct	0°	-	7×10^{22}
III Tethys	300,000	1.89	direct	1°	-	6.5×10^{23}
IV Dione	380,000	2.74	direct	0°	-	1.0×10^{24}
V Rhea	530,000	4.52	direct	0°	700	2.3×10^{24}
VI Titan	1,220,000	15.95	direct	0°	2,440	1.4×10^{26}
VII Hyperion	1,480,000	21.28	direct	0°	-	1.1×10^{23}
VIII Iapetus	3,560,000	79	direct	15°	-	1.1×10^{24}
IX Phoebe	12,900,000	550	retrograde	30°	-	-
X Janus	170,000	0.82	direct	0°	-	-
Uranus†						
I Miranda	190,000	2.52	direct	0°	-	1.2×10^{24}
II Ariel	270,000	4.14	direct	0°	-	5×10^{23}
III Umbriel	440,000	8.71	direct	0°	-	4×10^{24}
IV Titania	590,000	13.46	direct	0°	-	2.6×10^{24}
V Oberon	130,000	1.41	direct	-	-	1.1×10^{23}
Neptune						
I Triton	350,000	5.88	retrograde	20°	2,000	1.4×10^{26}
II Nereid	5,600,000	360	direct	28°	-	3×10^{22}

*Unlisted planets have no known satellites.
†Note that the rotation of Uranus is retrograde.
‡Tilt varies from 18° to 29°.
§Oblong shape; this "radius" equals one-half of the long dimension.

Index

Index

Index

Index